Calcified Tissues 1975

Proceedings of the XIth European
Symposium on Calcified Tissues

Edited by S. Pors Nielsen
E. Hjørting-Hansen

SPRINGER-VERLAG BERLIN HEIDELBERG GMBH

Published simultaneously as a supplement to
CALCIFIED TISSUE RESEARCH and by FADL Publishing Co.,
Blegdamsvej 84, DK-2100 Copenhagen Ø, Denmark
Composition by Sonja Søby

ISBN 978-3-662-27776-8   ISBN 978-3-662-29272-3 (eBook)
DOI 10.1007/978-3-662-29272-3
Softcover reprint of the hardcover 1st edition 1976

Sale through: FADL Publishing Co.,
              Blegdamsvej 84,
              DK-2100 Copenhagen Ø.,
              Denmark.

PREFACE

The XIth European Symposium on Calcified Tissues was held at Hotel Marienlyst, Elsinore, Denmark, May 25–29, 1975. The main topics were: 1. cellular calcium transport, 2. vitamin D, and 3. osteoporosis. Seventy-six of the submitted papers were selected for verbal presentation, and these papers are the basis of the articles found in this book. We hope that the book, which also appears as a supplement to CALCIFIED TISSUE RESEARCH has a form which will be adopted by organizers of future symposia of this series.

The Organizing Committee wishes to express its gratitude to the Danish Medical Research Council (Statens Lægevidenskabelige Forskningsråd), to Tuborgfondet, and to Danish and foreign industry for their generous financial support, without which the symposium would not have been held and without which the proceedings would not have appeared. Also, we wish to thank our secretarial staff, in particular Mrs. Karin Christensen for skilled and keen assistance before, during and after the symposium. We hope that the contents of the book confirm the view that the XIth European Symposium on Calcified Tissues, having an interdisciplinary character was important for future research in the field and for the treatment of patients with metabolic bone disease.

S. Pors Nielsen
E. Hjørting-Hansen

4

Organizing Committee

S. Pors Nielsen, Copenhagen (chairman)
Th. Friis, Copenhagen
E. Hjørting-Hansen, Copenhagen
O. Helmer Sørensen, Copenhagen
F. Kuhlencordt, Hamburg
H.J. Dulce, Berlin

List of Donators

A/S Alfred Benzon
Ciba-Geigy A/S
Ercopharm A/S, Organon
Gammatec
LEO Pharmaceutical Products
Noco A/S
Novo Industri A/S
Sandoz A/S
G.D. Searle A/S
Statens Lægevidenskabelige Forskningsråd
Tuborgfondet
The Upjohn Company

Exhibitions During the Symposium

AB Atomenergi, Studsvik
Claus Kettel (Orion Research Inc.)
Gammatec
Perkin-Elmer A/S
Sandoz A/S

6

CHAPTER V
Parathyroid Hormone

CHAPTER VI
Calcitonin

CHAPTER I
Cellular Transport of Calcium

Calcium Transport Processes and their Regulation in Endocrine Cells

N.A. THORN

In recent years much research work has been done on a function of calcium ion which is very different from that of calcium salts in supportive tissues.

It has turned out that the calcium ion, often in minute amounts, plays an essential role as a trigger or a messenger in stimulus-contraction coupling and in stimulus-secretion coupling.

In many endocrine cell systems the calcium ion functions as an intracellular messenger for hormone release, either alone or in combination with cyclic nucleotides (cAMP or cGMP) (8).

Some of the functions of calcium and the problems involved may be illustrated from our work on release of antidiuretic hormone (vasopressin) from the nerve endings in the neurohypophysis. This tissue in its function is a sort of mixture of nerve tissue and endocrine cells, since it receives and conducts impulses and releases nonapeptides from large secretory granules to the circulation. It has been established that the presence of calcium in the extracellular fluid is critical for release of this hormone. The main cellular transport processes for calcium are illustrated in Fig. 1. It has been reported by others that stimulation is associated with a cellular calcium uptake. Using refined methods for studying such calcium uptake (among them the lanthanum method of van Breemen et al. to distinguish intracellular calcium uptake from other uptake) we have not been able to demonstrate uptake in all situations in which we could stimulate release, and there are doubts that the "uptake" found previously in certain situations represents uptake of calcium that is essential in stimulus-secretion coupling (4). We have, however, shown that substances which in a number of other systems block calcium channels do block release in the neurohypophysis. This is the case with D600, a verapamil derivative (5) and diphenylhydantoin (2). We, therefore, conclude that a trigger amount

Institute of Medical Physiology C, University of Copenhagen.

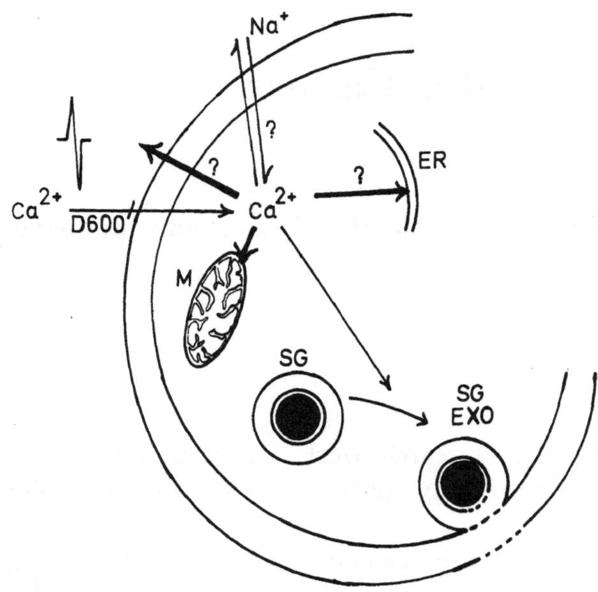

Figure 1. Mechanisms of Ca^{2+} entry on stimulation of secretion and removal of trigger Ca^{2+}. It is unclear whether Ca^{2+} is involved in the exocytosis itself. Thick arrows represent "active" transport of Ca^{2+}, thin ones passive flux. D600: D600 sensitive Ca^{2+} entry; ER: endoplasmic reticulum; M: mitochondrion; SG: secretory granule; SG EXO: exocytosis itself.

of calcium moves across the cell membrane and stimulates release.

Using somewhat artificial procedures it is possible to increase the intracellular calcium concentration and cause release of hormone without any apparent stimulation of the cell membrane. This can be done by means of the calcium ionophores A-23187 (6) and X537A (9).

Since these situations with increased release have in common an increase in intracellular calcium, it becomes an essential question how the intracellular calcium concentration is again brought to a resting state. This can be done by a temporary seqestration in subcellular structures, but it has to be combined with an eventual extrusion over the cell membrane in systems as the present one where at least part of the activator calcium comes from the extracellular pool.

We have demonstrated the presence of a strong ATP-dependent calcium accumulation in mitochondria from neurohypophyseal nerve endings. However, also a microsomal fraction accumulates calcium when ATP is present. This could represent a calcium extrusion over the cell membrane or a sodium to calcium exchange (7). The processes involved are being further characterized.

It should, however, be discussed whether there may be other ways of

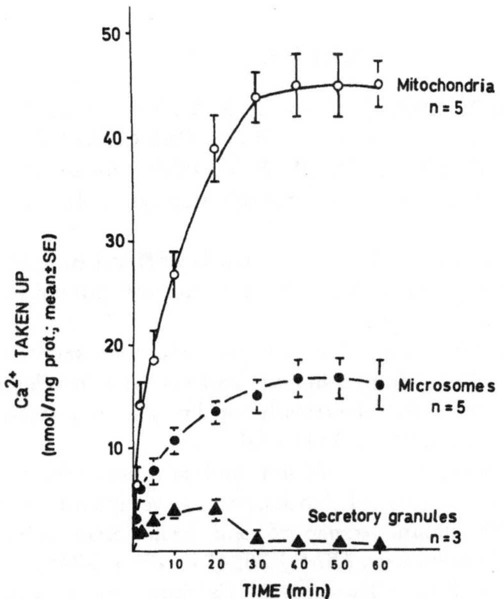

Figure 2. ATP-dependent Ca^{2+} uptake by mitochondrial, microsomal and secretory granule fraction isolated from ox neurohypophyses. The incubation was carried out in a K-rich medium with TES buffer, in the presence of 2 mM ATP and a regeneration system (7).

activating the hormone releasing machinery. One possibility is that the stimulus interferes with the internal balance of calcium transport e.g. by inhibiting mitochondrial calcium uptake by a sodium concentration higher than that present in the intracellular fluid in the resting tissue.

Such a phenomenon which was first demonstrated (1) in cardiac muscle mitochondria can also be demonstrated in the neurohypophysis (9). It would, however, seem that it cannot be dominant in the neurohypophysis under normal circumstances, since release can go on in a sodium free medium (3).

CONCLUSION

Calcium ions in trigger amounts are essential for activating a number of endocrine tissues to release their hormones. The cells are equipped with a number of processes for reestablishing the normal very low intracellular concentration of free calcium.

REFERENCES

1. Carafoli, E., Tiasso, R., Lugli, G., Croveth, F. & Kratzing, C.: The release of calcium from heart mitochondria by sodium. *J. Molec. Cell. Cardiol.* **6**, 361–371 (1974)
2. Guzek, J.W., Russell, J.T. & Thorn, N.A.: Inhibition by diphenylhydantoin of vasopressin release from isolated rat neurohypophyses. *Acta pharmacol. (Kbh.)* **34** 1–4 (1974)
3. Müller, J.R., Thorn, N.A. & Torp-Pedersen, C.: Effects of calcium and sodium on vasopressin release *in vitro* induced by a prolonged potassium stimulation. *Acta Endocr. (Kbh.)* **79**, 51–59 (1975)
4. Russell, J.T. & Thorn, N.A.: Calcium and stimulus secretion coupling in the neurohypophysis I. 45-calcium transport and vasopressin release in slices from ox neurohypophyses stimulated electrically or by a high potassium concentration. *Acta Endocr. (Kbh.)* **76**, 449–470 (1974a)
5. Russell, J.T. & Thorn, N.A.: Calcium and stimulus secretion coupling in the neurohypophysis II. Effects of lanthanum, a verapamil analogue (D600) and prenylamine on 45-calcium transport and vasopressin release in isolated rat neurohypophyses. *Acta Endocr. (Kbh.)* **76**, 471–487 (1974b)
6. Russell, J.R., Hansen, E.L. & Thorn, N.A.: Calcium and stimulus secretion coupling in the neurohypophysis III. Ca^{2+} ionophore (A-23187)-induced release of vasopressin from isolated rat neurohypophyses. *Acta Endocr. (Kbh.)* **77**, 443–450 (1974)
7. Russell, J.T. & Thorn, N.A.: Adenosine triphosphate dependent calcium uptake by subcellular fractions from bovine neurohypophyses. *Acta physiol. scand.* **93**, 364–377 (1975)
8. Thorn, N.A.: Role of calcium in secretory processes. In: *Secretory Mechanisms of Exocrine Glands.* Thorn, N.A. & Petersen, O.H. (eds.). The Alfred Benzon Symposium VII. Munksgaard, Copenhagen p 305–326, 1974
9. Thorn, N.A., Russell, J.T. & Robinson, I.C.A.F.: Factors affecting intracellular concentration of free Ca^{2+} ions in neurosecretory nerve endings. In: *Calcium Transport in Contraction and Secretion.* Carafoli, E., Clementi, F. & Margreth, A. (eds.), North-Holland, Amsterdam, pp 261–269, 1975

The Interaction of Divalent Cations, Hormones and Cyclic Nucleotides in the Control of Mitosis

A.D. PERRIS & J.I. MORGAN

A series of intraperitoneal calcium or magnesium chloride injections will stimulate mitotic activity in rat bone marrow and thymus tissue as will parathyroid extract by virtue of its ability to induce hypercalcaemia. Indeed, bone marrow mitosis is proportional to plasma calcium concentration. The reduction in cell division which thus accompanies hypocalcaemia can have dramatic consequences. Thymic and splenic atrophy and bone marrow hypoplasia all follow parathyroidectomy and there are impediments in liver regeneration after partial hepatectomy and in erythrocyte replenishment after haemorrhage (see reviews by Perris (6) and Whitfield *et al.* (9)). It has also been shown in a variety of natural situations that a calcium-dependent control of mitosis operates when there is a physiological requirement for enhanced cell division in rat haemopoietic and lymphoid tissue.

Thus, the altered patterns of mitosis which accompany changing body growth rates are paralleled by concomitant plasma ionised calcium fluctuations (6). A parathyroid-dependent hypercalcaemia also develops one or two days after haemorrhage in rats and this induces an increase in cell division in the bone marrow which thus speeds the restoration of the normal erythrocyte complement. Similarly in other heightened erythropoietic circumstances which occur after erythropoietin or cobaltous chloride administration or during pregnancy there again develops a parathyroid-dependent hypercalcaemia which stimulates bone marrow mitosis (6).

Circadian variations in the proliferative activity of thymus and bone marrow tissue also seem to be linked to plasma calcium concentration changes. Most vividly this is seen between 16.00 and 20.00 h just before the onset of darkness when there are parallel decreases in calcium levels

Department of Biological Sciences, The University of Aston in Birmingham.

and mitosis in normal but not aparathyroid rats (3). Cyclical variations are also seen in female rats during the oestrous cycle where plasma calcium concentrations reach a maximum during oestrous and bone marrow mitosis once again follows a similar pattern (Smith, Davis and Perris 1975, unpublished observations).

Finally it appears that the mitotic response to antigenic challenge is also a parathyroid-dependent event. Three days after rats are immunised with sheep red blood cells they become markedly hypercalcaemic and marrow mitosis is enhanced. Neither occurs in the parathyroidectomised rat (Edwards and Perris, 1975, unpublished observations). It is not yet established whether any natural fluctuations in magnesium homeostasis are likewise involved in the control of these or other mitotic events. The mechanism by which these two divalent cations might function as mitogens has been deduced from experiments using isolated cell suspensions.

When thymic lymphocytes are maintained in a tissue culture medium lymphoblasts will continue to synthesise DNA and enter the mitotic

Table I. Properties of different mitogens

Mitogen	Concentration	Ca dependent	Oestradiol blockade	Testosterone blockade	Mg dependent
Ca	1.8 mM	Yes	Yes	No	No
Mg	2.5 mM	No	No	Yes	Yes
Adrenaline	5×10^{-6}M	No	No	?	?
Isoprenaline	10^{-6}M	No	No	Yes	Yes
Glucagon	10^{-4}M	No	No	Yes	Yes
Dopamine	10^{-7}M	No	No	Yes	Yes
Acetylcholine*	10^{-12}M	Yes	Yes	No	No
Insulin	10^{-10}M	Yes	Yes	No	No
Histamine	10^{-13}M	Yes	Yes	No	No
Parathyroid hormone	10^{-9}M	Yes	Yes	No	No
High c-GMP	10^{-6}M	No	No	Yes	Yes
Low c-GMP	5×10^{-11}M	Yes	Yes	No	No
High c-AMP	10^{-7}M	No	No	Yes	Yes
Low c-AMP	10^{-14}M	Yes	Yes	No	No

Typically in the basal medium containing 0.6 mM Ca and 1.0 mM Mg the percentage of cells entering mitosis over a 6 h period would be 3.4±0.1. The addition of the different mitogens would increase this to approximately 6.0±0.3 per cent.

* Acetylcholine was tested in the presence of eserine (10^{-9}M) to prevent degradation by acetylcholinesterase.

sequence much as they do *in vivo*. Our standard medium contains 0.6 mM Ca and 1.0 mM Mg. Omission of either calcium or magnesium does not affect the basal level of mitotic activity. An increase in calcium concentration to 1.8 mM significantly increases the rate of entry of cells into mitosis whether magnesium is present at the usual concentration or is completely absent. In a similar way an increase in magnesium concentration to 2.5 mM stimulates mitosis in normal and calciumfree media (Table I). The distinct identity of these two agents is further shown by the ability of calcitonin and imidazole to block calcium's mitogenic action but to leave magnesium's unimpaired (5). The most vivid demonstration of their discrete mitogenic identities is the contrast between the effects of oestradiol and testosterone upon these divalent cations. The mitogenic action of calcium is blocked by oestradiol; testosterone and progesterone have no effect. In contrast magnesium's action is blocked by testosterone alone (Table I).

In addition to calcium and magnesium many other hormones are mitogenic in this cell type. Some of these compounds require the simultaneous presence of calcium. Since many of these compounds induce increases in intracellular cyclic $3'5'$ adenosine monophosphate (c-AMP) concentration it was thought that this cyclic nucleotide might be the ultimate mitogenic mediator in this cell type especially when it was shown that exogenous c-AMP can also stimulate cell division. The ability of mitogenic concentrations of calcium and some of the calcium-dependent and -independent hormones to increase thymocyte c-AMP concentrations supports this view. The central role of c-AMP in the triggering of DNA synthesis and cell division in this cell type has been reviewed by Whitfield *et al.*, (9).

Other workers, however, have suggested that c-AMP inhibits cell division and that cyclic $3'$-$5'$ guanosine monophosphate (c-GMP) is the nucleotide which triggers events culminating in mitosis. Thus, the initiation of DNA synthesis in and transformation of peripheral blood lymphocytes induced by phytohaemagglutinin and concanavalin A is preceded by an increase in c-GMP (2). Similarly when fresh serum, which initiates a new wave of DNA synthesis and mitosis, is added to quiescent $3 \, T_3$ cells there is a rapid increase in c-GMP and a concomitant decrease in c-AMP concentration (7).

Our approach to these apparent contrasting actions of the cyclic nucleotides was initially to study the mitogenic actions of hormones which classically antagonize each other because their actions are thought to be mediated by c-AMP on the one hand and c-GMP on the other. Acetylcholine, insulin and histamine may all use c-GMP as their intracellular mediators (1, 4) and all are mitogenic provided calcium is present (Table I). Parathyroid hormone clearly falls into the same calcium-dependent category. Oestradiol (which blocks calcium's mitogenic action) also blocks the action of all these calcium-dependent hormones but testosterone has no effect (Table I).

Adrenaline and glucagon (antagonists of acetylcholine and insulin) which are known to stimulate adenylate cyclase and increase intracellular c-AMP levels are also mitogenic but do not require calcium and cannot be blocked by oestradiol. Isoprenaline and dopamine are also calcium-independent mitogens and are not influenced by oestradiol (Table I). These compounds do, however, require magnesium ions and testosterone can block their mitogenic actions (Table I).

The clear cut separation between these two groups of mitogens, thought to have as their intermediaries c-AMP and c-GMP, led us to examine the effects of divalent cation omission and the effects of steroid addition on the mitogenic actions of the cyclic nucleotides themselves. Like other workers (8) we find that exogenous c-GMP has a biphasic action; mitotic stimulation is evident at 10^{-6} M and 5×10^{-11} M. Unlike these workers we find that the low concentration requires calcium whereas the high does not and it is also apparent that the high concentration requires magnesium whereas the low does not (Table I). Furthermore, high mitogenic concentrations of c-GMP are blockable by testosterone whereas low concentrations are inhibited by oestradiol (Table I). Cyclic-AMP also has a biphasic action. High (10^{-7} M) concentrations are mitogenic and are magnesium-dependent and testosterone-blockable. Low mitogenic levels (10^{-14} M) are calcium-dependent and can be blocked by oestradiol (Table I).

Assuming physiologically significant concentrations of c-AMP and c-GMP are approximately 10^{-7} M and 5×10^{-11} M respectively we imagine that the interaction of hormones, cyclic nucleotides and divalent cations might occur as illustrated in Fig. 1. Adrenaline, glucagon, isoprenaline and dopamine will each interact with their specific receptors at the cell surface linked to adenylate cyclase which becomes activated in the presence of magnesium. Thus, the intracellular c-AMP concentration will increase. Since c-AMP is only mitogenic if magnesium is present (Table I) we suggest that the nucleotide provokes the influx of magnesium ions to the cell which then become the ultimate mitotic initiators. Testosterone would then be envisaged to block the action of these mitogens by preventing the magnesium influx. In an analogous manner acetylcholine, histamine, insulin and parathyroid hormone (PTH) will increase intra-cellular c-GMP concentrations which may provoke an oestradiol-blockable influx of calcium ions (Fig. 1).

We believe that the interaction of hormones and cyclic nucleotides, and of

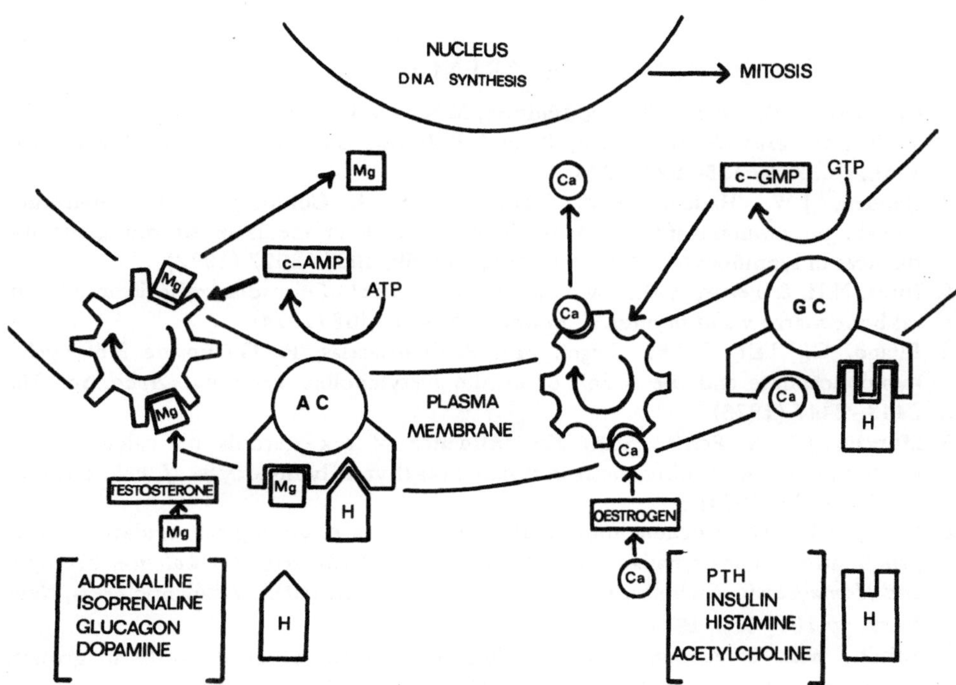

Figure 1. The mechanism of hormone — and divalent cation-induced mitogenesis. Two groups of hormones are linked via specific membrane receptors to guanylate cyclase (GC) or adenylate cyclase (AC) and cause an intracellular increase in c-GMP and c-AMP respectively. These nucleotides then provoke an increased influx of magnesium or calcium ions into the cell which in some way initiates DNA synthesis in a group of quiescent cells which ultimately proceed into mitosis.

course variations in the extracellular concentration of calcium or magnesium, will determine the intracellular divalent cationic climate. It is this environment which is so important in the physiological control of cell division in normal haemopoietic and lymphoid tissue, and it may be that derangements in the intracellular concentrations of calcium or magnesium is of major consequence in the economy of the malignant cell.

ACKNOWLEDGEMENTS

We thank The Royal Society, The Cancer Research Campaign, and the Science Research Council for their support of various aspects of this work.

REFERENCES

1. Goldberg, N.D., O'Dea, R.F. & Haddox, M.K.: Cyclic GMP. In: *Advances in cyclic nucleotide research*, Greengard, P. and Robison, G.A. (eds.). Raven Press, New York, vol. 3, pp 155–223, 1973
2. Hadden, J.W., Hadden, E.M., Haddox, M.K. & Goldberg, N.D.: Guanosine 3'-5'-cyclic monophosphate: A possible intracellular mediator of mitogenic influences in lymphocytes. *Proc. nat. Acad. Sci.* **69**, 3024–3027 (1972)
3. Hunt, N.H. & Perris, A.D.: Calcium and the control of circadian mitotic activity in rat bone marrow and thymus. *J. Endocr.* **62**, 451–462 (1974)
4. Illiano, G., Tell, G.P.E., Siegel, M.I. & Cuatracasas, P.: Guanosine 3'5' cyclic monophosphate and the action of insulin acetylcholine. *Proc. nat. Acad. Sci.* **70**, 2443–2447 (1973)
5. Morgan, J.I. & Perris, A.D.: The influence of sex steroids on calcium- and magnesium-induced mitogenesis in isolated rat thymic lymphocytes. *J. cell. Physiol.* **83**, 287–296 (1974)
6. Perris, A.D.: The calcium homeostatic system as a physiological regulator of cell proliferation in mammalian tissues. In: *Cellular Mechanisms for Calcium Transfer and Homeostasis.* Nicholas, G. Jr. & Wasserman, R.H. (eds.). Academic Press, New York, pp 101–131, 1971
7. Seifert, W.E. & Rudland, P.S.: Possible involvement of cyclic GMP in growth control of cultured mouse cells. *Nature (Lond.).* **248**, 138–140 (1974)
8. Whitfield, J.F., MacManus, J.P., Rixon, R.H. & Gillan, D.J.: The calcium-independent stimulation of thymic lymphoblast DNA synthesis by low cyclic GMP concentrations. *Proc. Soc. exper. Biol. (N.Y.).* **144**, 808–812 (1973 a)
9. Whitfield, J.F., Rixon, R.H., MacManus, J.P. & Balk, S.D.: Calcium, cyclic adenosine 3',5'-monophosphate, & the control of cell proliferation: A review. *In vitro, J. Amer. Tiss. Cult. Ass.* **8**, 257–278 (1973 b)

Control and Regulation of Calcium Homeostasis and Transport

A.B. BORLE & J.H. ANDERSON

Ten years ago, the prevalent view of cellular calcium homeostasis was very simple and modeled after the sodium transport scheme. It was accepted that the cytoplasmic calcium activity was low, around 10^{-6} M, that calcium influx occurs passively down its electrochemical gradient and that calcium efflux is an uphill metabolically dependent process. The implications of this model is that calcium transport is controlled and regulated by the plasma membrane permeability on one hand and by the activity of the calcium pump on the other. Furthermore, the cytoplasmic calcium activity is primarily regulated by events occurring at the plasma membrane. However, in the course of our studies on calcium metabolism and calcium transport in isolated cells, it became obvious that this model was too simple. For instance we discovered an intracellular compartment which is kinetically distinct from the cytoplasmic calcium pool. The magnitude of this compartment can increase or decrease and its calcium turnover rise and fall independently from the cytoplasmic calcium and from the calcium transport across the plasma membrane. On every available evidence, this compartment represents a calcium pool in the cell mitochondria. Then we asked ourselves the following question: Does this mitochondrial calcium pool passively reflect the fluctuations occurring in the cytoplasm and at the plasma membrane? Or, to the contrary, do mitochondria control and regulate cytoplasmic calcium activity and the cell calcium metabolism including calcium transport? Several observations suggested to us that the second possibility was more likely: First, the changes in this compartment often preceded any other cellular effects; and second, the magnitude of the changes was always greater than all other observed alterations. In addition, the calcium transport velocity and capacity of mitochondria is much larger than that of the plasma

Department of Physiology and Department of Pharmacology, University of Pittsburgh, School of Medicine, Pittsburgh.

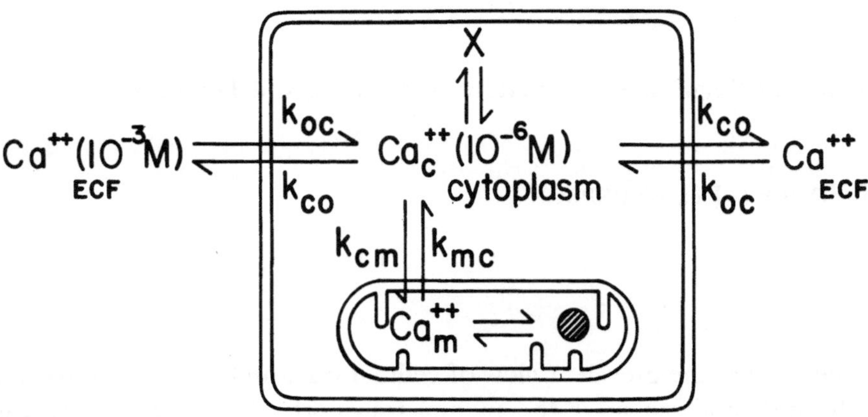

Figure 1. Model of cellular calcium distribution and exchange.

membrane, and the surface area of the mitochondria inner membrane is 30 to 100 times greater than the plasma membrane surface.

Consequently, it was reasonable to assume that the role of mitochondria in controlling and regulating cytoplasmic calcium activity must predominate over the transport of calcium in and out of the cell across the plasma membrane. On this ground, we proposed the following model of cellular calcium homeostasis (Fig. 1). In this model, cytoplasmic calcium activity is primarily controlled and regulated by calcium exchange across the mitochondrial membrane. In turn, calcium transport across the plasma membrane is determined by the cytoplasmic calcium activity (1). Furthermore, since calcium phosphate precipitates have been observed in many mitochondria in most tissue, one can assume that the calcium activity in the matrix of these mitochondria will be fixed by a constant of solubility product. The steady state control of cytoplasmic calcium activity can be expressed in very simple terms: Calcium influx into mitochondria, J_{cm}, is equal to the product of the rate constant of influx, k_{cm}, and the cytoplasmic calcium activity, Ca_c^{++}:

$$J_{cm} = k_{cm} \cdot Ca_c^{++} \tag{1}$$

Calcium efflux from mitochondria, J_{mc}, is equal to the product of the rate constant of efflux, k_{mc}, and the mitochondrial calcium activity, Ca_m^{++}:

$$J_{mc} = k_{mc} \cdot Ca_m^{++} \tag{2}$$

At steady state, influx equals efflux, so that

$$k_{cm} \cdot Ca_c^{++} = k_{mc} \cdot Ca_m^{++} \qquad (3)$$

In presence of calcium phosphate precipitates, the calcium activity of the mitochondrial matrix will be a function of the K_{sp} and of the phosphate concentration: $Ca_m^{++} = f(K_{sp}/P_i)$. Therefore, one can write:

$$Ca_c^{++} = \frac{Ca_m^{++} \left\{ f\dfrac{K_{sp}}{P_i} \right\} k_{mc}}{k_{cm}} \qquad (4)$$

If cytoplasmic calcium activity and calcium transport are controlled by the mitochondria, their hormonal and ionic regulation should also take place at the mitochondrial level. And, indeed, our studies in isolated mitochondria are consistent with this idea (2). For instance, it is well known that isolated mitochondria are able to accumulate great quantities of calcium and to lower the calcium of their bathing medium to very low levels. When cyclic AMP, the intracellular messenger of most polypeptide hormones, is added to the mitochondrial suspension, it stimulates calcium efflux and elevates the medium calcium 5 to 6 folds in a few seconds. When cyclic AMP is added before calcium, the mitochondria maintain the same elevated calcium concentration in the medium. The rise in medium calcium is proportional to the cyclic AMP concentration. However, the effective range of cyclic AMP is very narrow, from $5 \cdot 10^{-7}$ to $5 \cdot 10^{-6}$ M. When the concentration of free calcium is measured with a calcium electrode, the same effect can be observed in liver and in kidney mitochondria. The relative concentrations of calcium and phosphate markedly influence the effects of cyclic AMP. When the concentration of calcium and phosphate are reduced, the pattern of response shift from a sustained elevation of the medium calcium to a transient increase followed by reaccumulation of calcium by the mitochondria. The sustained response can be restored by increasing the calcium load when the phosphate concentration is kept low or by increasing phosphate while keeping the same calcium load. This suggests that a fixed calcium phosphate product must be reached before a sustained response can be achieved which could be due to the precipitation of calcium phosphate in the mitochondria matrix. If this hypothesis is correct, one can predict that in presence of a calcium phosphate precipitate and a sustained response to cyclic AMP, a rise in phosphate should depress the medium calcium and counteract the effect of the nucleotide. This is exactly what is observed in isolated

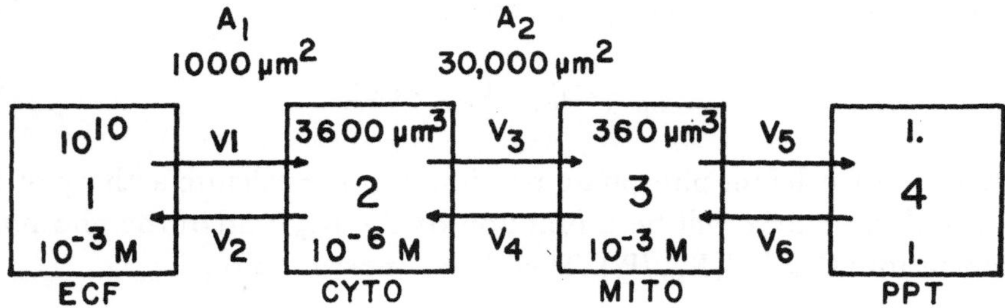

Figure 2. Computer model cellular calcium distribution and exchange.

mitochondria: A large increase in phosphate completely inhibits the effects of cyclic AMP; a more progressive rise produces a stepwise reduction in the medium calcium concentration.

The results obtained in liver and kidney mitochondria are in complete agreement with those observed in isolated kidney cells with cyclic AMP and phosphate. To test whether our interpretation of the results are theoretically and kinetically possible, we have built a computer model of cellular calcium metabolism. Fig. 2 represents one isolated cell $19\,\mu$ in diameter in an infinitely large extracellular compartment. The model consists of 4 compartments placed in series: ECF, cytoplasmic calcium activity, mitochondrial matrix calcium activity and the intramitochondrial calcium phosphate precipitate. The fluxes, rate constants, compartment sizes, surface areas, and the solubility product are all derived from our own data or from the literature. In this model, it is assumed that cyclic AMP increases the rate constant of efflux from mitochondria to cytoplasm, and that phosphate is passively distributed among the various compartments. Fig. 3 shows that the computer model precisely duplicates the results obtained in isolated mitochondria. In presence of a calcium phosphate precipitate, cyclic AMP increases the cytoplasmic calcium activity without affecting significantly the free calcium of the mito-chondrial matrix (Fig. 3-A). When the phosphate concentration is increased during the cyclic AMP stimulation, the cytoplasmic calcium activity immediately drops toward the control levels, and inhibits the effects of the nucleotide. This is due to an increased precipitation of calcium phosphate and to a drop in free calcium in the mitochondrial matrix (Fig. 3-B). Finally, in absence of calcium phosphate precipitate, cyclic AMP produces a transient rise in cytoplasmic calcium, but the new steady state is achieved with a depressed calcium activity of the mitochondrial matrix while the cytoplasmic calcium returns to normal (Fig. 3-C).

These results support the hypothesis that the cellular calcium homeostasis

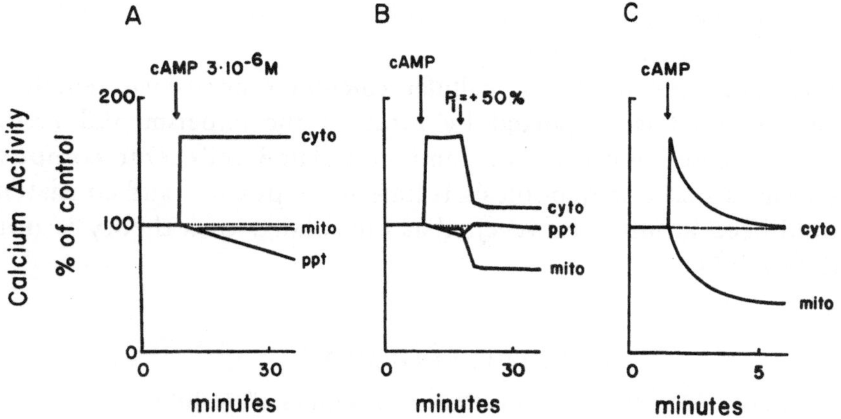

Figure 3. Computer simulation of the effect of cyclic AMP in the model shown in Fig. 9. Cyclic AMP = increase in the rate constant of Ca efflux from mitochondria. A: effect of cAMP in presence of mitochondrial calcium phosphate precipitate. B: inhibition by phosphate of cAMP action in presence of mitochondrial Ca precipitate. C: effect of cAMP in absence of mitochondrial Ca precipitate.

is controlled and regulated at the mitochondrial level. The presence of calcium phosphate precipitate has been observed in the mitochondria of most cells. It is clear, however, that not all mitochondria contain such precipitates. The fact that ultracentifugation of cell homogenates reveal the existence of heavy and light mitochondria which differ in their calcium content, suggests that there may be different populations of mitochondria. It is tempting to speculate that mitochondria which contain calcium phosphate precipitates have lost their ability to regulate the free calcium of their matrix, but they would become the main controller and regulator of cytoplasmic calcium activity, and of the cellular calcium homeostasis. On the other hand, mitochondria, free of precipitate, would influence the cytoplasmic calcium only transiently. However, they would be able to regulate the free calcium concentration of their matrix and the activity of intramitochondrial enzymes which are sensitive to calcium. According to our model, equation 3 becomes

$$Ca_m^{++} = \frac{Ca_c^{++} \cdot k_{cm}}{k_{mc}} \tag{5}$$

This equation predicts that cyclic AMP will lower the free calcium of the matrix by increasing, k_{mc} . Phosphate on the other hand would raise the free calcium of the matrix by increasing the electrical potential across the

26

mitochondrial membrane which, according to Lehninger, might be the driving force for calcium uptake (3).

In conclusion, the model of cellular calcium homeostasis which we have just presented is supported by most of the experimental results obtained in isolated mitochondria and in isolated cells. Our computer model confirms that our hypothesis is kinetically possible and consistent. One should keep in mind, however, that the solidity of a theory is not a proof of its validity.

ACKNOWLEDGEMENT

Supported by grant AM 07867 from the National Institutes of Health U.S.

REFERENCES

1. Borle, A.B.: Calcium metabolism at the cellular level. *Fed. Proc.* **32**, 1944—1950 (1973)
2. Borle, A.B.: Cyclic AMP stimulation of calcium efflux from kidney, liver and heart mitochondria. *J. Membrane Biol.* **16**, 221—236 (1974)
3. Lehninger, A.L.: Role of phosphate and other proton-donating anions in respiration-coupled transport of Ca^{2+} by mitochondria. *Proc. nat. Acad. Sci. (Wash.)* **71**, 1520—1524 (1974)

Calcium Binding Protein and Regulation of Calcium Transport

F. BRONNER, Y. CHARNOT, E.E. GOLUB & T. FREUND

Calcium absorption in the mammal results from two types of movements: The segmenting and peristaltic contractions of the small intestine which move calcium along the intestinal tract, and transmural movement which carries calcium across the intestinal epithelium. Regulation of calcium absorption could, therefore, involve alterations in intestinal contractions or events that affect transmural movement. In what follows we shall first describe studies that deal with the effect of intestinal movement on calcium absorption. We then deal with regulation of transmural movement, emphasizing the feedback regulation of calcium itself.

Earlier work (1, 8, 9) had shown that calcium absorption, studied in the whole animal, is an exponential function of calcium intake, i.e. it approaches a plateau value at high intakes. However, when calcium absorption is studied by an *in situ* loop method the results are quite different. In this technique an intestinal segment is tied at both ends. As a result, segmental and peristaltic movements have been interrupted. Under these conditions (15) calcium absorption is a linear function of the intraluminal calcium concentration. A logical explanation is that under normal absorptive conditions the intraluminal calcium does not remain in contact with the absorptive surfaces long enough for transmural movement to be completed. The amount of calcium that is absorbed is, therefore, less than would have been absorbed in the absence of movement along the intestinal tract. In other words, intestinal movement is a limiting factor for calcium absorption in the living animal.

Transmural calcium movement comprises two classes of concentration-dependent processes: Saturable and non-saturable. At low concentrations saturable processes predominate and some may require the input of metabolic energy to move calcium against an electrochemical gradient. The latter type of saturable process is often termed "active". However,

Department of Oral Biology, The University of Connecticut Health Center, Farmington, Connecticut.

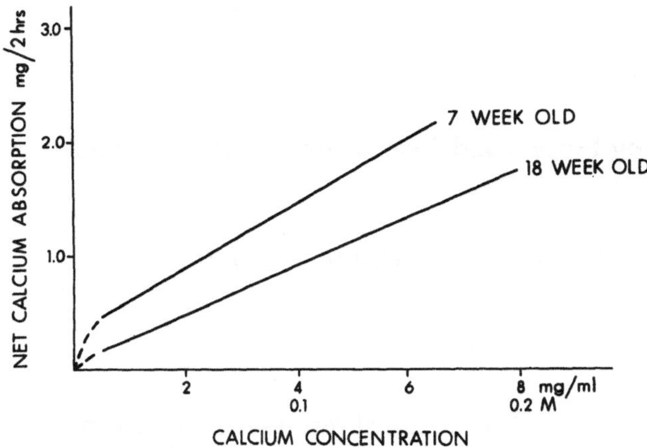

Figure 1. Net calcium absorption (*in situ* loop method (15), $\underline{y}$) as, a function of calcium concentration ($\underline{x}$) in 7-week-old and 18-week-old male Sprague Dawley rats. The animals had been on a semisynthetic diet (1.5% Ca, 1.5% P, III) and the equations describing the two age groups are

$$7\text{-wk-old: } \underline{y} = 0.34\ (\pm 12) + 0.28\ (\pm 0.03)\underline{x}$$
$$18\text{-wk-old: } \underline{y} = 0.06\ \pm(0.14) + 0.21\ (\pm 0.03)\underline{x}$$

with 95% confidence limits shown in parantheses. The data points are not shown. The dashed lines are drawn to indicate that the intercepts are apparent; no measurements were made in that concentration range. The apparent intercepts differ markedly and significantly. This indicates that the major difference in the two groups was due to the process represented by the apparent intercept (i.e. the active calcium transport component, see text).

Adapted from (15).

not all saturable processes necessarily constitute active transport in its strictest definition. Non-saturable processes are often called "passive".
Fig. 1 shows the relationship of saturable to non-saturable transport in the rat. The data were derived from rats at two ages, 7 and 18 weeks. As can be seen, the slopes of the lines were similar, but their intercepts differed. The slopes represent concentration-dependent behavior and may be thought of as reflecting calcium diffusion. The intercepts are only apparent, since there cannot be calcium absorption when the concentration is zero. Moreover, the intercept was virtually zero in the older group. The intercepts may, therefore, reflect active and saturable components of calcium absorption. As a result, the uptake curve may be considered to be the sum of two types of processes, passive diffusion and a saturable process that at least at low calcium concentrations is active. (See also (4)).

Table I.

Characteristics of Duodenal Calcium Binding Proteins in Rat

	Band 1	Band 2
MW (gel chromatography)	11,000	
Apparent Binding Constant, micromolar	1.5	
R_f	0.10	0.19
Specific Binding Activity, arbitrary units	6.0	7.4
Purified Peak B, Total Binding Capacity (nanomoles Ca bound/mg protein)		
High Ca Diet	174	
Low Ca Diet	233	
Purified Peak B, protein ratio, Band 1 / Band 2		
High Ca Diet	1.9	
Low Ca Diet	0.2	

Note: Purified Peak B refers to material that has been chromatographed twice.

Our knowledge of saturable calcium transport has grown greatly in recent years. This has been due to the discovery of the vitamin D-dependent calcium binding proteins (CaBP) in the mammalian and chicken duodenum (4, 14) and to advances in the understanding of vitamin D metabolism (6, 7). Our own work has centered on the function and regulation of CaBP in the rat (2, 4, 5).

Murine CaBP differs from the better known avian CaBP in several important ways. It is a much smaller molecule, MW = 11,000 daltons, and occurs in the rat as two molecules (Table I). Both have about the same molecular weight, but differ in charge, so that band 1 has a R_f of 0.1, whereas band 2 has about twice the mobility in our modified Davis system (4), i.e. R_f = 0.19. Both bands have similar specific calcium binding capacity (Table I), so that it is not necessary to determine the amount of each band separately when assaying for CaBP as a function of calcium status.

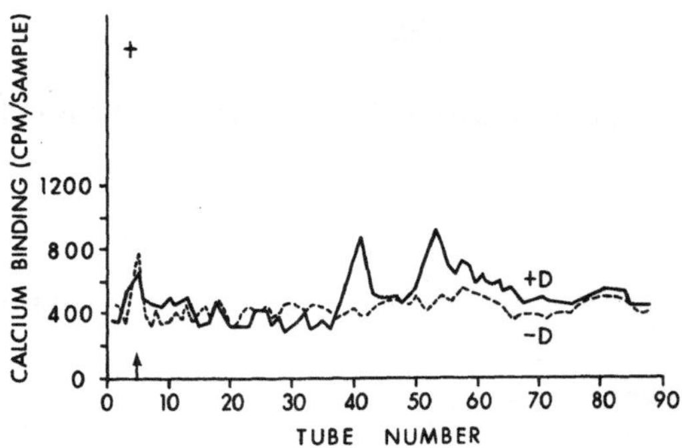

Figure 2. Electrophoretogram of rat intestinal calcium binding proteins. Intestinal scrapings from control and vitamin D-deficient animals were homogenized, centrifuged and the supernates fractionated on Sephadex G-50. The eluate constituting peak B, i.e. the second, vitamin D-dependent calcium binding peak, was then pooled and subjected to preparative, acrylamide gel electrophoresis. Note the absence of the two peaks in the material from the vitamin D-deficient animals. When the two peaks were rechromatographed on Sephadex G-50, they eluted in the appropriate place. The arrow indicates the migration front.

Fig. 2 shows that both bands are vitamin D-dependent. Some investigators have had difficulty demonstrating vitamin D dependence in the rat (3). In our hands the vitamin D dependence has been quantitative, i.e. the amount of CaBP (the sum of the activity of both bands) is a log-dose function of the amount of vitamin D (in the form of 25-hydroxyvitamin D_3) administered to vitamin D-deficient animals (Fig. 3). Moreover, when in similar animals we tested for active calcium transport, as determined by an everted gut technique, we found a similar, dose-dependent relationship (Fig. 4).

From these data it appears that there is a dose relationship between CaBP and active calcium transport, both varying in the same direction. Moreover, as shown by Schedl and colleagues (11), calcium absorption is depressed in rats made diabetic by alloxan treatment. These rats also have low levels of CaBP. Whatever the mechanism by which CaBP participates in saturable calcium transport, one can now state that when CaBP goes up, saturable calcium transport rises, and when CaBP levels drop, saturable calcium transport also drops.

An important characteristic of a regulatable process is the existence of a negative feedback mechanism. A major distinction between active and passive processes is that the former are directly subject to regulation, while the latter do not generally exhibit autocatalytic, or feedback — regulated behavior. In trying to determine whether CaBP plays a role in

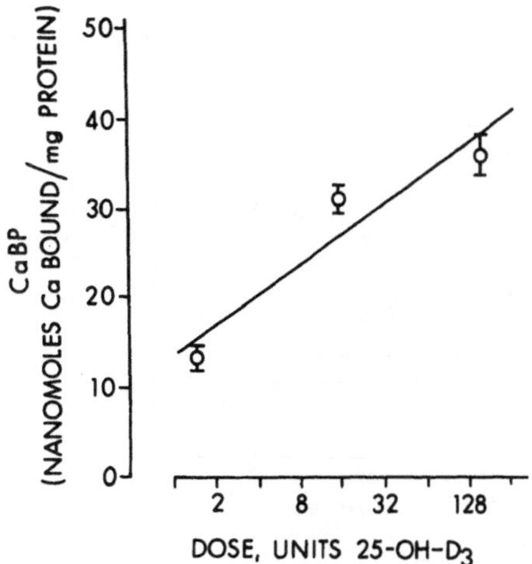

Figure 3. Intestinal calcium binding protein levels ($\underline{n}$; nanomoles Ca bound/mg protein) in hypocalcemic, vitamin D-deficient rats treated with varying doses of 25-hydroxy-vitamin D_3 30 h earlier. Each determination was made on a pooled sample from 6 rats.

The equation of the regression line is:

$$n = 13.2 + 3.5 \log_2 \text{dose}$$

where dose is units, administered intraperitoneally.

(From (2). Reproduced with permission from the American Journal of Physiology).

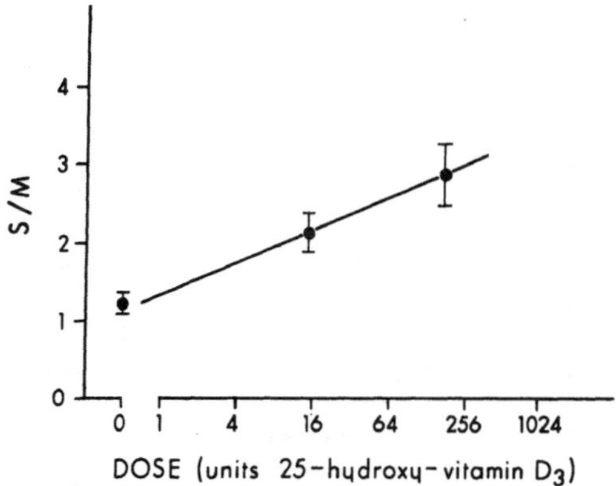

Figure 4. Serosal to mucosal ^{45}Ca ratio (S/M) in everted intestinal sacs from vitamin D-deficient, hypocalcemic rats.

15 male weanling (Sprague-Dawley) rats were placed on a vitamin D-deficient, semi-synthetic regimen (0.5% Ca, 0.5% P). Approximately 3 weeks later, when they had become hypocalcemic ($<$ 7 mg Ca/100 ml), 7 received 15 units and three 150 units 25-OH-D_3 by intraperitoneal injection. 30 h later the treated animals were sacrificed and everted sacs prepared from the proximal 5 cms of the duodenum according to Schachter and Rosen (10). The calcium concentration was 10^{-5}M and ^{45}Ca had been added to the buffer on either side. Measurements were made after 90 min incubation.

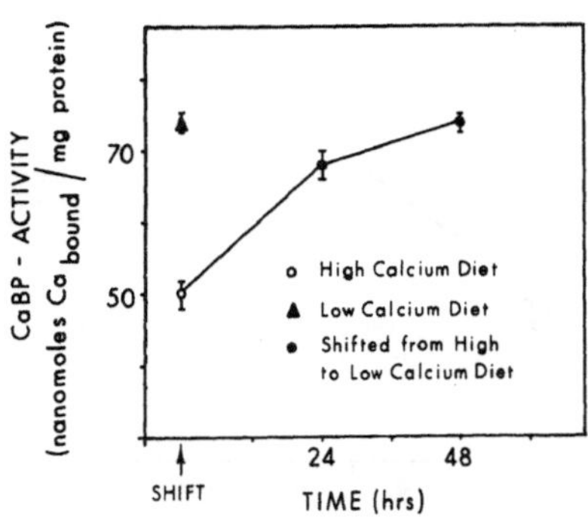

Figure 5. The effect of a dietary shift on calcium-binding protein (CaBP). 48 male Sprague-Dawley rats (BW = 125 g) were divided into two lots, 36 on diet III (1.5% Ca, 1.5% P), 12 on diet I (0.06% Ca, 0.2% P). Ten days later 12 animals on diet III and the animals on diet I were killed, while the remaining 24 animals on diet III were placed on diet I (shifted), and lots of 12 were killed 24 and 48 h later. Body weights of all animals were similar (185 g). CaBP analysis by quantitative, competitive binding assay on the purified material (4).

Adapted from (4).

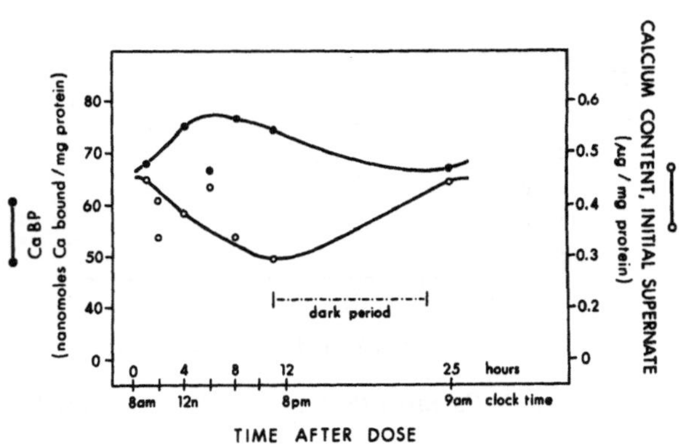

Figure 6. Diurnal variation of CaBP and calcium content of rat duodenum.

37 male Sprague-Dawley rats (BW = 191 g) that had been on a low-calcium (0.06% Ca, 0.2% P) semi-synthetic regimen for 3 weeks were killed in groups at the times indicated. Their CaBP was fractionated and analyzed (4). The calcium content refers to the calcium concentration of the initial supernate (S—100) from which CaBP was fractionated.

saturable calcium transport we searched for evidence of a negative feedback relationship between calcium absorption and CaBP.

Our evidence for such a feedback relationship is based on several kinds of experiments:

i) CaBP levels are higher in animals on low calcium diet than in animals on a high calcium diet (Table I; (4)).

ii) When animals are shifted from a high to low calcium diet, their CaBP rises in 24 h and reaches the steady state level of animals on the low calcium diet by 48 h (Fig. 5).

iii) Fig. 6 shows that when animals were sacrificed throughout the day and night, there was a characteristic diurnal variation in duodenal CaBP content, with CaBP high at times when food and calcium intake were low. Moreover, there was an opposite diurnal variation in the calcium content of the supernate from which CaBP was fractionated (See also (4)).

iiii) Finally, when we plotted all our CaBP data as a function of the calcium content of the supernate from which the CaBP was subsequently fractionated we found the negative, curvilinear relationship shown in Fig. 7. To appreciate that this relationship is not trivial, it must be emphasized that the CaBP content is determined after gel chromatography, when the calcium content of the pooled fraction that contains CaBP is similar in all diet groups. Moreover, the quantitative assay of CaBP (4) involves total

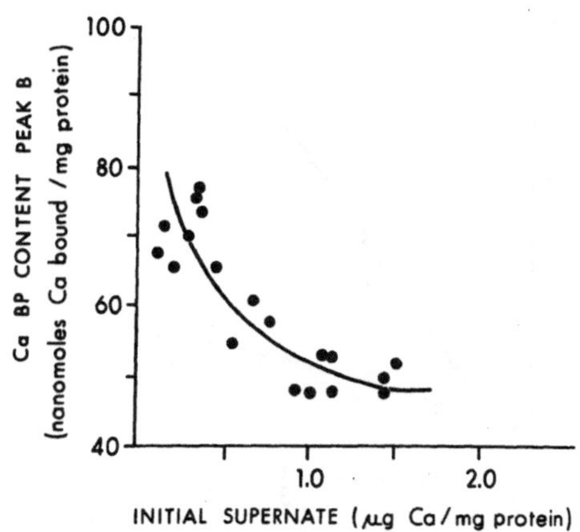

Figure 7. CaBP content of mucosal scrapings as a function of their calcium content.

Mucosal scrapings of male Sprague-Dawley rats were homogenized and centrifuged (100,000 g). The supernate (initial supernate) was analyzed for calcium and protein content and subjected to gel chromatography (4). Peak B of the eluate was then analyzed for CaBP. Each point is based on material from 6 rats. Eight different experiments were done over a span of 18 months. Peak B contains the vitamin D-dependent CaBP (4).

calcium concentrations of 1 to 100 μg Ca/ml, whereas the sample calcium content alone is below 1 μg Ca/ml. Finally, in separate experiments it was shown that the calcium content of duodenal tissue varies with calcium intake and is proportional to the calcium content of the supernate. It seems probable, therefore, that the supernate calcium concentration reflects calcium in transit.

The inverse relationship shown in Fig. 7 therefore is consistent with a negative feedback relationship between CaBP and absorbed calcium.

On the basis of these data we, therefore, propose the following set of relationships:

1) when CaBP goes up, calcium transport goes up;

2) when CaBP goes down, calcium transport goes down;

but, 3) when calcium transport goes up, CaBP goes down (negative feedback relationship).

Although these observations and considerations are consistent with a direct involvement of CaBP in saturable calcium transport, they do not shed light on how this may be brought about.

In recent months we have attempted to answer this question by studying a system of isolated duodenal cells (5). We obtained the cells by shaking minced duodenal tissue in an oxygenated medium that contains hyaluronidase. The harvested cells are resuspended and assayed for calcium uptake by a millipore filtration technique. Fig. 8 shows that the cells from

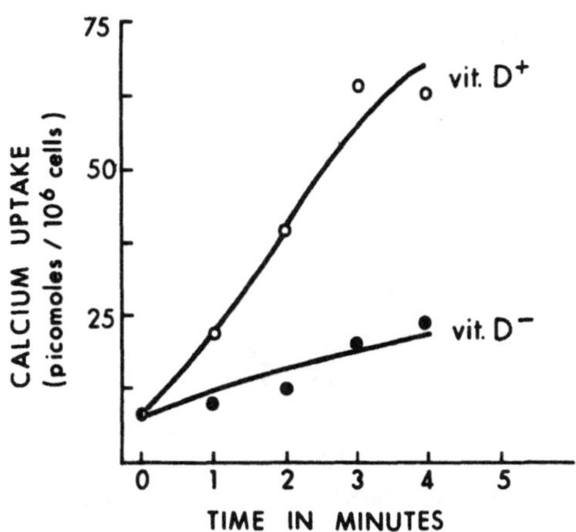

Figure 8. Calcium uptake of isolated intestinal cells from vitamin D-replete and vitamin D-deficient rats.

Male weanling rats were placed on a low calcium (I, 0.06% Ca, 0.2% P) semi-synthetic regimen with no or 2200 units vitamin D_2/kg. When the animals on the deficient diet had become hypocalcemic ($<$ 6 mg/100 ml plasma), both groups were killed, the duodenum excised and cells harvested and assayed for calcium uptake by millipore filtration.

vitamin D-replete animals took up 3x more calcium than cells from vitamin D-deficient animals. Moreover, when minced duodenal tissue from deficient animals was treated during the 90 min isolation with 1,25-dihydroxy-vitamin D_3, the isolated cells took up more than twice the amount of calcium than control cells (Fig. 9). Treatment with the immediate precursor, 25-hydroxy-vitamin D_3, failed to raise calcium uptake significantly. Thus, treatment of vitamin D-deficient tissue with the active vitamin D metabolite for only 90 min yielded cells whose calcium uptake was close to normal.

We then looked for CaBP and found CaBP in cells from replete animals (panel A, Fig. 10), were unable to detect CaBP in cells from deficient animals (panel B, Fig. 10), but detected CaBP in cells that had been treated with 1,25-dihydroxy-vitamin D_3 *in vitro* (panel C, Fig. 10). These measures have since been quantitated (16). Moreover, CaBP was not detected in cells from tissue that had been treated with the precursor, 25-hydroxyvitamin D_3.

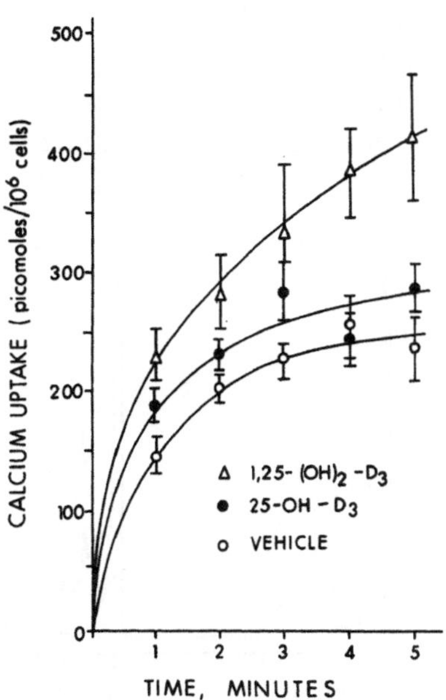

Figure 9. The effect of the addition of 1,25-dihydroxy-vitamin D_3 to the isolation medium on calcium uptake by rat intestinal cells.

Cells were harvested and assayed as described in the legend, Fig. 8, except that during the isolation procedure approximately one-third of the intestinal strips were exposed to isolation medium that also contained 4.7×10^{-8} M 1,25-$(OH)_2$-D_3 and one-third to medium that contained 4.7×10^{-8} M 25-(OH)-D_3. The remainder of the strips were placed in an isolation medium that contained a control volume of a 1:1 mixture of propylene glycol and ethanol ("vehicle").

From (16).

These *in vitro* experiments reveal how short is the time needed to induce CaBP and that the time of induction of increased Ca uptake parallels that period. As yet, we only have one time point, but our data, however preliminary, are fully consistent with a close, direct link between CaBP and intestinal calcium transport.

So far our presentation has dealt only with saturable calcium transport in relation to net uptake in the duodenum. Recent studies from the laboratory of Kimberg (12, 13) have produced evidence of the existence of regulation of the movement of calcium from blood to lumen, in the lower ileum. Early work from our laboratory (1) had shown that in the rat fecal endogenous calcium increases with increased calcium intake. This puzzling observation can now be explained by the work of Walling and Kimberg (12, 13). These investigators have shown that in animals on high calcium diets the unidirectional ileal flux from lumen to blood is markedly decreased, while the blood to lumen flux is virtually unchanged. As a result, net calcium movement is in the direction of the lumen. In high calcium diets, therefore, there occurs not only a marked decrease of the CaBP-related active calcium entry in the duodenum, but also a marked decrease in the ileal lumen-to-blood flux. The net result of a high calcium diet, therefore, is heavy reliance in the duodenum on passive, diffusion-dependent inward processes, whereas in the ileum there results net flux

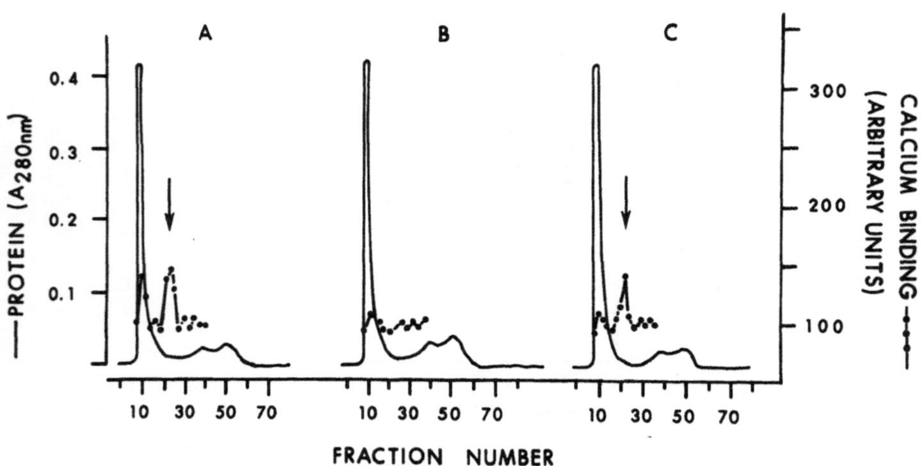

Figure 10. Elution profiles showing presence and absence of calcium binding protein. Isolated intestinal cells were homogenized, centrifuged (38,000 g, 60 min) and 2 ml of the supernate lyophilized. The sample was taken up in 0.5 ml elution buffer (0.02 M ammonium acetate, 1 mM mercapoethanol, pH 7.2) and chromatographed on a Sephadex G-50 solumn (1.5 x 30 cm). Protein was monitored at 280 nm and calcium binding determined by a competitive Chelex resin assay (4).

into the lumen. On a low calcium diet, on the other hand, CaBP livels rise, saturable duodenal calcium transport increases, whereas ileal net flux from blood to lumen decreases.

To this dual feedback effect of luminal calcium should be added an interesting third effect that is not understood. We have found (Table I) that the relationship of the two bands of CaBP is also calcium-dependent. As shown in Table I, in animals on a high calcium diet the ratio of band 1 to band 2 is 2.2, whereas in animals on the low calcium diet this ratio is 0.1, with virtually all of CaBP in the form of band 2.

SUMMARY

We have discussed the calcium-dependent feedback regulation of calcium absorption. This appears to involve two major mechanisms: a) The calcium-binding protein-related saturable calcium transport in the duodenum. This increases when calcium intake is lowered. b) The ileal calcium flux from lumen to blood which decreases when calcium intake is high. This is accompanied by an unmodified flux from blood to lumen so that net ileal flux into the lumen increases with high calcium intake.

ACKNOWLEDGEMENTS

These investigations have benefited from support by the National Institutes of Health (grants AM 14251 and AM 16408) and the University of Connecticut Research Foundation. Y. Charnot was holder of a World Health Organization travelling fellowship. Her permanent address is Université Claude Bernard, Lyon, France. T. Freund's current address is Fairleigh Dickinson University, New Jersey. We thank Janet Condino and Marilyn Reid for devoted technical assistance.

We thank Dr. A.W. Norman, University of California, Riverside, California, for making 1,25-dihydroxy-vitamin D_3 available to us.

REFERENCES

1. Bronner, F., Sammon, P.J., Stacey, R.E. & Shah, B.G.: Role of thyrocalcitonin in the regulation of the blood calcium level. *Biochem. Med.* 1, 261—279 (1967)
2. Bronner, F. & Freund, T.S.: Intestinal calcium binding protein: A new quantitative index of vitamin D-deficiency in the rat. *Amer. J. Physiol.* 229, 689—694 (1975)
3. Bruns, M.E. & Avioli, L.V.: The activity and the synthesis of calcium binding protein during vitamin D replacement in the rachitic rat. In: *Calcium Regulating Hormones.* Proc. 5th Parathyroid Conf., Oxford, 1974, edited by Talmage, R.V., Owen, M. & Parsons, J.A. (eds.). Excerpta Medica, Amsterdam, pp 336—345, 1975
4. Freund, T.S. & Bronner, F.: Regulation of intestinal calcium binding protein by calcium intake in the rat. *Amer. J. Physiol.* 228, 861—869 (1975)

5. Freund, T.S. & Bronner, F.: Vitamin D-dependent calcium uptake by isolated intestinal cells. *Fed. Proc.* **34**, 475 (1975)

6. Norman, A.W. & Henry, Helen: 1,25-dihydroxycholecalciferol — a hormonally active form of vitamin D_3. *Recent Progr. Hormone Res.* **30**, 431—480 (1974)

7. Omdahl, J.L. & DeLuca, H.F.: Regulation of vitamin D metabolism and function. *Physiol. Rev.* **53**, 327—372 (1973)

8. Sammon, P.J., Stacey, R.E. & Bronner, F.: Further studies in the role of thyrocalcitonin in calcium homeostasis and metabolism. *Biochem. Med.* **3**, 252—270 (1969)

9. Sammon, P.J., Stacey, R.E. & Bronner, F.: Role of parathyroid hormone in calcium bomeostasis and metabolism. *Amer. J. Physiol.* **218**, 479—485 (1970)

10. Schachter, D. & Rosen, S.M.: Active transport of ^{45}Ca by the small intestine and its dependence on vitamin D. *Amer. J. Physiol.* **196**, 357—362 (1959)

11. Schneider, L.E., Wilson, H.D. & Schedl, H.P.: Effect of alloxan diabetes on duodenal calcium-binding protein in the rat. *Amer. J. Physiol.* **227**, 832—838 (1974)

12. Walling, M.W. & Kimberg, D.V.: Active secretion of calcium by adult rat ileum and jejunum *in vitro*. *Amer. J. Physiol.* **225**, 415—422 (1973)

13. Walling, M.W. & Kimberg, D.V.: Calcium absorption or secretion by rat ileum *in vitro*: Effects of dietary calcium intake. *Amer. J. Physiol.* **226**, 1124—1130 (1974)

14. Wasserman, R.H. & Taylor, A.N.: Some aspects of the intestinal absorption of calcium with special reference to vitamin D. In: *Mineral Metabolism — An Advanced Treatise*, Comar, C.L. & Bronner, F. (eds.), Academic Press, New York, Vol. III, pp. 321—403, 1969

15. Zornitzer, A.E. & Bronner, F.: *In situ* studies of calcium absorption in rats. *Amer. J. Physiol.* **220**, 1261—1266 (1971)

16. Freund, T.S. & Bronner, F.: Stimulation *in vitro* by 1,25-dihydroxy-vitamin D_3 of intestinal cell calcium uptake and calcium-binding protein. *Science* **190**, 1300—1302 (1975)

Duodenal Calcium-binding-protein (CaBP) in the Sodium Deficient Growing Rat

M. THOMASSET, P. CUISINIER-GLEIZES & H. MATHIEU

During the past decade a specific calcium binding protein (CaBP) was detected in the intestinal mucosa of several species (6, 11). The exact role of this protein is not yet clearly defined. There is, however, a high direct correlation between CaBP and the duodenal capacity to transport calcium under definite conditions known to alter calcium absorption.

Indeed, vitamin D deficiency depletes CaBP while vitamin D (or metabolites) injections to vitamin D-deficient animals restores CaBP (6, 9) with corresponding changes in calcium transport. On the other hand, in the rat and in the chick, low calcium diets increase the production of CaBP (6, 12) and stimulate the synthesis of an active vitamin D metabolite (1,25-DHCC) (3). The calcium transport increases in parallel with the decrease in dietary calcium level and, therefore, with the decrease of the amount of absorbed calcium.

Recent investigations have demonstrated that sodium is required for the transport of calcium across the intestine (1, 2, 8).

An *in vivo* study from this laboratory (4) showed a decrease in the fractional intestinal calcium absorption in sodium deficient rats.

Since CaBP may play an important role in duodenal calcium absorption, we now report studies of CaBP in growing rats kept on adequate vitamin D and calcium intake and in which fractional calcium absorption decreased in response to a sodium deficiency.

MATERIALS AND METHODS

After a one week adaptation on a semi-synthetic diet, 4 weeks old Sherman male rats were divided in two groups. Group I was fed a low

Unite de Recherches sur le Metabolisme Hydro-Mineral, Inserm — U.120 — 44, Le Vesinet.

sodium diet (0.02%). Group II was *pairfed* and received a regimen adequate in sodium (0.37%) and served as controls. The diets supplied 0.48% Ca, 0.44% P, 0.33% K and 1 i.u. of vitamin D_3/g. The animals received deionized water *ad lib*. 8 days later all the rats were killed. The last 24 h urine and blood were collected. The first 15 cm of intestine were rapidly removed and everted. Mucosa from 6 animals was slitted and homogenized in an iced Tris buffer (0.013 M Tris-HC1, 0.12 M NaCl, pH 7.4) with the aid of a Potter-Elvehjem Teflon homogenizer. The homogenate was centrifuged at 100,000 g for 1 h. The supernate (S_{100}) was decanted, lyophilized, then redissolved and further fractioned by elution on Sephadex G-50 with 0.02 M ammonium acetate and 1 mM mercaptoethanol as an elution buffer.

Each fraction was assayed for calcium binding activity by a micro-modification of the competitive Chelex binding assay (10). Thus, peak B vitamin D-dependent was isolated and considered as CaBP partially purified.

To determine the apparent calcium binding constant (K_d) and the binding capacity ($\underline{n}$), the assay procedure was modified by varying the added calcium concentration (6).

Serum urine and S_{100} were analyzed for sodium and calcium by atomic absorption spectrophotometry and for phosphate by the colorimetric method (phosphomolybdate reduction). During the steps of CaBP isolation, protein was determined spectrophotometrically (7). In the serum, protein was evaluated by Lowry's method.

A similar experiment was made with 9 weeks old Sherman rats.

The *in vivo* percentage of calcium absorption was evaluated with the aid of chrome 51 as a non-absorbed reference substance. The general system of analysis and the calculation of net fractional absorption were described previously (5).

RESULTS

Table I lists the results. The drop of sodium urinary excretion was the indication of the state of sodium deprivation. In the serum, sodium and phosphorus were unchanged whereas calcium was slightly increased. The concomitant increase in protein concentration (6.7 v. 6.0 p < 0.001) suggested some degree of hemocencentration which could explain the slight rise of calcemia. In the urine solely, a rise of phosphaturia was observed.

TABLE I — EFFECT OF SODIUM DEFICIENCY (one week)

RATS		Ca					Ca BP		Na		P	
		INTAKE mg/day	FRACT. ABS. %	S 100 μg/g mucos	S mg/100ml	U mg/l	K_d μM	n nMCa/mgPr	S mEq/l	U mEq/24h	S mg/100 ml	U mg/24h
5 Weeks	LOW Na intake	39.4 ± 1.36	71.4 ± 4.76	16.0 ± 1.92	10.5 ± 0.08	30.2 ± 2.24	1.4 ± 0.13	27.1 ± 5.39	130.9 ± 1.34	0.002 ± 0.000	9.1 ± 0.22	8.5 ± 0.55
	CONTROLS	39.4 ± 1.36	82.6 ± 1.61	10.1 ± 1.40	10.0 ± 0.09	27.5 ± 2.58	1.2 ± 0.28	15.4 ± 4.08	130.1 ± 1.43	0.712 ± 0.0561	8.8 ± 0.19	11.4 ± 0.68
	p		< 0.05	< 0.05	< 0.001	n.s.	n s	< 0.01	n.s	< 0.001	n. s.	< 0.001
9 Weeks	LOW Na intake	55.0 ± 2.48	15.8 ± 4.26	34.8 ± 4.42	10.3 ± 0.06	22.0 ± 1.89	1.2 ± 0.14	18.2 ± 3.33	131.5 ± 0.91	0.005 ± 0.0000		
	CONTROLS	55.0 ± 2.48	29.9 ± 5.32	19.7 ± 3.47	10.3 ± 0.07	18.8 ± 1.52	0.8 ± 0.34	7.1 ± 1.36	133.0 ± 1.00	2.843 ± 0.1212		
	p		= 0.05	< 0.05	n.s.	n s.	n.s	< 0.02	n. s.	< 0.001		

Abbreviations : S : serum S_{100} : Supernate (100.000g)   n : binding capacity
U : urine K_d : apparent calcium binding constant   Fract; Abs : Fractional Absorption
Each data represents means of 36 5 weeks animals
and 24 9 weeks animals

In 5 weeks old rats the net fractional absorption of calcium was high, more than 80%, and decreased to 70% in sodium deprived animals. In 10 weeks old rats the calcium absorption coefficient fell to 30%. In addition, a low sodium diet imposed for one week reduced calcium absorption by one-half.

The analysis of supernate (S_{100}) showed a rise in calcium concentration (μg Ca/g mucosa) in each group of sodium deficient rats. The calcium concentration was higher in older rats that in younger ones and paralleled the increase in calcium intake as well.

The absorbance and calcium binding profiles are shown in Fig. 1. When material was derived from the sodium depleted animals, peak B profile (vitamin D-dependent) was similar to the one from the control animals. To characterize further the peak B material, the fractions of high activity were pooled. Material at the stage of purification yielded an apparent dissociation constant K_d of 1.2 (μM). When the material was derived from the younger animals on a low sodium diet the specific binding activity (n, nanomoles Ca bound per mg protein) was higher (n = 27) than the one from the control animals (n = 15).

Peak A, the void volume peak, vitamin D-non-dependent was characterized by a binding capacity of about 1. This parameter was unchanged in low sodium animals.

The different changes induced by a low sodium diet in 5 weeks old rats were also observed in 10 weeks old animals. However, in the control animals the specific binding activity diminished with age.

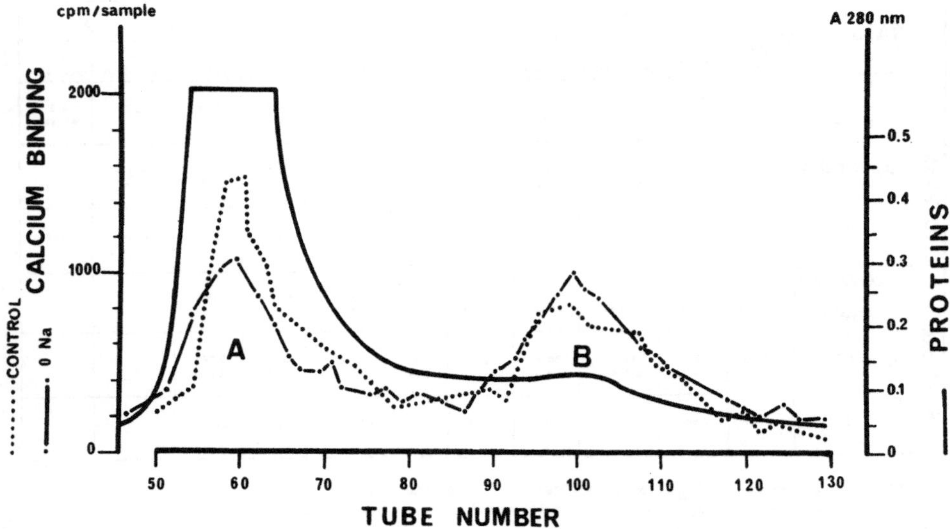

Figure 1. Sephadex G-50 elution profiles of duodenal mucosa supernates from sodium deficient (0 Na) and control animals.

DISCUSSION

It is apparent from the above results, that a low sodium diet imposed for one week leads to: A decrease in net fractional calcium absorption, an increase in supernate calcium concentration, and the maintenance of peak B (CaBP) with a slight enhancement of its binding activity.

Because calcium intake was adequate, the decrease in net fractional calcium absorption indicates an actual drop of the amount of calcium absorbed. It was surprising to find simultaneously, an enhancement of supernate calcium concentration which presumably reflects an accumulation of mucosal calcium. These two parallel findings suggest that the calcium extrusion is blocked at the cellular level in sodium deficient rats. This interpretation is in agreement with recent *in vitro* studies (1, 2) indicating that the step of calcium release at the basal lateral membrane needs sodium.

Up to now the CaBP regulation was studied under a variety of physiological conditions known to alter calcium absorption in consequence to calcium intake changes. These prior studies established firstly a direct correlation between calcium transport capacity and CaBP content (9), and secondly, an inverse relationship between mucosal concentration and CaBP (6). Such results were interpreted as reflecting a regulation of CaBP by the mucosal calcium concentration (6). In contrast, our results show parallelly an increase in mucosal calcium, and the maintenance of peak B (CaBP) with a slight enhancement of its binding activity. The increase in binding activity of CaBP is interpreted as the effect of the physiological regulation system in order to facilitate calcium transport across mucosal cells, in response to the decrease in calcium actually absorbed.

The present study suggests that calcium release, reflecting the actual amount of absorbed calcium, could be also involved in regulation mechanism of CaBP. Measurements of duodenal calcium influx and calcium efflux in our sodium deficient animals are required before a definitive statement concerning this possibility can be made.

Thus, the sodium deprived rat provides a new model system for examining the role of CaBP in duodenal calcium transport and for studying its regulation mechanism.

CONCLUSIONS

Sodium deficiency imposed for one week on growing rats with adequate vitamin D and calcium intake led to: — a decrease of fractional calcium absorption, — an increase of intracellular calcium from duodenal mucosa (S_{100}), — the maintenance of peak B (CaBP) and an enhancement of its calcium binding activity.

According to our results, the increase of calcium binding activity could be interpreted as one physiologic response of the regulating system of calcium homeostasis to the decrease in absorbed calcium.

REFERENCES

1. Birge, S.J., Gilbert, H.R. & Avioli, L.V.: Intestinal calcium transport: The role of sodium. *Science* **176**, 168–170 (1972)
2. Birge, S.J. & Gilbert, H.R.: Identification of an intestinal sodium and calcium dependent phosphatase stimulated by parathyroid hormone. *J. clin. Invest.* **54**, 710–717 (1974)
3. Boyle, I.T., Gray, R.W. & DeLuca, H.F.: Regulation by calcium of *in vivo* synthesis of 1,25-dihydroxycholecalciferol and 21,25-dihydroxycholecalciferol. *Proc. nat. Acad. Sci. (Wash.)* **68**, 2131–2134 (1972)
4. Cuisinier-Gleizes, P. & Mathieu, H.: Effect of low sodium diet on the intestinal absorption of calcium in rats. *Rev. franç. Étud. clin. biol.* **16**, 273–277 (1971)

5. Cuisinier-Gleizes, P., Desbuquois, B. & Mathieu, H.: Mesure de l'absorption intestinale apparente du calcium chez le rat à l'aide de chlorure de chrome radio actif comme marqueur. *Rev. franç. Étud. clin. biol.* **13**, 398–404 (1968)
6. Freund, T. & Bronner, F.: Regulation of intestinal calcium-binding protein by calcium intake in the rat. *Amer. J. Physiol.* **228**, 861–869 (1975)
7. Layne, E.: Spectrophotometric and turbidimetric methods for measuring protein. In: *Methods of Enzymology*, Colowick, S.P. and Kaplan, N.O. (eds.), Academic Press, New York **3**, 447–454 (1957)
8. Martin, D. & DeLuca, H.F.: Influence of sodium on calcium transport by the rat small intestine. *Amer. J. Physiol.* **216**, 1351–1359 (1969)
9. Taylor, A.N. & Wasserman, R.H.: Correlations between the vitamin D-induced calcium binding protein and intestinal absorption of calcium. *Fed. Proc.* **28**, 1834–1838 (1969)
10. Wasserman, R.H., Corradino, R.A. & Taylor, A.N.: Vitamin D-dependent calcium binding protein: Purification and some properties. *J. biol. Chem.* **243**, 3978–3986 (1968)
11. Wasserman, R.H. & Taylor, A.N.: Vitamin D_3-induced calcium binding protein in chick intestinal mucosa. *Science* **152**, 791–793 (1966)
12. Wasserman, R.H. & Taylor, A.N.: Vitamin D-dependent calcium binding protein: Response to some physiological and nutritional variables. *J. biol. Chem.* **243**, 3987–3993 (1968)

Transepithelial Calcium Transport Enhanced by Xylose and Glucose in the Rat Jejunal Ligated Loop

D. PANSU[1], M.C. CHAPUY[2], M. MILANI[2] & C. BELLATON[1]

Calcium absorption is increased by lactose and many other sugars such as xylose, mannose and arabinose, several polyalcools such as sorbitol, mannitol and xylitol and hexosamines for example glucosamine (7, 16). Lactose prevents tetany in parathyroidectomized dogs (3) and osteoporosis in lactating rats (6). The increased absorption was observed in humans where a positive balance was obtained in osteoporotic patients receiving 30 g of lactose per day (12). Some results suggest a direct effect without involving a synthesis of an intermediate carrier in the cell: calcium absorption is increased if the sugar and the mineral are present together in the lumen, no lag time is required to obtain an increased absorption (1, 11), the effect is not suppressed by actinomycin (4, 10, 11) it occurs in Vitamin D deficient rats (7). However, the exact mechanism is still unknown. The present study was undertaken to delineate the minimal calcium and xylose concentrations needed to observe an increase in calcium absorption and to characterize the kinetics of calcium absorption in the presence of sugars.

MATERIAL AND METHODS

Wistar rats of both sexes weighing 200—250 g were used. They received from weaning a standard diet that containing 0.84% Ca, 0.78% P and 400 I.U./100 g Vitamin D. After 24 h of fasting the animals were anesthetized by an intraperitoneal injection of sodium pentobarbital (4 mg/100 g). The small intestine was exposed, the jejunal loop was tied off 10 cm distal to the ligament of Treitz and 14—16 cm further distal from the first ligature.

1) Ecole Pratique des Hautes Etudes, Unité de Recherche Inserm U 45, Hôpital Edouard Herriot, Lyon.
2) Clinique de Rhumatologie, Hôpital Edouard Herriot, Lyon.

Exactly 3 ml of the test solution were injected. Rats were maintained under anesthesia during the experiment. After 2 h the final content of the loop was collected, the residual Ca and sugar were estimated. Histological control was made for some sacs.

In the first experiment, glucose or xylose concentration in test solutions varied from 0 to 200 mg/ml while calcium concentration varied from 0.1 mg/ml to 2 mg/ml. Isotonic saline was used as a control. Six rats were tested for each concentration. Keeping constant the concentration of xylose (200 mg/ml = 1.33 M) and glucose (200 mg/ml = 1.11 M) the calcium concentration was varied from 2 mg/ml to 10 mg/ml. Controls were made with two NaCl solution 9 mg/ml (0.3 M) and 36 mg/ml (1.2 M). Each solution contained polyethylene glycol 4,000 (PEG) 5 mg/ml as a volume marker. Two or three rats were tested for each concentration.

Calcium was measured by an EDTA method using the Patton and Reeder reagent as an indicator (13), PEG was measured by the turbidimetric method of Hyden (8) xylose by the method of Roe and Rice (14), glucose by the orthotoluidine method (9), and Na by flame photometry. Calculation was based on the assumption that PEG was unabsorbed. Water exchange was obtained with the aid of the following formula:

$$\text{Final Volume} = \text{Initial Volume} \times \frac{\text{Initial PEG concentration}}{\text{Final PEG concentration}} \quad \text{and}$$

$$\text{water exchange was expressed as the ratio} \frac{\text{Final Volume}}{\text{Initial Volume}} = \frac{V_f}{V_i}$$

This ratio is greater than unity when water is secreted into the lumen and less than unity when water is absorbed.

Calcium absorption was calculated as follows:

Ca Absorption = Initial Ca concentration $\times$ V_i $-$ Final Ca concentration $\times$ V_f

Similar calculations were made for xylose, glucose, and Na. Results were expressed in mg ± SD absorbed during the 2 h study.

RESULTS

The first experiment tested the minimal quantity of sugar needed to obtain an increased absorption of calcium. For a calcium concentration between 0.1 and 2 mg/ml, the absorption of Ca in the NaCl solutions was nearly twenty per cent of the initial quantity. In the presence of both sugars at concentrations lower than 200 mg/ml the Ca absorption did not

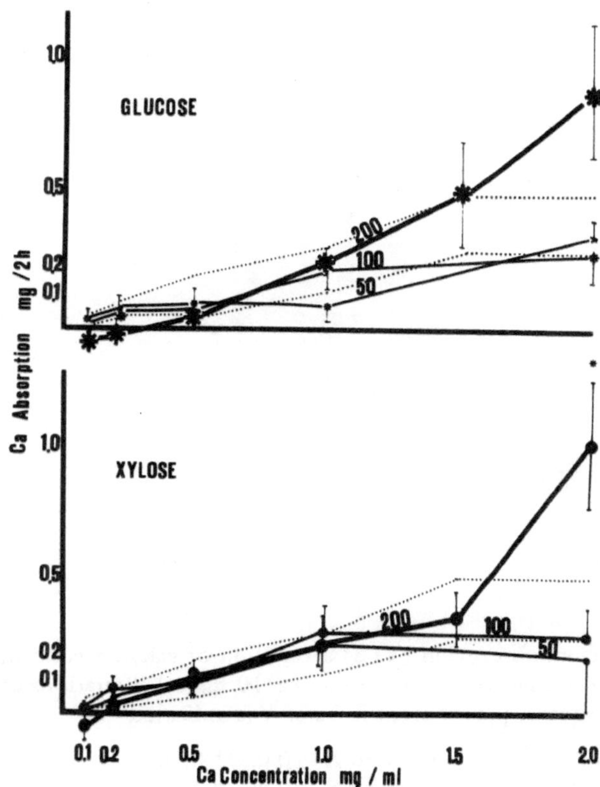

Figure 1. Calcium absorption in the presence of solutions containing glucose * at 50 mg/ml, 100 mg/ml and 200 mg/ml concentration, or xylose ● at 50 mg/ml, 100 mg/ml and 200 mg/ml concentration. Results are presented as mean ± SD. The results obtained for NaCl solutions are presented as a ⋯ area covering the mean ± SD. The Ca absorption is increased significantly (p < 0.05) for a Ca concentration = 2 mg/ml and a sugar concentration = 200 mg/ml.

differ from that observed with the NaCl solutions. When a 200 mg/ml sugar solution was used the Ca absorption increased significantly (p < 0.05) for the 2 mg/ml Ca concentration and attained 0.913 ± 0.284 mg/2 h in the glucose solution, 1.045 ± 0.233 mg/2 h in the xylose solution versus 0.415 ± 0.107 mg/2 h in the NaCl solution (Fig. 1).

The action of both xylose (200 mg/ml) and glucose (200 mg/ml) solutions on Ca absorption was tested for Ca concentrations between 2 mg/ml and 10 mg/ml and compared with the absorption of calcium observed in the presence of a 9 mg/ml NaCl solution and a 36 mg/ml NaCl solution. In the four conditions, the absorption of calcium was linearly correlated with the Ca concentration. The slope obtained for the three hypertonic solutions

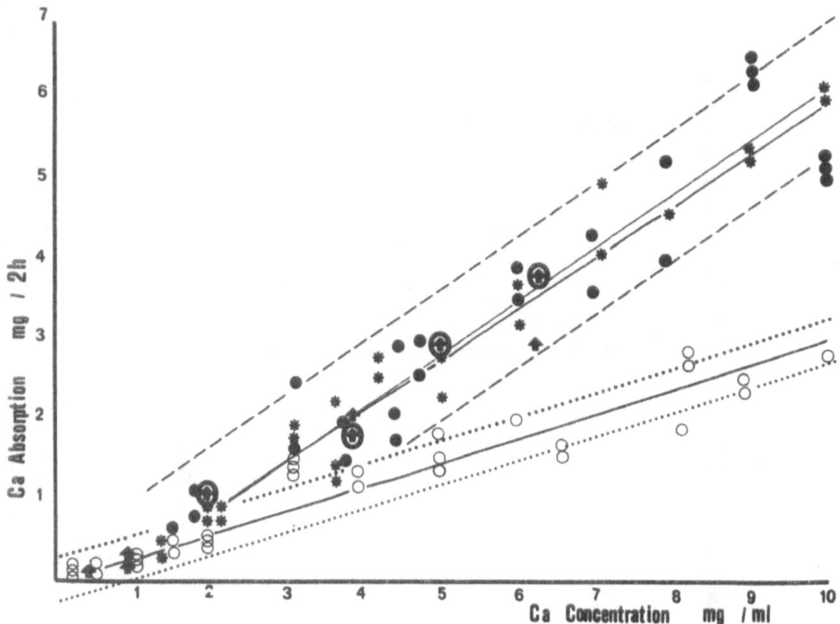

Figure 2. Net calcium absorption (Y) as a function of calcium concentration (X) for calcium concentration between 1.5 mg/ml and 10 mg/ml. Equations of the lines were derived by a least squares procedure, the Sd of Y is indicated

for 9 mg/ml NaCl	○ Y = 0.30 X − 0.03 (± 0.10)	n = 38	r = 0.97
for 36 mg/ml NaCl	◯ Y = 0.59 X − 0.26 (± 0.29)	n = 15	r = 0.97
for 200 mg/ml NaCl xylose	● Y = 0.66 X − 0.44 (± 0.33)	n = 24	r = 0.98
for 200 mg/ml glycose	* Y = 0.65 X − 0.35 (± 0.25)	n = 27	r = 0.96

The slope of the lines obtained in the hypertonic solution differed significantly from the slope obtained in the presence of the isotonic saline solution.

(xylose, glucose, NaCl 36 mg/ml) differed significantly from the slope obtained in the presence of NaCl 9 mg/ml (p < 0.01). As a result, the percentage of absorption in hypertonic solution rose beyond 60% whereas it did not reach 30% in isotonic NaCl solutions (Fig. 2).

During the experiment the morphological and histological status were controlled. The intestinal loops in contact with hypertonic solutions gradually increased in diameter and length. The histological control showed in all cases that the width and height of the villi were normal, however, there was an enlargement of the intercellular spaces and dilatation of the capillaries and lymphatic vessels. For Ca concentration higher than 4 mg/ml especially in NaCl hypertonic solutions, hemorragic dots were seen and a large mucous secretion appeared. In these cases the fixation for histological preparation was difficult due to the hypertonic status of the cells. After fixation the cells appeared clear and dilated by the hypotonic shock as well as by the test.

The absorption of sugars and the flux of water were measured for each experiment. No relationship between Ca absorption and either water exchange or sugar absorption was found. Numerical results are given for the experiment undertaken with the 2 mg/ml Ca solution, in this case the Ca solution was quite physiological and the mucosa did not show any signs of important damage. For the 9 mg/ml saline solution and the 50 mg/ml sugar solution the calculated osmolarity of the test solution was 430 mOsm/l in the glucose solution, 460 mOsm/l in the NaCl solution, and 480 mOsm/l in the xylose solution. The calcium absorption was similar when using either the solution containing xylose (0.29 ± 0.3 mg/2 h), or glucose (0.35 ± 0.07 mg/2 h), or NaCl (0.41 ± 0.10 mg/2 h). The water and sugar movement, however, was dependent on the sugar added. In the glucose solution the absorption of glucose (43.7 ± 10 mg/2 h) represented 87% of the initial quantity and occurred with water absorption ($\frac{Vf}{Vi}$= 0.99) and with a small secretion of Na (58 ± 25 μeg/2 h). On the other hand xylose was not absorbed during the experiment (1 ± 2 mg/2 h), and an extensive secretion of water occurred ($\frac{Vf}{Vi}$ = 2.30) with a secretion of Na$^+$ (−171 ± 30 μeg/2 h) out of the cell into the lumen. For the hypertonic solutions the calculated osmolarity was 1270 mOsm/l for the glucose solution, 1390 mOsm/l for the NaCl solution, and 1470 mOsm/l for the xylose solution. The absorption of calcium was similar in the three conditions and was nearly 50% of the initial quantity e.g. 1.05 ± 0.15 mg/2 h in the xylose solution, 1.04 ± 0.25 mg/2 h in the glucose solution, 1.05 ± 0.07 mg/2 h in the NaCl solution. A similar secretion of water into the lumen was observed in all three solutions with a ratio $\frac{Vf}{Vi}$ reaching 2.08 ± 0.11 for the xylose solution, 2.20 ± 0.25 for the glucose solution and 2.00 ± 0.34 for the NaCl solution. The percentage of absorption of both sugars was high and not equal for xylose and glucose with the absorption of 73 ± 6 mg/2 h (36.5%) of xylose and 80 ± 6 mg/2 h (40%) of glucose. The increased absorption of calcium was principally related to the hyperosmolarity which could be induced as well by sugars as by NaCl.

DISCUSSION

The absorption of calcium in NaCl solution as measured by the tied loop method is a linear function of the Ca concentration for concentrations between 2.5 and 250 mM as previously shown by Zornitzer and Bronner (17). These authors suggested that the apparent intercept of the slope with the abcissa is principally related to the saturable active component of calcium absorption. In this experiment the intercept was zero which suggested that active transfer did not occur at the jejunal site in anesthetized rats.

In the presence of d-xylose or glucose, calcium absorption was increased solely for Ca concentration greater than 2 mg/ml and indicated that the sugars were acting only on the passive transfer across the cell and did not modify the active transport. The absorption of Ca occurred at the same time as secretion of water previously found by Labat (10). The two events were not directly related as proven by xylose inducing a water efflux toward the lumen at low xylose concentration without enhancing the Ca absorption. The increased absorption of Ca was observed at the same time as an absorption of sugar. The effects were simultaneous but not related as proved by the similar effect induced by the hypertonic NaCl solution without sugars. This result does not confirm the hypothesis of Charley and Saltman (2) calling for the formation of a Ca-sugar complex which permeates.

The identical enhancement obtained with the use of an hypertonic saline solution proves that the high percentage of absorption was a non-specific consequence of high osmotic pressure within the lumen. The increased permeability for calcium occurred with an increased permeability for xylose and a secretion of water, the three mechanisms acting simultaneously to reduce the hyperosmolarity of the luminal content. The absence of increase of the Ca absorption rate in the hypertonic solutions containing calcium at a concentration lower than 2 mg/ml could be explained by the excessive dilution of the calcium in the lumen when the water is secreted. The identical efficiency of glucose and xylose in the tied loop was previously shown by Vaughan and Filer (15), the absence of effect observed *in vivo* with this sugar was explained by the rapid absorption of glucose in the higher part of the intestine. The high concentrations of Ca and sugar injected in the tied loop have been used by other workers to obtain a significant increase of the absorption, as an example xylose 100 mg/ml and Ca 5 mg/ml (15) lactose 150 mg/ml and Ca 5 mg/ml (11) lactose 160 mg/ml and Ca 2 mg/ml (5). Wasserman (16) succeeded in obtaining an increase in the absorption with a solution containing 37.5 mg/ml of xylose and 1 mg/ml of calcium, but NaCl was added in the test solution and the final osmolarity was near 800 mOsm/l.

SUMMARY

In the tied jejunal loop the sugars increased the Ca absorption when their concentration was sufficient to induce an osmotic pressure higher than 1,000 mOsm/l. Only the passive transfer was increased: the effect was observed for Ca concentration equal or higher than 2 mg/ml. The kinetics of the transfer was not saturable. The absorption of Ca occurred with an absorption of the sugar and a secretion of water. The increased permeability induced by the sugar could be related with the cellular and physicochemical mechanisms involved in the regulation of the osmolarity.

ACKNOWLEDGEMENT

Study supported by the Contrat No. 1908 A.T.P. de Physiologie et Pathologie du Tissu Calcifié, Centre National de la Recherche Scientifique, France.

REFERENCES

1. Chapuy, M.C., Milani, M., Pansu, D.: L'absorption du calcium en présence de glucose ou de xylose étudiée en anse isolée *in vivo* chez le rat. *Cah. méd. L.* 50, 681–686 (1974)
2. Charley, P. & Saltman, P.: Chelation of calcium by lactose its role in transport mechanisms. *Science* 139, 1205–1208 (1963)
3. Dragstedt, L.R. & Peacock, S.C.: Studies on the pathogenesis of tetany. The control and cure of parathyroid tetany by diet. *Amer. J. Physiol.* 64, 424–434 (1923)
4. Dupuis, Y. & Fournier, P.: Effets différents de la Vitamine D et du lactose sur le rat traité par l'actinomycine D en cours de carence fonctionnelle du calcium. *C.R. Soc. Biol. (Paris)* 163, 2530–2533 (1969)
5. Finlayson, B.: Lactose and intestinal absorption of calcium. *Investigative Urol.* 7, 433–441 (1970)
6. Fournier, P.: L'effet protecteur du lactose vis à vis du squelette de la ratte allaitante. *C.R. Acad. Sci. (Paris)* 238, 509–511 (1954)
7. Fournier, P. & Dupuis, Y.: Pouvoir antirachitique de composés divers dits de structure lactose, glucosamine, L xylose, mannitol. *C.R. Acad. Sci. (Paris)* 250, 3050–3052 (1960)
8. Hyden, S.A.: A turbidometric method for the determination of higher polyethylene glycols in biological materials. *Kgl. Lantbruks. Högskol Ann.* 22, 139–145 (1955)
9. Hyvarinen, A. & Nikkila, E.: Specific determination of blood glucose with o. toluidine. *Clin. chim. Acta* 7, 140–144 (1962)
10. Labat, M.L.: Effet de l'actinomycine D sur l'absorption intestinale du calcium provoquée par le lactose. *C.R. Acad. Sci. (Paris)* 272, 2005–2008 (1971)
11. Lengemann, J.W.: The site of action of lactose in the enhancement of calcium utilization. *J. Nutr.* 69, 23–27 (1959)
12. Pansu, D., Dupuis, Y., Bernard, J. & Fournier, P.: Lactose et utilisation du calcium chez l'homme. Observation relative à la femme âgée ostéoporotique. *C.R. Acad. Sci. (Paris)* 264, 2207–2210 (1967)
13. Patton, J. & Reeder, W.: New indicator for titration of calcium with ethylenedinitrilotetraacetate. *Anal. Chem.* 28, 1026–1030 (1956)
14. Roe, I.H. & Rice, E.W.: A photometric method for the determination of free pentoses in animal tissues. *J. biol. Chem.* 173, 507 (1948)

15. Vaughan, O.W. & Filer, L.J.: The enhancing action of certain carbohydrates on the intestinal absorption of calcium in the rat. *J. Nutr.* **71**, 10—14 (1960)
16. Wasserman, R.H. & Taylor, A.N.: Intestinal absorption of calcium. *Mineral Metabolism, Vol. III*, Comar, C.L. and Bronner, F. (eds.), Academic Press, New York, pp. 321—403, 1969
17. Zornitzer, A.E. & Bronner, F.: *In situ* studies of calcium absorption in rats. *Amer. J. Physiol.* **220**, 1261—1266 (1971)

Ca Transport of Sarcoplasmic Reticulum During Experimental Uremia

K.-W. HEIMBERG,[1] C. MATTHEWS & E. RITZ[2]

In recent years evidence has been provided that during uremia various active transport processes are impaired (8, 1). Kinetic alterations of the sarcoplasmic reticulum (SR) of rabbit skeletal muscle were observed by Ritz and Matthews (6). In the following, some of the altered kinetic parameters of this well established active transport system will be presented. It will be shown that they are paralleled by alterations of the protein and lipid moiety of the SR membrane. Some of these alterations are also found in non-sarcoplasmic membranes and lipoproteins.

METHODS

In the following inbred 5/6 nephrectomized rabbits are compared with sham operated and pair fed controls. 7 days after operation the rabbits were sacrificed. SR was prepared according to Hasselbach and Makinose (3). Kinetic measurements were performed according to the same authors (2).

Figure 1.

$$2\ Ca^{2+} + 1\ ATP \xrightarrow[\text{vesicle membrane}]{Mg^{2+}} 2\ Ca + 1\ P_i + 1\ ADP$$

"outside" "inside"

---

1) Max-Planck-Institut für medizinische Forschung, Department of Physiology, Heidelberg.
2) Medizinische Universitätsklinik, Heidelberg.

RESULTS

Fig. 1 presents a general scheme of the vesicular transport process: 2 moles Ca are shifted across the vesicle membrane from the outside to the inside. Concomitantly 1 mol ATP is split. An osmotic gradient of Ca from inside to outside is thus created. The osmotic potential of this gradient can be transformed back to a chemical potential by the reverse reaction. In general two kinds of Ca fluxes through the vesicle membrane can be distinguished: first an energy-linked active flux according to this scheme and second a passive Ca-flux due to diffusion following concentration gradients. The passive fluxes are very small compared to the active fluxes and only affect kinetic parameters determined near equilibrium of Ca uptake in the presence of a high Ca concentration gradient.

Table I. Ca Kinetics of Control and Uremic Vesicles.

	Control	Uremic
1a		
Storing capacity		
(nM Ca/mg protein)		
in the absence of oxalate	408 ± 296	572 ± 271
in the presence of oxalate	5910 ± 1683	4840 ± 1487
1b		
Concentrating ability		
Ca^{2+} concentration "outside" the vesicles	4.25 ± 2.89	6.98 ± 4.44
($\times 10^{-8}$ M)		
1c		
Initial rate of Ca uptake	1997 ± 884	1338 ± 474
(nM Ca/mg protein $\times$ min)		

All experiments were performed at room temperature with: 20 mM histidine pH 7.0, 5 mM ATP, 10 mM $MgCl_2$, 40 mM KCl, 5 mM phosphoenol-pyruvate, 0.05 mg pyruvate kinase/ml.

Using the Wilcoxon test for paired differences all results differ with $p \leqslant 0.01$.

Storing capacity and concentrating ability were determined after 20 min of Ca uptake, initial rate of Ca uptake was calculated from the first 30 sec of Ca uptake.

Table 1a: 0.01 mg protein/ml, 0.1 mM Ca.
Table 1b: 0.05 mg protein/ml, 0.1 mM Ca, 0.2 mM EGTA*, 5 mM oxalate.
Table 1c: 0.05 mg protein/ml, 0.1 mM Ca, 5 mM oxalate.

*EGTA: Ethylene glycol bis (β-aminoethyl ether)-N,N' tetraacetic acid

Table Ia compares the storing capacity of both preparations. This kinetic parameter describes the maximal amount of Ca which can be taken up in the presence of an excess of Ca and an optimal ATP concentration. In the absence of the Ca precipitating agent oxalate both preparations store about the same amount of Ca. In the presence of oxalate, however, the storing capacity of both preparations differs markedly: the control vesicles take up more Ca than the uremic. This result indicates a different sensitivity of both membranes to divalent anions.

Table Ib now presents the concentrating ability of both preparations. After cessation of net Ca uptake the free Ca concentration outside the uremic SR is higher than that of the control, indicating that the uremic membranes are not able to build up an osmotic Ca gradient that is as high as that of the controls. This result can be shown to be mainly due to an increased passive Ca permeability of the uremic preparations.

Table Ic: Storing capacity and concentration ability are kinetic parameters determined at equilibrium of Ca uptake. The initial rate of Ca influx discussed now, however, is a kinetic parameter determined at the start of the Ca uptake reaction. Because of the optimal Ca uptake conditions and the absence of Ca within the vesicles this influx can be regarded to represent a unidirectional flux not influenced by any active or passive fluxes in the opposite direction. It can be seen that the Ca influx rate of the uremic preparation decreases compared to that of the controls. This can result either from a decrease of the number of the Ca transporting sites of the membrane and/or from a decrease of an influx rate constant.

Gel electrophoretic separation of the vesicle protein shows a significant decrease of a protein with a molecular weight of about 100,000 dalton (control: 78.21 ± 14.66% of total protein; uremic: 63.88 ± 30.52% of total protein. Using the Wilcoxon test for paired differences both results differ with $p \leqslant 0.05$). This protein is generally accepted to represent the ATPase carrier complex. Its decrease allows the (deductive) conclusion that the decreased influx rate observed is at least in part due to a decreased number of transporting sites.

However, it is very uncertain whether this altered protein content of the membrane accounts for all the kinetic changes, because the lipids of the uremic membrane also change.

The phosphatide content of the uremic membranes decreases. For the control vesicles 48.41 ± 5.96 mg phosphatides/100 mg protein are found compared to 37.45 ± 9.47 mg phosphatides/100 mg protein for the uremic vesicles. (Using the Wilcoxon test for paired differences both results differ with $p \leqslant 0.01$). The decreased phosphatide content of the uremic membranes results in a decreased phosphatide/protein ratio of

Table II. Fatty Acids of Total Phosphatides.

Relative amount (%) of fatty acid in

Mean value and standard of deviation	Vesicles (n = 12)		Liver (n = 6)		Serum (n = 6)	
	Control	Uremic	Control	Uremic	Control	Uremic
$16:0$	22.9 ± 1.9	19.7 ± 2.3***	23.4 ± 5.4	21.9 ± 3.6*	20.3 ± 3.6	20.5 ± 3.5
$18:0$	10.0 ± 1.4	11.9 ± 1.7***	19.1 ± 5.2	20.8 ± 4.4*	18.2 ± 4.0	17.1 ± 1.8
$18:1\,\omega\,9$	14.4 ± 2.2	11.2 ± 1.4***	17.7 ± 10.1	11.5 ± 8.5*	9.1 ± 1.7	8.9 ± 1.5*
$18:2\,\omega\,6$	18.4 ± 2.8	15.2 ± 3.3	22.7 ± 6.7	25.6 ± 6.5	23.4 ± 7.9	24.8 ± 6.7
$20:4\,\omega\,6$	12.6 ± 2.2	16.7 ± 2.6***	5.8 ± 4.4	9.0 ± 3.3*	5.7 ± 1.9	7.9 ± 1.9
$22:5\,\omega\,6$	0.94 ± 0.53	1.2 ± 0.58***	0.78 ± 0.85	0.8 ± 0.5	0.28 ± 0.2	0.97 ± 0.39*
$22:5\,\omega\,3$	3.7 ± 1.1	4.9 ± 1.6*	1.2 ± 0.5	1.2 ± 0.9	0.75 ± 0.31	1.4 ± 0.6*

The asterisks indicate those fatty acids which show significant differences in the Wilcoxon test for paired differences.
*, **, *** indicating p $\leqslant$ 0.05, 0.02 and 0.01, respectively.
In the fatty acid column, the first number represents the number of carbon atoms, the second the number of double bonds, ω refers to the position of the first double bond, counting from the methyl end of the molecule.

these membranes. Since free and esterified cholesterol does not change it also results in a decreased phosphatide/cholesterol ratio.

Table II shows a change of the fatty acid pattern SR which is also observed in serum lipoproteins and liver serum. This may point to a generalized disturbance of fatty acid metabolism observed in a decrease of palmitic and oleic acid and an increase of stearic, arachidonic and higher unsaturated fatty acids.

CONCLUSION

In uremia, the membrane composition of the SR is altered: the ATPase carrier complex decreases, the phosphatide, but not the cholesterol content of the membranes decreases and the fatty acid pattern changes. All these lipid changes are known to affect the kinetic properties of lipoprotein enzymes (6, 7, 8). In uremia the altered lipoprotein interactions of the membrane are paralleled by an impairment of nearly all kinetic parameters of the SR.

REFERENCES

1. Baerg, H.D., Kimberg, D.V. & Gershon, E.: Effect of renal insufficiency on the active transport of calcium by the small intestine. *J. Clin. Invest.* **49**, 1288—1300 (1970)
2. Hasselbach, W. & Makinose, M.: Die Calciumpumpe der "Erschaffungsgrana" des Muskels und ihre Abhängigkeit von der ATP-Spaltung. *Biochem. Z.* **333**, 518—528 (1961)
3. Hasselbach, W. & Makinose, M.: Über den Mechanismus des Calciumtransportes durch die Membranen des sarkoplasmatischen Retikulums. *Biochem. Z.* **339**, 94—111 (1963)
4. Kimelberg, H.K. & Papahadjopoulos, D.: Effects of Phospholipid Acyl Chain Fluidity, Phase Transitions and Cholesterol on (Na^+ + K^+)-stimulated Adenosine Triphosphatase. *J. Biol. Chem.* **249**, 1071—1080 (1974)
5. Papahadjopoulos, D., Cowden, M. & Kimelberg, H.: Role of Cholesterol in Membranes: Effects on Phospholipid-Protein Interactions, Membrane Permeability and Enzymatic Activity. *Biochem. Biophys. Acta* **330**, 8—26 (1973)
6. Ritz, E., Bundschu, H.D. & Hasselbach, W.: Calcium Uptake by Sarcoplasmic Reticulum in Experimental Uremia. *IXth Europ. Symp. Calc. Tiss.* Baden near Vienna 1972, Czitober, H. & Eschberger, J. (eds.), Facta-Publication Vienna, pp. 361—364, 1973
7. Rottem, S., Cirillo, V.P., De Kruyff, B., Shinitzky, M. & Razin, S.: Cholesterol in Mycoplasma Membranes: Correlation of Enzymic and Transport Activities with Physical State of Lipids in Membranes of Mycoplasma Mycoides Var. Capri Adapted to Grow with Low Cholesterol Concentrations. *Biochem. Biophys. Acta* **323**, 509—519 (1973)

8. Welt, L.G., Sachs, J.R. & McManus, T.J.: An ion transport defect in erythrocytes from uremic patients. *Trans. Amer. Ass. Physic.* 77, 169–181 (1969)

CHAPTER II
Structure and Ultrastructure of Bone Cells. Functional Aspects

Morphology of Cell Calcium Homeostasis

W.L. DAVIS, J.L. MATTHEWS, R.V. TALMAGE & J.H. MARTIN

Calcium levels in the blood and in cells are maintained within narrow limits in normal animals, although variations occur in dietary constituents and absorption. This blood balance is maintained primarily by using the calcium reserves in skeletal tissues and modulating its loss in the kidney (15). It is thus essential to understand the underlying role of the cells of these tissues in effecting selective ion distribution, as calcium plays a major role in physiologic processes, and maintenance of appropriate distribution and concentration is mandatory (16). All cells regulate and utilize calcium in some activity, and the means employed by different cell types may be used to give clues to the mechanisms involved in regulating calcium ion distribution in mineralized systems. The following examples are selected to illustrate the diverse distribution and mechanisms of regulation that may by employed.

Ionized calcium is directly involved in excitation-contraction-coupling, whereby the depolarization of the sarcolemma and its associated T tubule system activates the release of calcium from specific closed membranous saccules, the sarcoplasmic reticulum (8, 4). Membranes of these saccules possess Ca/Mg, ATPase and other enzymes necessary to recoupe the free calcium after it has been temporarily coupled with select receptor proteins of the myofilaments. The intracellular distribution of partially contracted skeletal muscle fixed with K-pyroantimonate to demonstrate cell calcium localization is shown in Fig. 1. Muscle fibers with little or no sarcoplasmic reticulum use extracellular calcium, varying plasma membrane calcium permeability to elevate cytosolic calcium. Mitochondria assist this system to rapidly remove free calcium during the relaxation phase. Fig. 2.

Calcium itself influences plasma membrane permeability (9, 19) and has also been shown to regulate the flow of other ions between cells at special

Department of Microscopic Anatomy, Baylor College of Dentistry, Dallas, Texas.

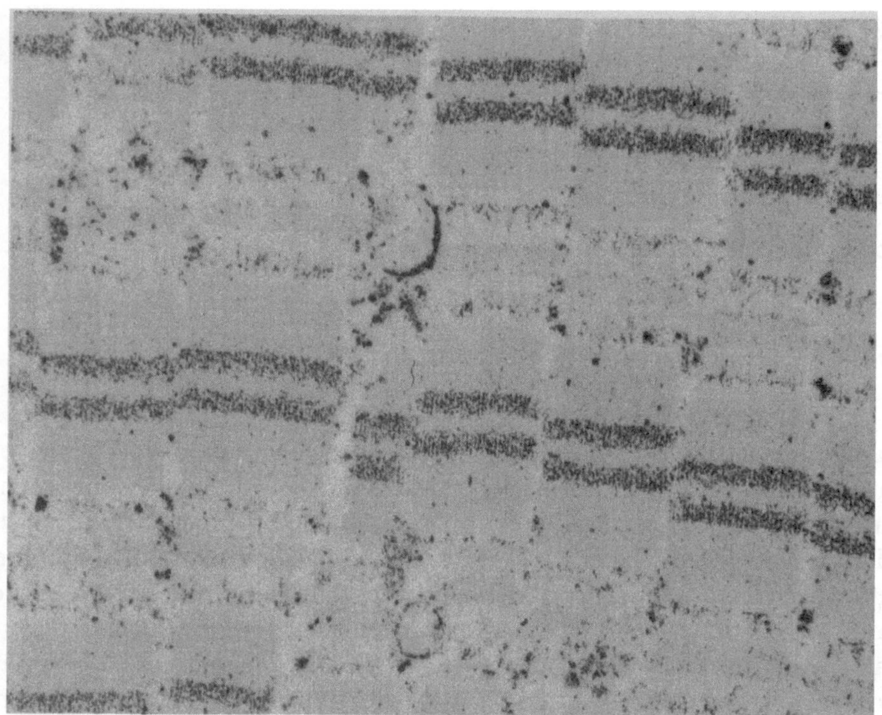

Figure 1. Electron micrograph of skeletal muscle fiber following glycerolation and fixation in K-pyroantimonate. Discrete bands of calcium are localized within myofilaments. Calcium is also seen in saccules of sarcoplasmic reticulum. (X 17,400)

cell-cell junctions. Calcium is not totally independent of other ions so that calcium quantities as well as its ratio to other ions such as sodium, magnesium, potassium, hydrogen ion, phosphorous, and others are related to the ultimate response (17).

Calcium serves to activate certain enzymes such as adenosine triphosphatase, succinic dehydrogenase and serves to inhibit others, participating as a co-messenger in events related to activation of adenyl-cyclase (18).

Calcium is also involved in excitation-secretion-coupling being involved in secretion of neurotransmitter substances, vasopressin, insulin, adrenal hormones, etc. (3, 7, 6). In this instance, calcium appears to "coat" the secretory vesicles, thereby influencing their "stickiness" to the plasma

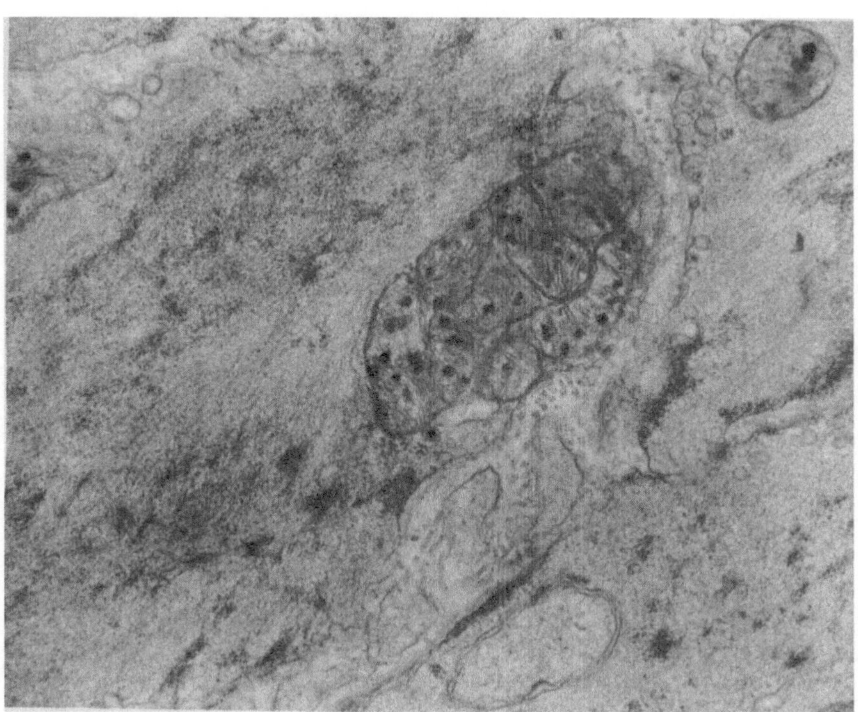

Figure 2. Electron micrograph of smooth muscle. Numerous mitochondrial granules are found. The mitochondria are close to the plasma membrane at which extensive pinocytotic transport activity is seen. (X 26,000)

membrane. Coupled with the movement of these secretory granules through the cytoplasm are microtubules whose activity and aggregation of subunits are also presumed to be calcium dependent (10).

Calcium also plays a special role in blood coagulation and clot retraction and is selectively stored in serotonin containing vesicles of blood platelets (13). Concentrations in these vesicles are influenced by serotonin levels and levels of adenine dinucleotides. Fig. 3 is an electron micrograph showing platelet granules whose electron density has been shown to be due to accumulated calcium by ashing, X-ray spectrometry, selective chelation, reloading experiments, and K-pyroantimonate staining. Special calcium containing vesicles are also present in fibrocytes of calciphylactic skin, avian shell glands, cartilage, mammary gland, and possibly intestine.

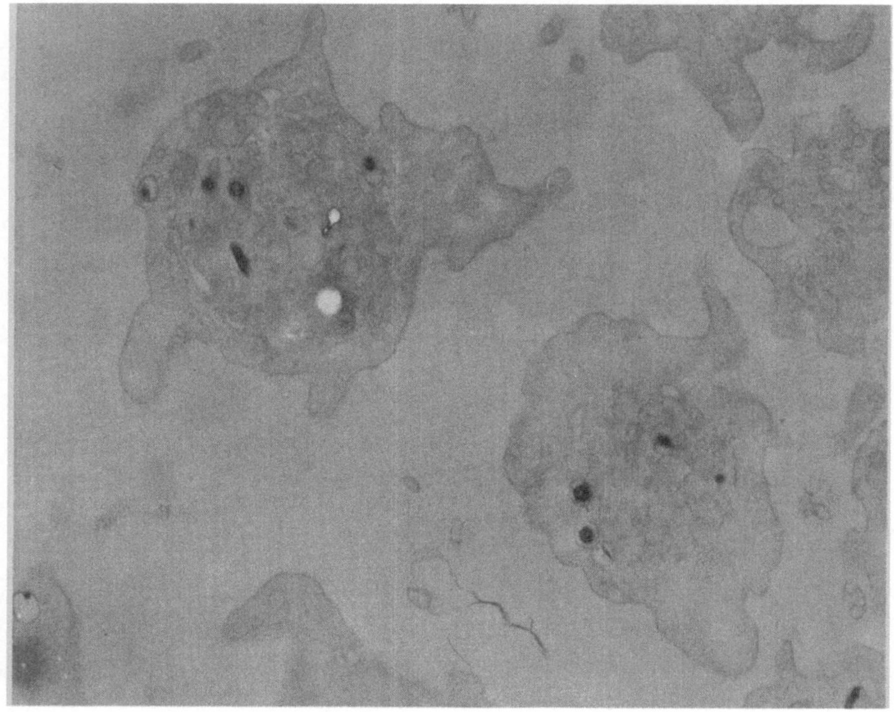

Figure 3. Electron micrograph of blood platelet in which several platelet granules are seen. The less numerous 5 HT granules contain calcium and appear as electron dense vesicles. (X 25,650)

Fig. 4 is an electron micrograph of calciphylactic skin showing the secretion of intracellular mineral packets into the collagenous matrix.
Nuclear calcium represents another physiologic compartment. Increased nuclear calcium is coupled with the initiative events of mitosis. We have observed nuclear mineralization in pathologic conditions such as damage from X-radiation of ganglia and streptococcal infections of kidney, other parts of the cell being undamaged.

Among the organelles involved in cell calcium regulation, the mitochondria play a primary role (11). These organelles in both intact and isolated systems have been shown to be capable of removing calcium from the cytosol and depositing it with phosphate in the mitochondrial matrix in the form of calcium phosphate salts. Fig. 5 is an electron micrograph of

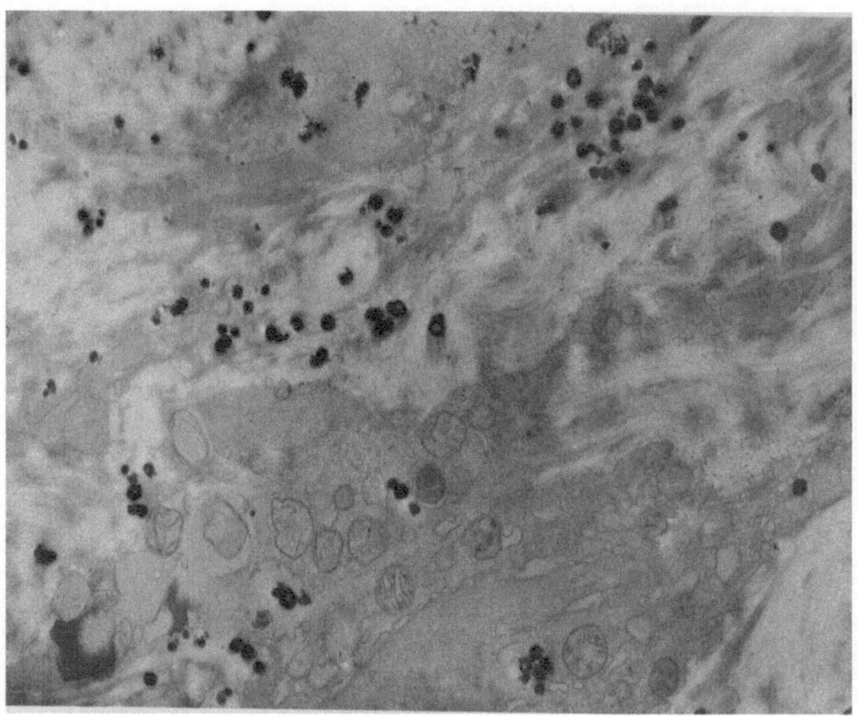

Figure 4. Electron micrograph of skin from calciphylactic rat skin. Calcium containing electron dense vesicles are seen forming with the fibrocyte and are also seen secreted into the matrix. (X 20,235)

bone showing mitochondrial granules in an osteocyte. These granules are "amorphous" in nature. Granules are usually observed in mitochondria of all cell types when the intracellular levels of ionic calcium exceed 10^{-6} molar concentration (1). Thus, mitochondrial granules are seen in cells associated with calcium transport; i.e., gut cells following vitamin D (20) active osteoclasts, osteocytes following parathyroid hormone, chondrocytes in the hypertrophic zone of the growth plate, cells of active avian shell gland, pathologic soft tissue calcification, cells of differentiating systems, etc. (14). Mitochondrial incorporation is coupled to steps of oxidative phosphorylation. Some investigators have suggested that mitochondria of bone and cartilage are capable of loading at oxygen levels different from those of other tissues, and it has been suggested that these

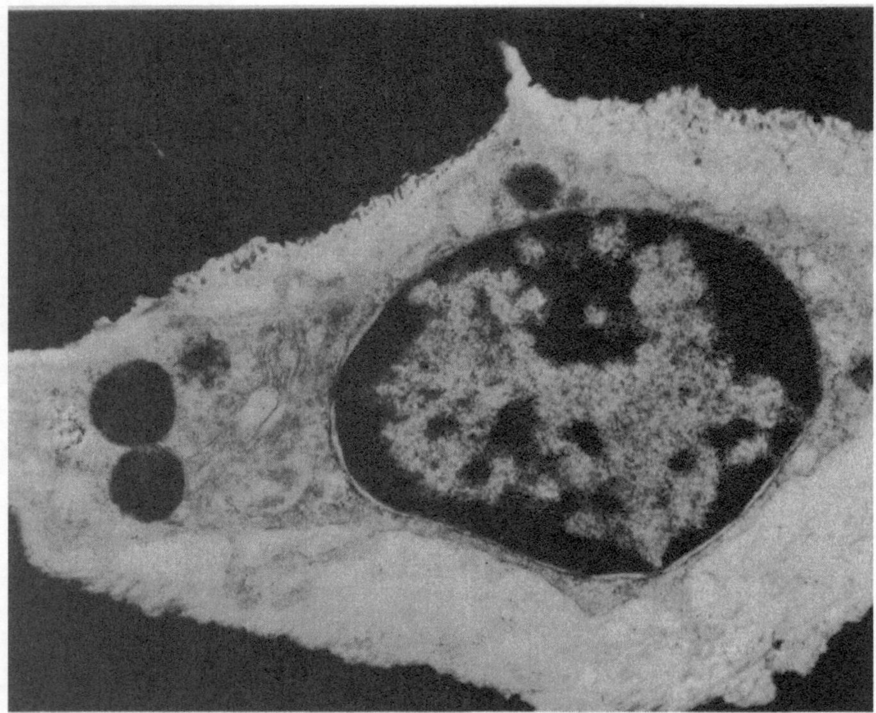

Figure 5. Electron micrograph of osteocyte from a rat pretreated with EHDP for four days at 5 mg dose. Following stimulation with PTH, mitochondria appear as electron dense bodies. (X 19,000)

mitochondrial mineral stores are utilized in the loading of matrix vesicles associated with mineralization (2, 12).

From these examples, numerous mechanisms are given that might be used by cells of bones and cartilage in regulating ion distribution. Among these are plasma membrane binding, vesicle compartments, mitochondria, nuclear compartments, protein binding, membrane permeability changes, microfilament and microtubule binding, and lipid binding. Examination of the ultrastructural localization of calcium in bone and cartilage cells in different physiologic states implicates all of these mechanisms in bone and cartilage cells. Following treatment with parathormone, (PTH), calcium entry into cells is enhanced. By using K-pyroantimonate staining, a stain that gives a reaction precipitate with cations, especially calcium (21),

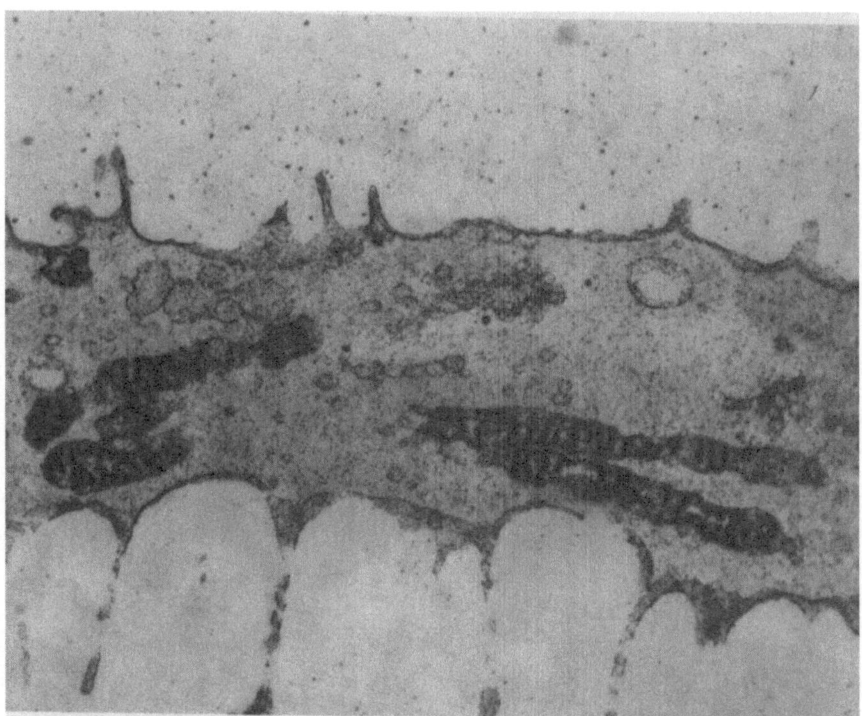

Figure 6. Electron micrograph of osmium-K-pyroantimonate fixed chondrocyte from rachitic chick growth plate. The plasma membrane, mitochondria, lipid droplets, and vesicles show calcium, i.e. pyroantimonate reaction product. (X 51,300)

increased product is seen within mitochondria, cytosol, and nucleus of osteocytes following PTH. Conversely, when calcitonin (SCT) is given, a depletion of intracellular ions is noted in osteoblasts, osteocytes, and osteoclasts, suggestive of selective efflux of calcium and phosphate. The latter is confirmed by the production of apatite crystals in the osteoid following incubation in aqueous media. Incubated specimens from non-treated animals do not show this effect (23). Pyroantimonate stains the lacunar contents in these animals. Pretreatment of animals with the diphosphate EHDP lessens the lacunar product when SCT is given and appears to block mitochondrial unloading when PTH is given. Fig. 5 is an electron micrograph of osteocytes stimulated with PTH following EHDP. When rachitic growth plates of chicks are treated with K-pyroantimonate,

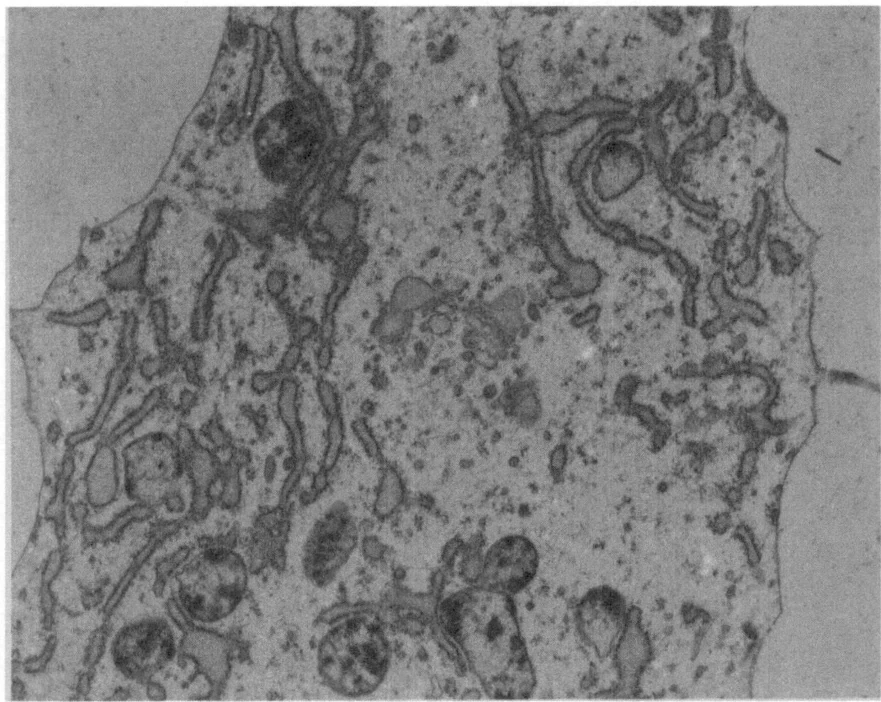

Figure 7. Electron micrograph of chondrocyte of rat growth plate. Granules are found in mitochondria. The distribution of mitochondrial mineral from this tissue differs markedly from the compartmentalization seen in Fig. 6. (X 11,250)

the mitochondria, plasma membrane, lipid inclusions, and Golgi vesicles show positive cation binding in the growth plate zones preceding matrix mineralization. These reactions are reduced in cells adjacent to matrix mineralization. Fig. 6 is an electron micrograph of a chondrocyte from rachitic chick growth plate. Lipid droplets, small cytoplasmic vesicles, and dispersed mitochondrial densities are reacted with K-pyroantimonate. Different patterns of calcium distribution in mitochondria are seen in rat chondrocytes. (Fig. 7).

By using scanning and transmission electron microscopic analysis of the endosteal lining of tubular bones, significant changes in cell morphology are observed in bones from animals stimulated with PTH, SCT, colchicine, cytochalasin B, and vinblastine. Fig. 8 is an electron

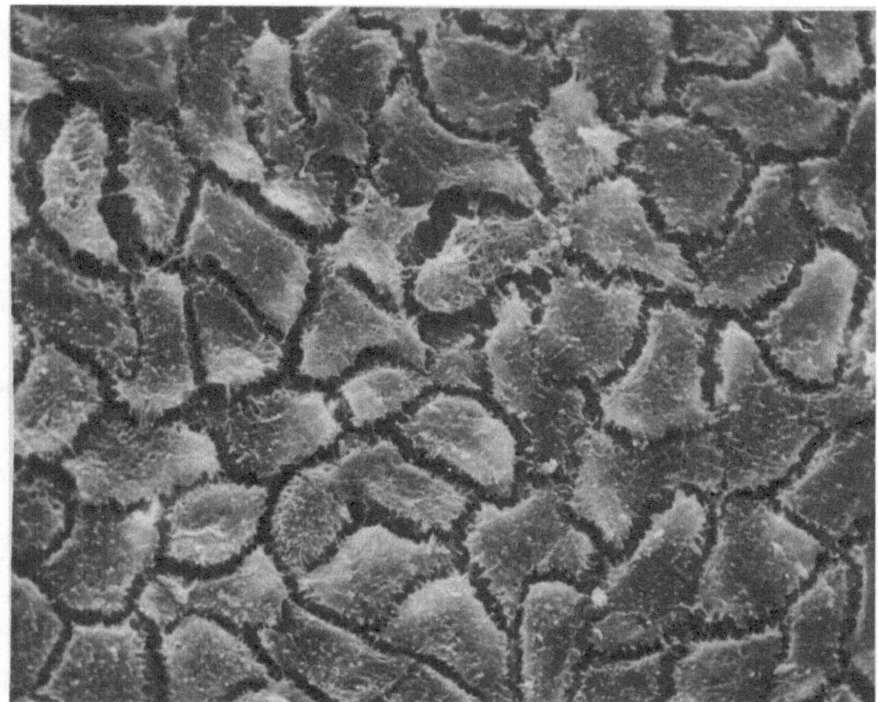

Figure 8. Scanning electron micrograph of endosteal lining cells of rat tibia. Osteoblasts show surface microvilli. Narrow spaces appear between adjacent cells. (X 2,500)

micrograph of the endosteal cells of normal rat bone. Dramatic changes in this morphology indicate the presence of microfilaments and microtubules within bone cells in regulating fluid flow. Retraction of microvilli of osteoclasts or enhancement of microvilli of osteoblasts, or varying the intercellular distances between lining cells can all influence the equilibrium of calcium compartmentalization. The organization of bone cells and their morphologic variations following different stimuli has led to a concept of a bone cell complex acting as a "functional membrane" (5, 22). Complete understanding of calcium regulatory phenomenons must consider all of these intracellular and extracellular mechanisms.

ACKNOWLEDGEMENT

This study was sponsored in part by U.S. Atomic Energy Commission # AT-40-1-40562 and National Institutes of Health AM-14717 and National Science Foundation OMS 74-13248.

REFERENCES

1. Borle, A.B.: Membrane transfer of calcium. *Clin. Orthop.* 52, 267 (1967)
2. Brighton, C.T. & Hunt, R.M.: Mitochondrial calcium and its role in calcification. *Clin. Orthop.* 100, 406 (1974)
3. Curry, D.L., Bennett, L.L. & Grodsky, G.M.: Requirements for calcium ion in insulin secretion by the perfused rat pancreas. *Amer. J. Physiol.* 214, 174 (1968)
4. Davis, W.L., Matthews, J.L. & Martin, J.H.: An electron microscopic study of myofilaments calcium binding sites in native, EGTA-chelated and calcium reloaded glycerolated mammalian skeletal muscle. *Calc. Tiss. Res.* 14, 139 (1974)
5. Davis, W.L., Matthews, J.L., Martin, J.H. & Talmage, R.V.: *The Endosteum as a Functional Membrane.* In: Calcium-regulating hormones. Talmage, R.V. and Owen, M. (eds.), Excerpta Med. (Amst.), pp. 275, 1975
6. Dean, P.M. & Matthews, E.K.: Electrical activity in pancreatic islet cells: Effect of Ions. *J. Physiol. (Lond.)* 210, 265 (1970)
7. Douglas, W.S.: Stimulus-secretion coupling: The concept and clues from chromaffin and other cells. *Brit. J. Pharmacol.* 34, 451 (1968)
8. Ebashi, S. & Endo, M.: Calcium ion and muscle contraction. *Progr. Biophys. Moles. Biol.* 18, 123 (1968)
9. Kanno, Y. & Loewenstein, W.R.: Intracellular diffusion. *Science* 143, 959 (1964)
10. Lacy, P.E.: Beta cell secretion from the standpoint of a pathologist. *Diabetes* 19 895 (1970)
11. Lehninger, A.L., Carafoli, E. & Rossi, C.S.: Energy-linked ion movements in mitochondrial systems. *Advanc. Enzymol.* 29, 259 (1967)
12. Martin, J.H. & Matthews, J.L.: Mitochondrial granules in chondrocytes. *Calc. Tiss. Res.* 3, 184 (1974)
13. Martin, J.H., Carson, F.L. & Race, G.J.: Calcium-containing platelet granules. *J. cell. Biol.* 60, 775 (1974)
14. Matthews, J.L., Martin, J.H., Kennedy, J.W. & Collins, E.J.: An ultrastructural study of calcium and phosphate deposition and exchange in tissues. In: Tissue Growth, Repair and Remineralization. Ciba Symposium. Sognnaes, R.F. & Vaughan, J.M. (eds.), Excerpta Med. (Amst.), pp. 187, 1973
15. Nordin, B.E.C. & Peacock, M.: Role of kidney in regulation of plasma calcium. *Lancet* (1969) ii, 1280
16. Rasmussen, H.: Cell communication, calcium ion, and cyclic adenosine monophosphate. *Science* 170, 404 (1970)
17. Rasmussen, H. & Bordier, P.: *The Physiological and Cellular Basis of Metabolic Bone Disease.* Williams & Wilkins, 1974
18. Robison, G.A., Butcher, R. & Sutherland, E.W.: *Cyclic AMP,* Academic Press, 1971

19. Romero, P.J. & Whittam, R.: The control by internal calcium of membrane permeability to sodium and potassium. *J. Physiol. (Lond.)* 214, 481 (1971)
20. Sampson, H.W., Matthews, J.L., Martin, J.H. & Kunin, A.S.: An electron microscopic localization of calcium in the small intestine of normal, rachitic, and vitamin-D-treated rats. *Calc. Tiss. Res.* 4, 305 (1970)
21. Simson, J.A.V. & Spicer, S.S.: Selective subcellular localization of precipitate using modifications of Kominicks pyroantimonate-osmium technique (Abstr.). *J. Histochem. Cytochem.* 22, 291 (1974)
22. Talmage, R.V.: Morphological and physiological considerations in a new concept of calcium transport in bone. *Amer. J. Anat.* 129, 467 (1970)
23. Talmage, R.V., Matthews, J.L., Davis, W.L., Martin, J.H. & Roycraft, J.H. Jr.: Calcitonin, phosphate and the osteocyte-osteoblast bone cell unit. In: *Calcium-regulating hormones.* Talmage, R.V. and Owen, M. (eds.), Excerpta Med. (Amst.), pp. 284, 1975

Effects of PTH and some Synthetic Fragments on Embryonic bone *in vitro*

P.J. GAILLARD, M.P.M. HERRMANN-ERLEE & J.W. HEKKELMAN

INTRODUCTION

Recent studies, employing the synthetic peptide bovine PTH 1—34 and analogues of this peptide have led to the conclusion that the amino terminal 1—34 or perhaps even the 1—27 residues of the hormone represent a structure sufficient for the expression of a series of biological effects: elevation of serum calcium; increase in urinary phosphate clearance and decrease of urinary calcium clearance; increase in urinary cyclic AMP; and the stimulation of adenyl cyclase activity in renal cortical membrane suspension *in vitro* (10, 14, 13, 16).

Considering that no tests have yet been reported as to whether the amino terminal residues are also sufficient to account for the effects of PTH on bone and other supporting tissues it was decided to investigate the effect of some hormone fragments on these tissues. In the present study (see also 8) the biological activities of bovine PTH were compared with those of some synthetic fragments comprising PTH 1—34, PTH 2—34 and PTH 3—34 using three *in vitro* bioassay systems.

MATERIAL AND METHODS

The PTH fragments were synthesized by Tregear (16).

Cyclic AMP was determined in calvaria of 1-day-old rats (6) using a protein binding assay (2).

Bone demineralization was determined by measuring the release of Ca and P_i into the medium from calvaria of 19-day-old rat embryos as described by Herrmann-Erlee and v.d. Meer (7).

Laboratorium voor Celbiologie en Histologie, der Rijksuniversiteit te Leiden, Academisch Ziekenhuis, Leiden.

The effects on bone, cartilage and connective tissue histology were studied in explants of 15-day-old embryonic mouse radii as described by Gaillard (3, 4, 5).

The activities of the peptides were estimated in a concentration range of 0.004—10 μg/ml. For each of the hormone preparations the dose-response relationships were studied at not less than three dose levels.

RESULTS

Table I shows the results of our estimations (II, V and VI) in comparison with those of other assays (I, II and IV). From the table it follows that PTH 1—34 is indeed active on bone. Moreover, it is evident that the absence of the amino acids 1 and 2 results in a complete inactivity. In addition, it is clear that the average values obtained with the fragment 1—34 are low as compared with the effects of the native hormone and that the average effects of the fragment 2—34 are again lower. Apart from this, relatively large differences appear to exist between the results of the different methods, which, however, do not seem to depend exclusively on the length of the incubation periods.

As regards the cyclic AMP estimations in calvaria (test number II) it is clear that the 1—34 fragment is about half as active as the native hormone. In this connection it is of interest to note that this result was obtained after the optimum period of incubation (15 minutes).

Concerning the values of the calcium release after 48 h incubation (test number V) it is evident that PTH 1—34 and the native hormone are equally active. It can be added that both preparations caused a half maximal effect at a concentration of about 5 x 10^{-9} M. However, in

Table I. Molar Activity as % of the Activity of Native Hormone.

	Method	Duration	Route	1-84	1-34	2-34	3-34
I.	Kidney ad. cycl.[1]	10'	in vitro	100	80	2	0.3
II.	Calvar. cAMP	15'	in vitro	100	45-60	8-30	0
III.	Chick hypercalc.[2]	1 h	i.v.	100	130	65	0.2
IV.	Rat hypercalc.[3]	1 h	s.c.	100	3	-	-
V.	Ca, P_i release	24 h)	in vitro	100	3-5	0.2-1	0
	calvaria	48 h)			100	0.4	-
VI.	Histol. radii	48 h	in vitro	100	50	0	-

1) Martin *et al.*, 1974
2) Tregear *et al.*, 1973
3) Munson, 1961

experiments which were terminated after *24* h of incubation the results appeared to be different. As would be expected the maximum amount of calcium mobilized was smaller but surprisingly the dose requirements also appeared to be strikingly different. The native hormone caused a half maximal effect at a $8 \times 10^{-10}\,M$ concentration, whereas a $2 \times 10^{-8}\,M$ solution of the 1—34 preparation was needed to match it. So far we are unable to explain this difference. However, in this connection it is of importance to realize that the two dose response curves were running parallel suggesting that the basic mechanisms of action may not be different.

In order to check whether the 1—34 fragment also causes an effect on the lactate flux it became clear that both the native hormone and the 1—34 fragment induced an enhanced lactate flux lasting at least one h after removal of the hormone. However, the 1—34 fragment showed a relatively low potency.

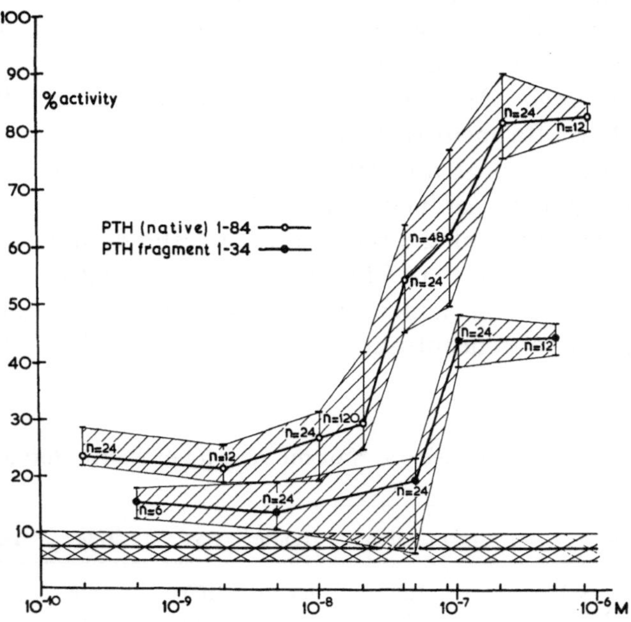

Figure 1. Dose response curves obtained from experiments with 48-h cultures of embryonic mouse radii. The presence of full blown reactions in the cartilage, the bone and the connective tissue as described (see 3) is indicated as a 100% activity. It is clear that the native hormone exhibits a characteristic S-shaped curve reaching a maximum of about 85%. The 1—34 fragment shows a maximum effect of about 45%. The values for the radii in control media without added hormone all lay between 5 and 10%.

Concerning the histology of the radii studied after 48 h of incubation it became clear that the 1—34 fragment is capable of influencing all three tissue compartments viz. the cartilage, the bone and a reticular type of connective tissue occurring inside the shaft. This implies that in this respect the 1—34 fragment and the native hormone acted in a similar way. However, quantitatively the effects turned out to be significantly different (see Fig. 1). From this figure it follows that all concentrations (2 x 10^{-10} M to 8 x 10^{-7} M) the PTH 1—34 fragment is about half as active as the native hormone. In this connection it is extremely interesting to note that also the maximal effect of the fragment is only 50% of that of the native hormone. Considering that the molecular weight of the 1—34 fragment is about half of that of the native hormone (4109 versus 9562) and accepting that at these concentrations in both instances all receptor sites will be "occupied" it is tempting to suggest differences in the stability of the hormone receptor complex and/or the hormone fragment itself.

In addition the histology of the radii grown in low concentrations of the 1—34 fragment (4.9 $x10^{-9}$ and 4.9 x 10^{-8} M) revealed some unexpected phenomena. In sharp contrast to the native hormone, the majority of the radii exhibited an *increased* number of active osteoblasts instead of a depression and a *increased* maturation of cartilage instead of an inhibition. In more than 60% of the explants even an ectopic osteoid formation was observed inside the bone shaft. This observation suggests that particularly at lower concentrations the PTH 1—34 fragment is capable of inducing *anabolic* effects. This finding is of particular interest because there is also good *in vivo* evidence that PTH can enhance bone formation when given chronically in low dose (15, 9, 13). Moreover, it is known that in generalized osteitis fibrosa cystica new, be it abnormal, bone occurs (1).

SUMMARY

The effects of PTH and some synthetic fragments on bone were studied and compared with the results of other authors using other assay methods. From the study it became clear that the bovine PTH 1—34 fragment is indeed active on bone. In addition it is concluded that in order to be active on bone the bovine PTH fragment must contain amino acids number 2 through 34. However, comparing the results of various assay methods relatively large differences in activity became evident illustrating the value of quantitatively comparing the properties of a series of active substances in the widest possible range of analytical bioassay systems. Finally it became evident that low concentrations of the bPTH 1—34 fragment possessed unexpected anabolic activities and even caused ectopic osteoid formation, substantiating the findings of others that PTH can enhance bone formation *in vivo* when given chronically in low dose.

REFERENCES

1. Ascenzi, A. & Marinozzi, V.: Biophysical study of von Recklinghausen's disease of bone. *Arch. Path. (Lisboa)* **72**, 297–309 (1961)
2. Brown, B.L., Albano, J.D.M., Erkins, R.P., Sgherzi, A.M. & Tampion, W.: A simple and sensitive saturation assay method for the measurement of adenosine 3',5'-cyclic monophosphate. *Biochem. J.* **121**, 561–562 (1971)
3. Gaillard, P.J.: The influence of parathyroid extract on the explanted radius of albino mouse embryos. II. *Proc. kon. ned. Akad. Wet.* **C 64**, 119–128 (1961 a)
4. Gaillard, P.J.: Parathyroid and bone in tissue culture. In: *The Parathyroids.* Greep, R.O. & Talmage, R.V. (eds.), Thomas, Springfield, Ill., 20–45 (1961 b)
5. Gaillard, P.J.: The influence of ascorbic acid on the effect of parathyroid extract on the histology of explanted mouse radius rudiments. I. *Proc. kon. ned. Akad. Wet.* **C 77**, 101–114 (1974)
6. Heersche, J.N.M., Marcus, R. & Aurbach, G.D.: Calcitonin and the formation of 3',5'-AMP in bone and kidney. *Endocrinology* **94**, 241–247 (1974)
7. Herrmann-Erlee, M.P.M. & van der Meer, J.M.: The effects of dibutyryl cyclic AMP, aminophylline and propranolol on PTE-induced bone resorption *in vitro.* *Endocrinology* **94**, 424–434 (1974)
8. Herrmann-Erlee, M.P.M., Heersche, J.N.M., Hekkelman, J.W., Gaillard, P.J., Tregear, G.W., Parsons, J.A. & Potts, J.T. Jr.: A comparison of the effects of bovine PTH and synthetic fragments of bovine PTH on bone *in vitro.* *Endocrine Research comm.* **3**, 1 (1976)
9. Kalu, D.N., Pennock, J., Doyle, F.H. & Foster, G.V.: Parathyroid hormone and experimental osteosclerosis. *Lancet* (1970), i, 1363–1366
10. Marcus, R. & Aurbach, G.D.: Bioassay of parathyroid hormone *in vitro* with a stable preparation of adenyl cyclase from rat kidney. *Endocrinology* **85**, 801–810 (1969)
11. Martin, T.J., Vakalis, N., Eisman, J.A., Livesey, S.J. & Tregear, G.W.: Chick kidney adenylate cyclase: Sensitivity to parathyroid hormone and synthetic human and bovine peptides. *J. Endocr.* **63**, 369–375 (1974)
12. Munson, P.L.: Biological assay of parathyroid hormone. In: *The Parathyroids,* Greep, R.O. & Talmage, R.V. (eds.), Thomas, Springfield, Ill., 94–113 (1961)
13. Parsons, J.A. & Reit, B.: Chronic response of dogs to parathyroid hormone infusion. *Nature* **250**, 254–257 (1974)
14. Parsons, J.A., Reit, B. & Robinson, C.J.: A bioassay for parathyroid hormone using chicks. *Endocrinology* **92**, 454–462 (1973)
15. Selye, H.: On stimulation of new bone-formation with parathyroid extract and irradiated ergosterol. *Endocrinology* **16**, 547–558 (1932)
16. Tregear, G.W., van Rietschoten, J., Greene, E., Keutmann, H.T., Niall, H.D., Reit, B., Parsons, J.A. & Potts, J.T. Jr.: Bovine parathyroid hormone: Minimum chain length of synthetic peptide required for biological activity. *Endocrinology* **93**, 1349–1353 (1973)

Structural and Ultrastructural Responses of Calcifying Cartilage to Parathyroid Hormone *"in vitro"*

E.H. BURGER & P.J. GAILLARD

INTRODUCTION

In vitro as well as *in vivo* studies have shown that parathyroid hormone (PTH) promotes the release of calcium from bone, while stimulating matrix resorption.

Besides, tissue culture studies indicate an inhibitory effect of the hormone on bone matrix production as could be concluded from the PTH-induced loss of active osteoblasts alongside the bone. These studies also show that in long bones not only the bone compartment but also the cartilage is influenced by PTH (3, 2).

We studied the effect of PTH on deposition of mineral in hypertrophic cartilage matrix, using embryonic cartilaginous bones, dissected at a stage of development just prior to calcification. During subsequent cultivation these "bones" were observed to start the deposition of mineral between the central hypertrophic chondrocytes. Parathyroid extract appeared to suppress both the intracellular and extracellular aspects of this phenomenon.

METHODS

The middle three metatarsal bones of 16-day-old Swiss albino mouse embryos were dissected and cultivated for 48 h in rollertubes, rotating at 8 r.p.h. The medium consisted of Hepes-buffered M199 with Hanks salts (Flow Lab.) supplemented with 15% heat inactivated human serum (Flow Lab.) and phosphate buffer pH 7.4 to a final concentration of 3 mM P_i.

Laboratorium voor celbiologie en histologie der Rijksuniversiteit te Leiden, Academisch Ziekenhuis, Leiden.

Medium pH was measured before and after cultivation and varied between 7.4 and 7.5 (at 37°C).

In experiments with PTE, corresponding left and right rudiments were cultivated with either PTE (E. Lilly and Co.) added to the medium or a similar amount of placebo solution.

In some experiments the necessity of viable cells for mineralization to take place was checked by killing the rudiments before cultivation. This was done by five times freezing (15 min at −20°C) and thawing (15 min at room temp.) in Hanks balanced salt solution.

For light-microscopic observations the bones were stained fresh with 10% glyoxal bis (2-hydroxyanil) (GBHA) for 1.5−2 h according to Kashiwa (6) and mounted "en bloc" in malinol. Controls were decalcified for 3 h in 5% EDTA in 5% mannitol solution. Length and width of the mineralized area in these total-mount preparations were measured with an eyepiece micrometer at 100x magnification. In addition, some rudiments were fixed with 100% ethanol and stained as 7μ thick sections according to Von Kossa. For electron-microscopy the combined-fixative method of Hirsch and Fedorko (4) was used without postfixation in acetic uranyl acetate, to prevent loss of calcium. After Epon embedding longitudinal sections were cut with glass knives from the central portion of the rudiment, which contained the hypertrophic cartilage. They were stained with Reynold's lead citrate and examined with a Siemens Elmiskop operating af 80 KV.

RESULTS

A. *Light microscopy*
In the *non*-cultivated cartilaginous rudiments, many GBHA-positive red granules were found inside the central hypertrophic chondrocytes. Treatment with EDTA for 3 h completely removed these granules from the cells, while treatment with the mannitol solution alone left them intact. After 2 h of decalcification with EDTA most of the granules could still be found within the cells, though their staining intensity was now much lighter. Inspection of younger bones showed that these granules had accumulated in the course of the preceding 24 h of embryonic life, in the same cells that showed the first signs of chondrocyte hypertrophy. No indication of matrix mineralization was found, neither in the GBHA, nor in the Von Kossa stained preparations. After two days of cultivation in control medium all explants showed a characteristic pattern of development: Between the central hypertrophic cells a Von Kossa-positive, refractive material had been deposited which stained brightly yellow in

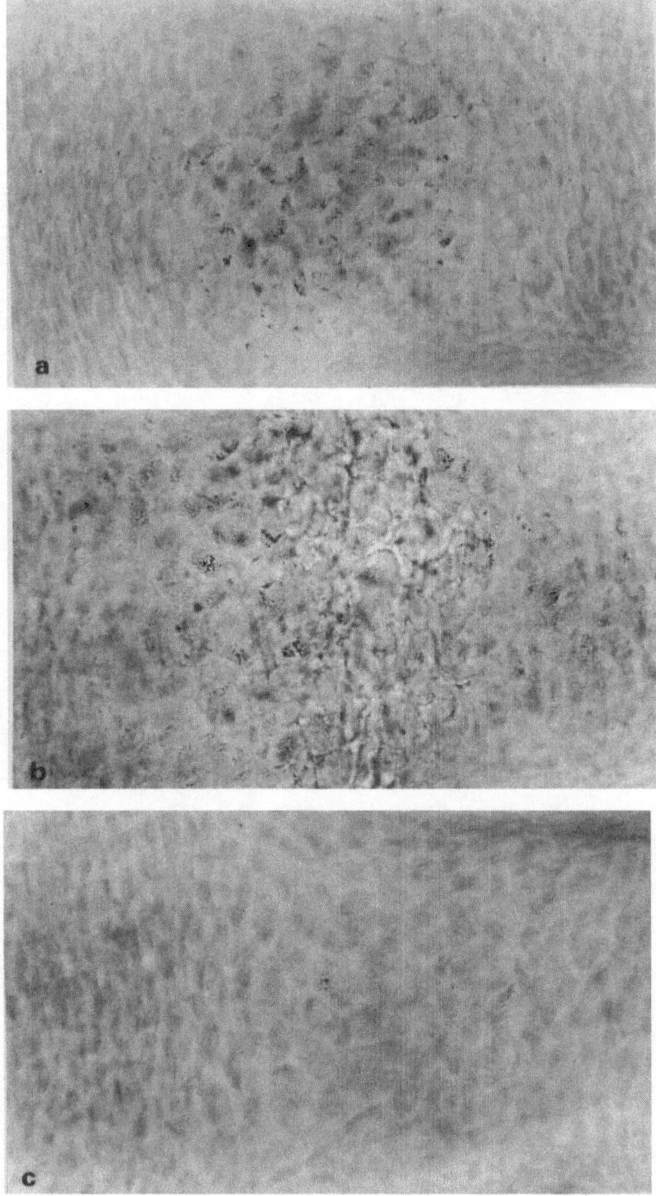

Figure 1. a-c. GBHA-stained total mount preparations; × 360.
(a) *Non*-cultivated control rudiment: note many GBHA-positive granules in the central hypertrophic chondrocytes.
(b) After 2 days of cultivation in control medium: a GBHA-negative, refractive material has been deposited between the central hypertrophic cells. New granules have appeared in the peripheral hypertrophic cells.
(c) After 2 days of cultivation with PTE (0.1 U/ml) the granules have disappeared from the cells, while no refractive material has been deposited.

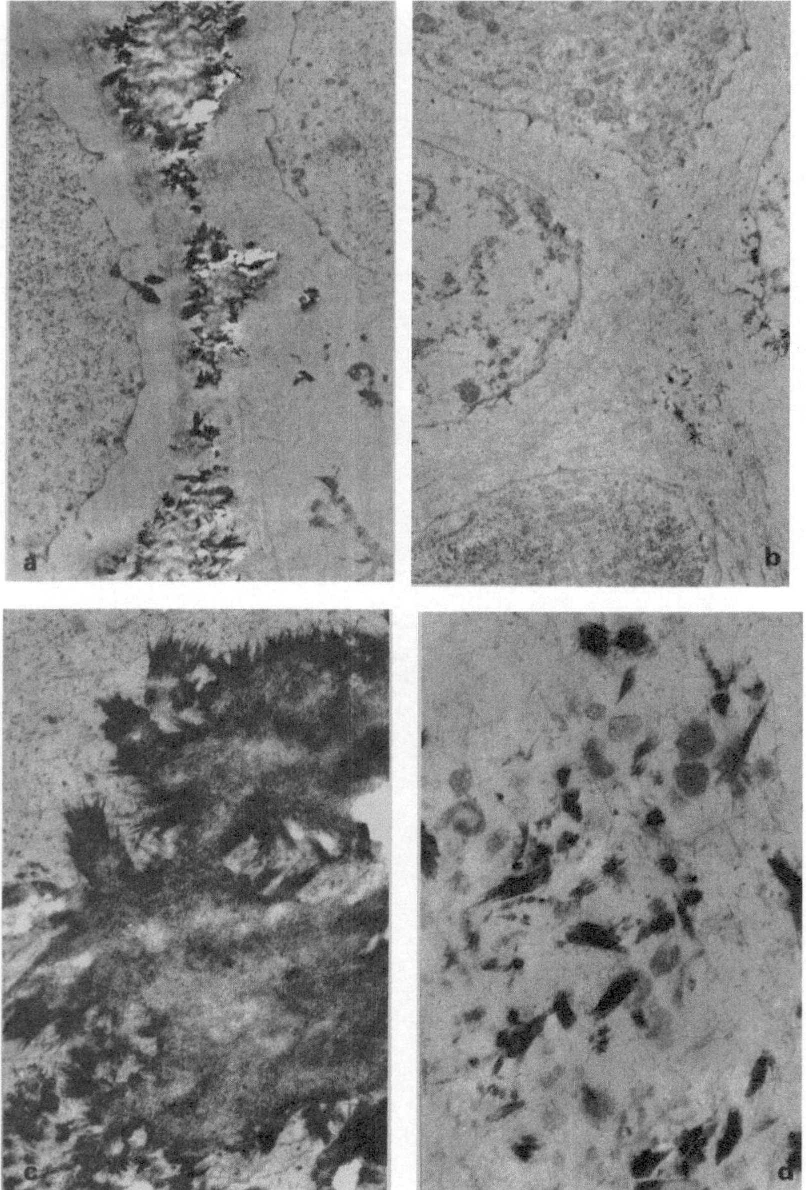

Figure 2. a-d. Electron micrographs.
(a) Survey micrograph of a control culture: note large mineralized areas between the hypertrophic cells; × 4500.
(b) PTE-treated culture: only very small patches of mineral can be found between the cells; × 4500.
(c) Control culture: detail of mineralized area coveres with small crystals; × 30000.
(d) PTE-treated culture: patches of needle shaped structures are closely associated with electron dense globules; × 30000.

the GBHA preparations. The central cells themselves were now devoid of GBHA-positive red granules, but new granules had appeared in the young hypertrophic cells on both sides of the mineralized area. The same pattern of events was found after *in vivo* development, in 17-day-embryos.

Cultivation in the presence of 0.1 U PTE per ml culture medium led to a total inhibition of this initial mineralization of cartilage matrix. Besides, the GBHA-positive granules had disappeared from the central hypertrophic cells while no new granules were formed in the maturing cells on both sides of this area (see Fig. 1 a-c). After treatment with a lower dose of PTE (0.01 U per ml culture medium) a partial inhibition was observed. The mineralized area was found to be 28% smaller than the corresponding controls, (P < 0.001, Student's t-test for paired observations), but the effect on the number of granules — scored in a semiquantitative way — was no longer significant.

B. *Electron microscopy*

In the *non*-cultivated rudiments many electron-dense, membrane bounded globules were found in the matrix between the hypertrophic cells. Elongated structures in the close vicinity of these globules suggested that they might have been formed at the tip of cell processes by budding. No indication of mineralization was found at this stage as yet.

After cultivation in control medium, large areas of the matrix between the hypertrophic cells were covered with small crystals (see Fig. 2 a, c). At the border of these areas the crystals were clustered, forming large needle shaped structures which protruded radially in the unmineralized matrix. Because of its hardness, cutting of this material with a glass knife was a problem.

Treatment with PTE (0.1 U/ml) prevented the formation of these large mineralized areas. Instead only small patches of an electron dense material were formed in or on the matrix globules (see Fig. 2 b, d). These patches contained the same needle shaped structures as were found protruding from the mineralized areas in the control cultures. Both in control- and PTE-treated cultures the hypertrophic cells themselves looked relatively healthy, with few organelles, but an intact cell membrane.

C. *Cultivation after freezing and thawing*

Five times freezing and thawing completely prevented the deposition of mineral during subsequent cultivation of the rudiments. Neither could any GBHA positive granules be found in the remnants of the cells after cultivation.

DISCUSSION

Kashiwa (5, 6, 7) has shown that the Schiff base glyoxal bis (2-hydroxy-anil) (GBHA) is able to localize labile cellular calcium. He described intracellular GBHA positive red granules in a variety of calcifying tissues such as bone, hypertrophic cartilage and calcifying tendon.

Observations on mouse calvaria led Aaron (1) to propose a mineral "loading" and "unloading" cycle for bone cells. She distinguished different types of osteocytes on the basis of their staining characteristics after GBHA treatment.

A similar sequence of "loading" and "unloading" could be observed in the hypertrophic chondrocytes in our material, both during *"in vivo"* and *"in vitro"* development. Also, matrix mineralization *"in vitro"* was shown to be a biological process, since killing the cells before cultivation prevented it. Besides, the studies reported here indicate that parathyroid hormone interferes with the process of cartilage mineralization by causing the cells to release their calcium, without using it for matrix mineralization. This suggests the possibility that in the intact animal parathyroid hormone raises serum calcium not only by stimulating mineral resorption but also by preventing mineral deposition.

REFERENCES

1. Aaron, J.E.: Osteocyte types in the developing mouse calvarium. *Calc. Tiss. Res.* 12, 259–279 (1973)
2. Burger, E.H.: Inhibition of chondrocyte hypertrophy and zone formation in explanted cartilaginous long bone rudiments by parathyroid extract. *Proc. Kon. Ned. Acad. Wet.* C77, 308–320 (1974
3. Gaillard, P.J.: The influence of parathormone on the explanted radius of albino mouse embryos. *Proc. Kon. Ned. Acad. Wet.* C63, 25–37 (1960)
4. Hirsch, J.G. & Fedorko, M.E.: Ultrastructure of human leukocytes after simultaneous fixation with glutaraldehyde and osmium tetroxide and "post-fixation" in uranyl acetate. *J. Cell. Biol.* 38, 615–627 (1968)
5. Kashiwa, H.K.: Calcium in cells of fresh bone stained with glyoxal bis (–2 hydroxyanil). *Stain Technol.* 41, 49–55 (1966)
6. Kashiwa, H.K.: Calcium phosphate in osteogenic cells. A critique of the GBHA and the dilute silver acetate methods. *Clin. Orthop. Rel. Res.* 70, 200–211 (1970)
7. Kashiwa, H.K. & Komorous, J.: Mineralized spherules in the cells and matrix of calcifying cartilage from developing bone. *Anat. Rec.* 170, 119–128 (1971)

A Bone Morphogenetic Polypeptide

M. R. URIST, H. NOGAMI & A. MIKULSKI

In this report, we demonstrate that bone matrix gelatine (BMG) prepared under specified conditions (17, 18) and implanted in diffusion chambers produces transfilter osteogenesis in a muscle pouch in allogeneic rats. We also report that a bone morphogenetic polypeptide BMP) can be isolated from a collagenase digest of ^{35}S-cysteine labelled BMG by column chromatography.

In the past, diffusion chambers have been employd in bone research almost exclusively for transplantation of living tissues. Goldhaber (5) transplanted calvarial bone inside and described appositional bone outside of a diffusion chamber. Heiple *et al.* (7) and Friedman *et al.* (4) placed mouse osteosarcoma inside of a millipore chamber and induced normal bone tissue differentiation on the outside. Buring and Urist (1) loaded chambers with demineralized bone matrix and minced muscle and produced intrachamber chondrogenesis; this in turn, induced transfilter osteogenesis in connective tissue lining a muscle pouch. In various other experiments with a wide selection of embryonic skeletal and non-skeletal tissues (2, 6, 10), the chamber was filled with living cells and transfilter cell differentiation occurred only when the membranes were single thickness and the pore sized ranged from 0.22 to about 0.45 μm. In order to investigate the influence of a non-viable chemically-defined preparation of bone matrix gelatin (BMG) implanted in a diffusion chamber, the experiments reported in this communication were performed with double as well as single (150 μm) walled chambers with membrane pore sizes beginning as low as 0.025 μm.

MATERIAL AND METHODS

A BMG that is insoluble in cold water but partially soluble in body fluids at 37°C, was prepared from *cortical* bone of Sprague-Dawley rats. The

Bone Research Laboratory, University of California, Los Angeles, California, 90024.

method of preparation and chemical analysis of BMG appears in a previous publication (18). Diffusion (Millipore) chambers were constructed of single and double thickness cellulose acetate membranes, pore sizes 0.025 and 0.45 μm, thickness 150 in μm^3. In double thickness chambers, 0.025 μm was placed on the inner and 0.45 μm on the outer surface. The chambers were loaded with either BMG or the residues of limited collagenase digestion of BMG and sealed with non-polar MF cement; 120 loaded chambers were implanted in muscle pouches in allogeneic adult rats for 4 weeks and examined by histochemical and EM methods.

Diced 500 mg quantities of BMG, prepared from bone labelled *in vivo* by daily injections of ^{35}S-cysteine (17, 19) were digested in 20 ml of 0.5 M Tris-HCl, pH 7.6 containing 4 mg of purified collagenase (9) with 0.005 M CaCl$_2$ and 200 units of penicillin streptomycin, over a period of 24 h at 37°C. The digestant solution was decanted and centrifuged at 15.000 rpm/20 min at 2°C; 5 ml of 0.5 M EDTA was added in order to bind Ca^{+2} and stop collagenase action; 1 g of chipped demineralized insoluble whole bone matrix was also added for 4 h at 2°C with stirring in order to remove free collagenase from the solution by absorption onto bone collagen. The opalescent supernatant was concentrated by lyophilization. Approximately 10 ml of the lyophilized product was equilibrated with 0.05 M Tris-HCl buffer pH 7.6, charged on Bio-Gel Agarose 0.5 M column, and eluted with the same buffer at room temperature. Three fractions were obtained in the following sequence: first, a distinctive peak eluting with void volumes of the column corresponding to a high molecular weight polymers; second, a broad fraction corresponding to a mixture of broad range of molecular sizes; third, a sharp peak of low molecular size. Fraction III was further purified by diethyl cellulose chromatography which removed hydroxyproline-containing tripeptides and other derivatives of the helical region of collagen molecules. The end product contained less than 0.03 μg OH proline/mg of Fraction III.

The biological activity of the three isolated fractions was determined by recombination with BMP-free BMG prepared from autolysed bone (16). Recombination was produced by incubating 25 mg of BMG in 2.0 ml of 0.1 M Tris-HCl buffer pH 7.4 containing 1 mg of each fraction separately for 24 h at 37°C. After incubation the recombinants were removed, lyophilized and implanted in a muscle pouch for three weeks in allogeneic adult rats.

RESULTS

Transfilter Osteoinduction by Bone Matrix Gelatin. In the interior of the chamber, BMG consisted of a conglomeration of fine granules, small filaments interspersed with ruthenium red-staining coarse granules, and remnants of old cross-banded collagen fibrils. Solubilized parts of the BMG consisted of similar fine and coarse granules suspended in interstitial fluid and mixed with packets and filaments of fibrin. The inner surface of the membrane was covered with a syrup of solubilized gelatin that appears in the EM as fine granules and filaments packed in layers.

The pores of the inner 0.025 μm pore size membrane were filled with irregularly-dispersed clumps of uranyl acetate staining granular material plastered against the cellulose acetate septa. Small and large electron-lucent fluid filled spaces occupied about half of the cross-sectional area each field. *No cross-banded collagen fibrils were ever found in the interior of the 0.025 μm pores.* The space between the membranes of double walled chambers was filled with densely packed layers of granular material. The pores of the outer membrane also were filled with a loosely-dispersed fine granular material but frequently contained isolated perfectly-formed 640 Å cross-banded renatured collagen *fibrils.* Ruthenium red-staining coarse granules were randomly distributed throughout the fine granular material and closely adherent to some of the cross-bands of the collagen fibrils.

New bone, densely stained *fibers*, formed by fibrous connective tissue cells, adhered to the outer surfaces of the membrane (Fig. 1). The pore openings on the outer surface of the membrane were covered with cartilage and new bone. Collagen *fibers* and ground substance extended from the deposits of new bone distances as long as 30 to 35 μm into the

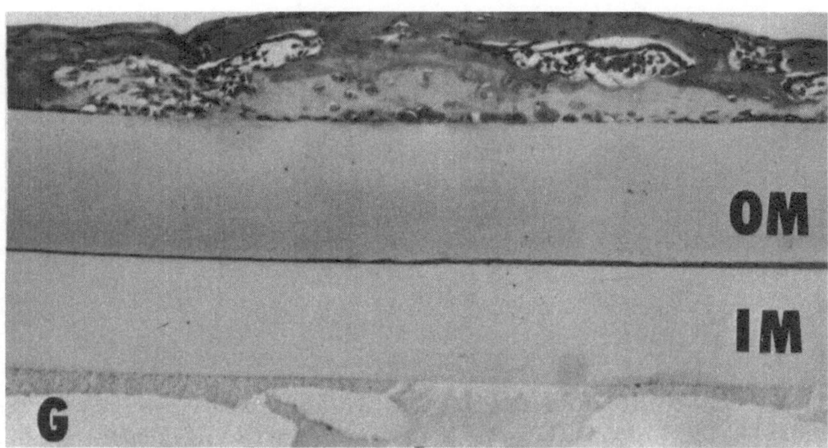

Figure 1. Photomicrograph of transfilter new bone (top) deposits developed on the exterior surface of the double walled chamber (thickness: 150 + 125 μm); pore size 0.45 μm outer membrane (OM) and 0.025 inner membrane (IM). Note: bone matrix gelatin (G) inside the chamber and juxtaposed new bone outside the chamber. X100

pore openings. Experiments with 20 control empty chambers, and 10 chambers filled with BMG produced from autolysed bone produced fibrous tissue only and neither cartilage nor bone.

Recombination of Fraction III with BMG Prepared from Autolysed Bone. Three fractions, obtained by chromatographic separation of a collagenase digest of BMG, contained radioactive cysteine. SDS gel showed a strong single band with a faint trailing material, and β-alanine gel electrophoresis showed the same strong band but with two very weak staining trailing collagen peptides (Fig. 2). The predominant amino acids were lysine, glycine, proline, and alanine with a small amount of cysteic acid. Tyrosine and methionine were absent. The total N by the Kjeldahl method was 160 μg/mg; total P was 0. Pentoses were present in trace quantities but hexoses were 0.

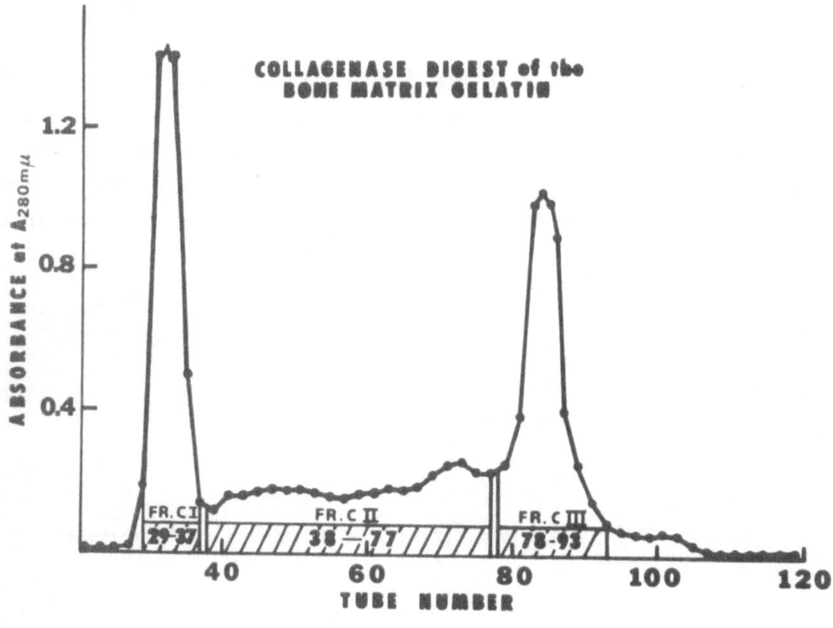

Figure 2. UV elution pattern measurements at 280 μm wave lengths (Bio-Gel Agarose column 0.5 M, 2.5 x 100 cm). From left to right: I, the first distinctive peak to elute with the void volume indicates a high molecular weight fraction; II, a broad fraction indicative of a wide range of molecular sized components; III, a sharp peak, the last to elute, delineates low molecular size components in the range less than 10,000 daltons. Note dense-staining single band obtained by SDS gel electrophoresis of Fraction III (right).

Autolysed BMG controls incubated in the phosphate buffer solutions alone or with Fraction I and II were resorbed and replaced with small round cells mixed with macrophages and fibroblasts. After incubation with Fraction III, autolysed inactive BMG produced appositional new bone formation. The new bone deposits were generally limited to external surfaces of the implants and suggested that recombination had occurred on the readily accessible surfaces of gelatin. The pH optimum for recombination of Fraction III was pH 7.4; pH 5.0 and 6.8 produced negative results.

DISCUSSION

BMG prepared under specified conditions (16) contains a hypothetical bone morphogenetic polypeptide (BMP). The helical portion of the collagen molecule itself is not BMP because bone matrix collagen divested of non-collagenous constituents by nine different enzymic and non-enzymic methods, produces fibrous tissue only (18, 21).

Either BMG or the residue of limited collagenase digestion of BMG, dissolves or disintegrates in interstitial fluids in the interior of a diffusion chamber at 37°C. The solubilized gelatin diffuses across a double walled chamber (inner wall 0.025 μm and outer wall, 0.45 μm pore sizes), transfers BMP to connective tissues proliferating near the pore openings on the outer surface of the membrane, and instigates cartilage and bone formation. The mechanism of action of BMP upon mammalian mesenchymal cell gene expression is not known, but the reaction that follows is one of the most consistently reproducible of all examples of extraskeletal bone formation (8, 12, 13).

BMP activity is physico-chemically separable from the helical or of the non-helical termini of the collagen molecule (14); it is irreversibly destroyed by radiation without altering the 640 Å cross-bands or insolubility of bone collagen (15), or by 0.1 N NaOH hydrolysis at 2°C without solubilization of hydroxyproline (16). BMP activity is unaffected by pepsin, collagenase, DNAase, RNAase, or phosphatases (16), but is associated with a trypsin-labile, polypeptides containing cysteine. Coupled with previous observations that it is extinguished reversibly by mercapto-ethanol or dithiothreitol-reduction and reoxidation (20); present evidence suggests that biologically active structure of BMP includes a disulfide bridge. The question arises whether BMP may be derived from a portion of the bone procollagen intermediate, e.g., registration peptides. These peptides do not contain hydroxyproline, an amino acid peculiar to the helical portion of the collagen molecule, or tyrosine, a component of the

N-termines, but they do contain cysteine residues which are *not* found in bone collagen. Research on bone procollagen synthesis is advancing rapidly (3); the fate and function of the cleaved bone registration peptides are under intensive investigation. It is interesting to speculate that BMP may be derived from some part of the bone tissue registration peptides. When purified BMP becomes available in significant quantities, interesting answers to the trenchant question about how extracellular substances could influence gene expression (11).

ACKNOWLEDGEMENT

This research was supported by grants-in-aid from NIDH (DE02103—11) and the Solo Cup Foundation.

REFERENCES

1. Buring, K. & Urist, M.R.: Effects of ionizing radiation on the bone induction principle in matrix of bone implants. *Clin. Orthop.* 55, 225—234 (1967)
2. Cooper, G.W.: Cartilage matrix induction transfilter cartilage induction. *Develop. Biol.* 12, 185—213 (1965)
3. Fessler, L.I., Burgeson, R.E., Morris, N.P. & Fessler, J.H.: Collagen synthesis: A disulfide-linked collagen precurser in chick bone. *Proc. nat. Acad. Sci. (Wash.)* 70, 2993—2996 (1973)
4. Friedman, B., Heiple, K.G., Vesely, J.C. & Honaoka, H.: Ultrastructural investigation of bone induction by an osteosarcoma using diffusion chambers. *Clin. Orthop.* 59, 39—58 (1968)
5. Goldhaber, P.: Osteogenetic induction across millipore filters *in vivo. Science* 133, 2065—2067 (1961)
6. Grobstein, C. & Dalton, A.J.: Kidney tubule induction in mouse metanephrogenic mesenchymal without cytoplasmic contact. *J. Exp. Zool.* 135, 57—66 (1957)
7. Heiple, K.G., Herndon, C.H., Chase, S.W. & Wattleworth, A.: Osteogenic induction by osteosarcoma and normal bone in mice. *J. Bone Jt. Surg.* 50A(2), 311—325 (1968)
8. Huggins, C., Wiseman, S. & Reddi, A.H.: Transformation of fibroblasts by allogenetic and xenogeneic transplants of demineralized tooth and bone. *J. exp. Med.* 132, 1250—1257 (1970)
9. Peterkofsky, B. & Diegelmann, R.: Use of a mixture of proteinase-free collagenases for the specific assay of radioactive collagen in the presence of other proteins. *Biochemistry* 10, 988—993 (1971)
10. Saxén, L.: Transmission and spread of kidney tubule induction. In: *Tissue Interactions in Carcinogenesis,* D. Tarin (ed.), Academic Press, pp. 49—80, 1972
11. Slavkin, H.C. & Gruelich, R. (eds.): *Extracellular Matrix Influences on Gene Expression,* Academic Press, p. 815, 1975
12. Urist, M.R.: Bone: Formation by autoinduction. *Science* 150, 893—899 (1965)

13. Urist, M.R.: The substratum for bone morphogenesis. *29th Symp. Soc. Develop. Biol. (Suppl.)* **4**, 125—163 (1971)

14. Urist, M.R., Earnest, F., Kimball, K.M., Dijulio, T.P. & Iwata, H.: Bone morphogenesis in implants of residues of radioisotope labelled bone matrix. *Calc. Tiss. Res.* **15**(4), 269—286 (1974)

15. Urist, M.R. & Hernandez, A.: Excitation transfer in bone — deleterious effects of cobalt 60 radiation-sterilization of bank bone. *Arch. Surg.* **109**, 486—493 (1974)

16. Urist, M.R. & Iwata, H.: Preservation and biodegradation of the morphogenetic property of bone matrix. *J. theor. Biol.* **38**, 155—168 (1973)

17. Urist, M.R., Iwata, H., Boyd, S.D. & Ceccotti, P.Q.L.: Observations implicating an extracellular enzymic mechanism of control of bone morphogenesis. *J. Histochem. Cytochem.* **22**, 88—103 (1974)

18. Urist, M.R., Iwata, H., Ceccotti, P.W.L., Dorfman, R.L., Boyd, S.D., McDowell, R.M. & Chien, C.: Bone morphogenesis in implants of insoluble bone gelatin. *Proc. nat. Acad. Sci. (Wash.)* **70**(12), 3511—3515 (1973)

19. Urist, M.R., Mikulski, A. & Boyd, S.D.: A chemosterilized antigen-extracted morphogenetic alloimplant. *Arch. Surg.* **110**, 416—428 (1975)

20. Urist, M.R., Mikulski, A. & Conteas, C.: Reversible extinction of the morphogen in bone matrix by reduction and oxidation of disulfide bonds. *Calc. Tiss. Res.*, In Press

21. Urist, M.R. & Strates, B.S.: Bone morphogenetic protein. *J. Dent. Res.* **50** (Suppl. 6), 1392—1406 (1971)

Effects of Ultimobranchialectomy (UBX) upon the Bone Catabolism and Anabolism and on some Aspects of the Calcium Metabolism Regulation in *Anguilla anguilla L.*

E. LOPEZ, J. PEIGNOUX-DEVILLE, F. LALLIER, E. MARTELLY-BAGOT & C. MILET

INTRODUCTION

In *Anguilla anguilla L.* the ultimobranchial body (UB) is situated between the oesophagus and the venous sinus; it presents a follicular structure (13). In higher vertebrates, UB is usually located at some distance from the heart and, in most mammals, it is embedded in the thyroid gland.

Rasquin and Rosenbloom (21) were the first to indicate that UB in fish might be related to the calcium metabolism. Then, later, Copp *et al.* (4) extracted calcitonin (CT) from the UB of dogfish and *Squalus suckleyi*; hypocalcemic activities were subsequently observed in ultimobranchial extracts from sharks (24) and teleosts (5). It was then obvious that CT was an important hypocalcemic factor in mammals. In teleosts, mammalian CT was first reported to have no hypocalcemic effect on *Fundulus heteroclitus* (18); subsequently other authors showed that it was effective in *Ictalurus melas* (16) and in *Anguilla japonica* (2). Pang and Griffith (19) have observed hypocalcemia and hypophosphatemia in fresh water — adapted *Anguilla rostrata*, injected with salmon calcitonin (SCT), but the same treatment was ineffective in sea water = adapted animals. Recently we have shown (Lopez *et al.*, in press) a decrease of serum calcium levels after injections of SCT in silver ♀ eel maintained in sea water and a marked glandular atrophy of the UB (20) but we failed to elicit an hypocalcemic response in trout maintained in tap water (13). It was also pointed out that hypercalcemia, induced by various experimental means, stimulated UB activity (13, 1, 11) and that UB activity was related to serum calcium fluctuations in *Salmo salar* (6); Dacke *et al.* (7) observed

Laboratoire de Physiologie générale et comparée, Muséum national d'Histoire naturelle. Laboratoire d'Endocrinologie comparée associé au CNRS, Paris.

that fish plasma CT responded to injected calcium.

In previous reports we have described some effects of exogenous CT on vertebral bone of some teleosts fish (14, 11, 15). It has also been noticed that ultimobranchialectomy could affect: serum calcium concentrations in teleost fish (1), bone in amphibians (22, 23) and in birds (3). The present work was carried out with the object to elucidate the effect of the absence of UB, in silver eel maintained in Ca^{++} rich tap water, on calcium and bone metabolism.

METHODS

Silver ♀ eels obtained from Peronne (Somme) were kept under laboratory conditions in aquariums for one month, in Ca^{++} rich tap water, $\theta = 18°-20°$. They were ranging in weight from 170 to 280 g. Two days before the operation, blood was collected from each animal, by means of heart puncture, to obtain, before the operation, the serum calcium level. The eels were divided into two groups and before the operation they were anesthetized with MS 222 from the Sandoz laboratory.

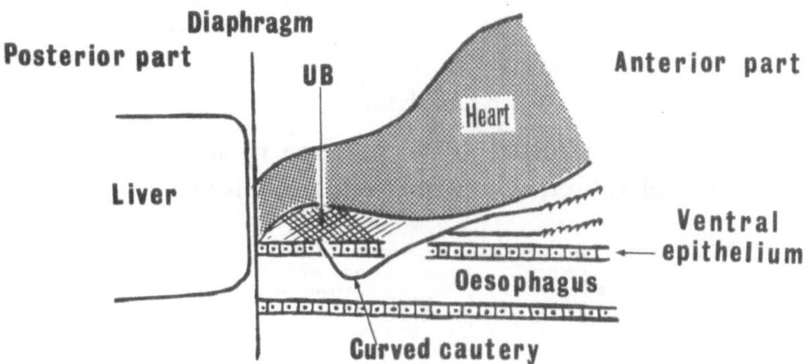

Figure 1.

1. A longitudinal incision was practised (1.5 cm) between the pectoral fins.
2. The heart was laterally pushed aside, so as to expose the oesophagus.
3. An incision (0.5 cm) was made in the ventral epithelium of the oesophagus.
4. A curved hot-wire cautery was passed through the ventral epithelium of the oesophagus and all the connective tissues, containing UB, located between the oesophagus and the venous sinus, was destroyed.

The first group (21 fish) was sham-operated, undergoing all the process of the operation. The second group (39 fish) was operated as described in Fig. 1. This operation, different from Chan's method (1970), was practised without any haemorrhage and without any damage to the neighbouring tissues such as heart and liver; the epithelium of the oesophagus was cicatrized approximately two days after the operation. We never observed mortality in operated eels. We noticed that, in the eel, the UB can be seen neither with the naked eye nor with a stereomicroscope. Some fish, both of the first and second groups, were sacrificed one week after the operation, then others two weeks and five weeks later. Blood was then collected from the ventral aorta. Serum calcium levels were determined by the atomic absorption method (Perkin-Elmer).

A histological control, with serial sections of the cauterized area, was done for each UBX animal. Stannius corpuscles were treated by classic histological methods. The vertebral bone of several fish of each group was studied, on thin sections of mineralized tissue (10 à 15 μ) and on microradiographs. The bone degree of mineralization was determined by means of a precise histophotometric method and the morphologic quantitation of bone turnover according to Jowsey *et al.*, (10). Osteocytic osteolysis was expressed as a percentage of the total number of osteocytic cavities. Details of these techniques have been published earlier (11).

RESULTS

Ultimobranchialectomy resulted in an increase in serum calcium levels, relative to the initial values, which reached a maximum after two weeks

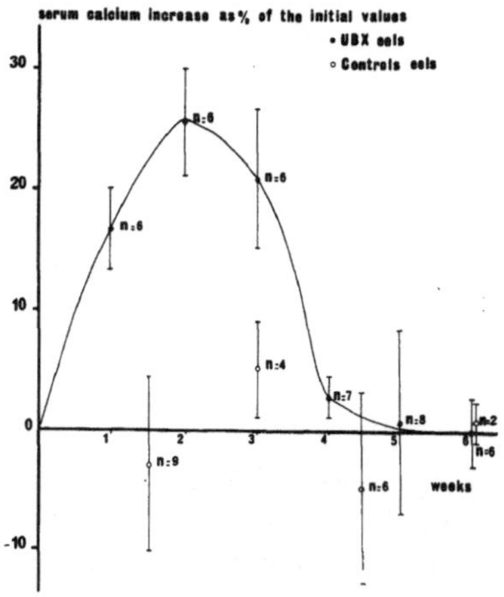

Figure 2.   Effect of ultimobranchialectomy (UBX) on serum calcium levels of the eel maintained in calcium rich tap water (80 mg/l)

and then subsequently declined to approximately physiological concentrations. Serum calcium concentrations in the controls did not vary by more than 10% and did not exceed the variation found between individuals (Fig. 2). We have observed in the vertebral bone several modifications which first appeared during the first week after the operation and remained unchanged two, three and five weeks later. In operated eels we noted principally some cytological changes in the osteoblasts. In bone of controls we observed an important number of active osteoblasts; they generally lined a new bone layer whose surface on microradiograph shows a low mineral density with a smooth surface, characteristic figures of bone apposition. In bone of UBX eels, osteoblasts were reduced in size, number and in the regularity of their placing along the bone surface; they were seen as completely flat cells lining bone inactive surfaces and a band of high mineral density is formed on microradiographs. The osteoblastic apposition was almost completely stopped (Table I).

Table I. Effects of Ultimobranchialectomy on the Various Bone Parameters.

Animals Silver ♀ eels	Mineral substance concentration g/cm³ (apatite + amorphous phosphate calcium)	osteoclastic resorption % surface	osteoblastic apposition % surface	osteocytic osteolysis %
Time after operation (weeks)				
1	1.02	11.2	5.2	3.9
1	1.01	13.8	4.8	3.2
2	1.10	14.1	4.5	3.8
2	1.08	14.5	4.9	3.8
3	1.12	13.2	5.3	3.4
3	1.01	13.5	4.8	4.5
5	1.19	14.3	4.5	3.6
5	1.08	12.8	5.2	4.1
5	1.13	12.1	4.1	4.5
mean ± SEM (a = SEM)	1.08 ± 0.022^a	13.27 ± 0.363	4.81 ± 0.132	3.86 ± 0.149
a = SEM F test	$p < 0.001$	$p > 5\%$	$p < 0.001$	$p > 5\%$
1	0.97	13.4	0.08	4.2
1	0.91	13.1	0.09	4.3
2	1.00	13.8	0.13	4.0
2	0.98	14.2	0.16	4.1
3	0.92	14.1	0.20	3.8
3	0.91	14.0	0.08	4.0
5	0.83	13.2	0.20	3.9
5	0.97	14.5	0.05	4.1
5	0.87	14.7	0.20	4.3
mean ± SEM	0.91 ± 0.022	13.88 ± 0.168	0.13 ± 0.018	4.05 ± 0.058

After ultimobranchialectomy we did not observe any changes in the number of osteoclasts therefore osteoclastic resorption surfaces were not increased; the degree of osteocytic osteolysis was not modified (Table I). We noticed that the degree of average mineralization of the intercellular substance of bone was noticeably decreased (without modification of the organic matrix) relative to the control values (Table I).

Stannius corpuscles were found to be very active glands in operated eels; initial spheric follicles were clumped together into sinuous strands. We have already described the formation of these strands after stimulation of the Stannius corpuscles (12). Between the cellular strands large blood vessels extended. Taking as a basis these observations compared to those we have already made concerning Stannius corpuscles, there is little doubt that these glands are in their active phase.

DISCUSSION

We noted, in previous studies, that CT (either mammalian CT or CT from fish) acted on the vertebral bone of fish (14, 15). Administration of exogenous CT may not necessarily reflect the physiological doses or reveal the total physiologic effect. The existence of UB as an individualized organ, readily accessible in fish, provides a very good means of evaluating its physiological significance.

Ultimobranchialectomy in the silver ♀ eel, maintained in Ca^{++} rich fresh water, provokes an important rise in serum calcium levels, this hypercalcemia is regulated during the third week after the operation. Such an increase in serum calcium concentrations was reported by Chan (1) in *Anguilla japonica* and by Robertson in *Rana pipiens* (22).

The two bone effects obtained, after the operation, in eels: demineralization of the bone organic matrix and inhibition of the osteoblastic apposition, can explain, at least partly, the observed hypercalcemia during the first week. The prolonged hypercalcemia to the second week can be the result of the effect of failing CT on kidney (1) and on calcium exchanges across the gills; indeed, preliminary results (Peignoux-Deville *et al.*, unpublished data) lead us to believe that CT (SCT), perfused in isolated UBX eel gills, acts on calcium influx and outflux and has a resulting effect in reducing the serum calcium concentration. The regulation of hypercalcemia after two weeks, in UBX eels, cannot be a parathormone response as Robertson has shown with the frog (22), because the parathyroid is absent in fish. In our experiment Stannius corpuscles of UBX eels are found to be very active glands, these glands were first shown to be hypocalcemic glands in teleost by Fontaine (8, 9);

we suggest that the hypercalcemia induced by ultimobranchialectomy, in eels, is regulated by Stannius corpuscles.

In eels the ultimobranchialectomy provokes an abolition of the osteoblastic apposition from the first week after the operation. Robertson (22) in *Rana pipiens* described a similar effect 12 weeks after the operation; we suppose that this variance is due to the fact that, in frogs, the first target organs for CT effect are paravertebral lime sacs and not the skeleton. In UBX eels we have noticed neither an increase of the number of osteoclasts as Robertson (22) did in frogs, nor an increase in the osteocytic osteolysis. These results are consistent with those exposed in our previous reports (14, 11, 15) in which we noted that exogenous CT reduces bone catabolism only when it is very high. If the bone resorption is not modified after ultimobranchialectomy we can point out an important demineralization of the bone organic matrix; reciprocally we have observed (Lopez *et al.*, in press) that salmon CT is able to enhance the degree of mineralization of the bone intercellular substance, in silver ♀ eels maintained in sea water. These results on the effect of CT on the degree of mineralization of vertebral bone, may be compared with those of Robertson (22, 23) noticing an effect of CT on the rate of calcium flux through the lime sacs epithelium of *Rana pipiens*. We have already suggested (15) that CT probably acts on the calcium exchanges at the site of the "bone lining cellular layer", which probably controls calcium ion transport (17); the existence of this cellular membrane (lining cells, osteoblasts, osteocytes), in fish with cellular bones, has been demonstrated (11). The role of the ultimobranchial gland, in silver eel, appears to be the release of a hormone which controls serum calcium, stimulates the production and activity of osteoblasts and regulates the bone degree of mineralization and, in some cases, reduces bone catabolism. We suggest that CT affects calcium transport both at the site of bone cells and gill epithelium.

REFERENCES

1. Chan, D.K.O.: Endocrine regulation of calcium and inorganic phosphate balance in fresh-water adapted teleost fish, *Anguilla anguilla* and *Anduilla japonica. Proc. 3rd int. Congr. Endocr.* Mexico City, 1968. Excerpta Med. Internat. Congress, Amsterdam, ser. 184, 709—716 (1970)

2. Chan, D.K.O., Chester-Jones, I. & Smith, R.N.: The effect of mammalian calcitonin on the plasma levels of calcium and inorganic phosphate in the European eel (*Anguilla anguilla L.*). *Gen. comp. Endocr.* 11, 243—245 (1968)

3. Copp, D.H.: Calcium regulation in birds. *Gen. comp. Endocr.*, suppl. 3, 441—447 (1972)

94

4. Copp, D.H., Cockroft, D.W. & Kueh, Y.: Calcitonin from ultimobranchial glands of dogfish and chicken. *Science* **158**, 924—925 (1967)
5. Copp, D.H., Cockroft, D.W., Kueh, Y. & Melville, M.: Calcitonin-ultimobranchial hormone. In: *Calcitonin.* Taylor, S. (ed.), Heineman, pp. 306—321, 1968
6. Deville, J. & Lopez, E.: Le corps ultimobranchial du saumon *Salmo salar L.* Etude histophysiologique à diverses étapes de son cycle vital en eau douce. *Arch. Anat. Micr. Morph. exp.* **59**, 393—402 (1970)
7. Dacke, C.G., Fleming, W.R. & Kenny, A.D.: Plasma calcitonin levels in fish. *Physiologist* **14**, 127 (abstract) (1971)
8. Fontaine, M.: Corpuscules de Stannius et régulation ionique (Ca, K, Na) du milieu intérieur de l'Anguille *(Anguilla anguilla L.). C. R. Acad. Sci. (Paris)* **259**, 975—878 (1964)
9. Fontaine, M.: Intervention des corpuscules de Stannius dans l'éguilibre phospho-calcique du milieu intérieur d'un Poisson téléostéen, l'Anguille. *C. R. Acad. Sci. (Paris)* **264**, 736—737 (1967)
10. Jowsey, J., Kelly, P.J., Riggs, B.L., Bianco, A.J., Scholtz, D.A. & Gershon-Cohen, J.: Quantitative microradiographic studies of normal and osteoporotic bone. *J. Bone Jt. Surg.* **47A**, 785—806 (1965)
11. Lopez, E.: *Étude morphologique et physiologique de l'os cellulaire des Poissons Téléostéens* (Thèse de Doctorat d'Etat, 1972). Mémoires du Muséum national d'Histoire naturelle (nouvelle série), série A., Zoologie, vol. LXXX, (1973)
12. Lopez, E. & Fontaine, M.: Réponse des corpuscules de Stannius de l'Anguille *(Anguilla anguilla L.)* à des blessures expérimentales. *C. R. Soc. Biol. (Paris)* **161**, n° 1, 36—39 (1967)
13. Lopez, E., Deville, J. & Bagot, E.: Etude histophysiologique du corps ultimo-branchial d'un Téléostéen *(Anguilla anguilla L.)* au cours d'hypercalcémies expérimentales. *C. R. Acad. Sci. (Paris)* **267**, 1531—1534 (1968)
14. Lopez, E., Chartier-Baraduc, M.M. & Deville, J.: Mise en évidence de l'action de la calcitonine porcine sur l'os de la truite *Salmo gairdnerii* soumise à un traitement déminéralisant. *C. R. Acad. Sci. (Paris)* **272**, 2600—2603 (1971)
15. Lopez, E. & Deville, J.: Effect of prolonged administration of synthetic salmon calcitonin (SCT) on vertebral bone morphology and on the ultimobranchial body (UB) activity of the mature female eel *(Anguilla anguilla L.). Proc. IXth Europ. Symp. Calc. Tiss.*, Vienna, Czitober, A. & Eschberger, J. (eds.), pp. 169—174, 1973
16. Louw, G.N., Sutton, W.S. & Kenny, A.D.: Action of thyrocalcitonin in the teleost fish Ictalurus melas. *Nature (Lond.)* **215**, 888—889 (1967)
17. Neuman, W.F.: The milieu interieur of bone: Claude Bernard revisited. *Fed. Proc.* **28**, 1846—1850 (1969)
18. Pang, P.K.T. & Pickford, G.E.: Failure of hog thyrocalcitonin to elicit hypocalce-mia in the teleost fish, *Fundulus heteroclitus. Comp. Biochem. Physiol.* **21**, 573—578 (1967)
19. Pang, P.K.T. & Griffith, R.W.: Unpublished data in calcitonin and ultimobranchial gland in fish. (P.K.T. PANG). Symposium. *J. exp. Zool.* **178**, 89—100 (1971)
20. Peignoux-Deville, J., Lopez, E., Lallier, F., Martelly-Bagot, E. & Milet, C.: Responses of the ultimobranchial body in eels *(Anguilla anguilla L.)* maintained in sea water and experimentally mature, to injections of synthetic salmon calcitonin. *Cell. Tiss. Res.* **164**, 73—83 (1975)

21. Rasquin, P. & Rosenbloom, L.: Endocrine imbalance and tissue hyperplasia in teleost maintained in darkness. *Bull. Amer. Mus. Nat. Hist. nat.* **104**, 363—425 (1954)
22. Robertson, D.R.: The ultimobranchial body of *Rana pipiens*. VIII: Effects of extirpation upon calcium distribution and bone cell types. *Gen. comp. Endocr.* **12**, 479—490 (1969 a)
23. Robertson, D.R.: The ultimobranchial body of *Rana pipiens*. X: Effect of glandular extirpation on fracture healing. *J. exp. Zool.* **172**, 425—441 (1969 b)
24. Urist, M.R.: Avian parathyroid physiology: Including a special comment on calcitonin. *Amer. Zool.* **7**, 883—895 (1967)

Number, Size and Arrangement of Osteoblasts in Osteons at Different Stages of Formation

G. Marotti, A. Zambonin Zallone & M. Ledda

The secretory activity of osteoblasts has been quantitatively investigated by means of different techniques. Autoradiographic studies on Glycine-H_3 and Thymidine-H_3 injected in young rabbits indicate that, on an average, a periosteal osteoblast produces 2860 μ^3 of matrix daily, during three days, before becoming either an osteocyte or a relatively inactive *lining* osteoblast (10). Using stereological methods on histological sections of human rib cortical bone, a mean value of 220 μ^3 of matrix daily produced per osteoblast was estimated (11). More recently, the 'secretory territories' of osteoblasts on rat parietal bone were precisely measured under the scanning electron microscope and a daily matrix production rate of approximately 470 μ^3 was derived (1).

It should be noted that these estimations were made assuming that the appositional growth progresses on the various osteogenetic surfaces of a given subject at the same constant rate with time, and, therefore, they cannot provide information on the changes that should occur on the metabolic activity of osteoblasts with different rates of bone accretion. In fact, by means of successive labelling the newly-formed bone layers with fluorescent substances (tetracyclines, alizarins), it has been demonstrated that the linear rate of lamellar bone apposition differs considerably not only in various skeletal regions, but even in adjacent areas of a given apposition front, on both the surfaces of trabecular bone (2, 7) and the wall of Haversian canals. Furthermore on the latter, the rate appears to decrease exponentially during osteon formation (3, 4, 5). Hence, what osteoblastic mechanism is involved in modulating the accretion rate of the osteogenetic surfaces?

In an attempt to answer this question, the structure of the osteoblastic

Institutes of Human Anatomy of the Universities of Sassari and Bari.

layer was analyzed in relation to the linear rate of concentric lamellar bone apposition (Marotti *et al.*, 1973, 1974a, b). Secondary osteonic bone was selected, because in forming Haversian systems, owing to the above-mentioned exponential decrement of the rate, which occurs during their completion, the degree of activity of the osteogenetic surface can be deduced from the caliber of the Haversian canal.

MATERIALS AND METHODS

Structural analyses and measurements of cell parameters (*vide infra*) were performed on serial semithin (1 μ thick) cross and longitudinal sections, obtained with the ultramicrotome from dog diaphyseal compact bone, embedded in araldite after prefixation in buffered paraformaldehyde and postfixation in osmium tetroxide. The sections were decalcified by floatation on 2% formic acid, stained with methylene blue and azur II, and photographed with oil-immersion planapochromatic objectives.

RESULTS

In large Haversian canals of osteons at the initial stage of formation, where the appositional growth rate is very high, the osteoblasts appear big, tightly packed and variously shaped. More frequently, however, the osteoblastic layer bears some resemblance to a pseudostratified columnar epithelium, whose basement membrane corresponds to the osteoid border; consequently, the cell surface of each osteoblast in contact with the preosseous matrix appears relatively narrow (Fig. 1 a). On the contrary, in progressively smaller canals, in parallel to the diminution of the radial growth rate, the osteoblasts become smaller, separated the one from the other by larger intercellular spaces and flattened so as to adhere to the osteoid border with a relatively broad surface (Fig. 1 b).

To evaluate on a quantitative basis these structural changes, the following parameters were recorded from the osteoblastic layer, in osteons at different stages of formation: *a*) the area of the cell surface facing the osteoid border [viz. the 'secretory territory' according to Jones (1)] *b*), the volume of each osteoblast and *c*) the number of osteoblasts per unit length of the osteogenetic front.

As an example the data obtained, at the mid-shaft level of the tibia of a 3-months-old dog, from three growing osteons, enclosing Haversian canals of 13000, 3000 and 900 μ^2 respectively, can be seen in Fig. 2. According to previous measurements based on alizarin labelling (5), the mean radial growth rate of these osteons corresponds approximately to 1.8, 0.9 and

0.4 μ per day. From the comparison of the mean values reported above each histogram, it appears that the larger the Haversian canal (viz. the higher the appositional rate) the bigger the volume and the smaller the 'secretory territory' of the osteoblasts. The behaviour of the latter parameter also indicates that the cell density of the osteoblastic layer decreases with the narrowing of the Haversian canal. In fact, the mean

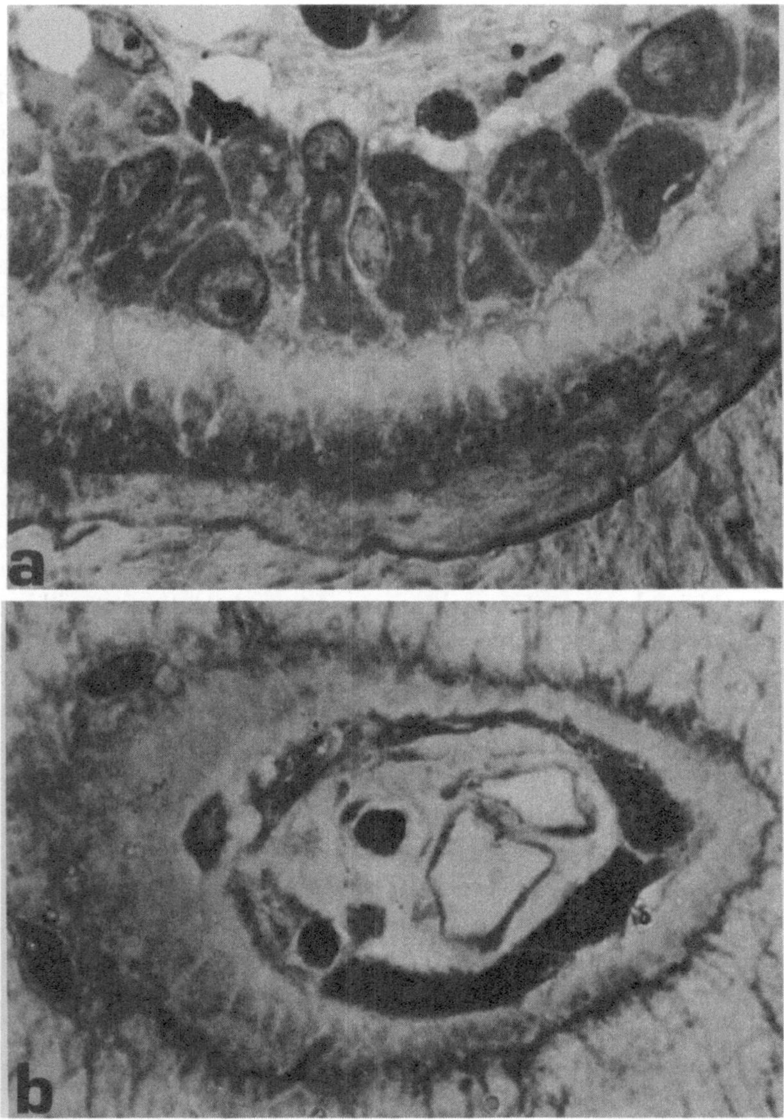

Figure 1. Cross sections of the tibial cortex of a 3-months-old dog. Microphotographs at the same magnification (2000 x) of the osteogenetic surfaces in osteons at the initial stage (*a*) and towards the end (*b*) of their deposition (see text).

number of osteoblasts lining a 100 μ length of osteogenetic front is 13, 9, 6 in the canal of 13000, 3000 and 900 μ^2 respectively.

The mean value for the 'secretory territories' when multiplied by the daily linear appositional growth rate gives, according to Jones (1), the volume of matrix secreted, on an average, by each osteoblast in one day. The following data can be approximately derived from the three growing osteons analyzed in this paper:

Haversian canal, μ^2	linear rate, μ/day	secretory territory, μ^2	matrix secreted by each osteoblast, μ^3/day
13000	1.8	100	180
3000	0.9	114	103
900	0.4	224	90

It is to note that, in all the osteoblastic layers so far analyzed, the ratio between the means of the cell volume and the amount of matrix daily produced per cell ranges from 5 to 7. Thus it seems that each osteoblast

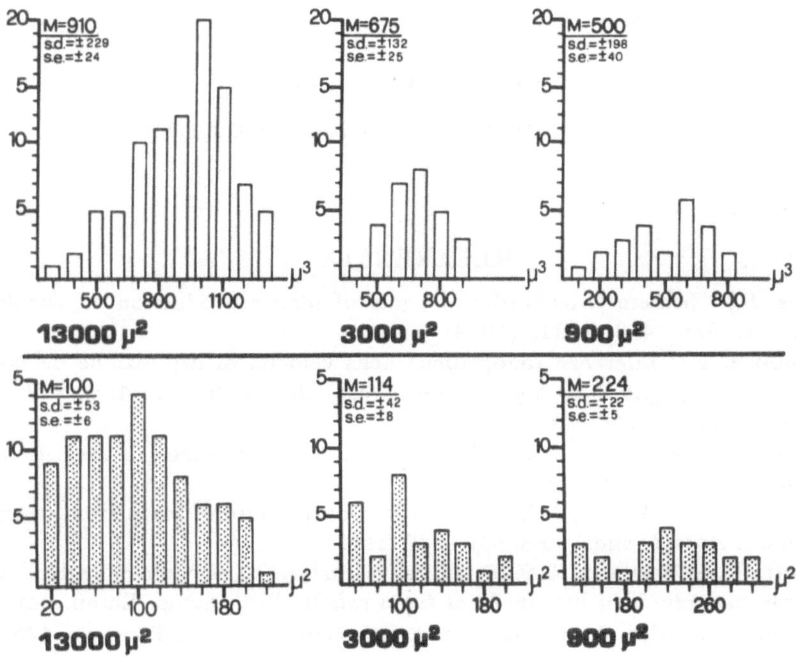

Figure 2. Frequency histograms of the volume (white bars) and secretory territory (dotted bars) of osteoblasts, measured in three growing osteons with Haversian canal of different size. The mean canal area is reported below each histogram. Horizontal axes: the cell volume in μ^3 and the secretory territory in μ^2 respectively. Vertical axes: the frequencies in absolute value. M = mean; s.d. = standard deviation; s.e. = standard error.

secretes an amount of matrix proportional to its protoplasmic volume, namely to the amount of organelles involved in glycoprotein and protein synthesis. However, more data are needed to confirm this assumption.

CONCLUSIONS

The present results indicate that the appositional growth rate of the osteogenetic surfaces seems to be related to two cell parameters: a) the secreting activity of each osteoblast, which in turn appears to be a function of both cell volume and extent of the 'secretory territory', and b) the number of osteoblasts per unit surface of the osteogenetic front. The observed marked variability of the shape and size of osteoblasts, in relation to simply physiological variations of the appositional growth rate, suggests that the analyses of the structural responses of osteoblasts to hormones, vitamins etc. or to pathological conditions must be carried out with care.

ACKNOWLEDGEMENT

This work was supported by the Italian National Research Council.

REFERENCES

1. Jones, S.J.: Secretory territories and rate of matrix production of osteoblasts. *Calc. Tiss. Res.* **14**, 303–315 (1974)
2. Lozupone, E.: Differenze topografiche nella velocità di deposizione del tessuto osseo nella spugnosa di ossa lunghe di cani di età diversa. *Arch. ital. Anat. Embriol.* **suppl. 78**, 54 (1973)
3. Manson, J.D. & Waters, N.E.: Maturation rate of the osteon of the cat. *Nature (Lond.)* **200**, 489–490 (1963)
4. Manson, J.D. & Waters, N.E.: Observations on the rate of maturation of the cat osteon. *J. Anat. (Lond.)* **99.3**, 539–549 (1965)
5. Marotti, G. & Camosso, M.E.: Quantitative analysis of osteonic bone dynamics in the various periods of life. In: *"Les tissus calcifiés"*, V. Symp. Europ. Bordeaux, Milhaud, G., Owen, M. & Blackwood, H.J. (eds.), SEDES, Paris, pp. 423–427, 1968
6. Marotti, G., Ledda, M. & Zambonin Zallone, A.: Dati quantitativi sulle dimensioni e sulla densità degli osteoblasti di superfici ossee a diversa attività osteogenetica. *Studi Sassaresi* **52**, 184–186 (1974 *a*)
7. Marotti, G., Lozupone, E., Favia, A. & Lattanzi, V.: Variazioni topografiche e relative all'età della velocità di apposizione del tessuto osseo sulle trabecole della spugnosa. *Boll. Soc. ital. Biol. sper.* **45**, 1017–1021 (1969)

8. Marotti, G., Zambonin Zallone, A., Del Rio, N. & Ledda, M.: Disposizione, forma e dimensione degli osteoblasti durante le varie fasi della costruzione degli osteoni. *Arch. ital. Anat. Embriol.*, **suppl.** **78**, 55 (1973)
9. Marotti, G., Zambonin Zallone, A. & Ledda, M.: Analisi della struttura degli osteoblasti in rapporto alla velocità di deposizione della matrice ossea. *Studi Sassaresi* **52**, 171—183 (1974 *b*)
10. Owen, M.: Cell population kinetics of an osteogenic tissue, I. *J. Cell. Biol.* **19**, 19—32 (1963)
11. Schen, S., Villanueva, A.R. & Frost, H.M.: Number of osteoblasts per unit area of osteoid seam in cortical human bone. *Canad. J. Physiol. and Pharmacol.* **43**, 319—325 (1965)

Cytochemical and Ultrastructural Characteristics of Human Osteoblasts in Relation to General Skeletal Growth Activity

INGER KJÆR & M.E. MATTHIESSEN

INTRODUCTION

Histological studies of human bone growth have been performed on biopsy and autopsy material by using the following methods: 1) measurements of osteoid seams (6, 18), 2) counting of osteoblasts per unit area of osteoid seam (15), 3) determination of osteoid volume (13), 4) microradiographic determinations of bone formation activity (8). The possibility of histochemical and ultrastructural assessment of bone growth activity has so far not been much utilized, possibly because of difficulties in tissue preparation. A previous study has shown that demonstration of morphological variations in human osteoblasts might provide a basis for diagnosis of pathological bone tissue (10). Since the activity of the alkaline phosphatase indicates the ossification intensity of the osteoblast and preosteoblast (5, 1, 2) determination of enzyme activity must be a useful method for determining interindividual growth differences (9, 11). The object of the present study was to relate the periodic variations in growth rate of normal growing individuals to the ultrastructure and cytochemistry of the osteoblasts.

MATERIAL AND METHODS

Material

The investigation was based on bone tissue from 30 human fetuses, 15 normal growing children and 5 adults.

The fetal material was derived from the bony primordium of the jaws,

Institute of Orthodontics, The Royal Dental College, Copenhagen and Anatomy Department A, University of Copenhagen.

which were dissected free immediately after the fetus had been removed from the uterine cavity in connection with legally approved abortion. The non-fetal material was made up of biopsies comprising periosteum and underlying bone, removed from the buccal alveolar lamella in the premolar region of the jaws. The biopsies were obtained in connection with dental extractions performed because of caries or for orthodontic purposes. Biopsies were taken only from individuals with clinical normal periodontal appearance.

Methods

The part of the material which was intended for the study of enzyme activity was fixed for half an hour in ice-cold formolcalcium (14) and then transferred to ice-cold gumarabic sucrose. The material was not decalcified. 10 μ sections were cut on a Pearse-Slee cryostat. Naphtol-AS-Bl phosphate method for alkaline phosphatase (4) was used. Incubation without substrate was used as control.

Immediately after removal the specimens for electron microscopic studies were placed in 2.5% glutaraldehyde in 0.1 M cacodylate buffer (pH 7.4) at 4°C and postfixed in 2% osmium tetroxide. The specimens were then dehydrated in ethanol and embedded in Epon without decalcification. Areas suitable for electron microscopy were located on 1 μ sections which were stained with toluidine blue. The thin sections (500 Å) were contrasted with uranyl acetate and lead citrate. From each individual 100 sections containing mitochondria were studied.

Individual postnatal skeletal maturation was assessed on the basis of roentgenograms of the hand (7). Based on the assessment of skeletal maturation the sample was divided in infantile, juvenile and adolescent groups according to Björk & Helm (3).

RESULTS AND DISCUSSION

The cytoplasm of the osteoblasts was ultrastructurally characterized by a well-developed granular endoplasmic reticulum, Golgi complexes, vesicles, and numerous mitochondria. In part of the material the osteoblasts contained mitochondria with electron-dense granules. The granules were about 600 Å in diameter, and corresponded in appearance to previously described mitochondrial granules in osteoblasts from rats (12).

In the fetal material sections through the mitochondria revealed an average number of 10 granules per mitochondrial section but minor differences in the number of mitochondrial granules could not be related to the size of the fetuses.

In the postnatal material from the younger individuals with the highest skeletal growth activity, sections disclosed 1 or 2 granules per mitochondrial section, but usually the postnatal material did not reveal mitochondrial granules.

The difference in intensity between prenatal and postnatal bone growth (16, 17, 3) was roughly reflected in the present study by the number of mitochondrial granules.

The periosteal activity of alkaline phosphatase was estimated on the basis of the period of time from the start of incubation of biopsy specimens at room temperature until occurrence of distinct red colour reaction (Fig. 1). The length of this period was considered indicative of periosteal growth activity. Tests were repeated on series of sections to avoid variations because of differences in section thickness. Prenatal osteoblasts showed instantaneous reaction. Postnatal osteoblasts showed reaction times varying from 5 seconds to 4 minutes.

The following are data on mean reaction times and osteoblast activity indexes, O.A.I., defined as $\frac{1 \times 1000}{\text{reaction time}}$.

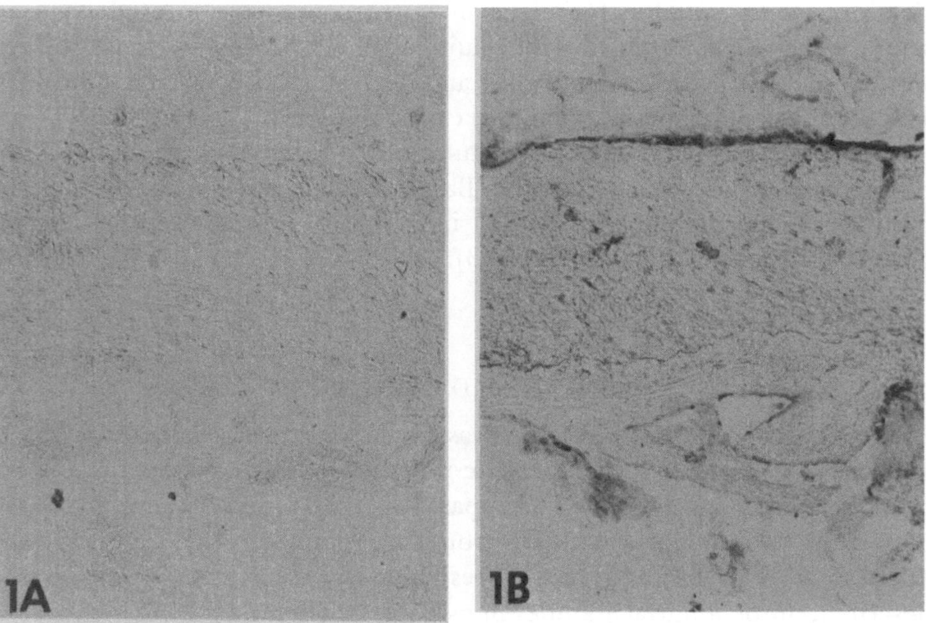

Figure 1. Buccal alveolar margin from second lower premolar of girl (13 years) with a skeletal maturation stage corresponding to puberal maximum.

Sections incubated for alkaline phosphatase (orig. mag. x 64), A) after 10 sec (no reaction) B) after 15 sec (reaction).

	N	Reaction time	Osteoblast activity index O.A.I.
Prenatal period	25	instantaneous	
Postnatal period			
Infantile period	2	10 sec.	100
Juvenile period	6	46 sec.	22
Adolescent period	7	19 sec.	53
Adult period	5	114 sec.	9

In Fig. 2 the osteoblast activity indexes are related to postnatal periodic variations in growth rate.

The graphical comparison of alkaline phosphatase activity and growth rate seems to indicate the presence of a correlation.

The question to what extent a high activity of alkaline phosphatase in the periost originates from high enzyme activity in the osteoblasts or from an increase of the number of osteoblasts in the periosteum is unanswered in the present report. As the ultrastructural study shows differences in cell morphology which could be related to growth activity it seems desirable to apply electron microscopy and enzyme histochemistry in the study of bone growth.

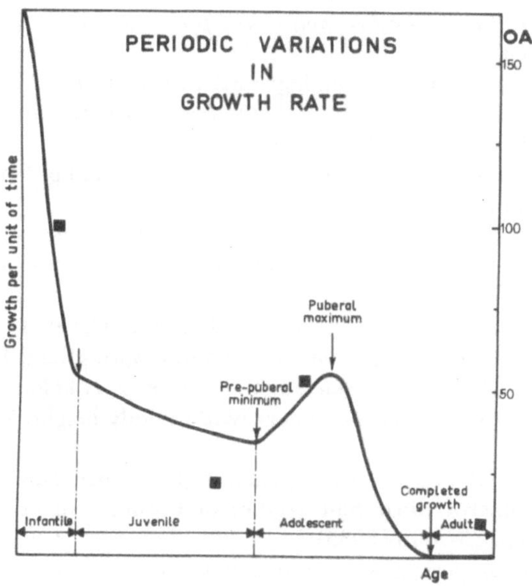

Figure 2. Stages of skeletal maturation defined by ages at periodic margin changes in rate of growth (left), (after Björk & Helm (3)) related to square plots for osteoblast activity index, O.A.I., (right).

SUMMARY

In prenatal and postnatal human bony tissue osteoblasts were studied for differences in cell morphology and activity of alkaline phosphatase in relation to general skeletal growth activity.

Ultrastructurally observed mitochondrial granules revealed differences in number of granules per mitochondrial section which could be related to variations in growth rate of normal growing individuals.

Activity of alkaline phosphatase was quantitated according to time of incubation of the sections before distinct visual reaction. Prenatal osteoblasts showed an instantaneous reaction. Postnatal osteoblasts showed reaction periods of time varying from 5 seconds to 4 minutes at room temperature. The measurement of osteoblast activity enzymehistochemically was expressed in an osteoblast activity index (O.A.I.) which seems to be correlated to general skeletal growth activity assessed on the basis of radiographic determination of general skeletal maturation.

ACKNOWLEDGEMENT

This work was supported by a grant from the Danish Medical Research Council.

REFERENCES

1. Andersen, H.: *Histochemical studies on the development of joints and bones in the human foetus during the first half of the prenatal period.* Thesis, University of Copenhagen, 1970

2. Anderson, H.C.: Calcium-accumulating vesicles in the intercellular matrix of bone. In: *Hard Tissue Growth, Repair and Remineralization,* Ciba Foundation Symposium 11, pp. 213–246, 1973

3. Björk, A. & Helm, S.: Prediction of the age of maximum puberal growth in body height. *Angle Orthodont.* 37, 134–143 (1967)

4. Burstone, M.S.: The relationship between fixation and techniques for the histochemical localization of hydrolytic enzymes. *J. Histochem. Cytochem.* 6, 322–339 (1958)

5. Cabrini, R.: Histochemistry of ossification. *Int. Rev. Cytol.* 11, 283–306 (1961)

6. Frost, H.M.: *Bone remodelling dynamics.* Thomas, Springfield, 1963

7. Helm, S., Siersbæk-Nielsen, S., Skieller, V. & Björk, A.: Skeletal maturation of the hand in relation to maximum puberal growth in body height. *Tandlægebladet* 12, 1223–1234 (1971)

8. Jowsey, J., Kelly, P.J., Riggs, B.L., Bianco, A.J., Scholz, D.S. & Gershon-Cohen, J.: Quantitative microradiographic studies of normal and osteoporotic bone. *J. Bone Jt. Surg.* 47, 785–806 (1965)

9. Kjær, Inger.: Human periosteal growth activity determined enzyme-histochemically. *J. Dent. Res.* 51, 872 (1972)

10. Kjær, Inger & Matthiessen, M.E.: Mitochondrial granules in human osteoblasts with a reference to one case of osteogenesis imperfecta. *Calc. Tiss. Res.* 17, 173–176 (1975)

11. Kjær, Inger & Prydsø, U.: Enzyme-histochemical determination of human bone growth activity in postmortem material. *J. Dent. Res.* **52**, 1293—1296 (1973)

12. Matthews, J.L., Martin, J.H., Kennedy, J.W. & Collins, E.J.: An ultrastructural study of calcium and phosphate deposition and exchange in tissues. In: *Hard Tissue Growth Repair and Remineralization,* Ciba Foundation Symposium 11, pp. 187—211, 1973

13. Merz, W.A. & Schenk, R.K.: A quantitative histological study on bone formation in human cancellous bone. *Acta Anat. (Basel)* **76**, 1—15 (1970)

14. Pearse, A.G.E.: *Histochemistry — theoretical and applied.* Vol. 1, 3rd ed., Churchill, London, 1968

15. Schen, S., Villanueva, A.R. & Frost, H.M.: Number of osteoblasts per unit area of osteoid seam in cortical human bone. *Canad. J. Physiol. Pharmacol.* **43**, 319—325 (1965)

16. Shepard, T.H.: Growth and development of the human embryo and fetus. In: *Endocrine and Genetic Diseases of Childhood,* Gardner, L.I. (ed.), Saunders, Philadelphia and London, pp. 1—6, 1969

17. Tanner, J.M., Whitehouse, R.H. & Takaishi, M.: Standards from birth to maturity for height, weight, height velocity and weight velocity: British children, 1965. *Arch. Dis. Childh.* **41**, 454—471 (1966)

18. Woods, C.G., Morgan, D.B., Patterson, C.R. & Gossmann, H.H.: Measurement of osteoid in bone biopsy. *J. Path. Bact.* **95**, 441—447 (1968)

Ca^{2+}-ATPase in Hard Tissue Forming Cells

A. LINDE, G. GRANSTRÖM & B.C. MAGNUSSON

In previous studies (4) it has been shown that the non-specific alkaline phosphatases (APase) present in bone, calcifying cartilage, odontoblasts and the enamel organ have several properties in common. The ATPases concerned in the formation of different hard tissues thus seem to be the same isoenzyme. Inorganic pyrophosphatase (PP$_i$ase) activities in odontoblasts and the cells of the enamel organ have very similar properties (5, 3).

Levamisole and its analogue R 8231 were recently shown to be very potent, specific non-competitive inhibitors of APase (9, 1). These agents almost completely inhibit APase and PP$_i$ase in active odontoblasts (7) and the cells of the enamel organ. Our earlier results thus support previous suggestions (2, 10) that APase and PP$_i$ase occurring at different loci of calcification are the same enzyme.

In histochemical studies of phosphatase activities in mouse jaws (8) the ATP degrading activity in osteoblasts, odontoblasts and the cells of the stratum intermedium was decreased to only a limited extent in the presence of levamisole or R 8231, while the activity in striated muscle fibers and cells of the blood vessel walls was unaffected. These results caused us to undertake a more detailed study of the ATP splitting enzyme activity in odontoblasts from the rat incisor, a model previously used for the study of biological calcification.

The preliminary results of these studies indicate that the ATP degradation in odontoblasts is mediated by two different enzymes, APase and an ensyme tentatively named "Ca-ATPase". The former is inhibited by levamisole or R 8231.

MATERIAL AND METHODS

Odontoblasts, active in dentine formation, were dissected out as previously described (6) and homogenized. The ATP splitting enzyme activity was

Laboratory of Oral Biology, Department of Histology and Department of Oral Histopathology, University of Göteborg.

normally determined by incubating portions of the homogenate in the presence of 3 mM Na_2-ATP in 0.1 M glycine-NaOH buffer, pH 9.8 at 37°C for 120 min. The liberated phosphate was determined colorimetrically (5). When R 8231, ouabain and ruthenium red were included, they were at a concentration of 0.1 mM in the incubation medium.

RESULTS

Both APase and "Ca-ATPase" showed an optimal ATP-splitting activity at pH 9.8, both in the presence of Ca^{2+} in equimolar concentrations with ATP and without any addition of Ca^{2+}. The optimal pH for the total ATP-splitting enzyme activity was independent of the ATP concentration. The total activity, under the assay conditions, amounted to 13.0 ± 2.7 (n = 9) μmol liberated phosphate per min and mg DNA. The degradation of ATP by APase was linear with incubation times up to 45 min, while the "Ca-ATPase" had a linear relationship for at least 300 min (Fig. 1a). Both APase and "Ca-ATPase" degradation of ATP were shown to be

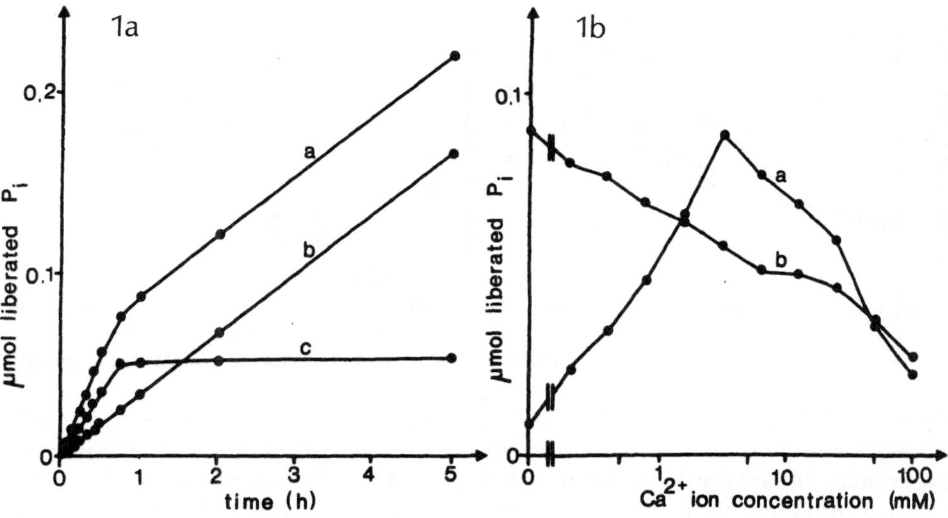

Figure 1a. Total ATP degradation (a), "Ca-ATPase" activity (b) and APase mediated ATP degradation (c) as a function of incubation time. The "Ca-ATPase" had a linear relationship for at least 5 h, while the APase was linear for only 45 min.

Figure 1b. Effect of varying concentrations of Ca^{2+} ions in the incubation medium on "Ca-ATPase" activity (a) and APase mediated ATP degradation (b). Note that the abscissa in logarithmic. The APase activity was decreased at all concentrations. The "Ca-ATPase" showed a maximal activation at a 3 mM Ca^{2+} concentration.

completely independent of the Na$^+$ or K$^+$ concentration in the incubation medium (Tris-HCl buffer). Furthermore, neither ouabain nor ruthenium red had any influence on the enzyme activities. Both enzyme activities were found to be very sensitive to heating at 56°C.

The "Ca-ATPase" was activated by the addition of Ca^{2+} ions (Fig. 1b). Maximal activation was obtained at a Ca^{2+}: ATP ratio of 1:1. The APase, on the other hand, was inhibited at all Ca^{2+} concentrations. When Mg^{2+} ions were added at different concentrations both "Ca-ATPase" and APase revealed activation with a maximum at equimolar concentration with the substrate. The maximal activity of "Ca-ATPase" in the presence of Mg^{2+} ions was found to be somewhat higher than that in the presence of Ca^{2+} ions. When Ca^{2+} and Mg^{2+} ions were added together in a 1:1 proportion, maximal activity was obtained when the combined ion concentration was twice that of the ATP. This activity was almost equal to that obtained with Mg^{2+} alone (Fig. 2a).

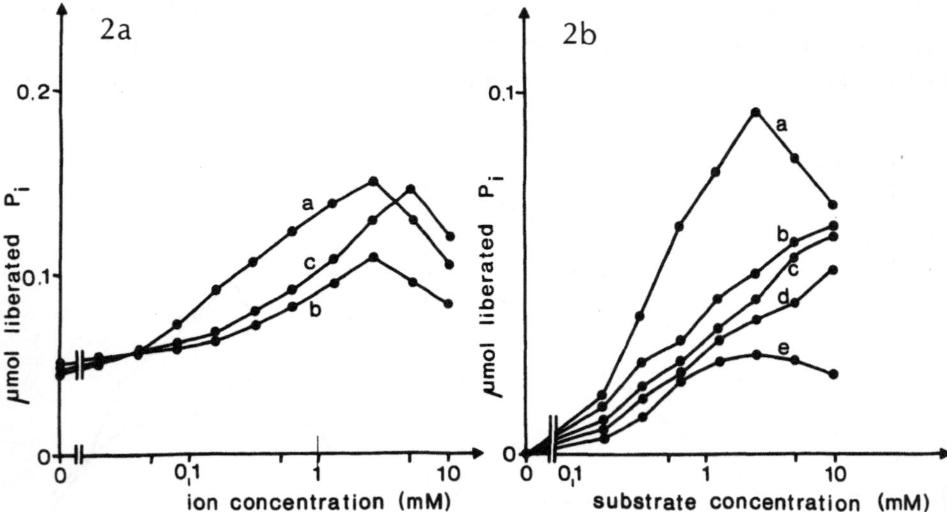

Figure 2a. Effect of varying concentrations of Mg^{2+} ions (a), Ca^{2+} ions (b) and Ca^{2+} and Mg^{2+} ions in a 1:1 ratio (c) on the "Ca-ATPase" activity. Note that the abscissa is logarithmic. The maximal activation with Mg^{2+} and Ca^{2+} ions alone was obtained at 3 mM concentrations, the activation by Mg^{2+} ions being stronger than that by Ca^{2+} ions. When Ca^{2+} and Mg^{2+} ions were present together the maximal "Ca-ATPase" activity was reached when the combined ion concentration reached 6 mM.

Figure 2b. Ability of "Ca-ATPase" to degrade different nucleotide substrates at different substrate concentrations in the presence of 3 mM Ca^{2+} ions. Note that the abscissa is logarithmic. The highest activity was obtained with ATP as a substrate (a). Both with ATP and AMP (e) an optimal activity was reached with a 3 mM substrate concentration. No optimal substrate concentration was obtained with ADP (b), GTP (c) or ITP (d) as substrates.

Both "Ca-ATPase" and APase mediated ATP degradation was completely inhibited by the addition of 0.3 mM EDTA or 1.25 mM EGTA, but the original activities could be recovered by the addition of either Ca^{2+} or Mg^{2+} ions.

The ability of "Ca-ATPase" to degrade other substrates such as ADP, AMP, GTP and ITP was investigated (Fig. 2b). The highest activity in the presence of 3 mM Ca^{2+} was found with ATP. An optimal activity was reached with a Ca^{2+}: substrate ratio of 1:1. The same result was found with AMP, although the activity was 70% less. ADP, GTP and ITP were also degraded by the "Ca-ATPase", albeit at a lower rate and without any optimal substrate concentration.

CONCLUSIONS

We have shown that in active odontoblasts of the rat incisor tooth, used as a model system for hard tissue forming cells, two enzymes capable of degrading ATP exist. One can be inhibited by levamisole and R 8231 and is probably identical with non-specific alkaline phosphatase (APase). The activity of the other enzyme, tentatively named "Ca-ATPase", is dependent on the presence of Ca^{2+} or Mg^{2+} and is activated by these ions. The "Ca-ATPase" is unaffected by ouabain and ruthenium red. It may be speculated that the "Ca-ATPase" is concerned with the transmembraneous transport of Ca^{2+} ions to be mineralization front.

REFERENCES

1. Borgers, M.: The cytochemical application of new potent inhibitors of alkaline phosphatase. *J. Histochem. Cytochem.* 9, 812–824 (1973)
2. Eaton, R.H. & Moss, D.W.: Partial purification and some properties of human bone alkaline phosphatase. *Enzymologia* 35, 31–39 (1968)
3. Fredén, H., Linde, A. & Magnusson, B.C.: Inorganic pyrophosphatase in isolated enamel organ and odontoblasts from the rat incisor. *Acta odont. scand.* 33, (1975). In Press.
4. Granström, G. & Linde, A.: A comparative study of alkaline phosphatase in calcifying cartilage, odontoblasts and the enamel organ. *Calc. Tiss. Res.* (1975a). In Press.
5. Granström, G. & Linde, A.: Determination of inorganic pyrophosphatase in the odontoblast layer of rat by a radiochemical method. *Scand. J. Dent. Res.* (1975b). In Press.
6. Linde, A.: A method for the biochemical study of enzymes in the odontoblastic layer during dentinogenesis. *Arch. oral. Biol.* 17, 1209–1212 (1972)

7. Linde, A. & Magnusson, B.C.: Inhibition studies of alkaline phosphatases in hard-tissue-forming cells. *J. Histochem. Cytochem.* (1975). In Press.
8. Magnusson, B.C. & Linde, A.: Alkaline phosphatase, 5′-nucleotidase and ATPase activity in the molar region of the mouse. *Histochem.* 42, 221—232 (1974)
9. Van Belle, H.: Kinetics and inhibition of alkaline phosphatases from canine tissues. *Biochim. Biophys. Acta* 289, 158—168 (1972)
10. Wöltgens, J.H.M., Bonting, S.L. & Bijvoet, O.L.M.: Relationship in inorganic pyrophosphatase and alkaline phosphatase activities in hamster molars. *Calc. Tiss. Res.* 5, 333—343 (1970)

Nuclear Inclusions in Osteoclasts in Paget's Bone Disease

A. REBEL, K. MALKANI, M. BASLÉ & CH. BREGEON

Osteoclasts, relatively rare among bone cells in normal tissue as in most pathological conditions, are abundant in the case of Paget's bone disease. Electron microscopy shows several cytological differences between these and normal osteoclasts (6). The most striking one is the presence of nuclear inclusions which appear to be specific to osteoclasts in this particular context (5). In structure, the inclusions are comparable to those associated with various viral infections and this suggests the possible action of an exogenous agent in Paget's bone disease.

MATERIAL AND METHODS

Iliac crest biopsies were taken from affected zones in 12 patients with Paget's bone disease before and after treatment with calcitonin; 13 other patients were examined at various stages of treatment. In all, the 37 biopsies provided well over a hundred samples for electron microscopy.

Undecalcified fragments of bone tissue were fixed at $4°C$ in 4.25% glutaraldehyde with a phosphate buffer at pH 7.2. After post-fixation in 2% osmic acid, the fragments were embedded in epon. Thin sections, cut with diamond knives, were double-stained with uranyl acetate and lead citrate according to Reynolds.

RESULTS AND DISCUSSION

The osteoclasts in Paget's bone disease are extraordinarily large and multinucleated. Almost all the osteoclasts observed, in biopsies from

Department of Histology and Department of Rheumatology, Faculty of Medicine and Pharmacy, University of Angers, Angers.

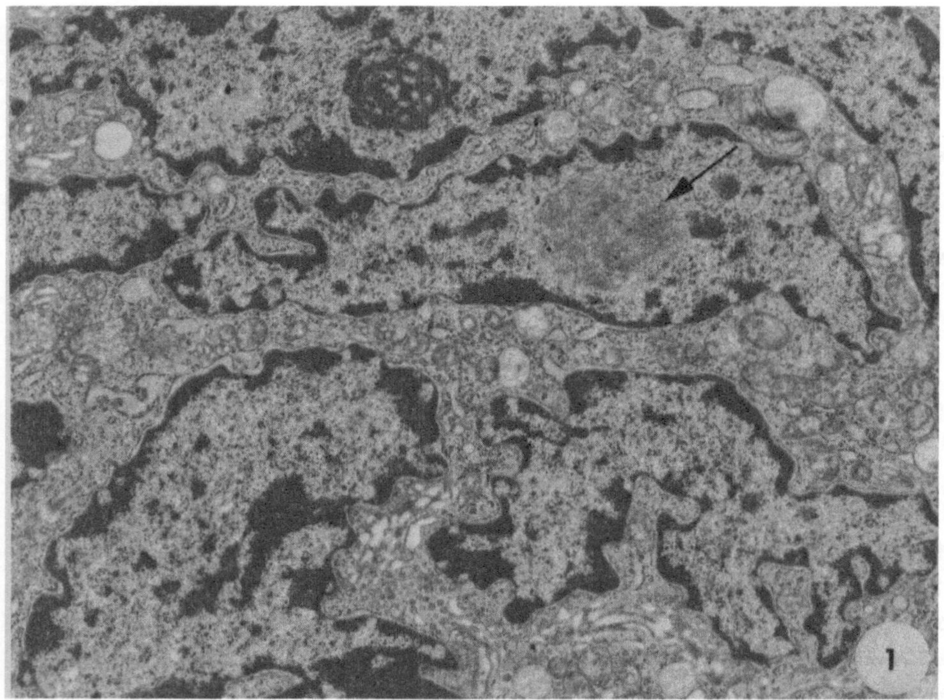

Figure 1. Typical low magnification view of part of an osteoclast in Paget's bone disease. One of the nuclei (arrow) contains the characteristic inclusion body easily dinstinguished from the surrounding nucleoplasm. (× 7000)

untreated as well as from calcitonin treated cases, show at least one nucleus containing characteristic inclusion bodies quite distinct in electron density from the nuclear chromatin and the nucleoli (Fig. 1). No such bodies are to be found in the nuclei of other bone cells.

The inclusions are generally in the form of striated cylindrical filaments about 15 nm thick and of indeterminate length. Transverse sections of the filaments show electron-dense structures about 5 nm in diameter surrounding a clear center of roughly the same dimensions (Fig. 2). The filaments often appear loosely organized and randomly distributed, but sometimes constitute close-packed paracrystalline arrays.

Selected high magnification micrographs of the inclusions were subjected to photographic integration using Markham's technique (3). The results of linear analysis suggest that the striated appearance of the filaments is due to an alternate light and dark helicoidal banding about 5 nm wide. Rotational analysis, on the other hand, gives no clear-cut information on the symmetry of the transverse sections of the filaments, but this is only to be expected since the dimensions of the dense peripheral elements are

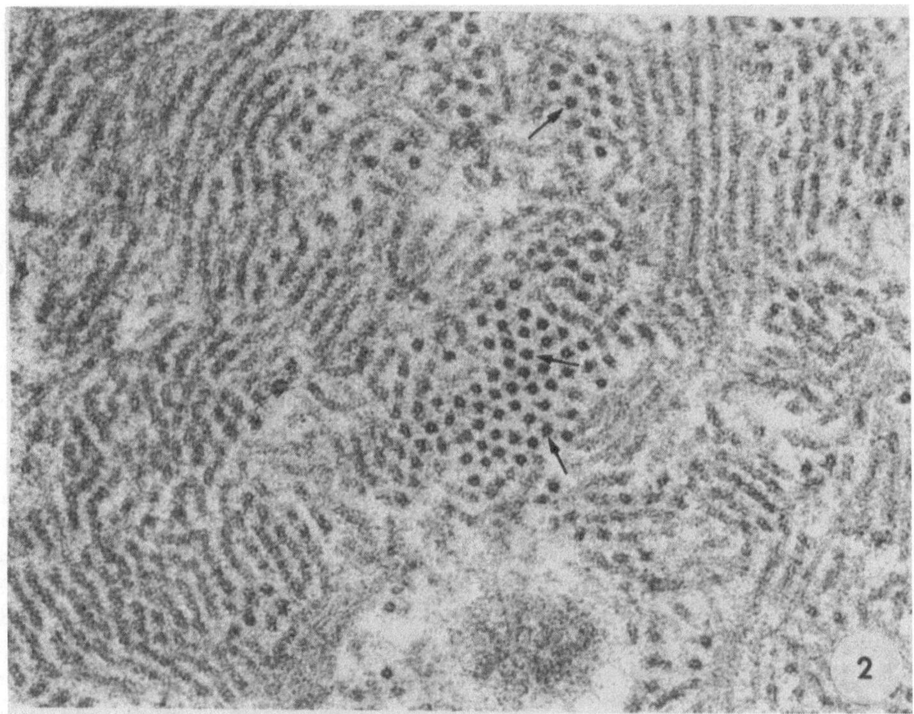

Figure 2. At higher magnification, the nuclear inclusion is seen to be made up of striated filament-like structures more or less randomly disposed in this particular case. The filaments are about 15 nm in diameter, and transverse sections show dense peripheral structures around a clear center (arrows). (x 100.000)

relatively small compared to the overall thickness of the sections used.

Some of the nuclear inclusions are separated from the rest of the nucleoplasm by an electron-transparent zone, while others are interspersed with and surrounded by much grosser filament-like structures whose electron-density rather resembles that of the nucleoli.

In a few cases, osteoclast nuclei show that what appears to be the massive passage of the filaments making up the inclusion bodies across gaps in the nuclear envelope into the cytoplasm.

To date, cytochemical and immunological methods using cryo-ultra-microtomy have, unfortunately, not produced very satisfactory results, but the application of Bernhard's staining technique (1) suggests that RNA rather than DNA is an essential component of the nuclear inclusions.

The specific nuclear structures described could be caused by some dysfunction of the osteoclast in Paget's bone disease, but so far, to our knowledge, no nuclear changes of this type have ever been induced by any experimental interference with cellular activity. On the other hand, the close morphological resemblance of the inclusion bodies to those observed in various cases of viral attack (2, 4, 7) leads to the hypothesis of the possible action of an exogenous agent in Paget's bone disease.

ACKNOWLEDGEMENT

This work was supported by a grant from the French Medical Research Foundation. We thank Miss A. Le Patezour and Mr R. Filmon for their skilfull technical assistance.

REFERENCES

1. Bernhard, W.: A new staining procedure for electron microscopical cytology. *J. Ultrastruct. Res.* 27, 250—265 (1969)
2. Dreyer, E.O., Muldiyarov, P.Y., Nassonova, V.A. & Alekberova, Z.S.: Endothelial inclusions and "nuclear bodies" in systemic lupus erythematosus. *Ann. rheum. Dis.* 32, 444—449 (1973)
3. Markham, R., Frey, S. & Hills, G.: Methods for the enhancement of image detail and accentuation of structure in electron microscopy. *Virology* 20, 88—102 (1963)
4. Raine, C.S., Feldman, L.A., Sheppard, R.D., Barbosa, L.H. & Bornstein, M.B.: Subacute sclerosing panencephalitis virus. *Lab. Invest.* 31, 42—53 (1974)
5. Rebel, A., Malkani, K., Baslé, M. & Bregeon, Ch.: Anomalies nucléaires des osteoclastes de la maladie osseuse de Paget. *Nouv. Presse méd.* 3, 1299—1301 (1974)
6. Rebel, A., Malkani, K., Baslé, M. & Bregeon, Ch.: Osteoclast ultrastructure in Paget's bone disease. *Calc. Tiss. Res.* (1975). (In Press).
7. Zu Rhein, G.M.: Association of papova-virions with a human demyelinating disease (Progressive multifocal leukoencephalopathy). *Progr. med. Virol.* 11, 185—247 (1969)

Transmission Electron Microscopy of Ion Erosion Thinned Hard Tissues

A. BOYDE & J.B. PAWLEY

INTRODUCTION

Material is removed from surfaces undergoing bombardment with non-reactive noble gas ions. Thus, this process can be used to thin down thin samples of hard materials in order that they may be examined in the transmission electron microscope (TEM) (6, 7, 1). However, the rate of removal of material depends upon several factors which may vary over very small areas; these factors include the angle of incidence of the ion beam with respect to the surface and with respect to the crystallographic orientation of the surface features. Regions having different compositions, for example, in the case of the hard tissues, differences in the proportion of calcium phosphate to organic matrix, also lead to differences in the rate of removal of superficial layers. An ion bombarded surface will therefore be etched to produce surface relief (or variations in the thickness of a section) if there are local variations in composition or orientation. Boyde and Stewart (3, 4), Stewart and Boyde (15) and Boyde, Switsur and Stewart (5) thus reported etching of enamel prisms due to variations in crystallographic orientation across prisms, a lack of etching in the uniformly orientated outer surface zone of the enamel and etching due to compositional variations in dentine, where the more highly mineralized peritubular dentine was removed at a lower rate than the intertubular dentine.

The etching effects due to ion beam erosion — which is also currently described as ion beam thinning, machining, micromilling, etching or sputtering — can be reduced by a) using a very low, grazing angle of incidence of the beam with respect to the surface under bombardment which also reduces the rate of removal of material and/or b) by varying the angle of incidence of the beam by rotating and tilting the sample

Department of Anatomy and Embryology, University College London, London.

continuously during the sputtering process. In spite of such attempts it is not possible to produce uniformly thick samples for TEM but the samples are nevertheless valuable for this purpose because they are thin enough to be examined in parts and the surrounding thicker regions might be used as a support.

Ion beam thinning has the advantage over conventional ultramicrotome procedures that the sample is not physically deformed due to its being bent through a large angle as the section passes over the edge of the knife. All knife sections have undergone elastic (and plastic) deformation. In the case of the adult hard tissues it is unavoidable that an element of plastic (permanent) deformation shall have occurred and this is significant because it may be the result of slip or fracture within the crystalline component. It is hardly possible, therefore, to rely upon ultramicrotomed, ultra-thin sections of hard tissue for the full interpretation of the morphology of the mineral component.

A further objection which can be raised to conventional preparative procedures used for TEM of hard tissues is that the sections (or the tissues) must at some time come into contact with watery solutions in a non-physiological environment. This may lead to the removal of, or addition to, parts of the mineral phase. This objection is not found with the ion beam thinning procedure.

MATERIALS AND METHODS

Human teeth and human and bovine bone have been used in the present studies. Ground sections were prepared using a rotary diamond saw with water cooling. The sections were prepared as for light microscopy by grinding on successively finer grades of abrasive using a LogiTech model PM−2 lapping machine (1). During the grinding and polishing stages of specimen preparation the specimens were in contact with non-aqueous solutions and at all other stages were kept dry.

In early stages of this work we found that a "frothy" structure resulted in otherwise apparently intact enamel crystallites. Towe and Thompson (16) reported a similar appearance in molluscan shell carbonates following ion beam thinning and considered whether these localized irregularities in crystal structure resulted from the effects of the electron beam in the TEM. We attempted to reduce or prevent the occurrence of this artefact by maintaining the specimens at a low temperature during ion bombardment. In the Edwards IBMA1 ion beam machining apparatus (as in many similar devices) the sections are thinned with two opposing ion beams simultaneously, one impinging on each side of the ca. 15° tilted specimen. This procedure undoubtedly results in a great elevation of the temperature of the central thinned portion of the sample. We constructed a solid copper specimen carrier to which one side of the specimen was attached using dental sticky wax. We had originally envisaged that it would be possible to attach the specimens to the liquid nitrogen-cooled holder with some frozen water, but the sublimation of this ice always occurred before the completion of the thinning procedure resulting in the detachment and loss of the samples. Cooling of the rotating specimen support was provided by wheels running on its rim whose axles in turn were cooled through contact with a liquid nitrogen filled copper probe. We had no direct method of measurement of the sample temperature during machining with this additional precaution, but that it was well below $0°C$ was established from the fact that the samples could be attached with ice as an adhesive for considerable periods.

With the one-sided erosion procedure we used argon ions (4 to 5 keV) with a beam current measured at an intercepting probe of $30-100 \mu A$. After a central hole was bored in the sample it was removed and placed in a Philips' EM 300 TEM with liquid nitrogen cooling of the specimen chamber space. Again, there was no means of monitoring the sample temperature in the TEM. Micrographs were recorded at a variety of magnifications at 80 or 100 kV. A tilting stage was used and images of the same area prepared at a variety of orientations in order to allow a stereoscopic evaluation of the orientation of particles. The magnification of the microscope was calibrated using a diffraction grating replica.

RESULTS

Reproducible fine structural details were found in the TEM pictures of sample areas which were sufficiently thinned. The contrast in these micrographs is entirely due to the calcium phosphate (hydroxyapatite) mineral component, but may arise in several ways which are important in interpreting the present results. The simple mass distribution (electron density) effect is probably the single most important factor in interpreting micrographs of bone and dentine, where the mineral component seems effectively to act as a negative stain for the oriented microfibrillar components of the matrix. All prior and alternative explanations of this effect have been couched in terms of describing the distribution and morphology of "crystallites", or "amorphous" calcium phosphate. Electron contrast is also affected by variations in the perfection of crystallinity as considered by Reimer (13) and Höhling *et al.* (9) and by electron diffraction effects giving rise to the so-called Bragg reflections. The latter are only found in larger and more perfect crystals such as the enamel crystals. A variety of mechanisms can give rise to contrasts interpretable as lattice planes in enamel crystals. Overlapping crystal lattice plane images can give rise to Moiré patterns.

Bragg reflections are found within very narrow hands on particular enamel crystals which happen to diffract the beam outside the objective aperture. It would seem that most enamel crystals are slightly strained or bent. A single dark band due to a Bragg reflection can be made to travel along a given crystal by tilting the specimen with respect to the electron beam. This effect can be used to trace the continuity of single enamel crystals. Enamel crystals are at least several microns long and are probably much longer (Fig. 1). Bragg reflections in adjacent crystals or groups of crystals may travel in opposing directions when the sample is tilted.

In dentine and bone the most interesting findings concerned the morphology and size features of the mineral component. Dentine is a particularly suitable tissue to study because of its more homogeneous composition and the random feltwork arrangement of the rather uniformly diametered collagen fibrils (Fig. 2). Thus, it is usually possible to find both longitudinal and cross-sectioned fibrils within a given field of view. In cross sections, a common appearance is that of a series of electron-dense circles showing no very regular packing pattern, with the closest distance between centres about 38Å. In longitudinal sections of the fibrillar components the most common appearance is that of parallel

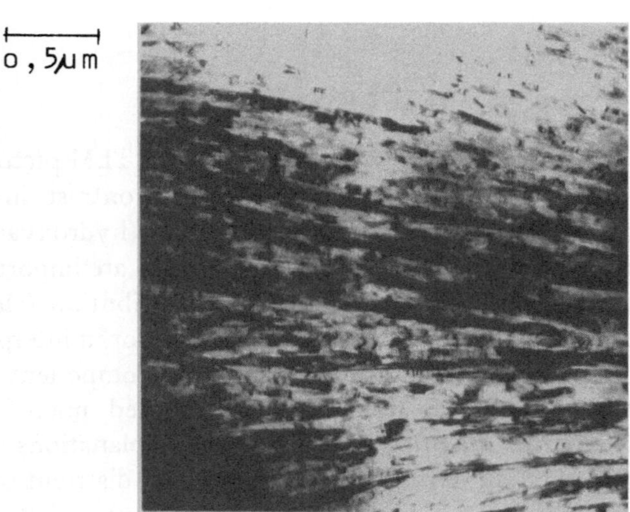

Figure 1. Ion beam thinned human permanent dental enamel showing long thin parallel enamel crystallites. The transverse striations are Bragg reflections and do not indicate imperfections in the crystal structure. These Bragg reflections move when the sample is tilted in the electron beam. Magnification 50,000X.

640Å

Figure 2. Ion beam thinned human permanent dentine. Circular and longitudinal profiles of the mineral component can be discerned (see text). Magnification 364,000X.

pairs of dark lines. Thus, the mineral component seems to be deposited around presumed microfibrils and in very close relationship to them. In the case of the very closely packed parallel dark lines the closest distance between their centres was again about 38Å. It should be emphasized, however, that there is a wide spread in the lateral separation distance between electron-dense features. Rarely was any evidence of a 640Å striation across the long axis of the fibrils to be seen. The lengths of individual electron-dense features were commonly greater than 640Å.

The so-called "frothy" (16) appearance in the enamel crystals was present in all samples, including those which had been prepared with the cooling procedure described above.

DISCUSSION

The fact that the "frothy" appearance could not be eliminated and that it was present in any given field of enamel as soon as it was moved under the electron beam indicates that it may be due to the ion beam machining procedure. It is not yet clear whether it is due to thermal damage or to some other, as yet unexplained, irradiation damage. It should be borne in mind that the argon ions enter the surface layers of a material under bombardment. Thus, ion implantation artefacts might be awaited. The significance of the "frothy" effect artefact (16) is probably limited because individual Bragg reflections and crystal lattice lines pass across the electron-lucent "pebble-like" features without interruption. Towe and Thompson (16) inclined to the belief that the "frothy" artefact is due to heating and to the presence of trapped water within crystals and they mention the presence of a similar appearance in micrographs of ultramicrotomed enamel crystals published by Rönnholm (14).

The present results confirmed impressions gained from carbon-replica studies of fractured enamel that crystal length in enamel is very large. The subdivision into ca. 1600Å units (14) is, therefore, to be attributed to the procedure of cutting enamel sections with a knife. The very small spaces between enamel crystals have been measured as ~ 50Å, confirming the findings of Orams *et al.* (12). However, it is not certain that these inter-crystalline spaces may not be altered in size due to translocation of mineral material as a sputtering-induced artefact. One should, therefore, treat these values with some reserve until further experiments have eliminated this possibility. Intercrystalline spaces at prism boundaries are irregular, and much larger spaces may be found in this location. Since, however, the prism boundaries may thin preferentially, measurements made in this region should be treated with extreme reservation.

The present findings also indicate that the length of the individual mineral particles in bone and dentine are greater than has commonly been reported to date. They also support the suggestion that the mineral is deposited in a rather specific relationship to the collagen component of the matrix and that it shrouds every individual microfibril. The micro-fibrillar concept of collagen organization is still under intensive investigation and speculation, but the present findings would support the five-stranded microfibril model of Miller and Parry (11) and would not appear to lend support to the octafibril model of Hosemann *et al.* (10). The results indicate that mineral is not located within the centres of the microfibrils or that, if it is, it is present in very significantly smaller amounts. For descriptive purposes, the mineral particles in dentine and bone may be regarded as needle-like (as most previous descriptions of the morphology of the mineral component in calcified collagenous tissues) or as plates corresponding to the mineral distributed between sheets of parallel microfibrils (see Fig. 9 of Höhling, Ashton and Köster (8)) or as tubes surrounding individual microfibrils. The continuity of the mineral phase between microfibrils where they are closely packed and the great length of the mineral needles, plates or tubes shrouding the microfibrils provides an explanation for the residual strength of bone and dentine from which the organic matrix has been extracted with 1, 2 ethane diamine or sodium hypochlorite. Such "anorganic" specimens appear to retain the original morphology and dimensions of the intact tissues and have formed the basis of extensive SEM studies (2).

ACKNOWLEDGEMENTS

This work has been supported by the Medical Research Council. We are grateful for the expert technical assistance of Elaine Bailey and Philip Reynolds. We have enjoyed stimulating discussions with Professors H.J. Höhling and E.P. Katz.

REFERENCES

1. Boyde, A.: Transmission Electron Microscopy of Ion Beam thinned Dentine. *Cell Tiss. Res.* 152, 543–550 (1974)

2. Boyde, A. & Jones, S.J.: In: *Principles and Techniques of Scanning Electron Microscopy*, Vol. 2, Hayat, M.A., ed., Van Nostrand-Reinhold, New York, pp 123–149 (1974)

3. Boyde, A. & Stewart, A.D.G.: A study of the etching of dental tissues with argon ion beams. *J. Ultrastruct. Res.* 7, 159–172 (1962)

4. Boyde, A. & Stewart, A.D.G.: In: *Proc. 5th Int. Conf. for Electron Microscopy*, Breese, S.S., ed., Academic Press, New York, paper QQ9 (1962)

5. Boyde, A., Switsur, V.R. & Stewart, A.D.G.: An assessment of two new physical methods applied to the study of dental tissues. In: *Proc. 9th ORCA Congr.*, Pergamon Press, Oxford, pp. 185–193 (1963)

6. Grove, C.A., Judd, G. & Ansell, G.S.: Determination of hydroxyapatite crystallite size in human dental enamel by dark-field electron microscopy. *J. Dent. Res.* 51, 22–29 (1972)

7. Hamilton, W.J., Judd, G. & Ansell, G.S.: Ultrastructure of human enamel specimens prepared by ion micro-milling. *J. Dent. Res.* 52, 703–710 (1973)

8. Höhling, H.J., Ashton, B.A. & Köster, H.D.: Quantitative Electron Microscopic Investigations of mineral nucleation in collagen. *Cell Tiss. Res.* 148, 11–26 (1974)

9. Höhling, H.J., Scholz, F., Boyde, A., Heine, H.G. & Reimer, L.: Electron microscopical and laser diffraction studies of the nucleation and growth of crystals in the organic matrix of dentine. *Z. Zellforsch.* 117, 381–393 (1971)

10. Hosemann, R., Dreissig, W. & Nemetschek, Th.: Schachtelhalm-structure of the octafibrils in collagen. *J. molec. Biol.* 83, 275–280 (1974)

11. Miller, A. & Parry, D.A.D.: Structure and packing of microfibrils in collagen. *J. molec. Biol.* 75 441–447 (1973)

12. Orams, H.J., Phakey, P.P., Bachinger, W.A. & Zybert, J.J.: Visualisation of micropore structure in human dental enamel. *Nature* 252, 584–585 (1974)

13. Reimer, L.: Änderung des elektronenmikroskopischen Bildkontrastes beim Übergang amorph-kristallin und flüssig-kristallin. *Naturwissenschaften* 49, 1 (1962)

14. Rönnholm, E.: The amelogenesis of human teeth as revealed by electron microscopy, II. The development of the enamel crystallites. *J. Ultrastruct. Res.* 6, 249–303 (1962)

15. Stewart, A.D.G. & Boyde, A.: Ion etching of dental tissues in a scanning electron microscope. *Nature* 196, 81–82 (1962)

16. Towe, E.H. & Thompson, G.R.: The structure of some bivalve shell carbonates prepared by ion-beam thinning. *Calc. Tiss. Res.* 10, 38–48 (1972)

Rabbit Ear Chamber Bone Cultures, a Novel *in vivo* System for Bone Tissue Investigation

E. SUDMANN

As a technique of bone research, microscopy of living bone tissue is unique in giving instant visualization and documentation of both bone resorption and bone formation not only at tissue, but also at cellular levels. This technique would thus seem to be a promising approach to a number of problems in osteology and orthopaedic surgery. There has, however, been surprisingly little progress in this area of research since Sandison's (2) classical contribution.

The purpose of the present paper is to review our vital microscopical studies in more than 100 ear chambers inserted in rabbits of both sexes, 5—12 months old. Included in this series are animals and ear chambers used for various bone tissue experiments described in detail elsewhere (3, 4, 5, 6, 7).

TECHNICAL CONSIDERATIONS

An ear chamber for bone tissue investigations should have two communicating compartments; a slit-shaped optical part (0.1—0.2 mm) for flake grafts of cortical bone and a shallow part (0.7—0.8 mm) for thicker cortical cancellous grafts. A novel ear chamber was made along these lines (3).

Bone grafts for initiating bone resorption and formation within the chambers were prepared from a lumbar spinous process.

Stereometric analysis of bone resorption and formation was done *in vivo* by measuring the bone area within the optical part of the chambers by a new method for stereometric analysis in microscopy called mixed-image planimetry. This method proved to be simpler and more

Department of Orthopaedic Surgery, University of Tromsø.

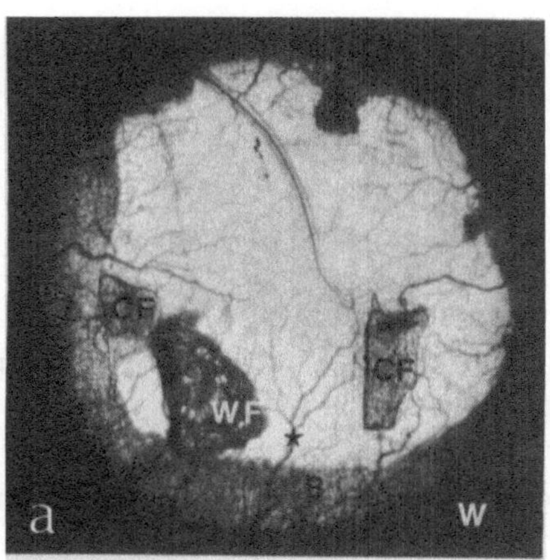

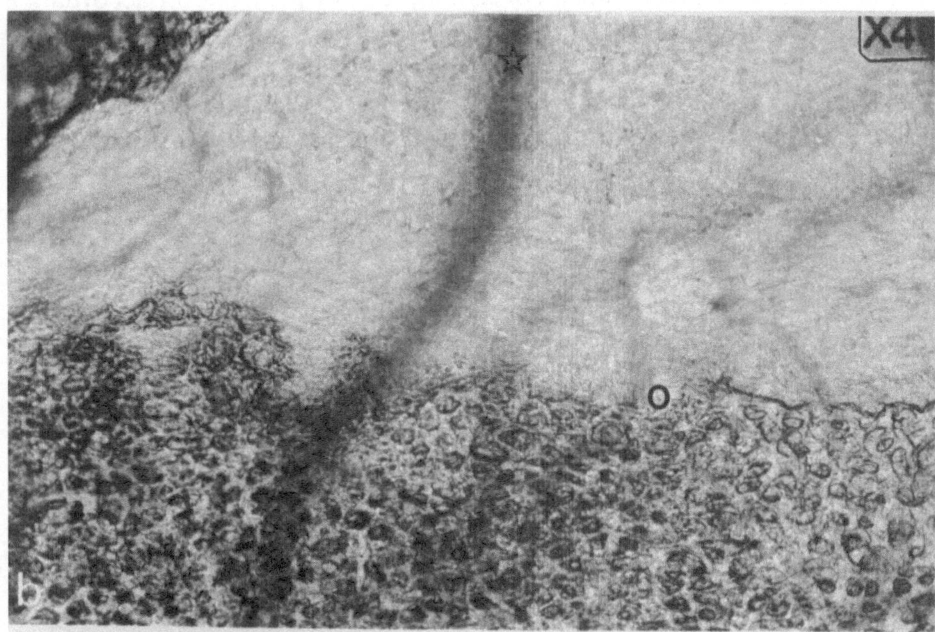

Figure 1 (a-b). Vital photomicrographs of bone culture in optical part of an ear chamber 36 days after autografting 3 cortical flake grafts (CF) to the optical part and 9 cortical cancellous grafts, with intact bone marrow, to the shallow, circular non-optical part, the well. Asterisk indicates same blood vessel in both photomicrographs.

Figure 1 a. One of the CF grafts is completely, the two others are partly resorbed. New (primitive) woven-fibred bone (WF) developed on one of the grafts, and bone (B) is growing into the optical part from the well (W). Field width, 6 mm. Objective 1x/0.04.

Figure 1 b. Part of (differentiated) woven-fibred bone flake growing directly underneath the cover glass. O = osteoid tissue. Field width, 0.9 mm. Objective 10x/0.25.

precise than any of the usual methods for area quantitation in microscopy and a precision better than 1 per cent was easily achieved (5).

BONE RESORPTION AND FORMATION

Spontaneous bone formation was not observed in any ear chamber. On the contrary, the technique of bone grafting for initiating bone resorption and osteogenesis proved essential (4).

Allografts of cortical bone flakes were very sparsely resorbed. Allografts of cortical cancellous bone did not initiate osteogenesis. In contrast, the bone resorption and formation initiated by grafting autologous bone to the chambers were remarkably consistent, and in principle in accordance with that of orthotopic bone (7) (Fig. 1).

The imbalance in favour of bone resorption seen in bone cultures *in vitro* was not seen in rabbit ear chamber bone cultures. Vital microscopical studies of the bone cultures was therefore possible for half a year or longer.

DRUG EXPERIMENTS

Indomethacin in a daily dosage of 10 mg/kg body weight *per os* for 3 weeks inhibited primitive (disordered) osteoclastic and osteoblastic activity and retarded the healing of fractures within the chambers (6). These findings were later reproduced in rats and in 1 patient where indomethacin caused non-union of (macro) fractures in orthotopic bone tissue (1, 8).

Indomethacin may delay the healing of microfractures in the human hip-joint causing drug-induced necrosis of the femoral head (6).

CONCLUSIONS

We find rabbit ear chamber bone cultures a remarkably convenient *in vivo* method for dynamic studies related to bone resorption and formation.

Indomethacin inhibited both osteoclastic and osteoblastic activity in ear chambers and retarded the healing of fractures in the bone cultures. Indomethacin may thus delay fracture healing in patients.

REFERENCES

1. Rø, J., Sudmann, E. & Marton, P.F.: Effect of indomethacin on fracture healing in rats. To be published
2. Sandison, J.C.: A method for the microscopic study of the growth of transplanted bone in the transparent chamber of the rabbit's ear. *Anat. Rec.* **40**, 41—49 (1928)
3. Sudmann, E.: A novel titanium rabbit ear chamber for vital microscopy of bone tissue. *Acta orthop. scand.* Suppl. 160, (1975a). In press
4. Sudmann, E.: Osteogenesis in rabbit ear chambers. *Acta orthop scand.* Suppl. 160, (1975b). In press
5. Sudmann, E.: Mixed-image planimetry. Assessment of a new method for sterometric analysis in microscopy. *Acta orthop. scand.* Suppl. 160, (1975c). In press
6. Sudmann, E.: Effect of indomethacin on bone remodelling in rabbit ear chambers. *Acta orthop. scand.* Suppl. 160, (1975d). In press
7. Sudmann, E. & Marton, P.F.: Bone remodelling in rabbit ear chambers. A vital microscopical and histological study. *Acta orthop. scand.* Suppl. 160, (1975e). In press
8. Sudmann, E. & Hagen, T.: Indomethacin-induced delayed fracture healing. Arch. orthop. Unfall-Chir. In press

CHAPTER III
Physiology and Biochemistry of Vitamin D

The Metabolism and Function of 1α-Hydroxyvitamin D_3

H.F. DeLuca, S.A. Holick & M.F. Holick

In 1968 25-hydroxyvitamin D_3 (25-OH-D_3) was isolated in pure form and chemically identified (4). During the ensuing years it was demonstrated that this hydroxylation occurs in the liver prior to the initiation of intestinal calcium transport or bone calcium mobilization (20, 5). Following this initial demonstration that vitamin D is not active as such but must be metabolically altered before it can function, it was soon learned that a metabolite of vitamin D more polar than 25-OH-D_3 appeared in the target tissues prior to response (19, 17, 9). This conversion was demonstrated by Fraser and Kodicek to occur in the kidney (8). In 1970 and 1971 this compound was isolated in pure form and identified as 1,25-dihydroxyvitamin D_3 (1,25-(OH)$_2 D_3$) (12). It was subsequently synthesized to confirm its structure and to provide evidence for the 1α configuration for the hydroxyl placed on ring A (22). During the course of this synthesis an important new analog was prepared, namely 1α-hydroxyvitamin D_3 (1α-OH-D_3) (13) (Fig. 1). Perhaps not surprising in retrospect is the fact that this compound proved to be almost equally as active as 1,25-(OH)$_2 D_3$ in the chick (10) but approximately one-half as active as the natural metabolite in the rat (14). The time course of onset of responses to the 1α-OH-D_3 was remarkably similar to the 1,25-(OH)$_2 D_3$ (13) and in the case of the chick (10), virtually identical. Of great interest was the fact that the 1α-OH-D_3 was much more effective when given in oil solutions orally to the rat than was the corresponding 1,25-(OH)$_2 D_3$, leading to the idea that perhaps the 1α-OH-D_3 in some species is more easily absorbed (14). Its ease of synthesis (13, 2, 16) and the relative inexpensiveness of the starting material led to its application to such diseases as chronic renal failure and hypoparathyroidism in which there is a clear indication of impared ability to carry out the 1-hydroxylation of 25-OH-D_3. This has met with marked

---

Department of Biochemistry, College of Agricultural and Life Sciences, University of Wisconsin-Madison, Madison, Wisconsin.

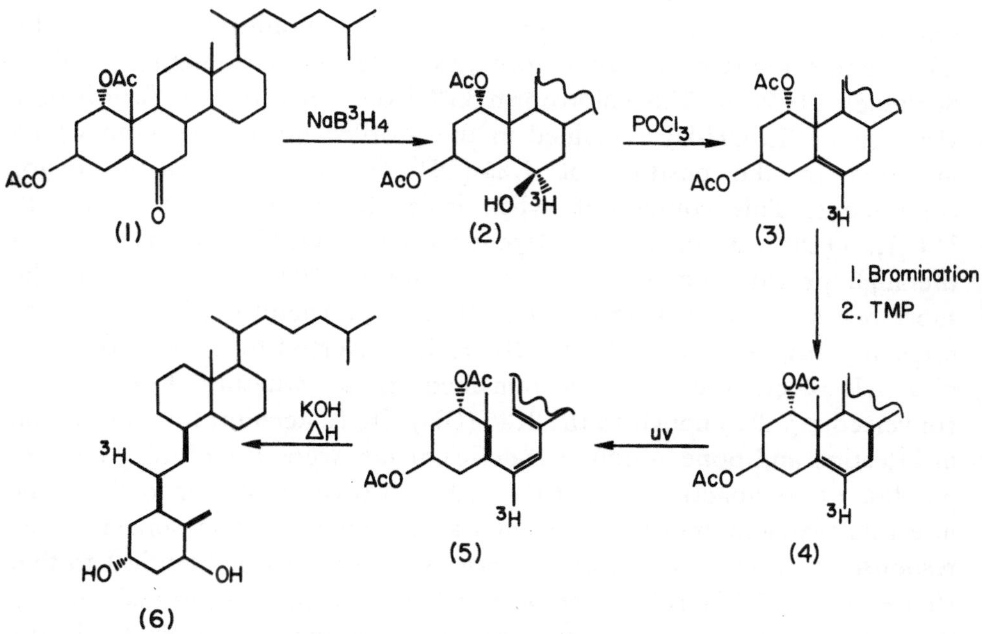

Figure 1. Chemical Synthesis of [6-³H]-1α-hydroxyvitamin D₃.

success in virtually every clinical center (6, 1, 7, 21). However, the surprising fact that the 1α-OH-D₃ produced a response almost identical in time course to the 1,25-(OH)₂ D₃ in animals led to the suggestion that 1α-OH-D₃ might function without the necessity of 25-hydroxylation. On the other hand it also seemed possible that the 1α-OH-D₃ might be hydroxylated on carbon 25 very rapidly. Although work by Zerwekh *et al.* (26) indicated that 25-hydroxylation of 1α-OH-D₃ does take place in the chick, the methods used to detect 1,25-(OH)₂ D₃ were indirect and not conclusive as they themselves stated.

During the past year we have been successful in preparing ³H-1α-OH-D₃ (13), namely the 6-keto-1α-hydroxycholesteryl acetate was reduced with sodium borotritide to introduce the label. 6∞-³H-1α-hydroxy-previtamin D₃ diacetate was isolated in pure form on silver nitrate impregnated silicic acid solumns. Saponification yielded the 6-³H-1α-OH-D₃ (15). This radioactive compound was shown to be identical with crystalline 1α-OH-D₃ obtained from Leo Pharmaceutical Products and from the Upjohn Company in biopotency and on cochromatography in a variety of systems. This includes a high pressure liquid chromatographic system which resolves all known metabolites of vitamin D (Jones and DeLuca, submitted for publication). In an initial experiment ³H-1α-OH-D₃ was administered to rachitic rats at a dose of 12.5 ng/day for 7 days. Eighteen hours after the last dose the animals were killed to examine their degree of calcification in the rachitic epiphyseal plate of the radii and ulnae. At the same time other tissues were removed, extracted and chromatographed on Sephadex LH-20 in 65% chloroform-35% Skellysolve B (11). Surprisingly virtually no 1α-OH-D_3 remained in the profile and a compound which migrated in the position of $1,25$-$(OH)_2 D_3$ appeared as the major component. This compound, when isolated, cochromatographed with ^{14}C-$1,25$-$(OH)_2 D_3$ on three different chromatographic systems including the high pressure liquid Zorbax SIL columns, demonstrating that the substance produced was in fact $1,25$-$(OH)_2 D_3$. Therefore, it was clear as originally suggested that the 1α-OH-D_3 is converted to $1,25$-$(OH)_2 D_3$ *in vivo*. However, the question remained as to whether 1α-OH-D_3 is converted rapidly enough to the $1,25$-$(OH)_2 D_3$ to account for its function in intestine and bone. Vitamin D-deficient rats were, therefore, given 325 pmoles of radioactive 1α-OH-D_3 either intravenously or orally. The intestinal calcium transport system was examined first inasmuch as the response to 1α-OH-D_3 had been previously shown to be identical to that elicited by $1,25$-$(OH)_2 D_3$. The plot of the response of intestinal calcium transport of these rats and the appearance of the $1,25$-$(OH)_2 D_3$ in the intestine following an oral or an intravenous dose is shown in Fig. 2. There is little doubt that $1,25$-$(OH)_2 D_3$ appears very rapidly in this target tissue and rapidly enough to account for the biological response. It is of importance to note that in the rat using a variety of 1α-OH-D_3 compounds prepared in our laboratory and prepared by the pharmaceutical companies by the method of Barton *et al.* (2) show identical responses, namely that the intestine shows a maximum calcium transport response at 6 h after an intravenous dose. We have been unable to confirm the report of Pechet and his colleagues (24) claiming a 30-min response of intestine to 1α-OH-D_3 and $1,25$-$(OH)_2 D_3$. Therefore, it is evident that the conversion of 1α-OH-D_3 to $1,25$-$(OH)_2 D_3$ occurs rapidly enough to account for the intestinal calcium transport response. Similar results have been obtained with the bone.

It is of some interest that 1α-OH-D_3 disappears very rapidly from the blood and intestine following an intravenous dose of 1α-OH-D_3. In fact, virtually all of the 1α-OH-D_3 has disappeared by 12 h, which indicates that this substance is very rapidly cleared from the body and is converted to the $1,25$-$(OH)_2 D_3$.

The rapidity with which $1,25$-$(OH)_2 D_3$ appears in the tissues *in vivo* suggested that the 25-hydroxylation must in fact be a very rapid reaction. Such has been suggested by Zerwekh and his colleagues using an indirect measurement of $1,25$-$(OH)_2 D_3$ (26). A comparison of the *in vitro* conversion of vitamin D_3 to 25-OH-D_3 and of 1α-OH-D_3 to the $1,25$-$(OH)_2 D_3$ has been made. Livers were taken from vitamin D-defi-

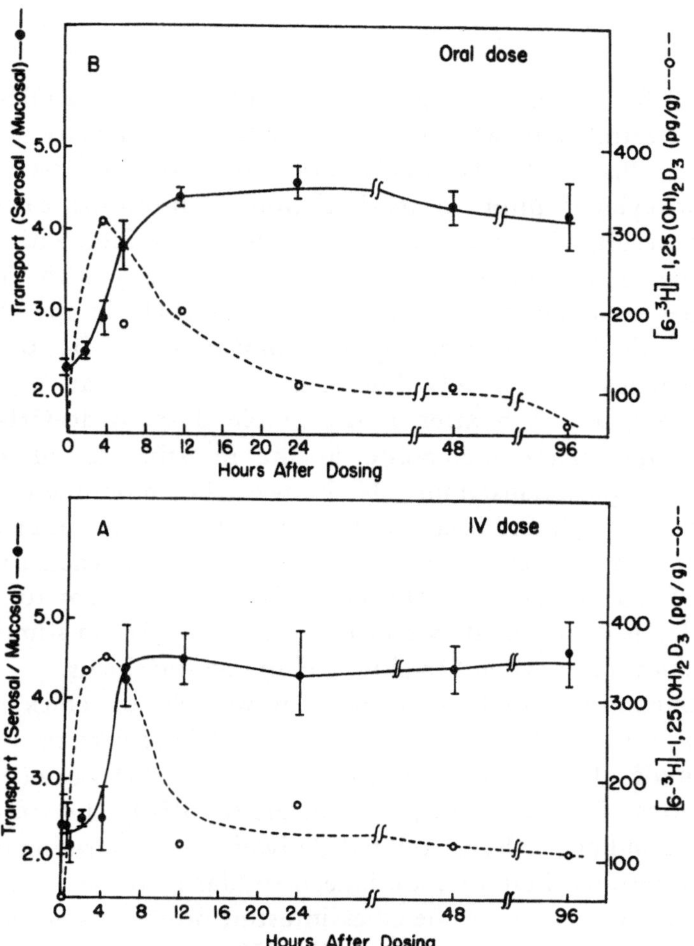

Figure 2. Response of intestinal calcium transport and appearance of $[6\text{-}^3\text{H}]\text{-}1,25\text{-}(OH)_2D_3$ in intestine of vitamin D-deficient rats given a single dose of 325 pmoles $[6\text{-}^3\text{H}]\text{-}1\alpha\text{-OH-}D_3$ orally or intravenously.

cient rats, homogenized and incubated as described by Bhattacharyya and DeLuca (3) utilizing an endogenous generation of NADPH. Identical conversions of vitamin D_3 and $1\alpha\text{-OH-}D_3$ by this *in vitro* preparation to the corresponding 25-hydroxylated derivatives was found. The rate and time course of this *in vitro* conversion was identical for both compounds, suggesting that this enzyme which hydroxylates vitamin D is also the one which hydroxylates $1\alpha\text{-OH-}D_3$. However, the fact that the $1,25\text{-}(OH)_2 D_3$ is seen very rapidly in the intestine and because of the report by Haussler's group that chick intestine is capable of 25-hydroxylation, we examined the question of whether rat intestine could carry out the 25-hydroxylation of $1\alpha\text{-OH-}D_3$. Rat intestine was, therefore, isolated and the mucosa was homogenized and incubated as described by Tucker *et al.* (25). For comparison similar homogenates of rachitic or vitamin D-deficient rat intestine were incubated with vitamin D_3. No 25-hydroxylated products could be detected for either substrate and certainly not for 1α-OH-D_3, suggesting that intestinal preparations do not have this capacity. It is, therefore, apparent that 1α-OH-D_3 likely proceeds to the liver where it becomes 25-hydroxylated prior to its initiation of intestinal calcium transport. These results, however, would appear to conflict with the results which have been obtained by Zerwekh *et al.* (26) with chick intestine. We have examined the metabolism of 1α-OH-D_3 in the chick and have learned that chicks very rapidly convert 1α-OH-D_3 to the $1,25$-$(OH)_2 D_3$, perhaps more rapidly than in the case of the rat. A good example is the fact that 1 h after a 325 pmole dose of 1α-OH-D_3, $1,25$-$(OH)_2 D_3$ is the major metabolite in the intestine found at a concentration of 317 picograms/gram. This is twice the amount accumulated in rat intestine in 1 h and is certainly sufficient to initiate intestinal calcium transport. The chicken likely has a greater ability or capacity to convert the 1α-OH-D_3 to the $1,25$-$(OH)_2 D_3$. This may account for the fact that 1α-OH-D_3 is approximately equal to $1,25$-$(OH)_2 D_3$ in stimulating intestinal calcium transport and calcification in rachitic chicks.

We have examined more closely the mechanism whereby the chicks will carry out the 25-hydroxylation of 1α-OH-D_3. Chick liver, like rat liver, when incubated under the conditions of Bhattacharyya and DeLuca (3) or under the conditions of Tucker *et al.* (25), convert 1α-OH-D_3 and vitamin D_3 to the corresponding 25-hydroxylated derivatives at approximately equal rates. These results, therefore, are indeed similar to those obtained in the rat. Unlike the rat, however, the chick intestine possesses significant ability to hydroxylate 1α-OH-D_3 to the $1,25$-$(OH)_2 D_3$ as suggested by Zerwekh *et al.* (26). The results shown in Fig. 3 demonstrate that chick intestinal homogenates will hydroxylate both vitamin D_3 and 1α-OH-D_3 to 25-OH-D_3 and $1,25$-$(OH)_2 D_3$, respectively. Furthermore, heating of the chick intestinal mucosal homogenate for 10 min at $100°C$ completely eliminates this 25-hydroxylating· ability. Apparently chicks have in addition to hepatic 25-hydroxylating system a similar capability in intestine. It is, therefore, nor surprising that 1α-OH-D_3 acts just as rapidly as $1,25$-$(OH)_2 D_3$ in initiating intestinal calcium transport when given by the oral route as found by Haussler and his colleagues (10) and as has been demonstrated in our laboratory (13).

The present results cannot exclude completely the possibility that 1α-OH-D_3 might act directly at high concentrations found initially in the intestine following dosage. Although this possibility in our view seems remote, it is likely that this cannot be excluded until further experiments with appropriate analogs are carried out.

The present results give considerable confidence to the clinical investigators who utilize 1α-OH-D_3 in the treatment of renal osteo-

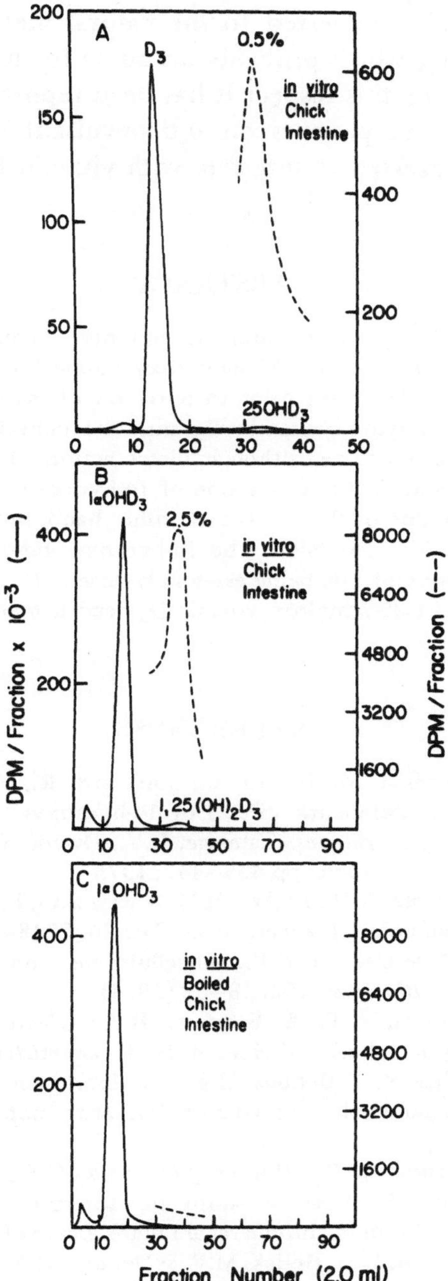

Figure 3. *In vitro* 25-hydroxylation of [6-³H]-1α-OH-D₃ by chick intestinal homogenates. The incubation mixture of Tucker *et al.* was used with either [³H-1,2]-vitamin D₃ or [6-³H]-1α-OH-D₃ as substrates. The incubation period was 2 h and chromatography was by Sephadex LH-20 (65:35, chloroform-Skellysolve B).

dystrophy and hypoparathyroidism inasmuch as it is evident that this compound is rapidly converted to the natural metabolite or hormonal form of vitamin D_3, which probably accounts for most, if not all, of its biological activity. In this respect it has been reported that 1α-OH-D_3 is much less effective in patients on anticonvulsant therapy, a treatment which has been suggested to interfere with vitamin D_3-25-hydroxylation (23).

ABSTRACT

High activity $[6\text{-}^3H]\text{-}1\alpha$-hydroxyvitamin D_3 has been chemically synthesized and shown to be converted rapidly to 1,25-dihydroxyvitamin D_3 in both rats and chicks. This rapid conversion takes place prior to initiation of intestinal calcium transport which suggests that 1α-hydroxyvitamin D_3 must be converted to 1,25-dihydroxyvitamin D_3 before it functions, although direct action of 1α-hydroxyvitamin D_3 cannot be totally excluded. The conversion of 1α-hydroxyvitamin D_3 takes place in the liver and not intestine of the rat. On the other hand, both liver and intestine of chicks can carry out this conversion. The 1α-hydroxyvitamin D_3 rapidly disappears from blood and intestine of rats being essentially absent 12 hours after a 325 pmole dose, while the formed 1,25-dihydroxyvitamin D_3 persists much longer.

REFERENCES

1. Balsan, S., Garabedian, M., Pezant, E., Sorgniard, R., Holick, M.F. & DeLuca, H.F.: Biologic and therapeutic effects of 1α-hydroxyvitamin D in children. In: *Calcium Regulating Hormones*. Talmage, R.V., Owen, M. & Parsons, J.A. (eds.), Excerpta Medica, Amsterdam, pp 439—447, 1975

2. Barton, D.H.R., Hesse, R.H., Pechet, M.M. & Rizzardo, E.: A convenient synthesis of 1α-hydroxyvitamin D_3. *J. Amer. chem. Soc.* 95, 2748—2749 (1973)

3. Bhattacharyya, M. & DeLuca, H.F.: Subcellular location of rat liver calciferol-25-hydroxylase. *Arch. Biochem.* 160, 58—62 (1974)

4. Blunt, J.W., DeLuca, H.F. & Schnoes, H.K.: 25-Hydroxycholecalciferol. A biologically active metabolite of vitamin D_3. *Biochemistry* 7, 3317—3322 (1968)

5. Blunt, J.W., Tanaka, Y. Y DeLuca, H.F.: The Biological Activity of 25-Hydroxycholecalciferol, a metabolite of vitamin D_3. *Proc. nat. Acad. Sci. (Wash.)* 61, 1503—1506 (1968)

6. Chalmers, T.M., Davie, M.W., Hunter, J.O., Szaz, K.F., Pelc, B. & Kodicek, E.: 1α-Hydroxycholecalciferol as a substitute for the kidney hormone 1,25-dihydroxycholecalciferol in chronic renal failure. *Lancet* (1973), ii, 696—699

7. Chan, J.C.M., Oldham, S.B., Holick, M.F. & DeLuca, H.F.: 1α-Hydroxyvitamin D_3 in chronic renal failure. *J. Am. Med. Assoc.* 234, 47—52 (1975)

8. Fraser, D.R. & Kodicek, E.: Unique biosynthesis by kidney of a biologically active vitamin D metabolite. *Nature* 228, 764—766 (1970)

9. Haussler, M.R., Myrtle, J.F. & Norman, A.W.: The association of a metabolite of vitamin D_3 with intestinal mucosa chromatin *in vivo*. *J. Biol. Chem.* 243, 4055—4064 (1968)

10. Haussler, M.R., Zerwekh, J.E., Hesse, R.H., Rizzardo, E. & Pechet, M.M.: Biological activity of 1α-hydroxycholecalciferol, a synthetic analog of the hormonal form of vitamin D_3. *Proc. nat. Acad. Sci. (Wash.)* **70**, 2248–2252 (1973)

11. Holick, M.F. & DeLuca, H.F.: A new chromatographic system for vitamin D_3 and its metabolites: Resolution of a new vitamin D_3 metabolite. *J. Lipid Res.* **12**, 460–465 (1971)

12. Holick, M.F., Schnoes, H.K., DeLuca, H.F., Suda, T. & Cousins, R.J.: Isolation and identification of 1,25-dihydroxycholecalciferol. A metabolite of vitamin D active in intestine. *Biochemistry* **10**, 2799–2804 (1971)

13. Holick, M.F., Semmler, E.J., Schnoes, H.K. & DeLuca, H.F.: 1α-hydroxy derivative of vitamin D_3: A highly potent analog of 1α,25-dihydroxyvitamin D_3. *Science* **180**, 190–191 (1973)

14. Holick, M.F., Kasten-Schraufrogel, P., Tavela, T. & DeLuca, H.F.: Biological activity of 1α-hydroxyvitamin D_3 in the rat. *Arch. Biochem.* **166**, 63–66 (1975)

15. Holick, M.F., Holick, S.A., Tavela, T., Gallagher, B., Schnoes, H.K. & DeLuca, H.F.: Synthesis of [6-^{3}H]-1α-hydroxyvitamin D_3 and its metabolism *in vivo* to [^{3}H]-1α,25-dihydroxyvitamin D_3. *Science* **190**, 576–578 (1975)

16. Kaneko, C., Yamada, S., Sugimoto, A., Eguchi, Y., Ishikawa, M., Suda, T., Suzuki, M., Kakuta, S. & Sasaki, S.: Synthesis and biological activity of 1α-hydroxyvitamin D_3. *Steroids* **23**, 75–92 (1974)

17. Lawson, D.E.M., Wilson, P.W. & Kodicek, E.: Metabolism of vitamin D. A new cholecalciferol metabolite, involving loss of hydrogen at C-1, in chick intestinal nuclei. *Biochem. J.* **115**, 269–277 (1969)

18. Maclaren, N. & Lifshitz, F.: Vitamin D-dependency rickets in institutionalized, mentally retarded children on long term anticonvulsant therapy. II. The response to 25-hydroxycholecalciferol and to vitamin D_3. *Pediat. Res.* **7**, 914–922 (1973)

19. Ponchon, G. & DeLuca, H.F.: Metabolites of vitamin D_3 and their biological activity. *J. Nutr.* **99**, 157–167 (1969)

20. Ponchon, G., Kennan, A.L. & DeLuca, H.F.: "Activation" of vitamin D by the liver. *J. clin. Invest.* **48**, 2032–2037 (1969)

21. Reade, T.M., Scriver, C.R., Glorieux, F.H., Nogrady, B., Delvin, E., Poirier, R., Holick, M.F. & DeLuca, H.F.: Response to crystalline 1α-hydroxy vitamin D_3 in vitamin D dependency. *Pediat. Res.* **9**, 593–599 (1975)

22. Semmler, E.J., Holick, M.F., Schnoes, H.K. & DeLuca, H.F.: The synthesis of 1α,25-dihydroxycholecalciferol. A metabolically active form of vitamin D_3. *Tetrahedron Letters* **40**, 4147–4150 (1972)

23. Stamp, T.C.B., Round, J.M., Rowe, D.J.F. & Haddad, J.G.: Plasma levels and therapeutic effect of 25-hydroxycholecalciferol in epileptic patients taking anticonvulsant drugs. *Brit. Med. J.* **4**, 9–12 (1972)

24. Toffolon, E.P., Pechet, M.M. & Isselbacher, K.: Demonstration of the rapid action of pure crystalline 1α-hydroxy vitamin D_3 and 1α,25-dihydroxy vitamin D_3 on intestinal calcium uptake. *Proc. nat. Acad. Sci. (Wash.)* **72**, 229–230 (1975)

25. Tucker, G., Gagnon, R.E. & Haussler, M.R.: Vitamin D_3-25-hydroxylase: Tissue occurrence and apparent lack of regulation. *Arch. Biochem.* **155**, 47–57 (1973)

26. Zerwekh, J.E., Brumbaugh, P.F., Haussler, D.H., Cork, D.J. & Haussler, M.R.: 1α-Hydroxyvitamin D_3. An analog of vitamin D which apparently acts by metabolism to 1α,25-dihydroxyvitamin D_3. *Biochemistry* **13**, 4097–4102 (1974)

The Regulation of Vitamin D Metabolism

I. MacIntyre, K.W. Colston & I.M.A. Evans

INTRODUCTION

Vitamin D_3 (cholecalciferol) undergoes two hydroxylations before exerting its biological effect. The first is in the liver to 25-hydroxycholecalciferol (25-OH-D_3), the major circulating form of vitamin D_3 (1), the second in the kidney to 1,25-dihydroxycholecalciferol (1,25-$(OH)_2 D_3$) (7), the most active metabolite of vitamin D_3 or to 24,25-$(OH)_2 D_3$, a less active metabolite whose action is not fully understood. Factors which influence the conversion of 25-OH-D_3 to these metabolites and which affect the activity of the renal 1- and 24-hydroxylase enzymes will be discussed and a hypothesis for the manner in which vitamin D_3 metabolism is regulated will be presented.

REGULATION OF VITAMIN D METABOLISM

1. Feed-back Regulation of Vitamin D_3

In states of vitamin D deficiency there is a marked increase in the activity of the renal 1-hydroxylase enzyme. Utilizing a standardised assay in chick kidney homogenates (3) capable of measuring both 1- and 24-hydroxylase activity, the influence of vitamin D deficiency on the activity of these renal enzymes has been studied. Chicks fed a vitamin D deficient diet from the day of hatching show a marked increase in 1-hydroxylase activity after 4 days. This stimulation of enzyme activity reaches a maximum after 2 to 3 weeks. Administration of vitamin D_3 or one of its metabolites causes a marked stimulation of 24-hydroxylase activity and a disappearance of 1-hydroxylase activity. 6.25 nmol cholecalciferol (2.5 μg) given subcutaneously to vitamin D deficient chicks induces the appearance of 24-hydroxylase activity after 24 h, accompanied by a fall in

Endocrine Unit, Royal Postgraduate Medical School, London.

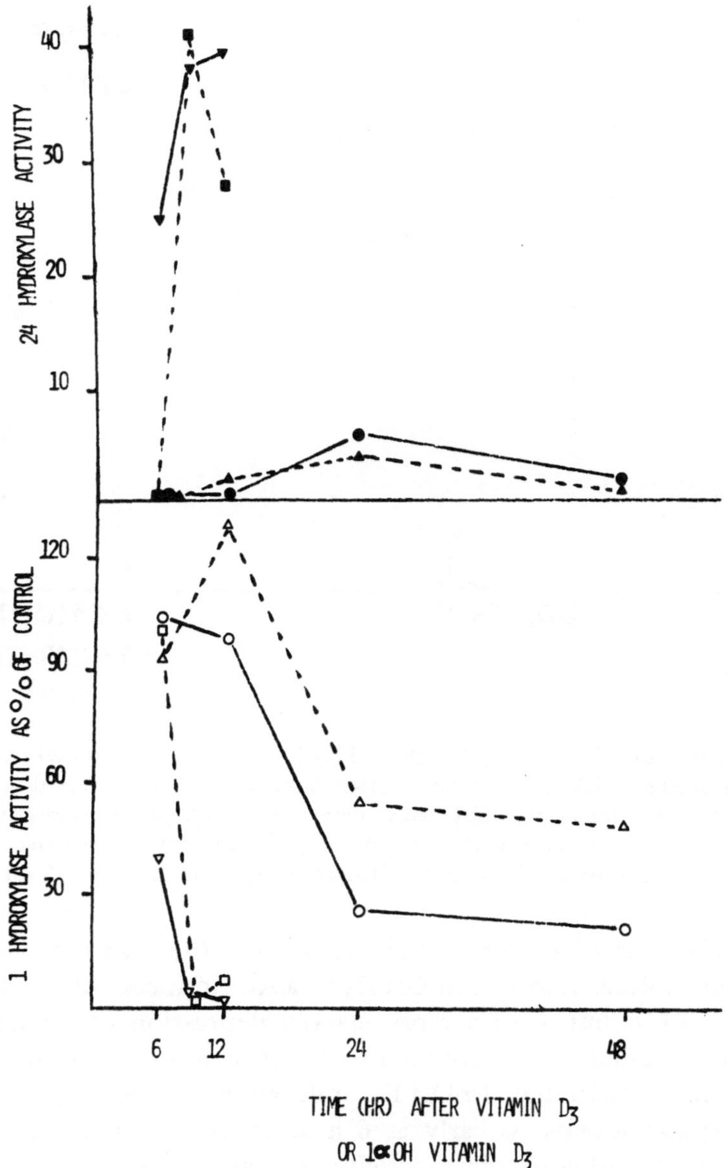

Figure 1. Effect of vitamin D_3 and 1α-OH-D_3 on 1- and 24-hydroxylase activity as assayed in chick kidney homogenates.

△ ▲ 6.25 nmol D_3		○ ● 6.25 nmol 1α-OH-D_3
□ ■ 62.5 nmol 1α-OH-D_3		▽ ▼ 625 nmol 1α-OH-D_3

Closed symbols represent 24-hydroxylase activity expressed as % of the substrate ^{3}H-25-OH-D_3 converted to $24,25$-$(OH)_2D_3$ per chick kidney per 15 min.

Open symbols represent 1-hydroxylase activity expressed as a % of that in ethanol treated controls. Each point represents the average of duplicate estimations for each of 3 birds.

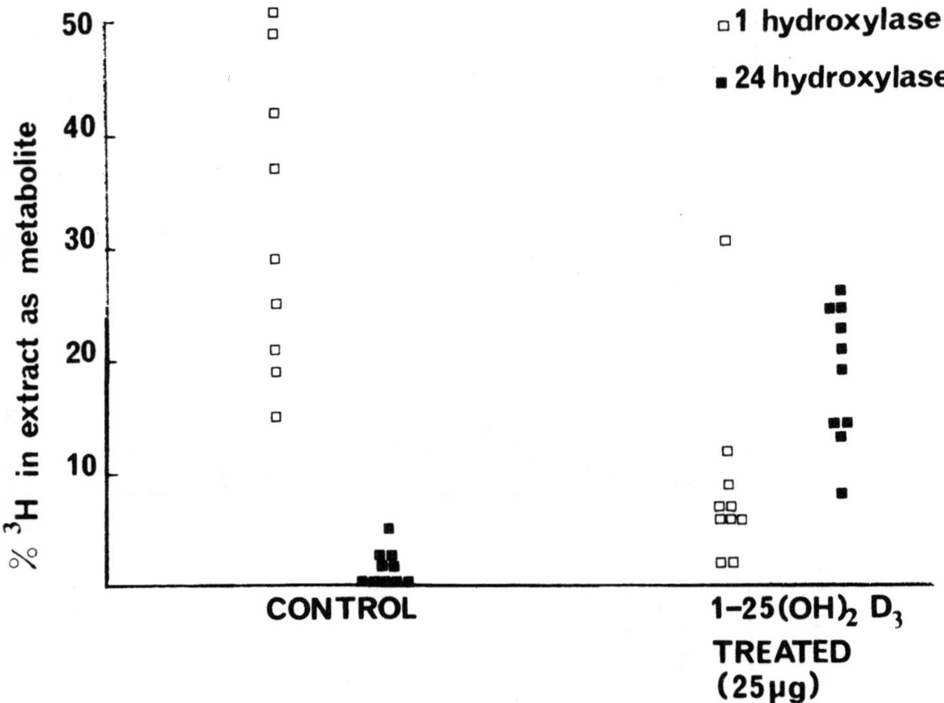

Figure 2. Effect of 62.5 nmol (25 µg) 1,25-(OH)$_2$D$_3$ on the 1- and 24-hydroxylase enzymes assayed in chick kidney homogenates. Enzyme activity expressed as % tritium in homogenate extract as the dihydroxy metabolite. Each point represents a single estimation of enzyme activity (3 estimations per bird, three birds per treatment group). The concentration of substrate (^{3}H-25-OH-D$_3$) was 30 ng per 3 ml incubate.

1-hydroxylase activity. An equal weight of the synthetic analogue 1α-hydroxycholecalciferol (1α-OH-D$_3$) also induces 24-hydroxylase activity at 24 h but with a more marked depression of 1-hydroxylase activity. More rapid effects are seen with larger doses of 1α-OH-D$_3$, such that 625 nmol (250 ng) 1α-OH-D$_3$ induced a dramatic appearance of 24-hydroxylase activity as early as 6 h after treatment. In all cases the appearance of 24-hydroxylase activity was accompanied by a marked decrease in 1-hydroxylase activity (Fig. 1). This induction of 24-hydroxylase activity was also seen with administration of 62.5 nmol (25 µg) the natural metabolit, 1,25-(OH)$_2$D$_3$ after 9 h (Fig. 2).

2. The Influence of Parathyroid Hormone

In the last few years there have been differing opinions concerning the influence of parathyroid hormone on vitamin D metabolism, but it has now been shown that the hormone is not essential for the production of 1,25-(OH)$_2$D$_3$. Garabedian (10) found a decrease in 1,25-(OH)$_2$D$_3$ in the blood of thyroparathyroidectomized rats and observed that administration of parathyroid hormone reversed this effect. However, operatively induced hyperphosphataemia rather than absence of parathyroid hormone *per se* may have caused this decreased production of $1,25\text{-}(OH)_2 D_3$. This is supported by the fact that glucose, which lowers serum phosphorus, reserves the inhibitory effect of thyroparathyroidectomy on $1,25\text{-}(OH)_2 D_3$ production (10, 13). Further, thyroparathyroidectomy does not prevent increased production of $1,25\text{-}(OH)_2 D_3$ in response to a low calcium, low phosphorus diet (11, 6). Finally, parathyroid hormone is not essential for the maintenance of the high levels of 1-hydroxylase activity seen in states of vitamin D deficiency (9). Nevertheless, parathyroid hormone probably plays an important role in vitamin D metabolism, although its action is likely to depend on the levels of calcium and phosphorus in the plasma. Thus, parathyroid hormone reverses the effect of parathyroidectomy when plasma calcium is low and phosphorus high, but the hormone may have the opposite action when plasma calcium is already high, or phosphorus already low (8).

3. Regulation by Intracellular Calcium Concentration

A possible acute regulator of $1,25\text{-}(OH)_2 D_3$ production is the concentration of free calcium ions within the renal 'D' cell which contains the 1- and 24-hydroxylase systems. In experiments with chick kidney homogenates, the rate of 1 hydroxylation responded instantaneously and inversely to changes in the concentration of this ion (3). The maximum effect of calcium ions on the rate of 1-hydroxylation was seen with this ion over the concentration range $0-0.5$ mM. The upper limit of this range is much higher than the accepted range for the cytoplasmic concentration of calcium of $0.1\ \mu M$ (12). However, the site of the inhibitory action of calcium ions on the 1-hydroxylase system is not yet known and inhibition may take place in an intracellular region such as the mitochondria. No such inhibition of enzyme activity with calcium was seen with the 24-hydroxylase system (4).

DISCUSSION

The factors regulating the metabolism of vitamin D are complex. Serum calcium (2) and phosphorus, intracellular calcium and vitamin D itself have all been shown to influence $1,25\text{-}(OH)_2 D_3$ production. Vitamin D regulates its own metabolism in a feed-back manner such that the 1-hydroxylase enzyme is stimulated in states of vitamin D deficiency while administration of vitamin D or one of its metabolites induces

140

24-hydroxylase activity and decreases 1-hydroxylase activity. There is evidence that this induction of 24-hydroxylase activity may be via transcription and new protein synthesis (5). This feed-back control of vitamin D_3 metabolism is likely to be one major means of long term regulation.

Serum phosphorus and calcium also have some influence on 1,25-$(OH)_2 D_3$ production, although it is not yet known how this effect is mediated. Parathyroid hormone is not essential for 1-hydroxylase activity, but under some dietary conditions it may stimulate 1,25-$(OH)_2 D_3$ production. It is likely that all these factors interact in a complex manner to regulate the activity of the renal 1- and 24-hydroxylase enzymes. Thus the activity, and perhaps the synthesis of the two renal hydroxylases may be regulated by vitamin D status, serum levels of phosphorus and calcium and intracellular calcium concentration.

ACKNOWLEDGEMENTS

We thank the Medical Research Council for generous financial assistance. We also thank Professor B. Lythgoe, Department of Chemistry, University of Leeds for the gift of synthetic 1α-hydroxy-D_3 and Dr. M.R. Uskokovic, Hoffman-La-Roche Inc., New Jersey for the gift of 1,25-$(OH)_2 D_3$. I.M.A.E. is in receipt of a Medical Research Council Clinical Research Fellowship.

REFERENCES

1. Blunt, J.W., DeLuca, H.F. & Schnoes, H.K.: 25-hydroxycholecalciferol. A biological active metabolite of vitamin D_3. *Biochemistry* 7, 3317–3322 (1968)
2. Boyle, I.T., Gray, R.W. & DeLuca, H.F.: Regulation by calcium of *in vivo* synthesis of 1,25-dihydroxycholecalciferol and 21,25-dihydroxycholecalciferol. *Proc. nat. Acad. Sci. (Wash.)* 68, 2131–2134 (1971)
3. Colston, K.W., Evans, I.M.A., Galante, L., MacIntyre, I. & Moss, D.W.: Regulation of vitamin D metabolism: factors influencing the rate of formation of 1,25-dihydroxycholecalciferol by kidney homogenates. *Biochem. J.* 134, 817–820 (1973)
4. Colston, K.W., Evans, I.M.A. & MacIntyre, I.: The interaction of vitamin D and calcium. In Press
5. Evans, I.M.A., Colston, K.W., Galante, L. & MacIntyre, I.: Feed-back regulation of 25-hydroxycholecalciferol metabolism by vitamin D_3. *Clin. Sci. Molec. Med.* 48, 227–230 (1975)
6. Favus, M.J., Walling, M.W. & Kimberg, D.V.: Effects of dietary restriction and chronic thyroparathyroidectomy on the metabolism of (^{3}H) 25-hydroxy vitamin D_3 and active transport of calcium by rat intestine. *J. clin. Invest.* 53, (4) 1139–1148 (1974)

7. Fraser, D.R. & Kodicek, E.: Unique biosynthesis by kidney of a biologically active vitamin D metabolite. *Nature* **228**, 764—766 (1970)

8. Galante, L., MacAuley, S.J., Colston, K.W. & MacIntyre, I.: Effect of parathyroid extract on vitamin D metabolism. *Lancet* (1972), i, 985—988 (1972)

9. Galante, L., Colston, K.W., Evans, I.M.A., Byfield, P.G.H., Matthews, E.W. & MacIntyre, I.: The regulation of vitamin D metabolism. *Nature* **244**, 438—440 (1973)

10. Garabedian, M., Holick, M.F., DeLuca, H.F. & Boyle, I.T.: Control of 25-hydroxy-cholecalciferol metabolism by parathyroid glands. *Proc. nat. Acad. Sci. (Wash.)* **69**, 1673—1676 (1972)

11. Larkins, R.G., MacAuley, S.J., Colston, K.W., Evans, I.M.A., Galante, L. & MacIntyre, I.: Regulation of vitamin D metabolism without parathyroid hormone. *Lancet* (1973), ii, 289—291

12. Rasmussen, H.: Cell communication, calcium ion and cyclic adenosine mono-phosphate. *Science* **170**, 404—412 (1970)

13. Tanaka, Y. & DeLuca, H.F.: The control of 25-hydroxy vitamin D metabolism by inorganic phosphorus. *Arch. Biochem.* **154**, 566—574 (1973)

142

The Effects of Vitamin D Metabolites and their Analogues on the Secretion of Parathyroid Hormone

A.D. CARE, R.F.L. BATES, D.W. PICKARD, M. PEACOCK[1], S. TOMLINSON[2], J.L.H. O'RIORDAN[2], E. BARBARA MAWER[3], CAROL M. TAYLOR[3], H.F. DELUCA[4] & A.W. NORMAN[5]

1,25-dihydroxycholecalciferol (1,25-DHCC) is the major biologically active metabolite of vitamin D_3 in the small intestine and bone, whereas roles for 24,25-dihydroxycholecalciferol and 25,26-dihydroxycholecalciferol have yet to be assigned. All three metabolites have been detected in plasma, the approximate concentration in a normal individual being 60 pg/ml for 1,25-DHCC (4), 0—4 ng/ml for 24,25-DHCC (13) and 200—1000 pg/ml for 25,26-DHCC (Mawer & Taylor, unpublished results). The principal circulating metabolite is 25-hydroxycholecalciferol (25-HCC) the normal concentration of which in healthy humans in Britain is 12 ng/ml, with a range of 4 to 33 ng/ml (12).

It is now generally accepted that the administration of parathyroid hormone (PTH) to normal or hypocalcaemic animals leads to increased renal synthesis of 1,25-DHCC (7). Recent work has shown that the renal 25-HCC-1-hydroxylase activity is directly related to the ratio of PTH to cholecalciferol (CC) status such that the rate of biosynthesis approaches the basal level either in the absence of PTH or in the presence of adequate CC (8).

Circumstances have been noted in which the secretion of PTH was not in accordance with the value to be expected from the prevailing concentration of plasma ionized calcium (6, 11, 3). A direct action of vitamin D metabolites on the parathyroid gland might explain these

Department of Animal Physiology and Nutrition, University of Leeds.
[1]) M.R.C. Mineral Metabolism Unit, The General Infirmary, Leeds.
[2]) Department of Medicine, Middlesex Hospital, London.
[3]) Division of Metabolism, University Department of Medicine, Manchester.
[4]) Department of Biochemistry, University of Wisconsin, Madison.
[5]) Department of Biochemistry, University of California, Riverside.

findings. Preliminary work carried out in goats with biosynthetic 24,25-DHCC has indeed demonstrated an inhibitory effect on PTH secretion (2).

METHODS

A superior parathyroid gland was isolated *in situ* in 22 anaesthetized goats and perfused under controlled conditions of composition, temperature, flow rate and pressure using a fluid based on tissue culture medium 199 (BDH Chemicals) as described for blood by Care, Sherwood, Potts & Aurbach (5). Sodium bicarbonate was added to medium 199 to give a 25 mM bicarbonate solution, pH 7.4, after gassing with $5\% \, CO_2 : 95\% \, O_2$ for one h. Dextran 70 (Glaxo) was added to make a 5% solution (w/w). This solution was pumped from a reservoir at a constant rate and to it was added analogous plasma to give a final plasma dilution of 10%. This plasma was used as the carrier medium for the vitamin D metabolite under test, the addition of the metabolite being omitted during the initial and final control periods. Towards the end of each experiment it was usual to test the perfused gland for a normal response of PTH secretion to hypercalcaemia and hypocalcaemia (induced by the addition of EDTA to the perfusing fluid). The rate of collection of the venous effluent was always recorded and compared with the rate of influx to monitor quantitative perfusion of the gland. The concentration of calcium in this fluid was measured, immediately after collection, by EDTA titration using a photoelectric titrator (Evans Electro Selenium Ltd) to detect the end-point. The concentration of PTH in the gland effluent was measured, in those experiments which involved 24,25-DHCC & 25,26-DHCC, by a conventional radioimmunoassay using an antiserum to bovine PTH. In this system, percentage bound vs dilution curves were set up using the dilutions of goat parathyroid venous effluent and human PTH to establish that the curves were parallel over the range in which the estimations of PTH concentration were made. For the remaining experiments, PTH was measured using a radioimmunometric assay and an antiserum to bovine PTH (1). One experiment was assayed by both methods and a similar result obtained.

RESULTS

24,25-Dihydroxycholecalciferol (a) Biosynthetic. This was prepared *in vitro* by hydroxylation of 25-HCC at the 24 position using a kidney homogenate from chicks treated with 1-hydroxyethyl-1, 1-diphosphonate

(14). In all of six technically satisfactory experiments 24,25-DHCC at concentrations ranging from 0.5—1 ng/ml caused a significant reduction in the secretion rate of PTH. (b) Synthetic. Both isomers, 24r,25-DHCC and 24s,25-DHCC, have been tested. In one of four experiments a significant decrease in PTH secretion rate was found using 24r,25-DHCC in the range 1.2—4.0 ng/ml. In three other experiments using 24r,25-DHCC (0.45—6.2 ng/ml) this result could not be confirmed. However, in two experiments carried out with 24s,25-DHCC, although no effect was noted with concentrations 0.8—0.9 ng/ml, when these were raised to 3.1 and 8.3 ng/ml respectively, a reduction in PTH secretion rate resulted.

1,25-Dihydroxycholecalciferol Five technically satisfactory experiments have been carried out using synthetic 1,25-DHCC at concentrations of 50—125 pg/ml. In two of these, significant stimulation of PTH secretion was observed. At high concentrations (25 ng/ml) 1,25-DHCC was without effect on PTH secretion.

25,26-Dihydroxycholecalciferol In one experiment, synthetic 25,26-DHCC (180 pg/ml) caused a significant reduction in PTH secretion but a concentration of 9 ng/ml was without effect.

1α-Hydroxycholecalciferol A concentration range of 170 pg/ml — 10 ng/ml was used during the course of three experiments. 1α-HCC was without significant effect on PTH secretion rate throughout this range of concentration.

25-Hydroxycholecalciferol In one experiment, 25-HCC (100 ng/ml) caused no significant change in PTH secretion. This concentration of 25-HCC has been found in the cow following the intramuscular injection of the usual recommended dose of vitamin D for the prevention of parturient hypocalcaemia (250 mg).

DISCUSSION

The concentrations of 25-HCC (14—53 ng/ml) in the plasma of the goats used for these experiments were all within the normal range so that vitamin D deficiency could not account for the variation observed in the response to a given metabolite in different animals. Nevertheless, effects of vitamin D metabolites on PTH secretion have been demonstrated, although these were usually less marked than those produced by changes in the calcium ion concentration in the perfusion fluid. The effectiveness of a vitamin D metabolite in a particular experiment may depend on the individual relationship between calcium ion concentration and PTH secretion rate for that gland. Since this relationship is not linear (10) it is clear that the effect of a metabolite would depend upon the part of the sigmoid curve which was in use during the experiment.

It is known that during hypercalcaemia or when high calcium diets are fed, 24,25-DHCC is the predominant derivative of 25-HCC. It is proposed that 24,25-DHCC serves to increase the inhibitory effect of hypercalcaemia on PTH secretion and thus quicken the restoration of calcium homeostasis.

In addition to bone and small intestinal mucosa, the parathyroid gland has recently been identified as a localization site for 1,25-DHCC (9). This observation provides circumstantial support for our finding that 1,25-DHCC can stimulate PTH secretion. Since 1,25-DHCC production is enhanced in hypocalcaemic conditions, stimulation of PTH release not only by the hypocalcaemia but also by 1,25-DHCC, would be expected to bring a return to normocalcaemia more quickly.

It is concluded that 24,25-DHCC and 1,25-DHCC alter PTH secretion in such a way as to increase the rate at which calcium homeostasis is restored.

REFERENCES

1. Addison, G.M., Hales, C.N., Woodhead, J.S. & O'Riordan, J.L.H.: Immunoradiometric assay of parathyroid hormone. *J. Endocr.* **49**, 521—530 (1971)
2. Bates, R.F.L., Care, A.D., Peacock, M., Mawer, E.B. & Taylor, C.M.: Inhibitory effect of 24,25-dihydroxycholecalciferol on parathyroid hormone secretion in the goat. *J. Endocr.* **64**, p 6 (1974)
3. Bordier, P.: Studies on the effectiveness of vitamin D metabolites and analogues in nutritional and renal osteomalacia. In: *Vitamin D and Problems related to uremic bone disease.* Norman, A.W., Grigoleit, H.G., von Herrath, D. & Ritz, E. (eds.) de Gruyter, Berlin, pp 133—148, 1975
4. Brumbaugh, P.F., Haussler, D.H., Bressler, R. & Haussler, M.R.: Radioreceptor assay for $1\alpha,25$-dihydroxy vitamin D_3. *Science* **183**, 1089—1091 (1974)
5. Care, A.D., Sherwood, L.M., Potts, J.T. & Aurbach, G.D.: Evaluation by radioimmunoassay of factors controlling the secretion of parathyroid hormone. Perfusion of the isolated parathyroid gland of the goat and sheep. *Nature (Lond.)* **290**, 55—57 (1966)
6. Fischer, J.A., Binswanger, U., Fanconi, A., Illig, R., Baerlocher, K. & Prader, A.: Serum parathyroid hormone concentrations in vitamin D deficiency rickets of infancy: effects of intravenous calcium and vitamin D. *Horm. Metab. Res.* **5**, 381—385 (1973)
7. Garabedian, M., Holick, M.F., DeLuca, H.F. & Boyle, I.T.: Control of 25-hydroxycholecalciferol metabolism by parathyroid glands. *Proc. nat. Acad. Sci. (Wash.)* **69**, 1673—1676 (1972)
8. Henry, H.L., Midgett, R.J. & Norman, A.W.: Regulation of 25-hydroxy vitamin D_3-1-hydroxylase *in vivo. J. biol. Chem.* **249**, 7584—7592 (1974)

9. Henry, H.L. & Norman, A.W.: Studies on the mechanism of action of calciferol VII. Localization of 1,25-dihydroxy-vitamin D_3 in chick parathyroid glands. *Biochem. Biophys. Res. Commun.* **62**, 781–788 (1975)

10. Mayer, G.P.: Effect of calcium and magnesium on parathyroid hormone secretion rate in calves. In: *Calcium Regulating Hormones.* Talmage, R.V. (ed.) Excerpta Medica, Amsterdam, pp 122–124 (1975)

11. Peacock, M., Nordin, B.E.C., Gallacher, J.C. & Varnavides, C.: Action of 1α-hydroxy vitamin D_3 in man. In: *Vitamin D and Problems Related to Uremic Bone Disease.* Norman, A.W., Schaefer, K., Griloleit, H.G., von Herrath, D. & Ritz, E. (eds.) de Gruyter, Berlin. pp 611–618 (1975)

12. Preece, M.A., O'Riordan, J.L.H., Lawson, D.E.M. & Kodicek, E.: A competitive protein-binding assay for 25-hydroxycholecalciferol and 25-hydroxyergocalciferol in serum. *Clin. chim. Acta* **54**, 235–242 (1974)

13. Stanbury, S.W.: Vitamin D metabolism in adults in health and disease. In: *Calcium Regulating Hormones.* Talmage, R.V. (ed.), Excerpta Medica, Amsterdam, pp 431–438, 1975

14. Taylor, C.M., Mawer, E.B. & Reeve, A.: The metabolism of cholecalciferol (Vitamin D_3) in 1-hydroxyethyl-1,1-diphosphonate-treated chicks. *Biochem. Soc. Trans.* **1**, 596–599 (1973)

The Effects of Hydroxylated Derivatives of Vitamin D_3 and of Extracts of Solanum Malacoxylon on the Absorption of Calcium, Phosphate and Water from the Jejunum of Pigs

J. Fox & A.D. Care

The prolonged ingestion by cattle of the leaves of the plant Solanum malacoxylon (SM) causes massive soft tissue calcification of large blood vessels, heart, lungs and tendons. There is also marked erosion of cartilage of the appendicular skeleton. This condition, known in the Argentine as "enteque seco", can be a cause of considerable economic loss (10). It has been shown with rabbits that the oral or subcutaneous administration of an aqueous extract of SM leaves increases plasma concentration of calcium and inorganic phosphate (5). These effects were shown to result from increased absorption of calcium and phosphate from the digestive tract and raised bone resorption rate. They were similar to those observed following vitamin D administration, but quicker in both onset and decay. The stimulatory effect of SM leaves on dietary calcium absorption has also been clearly shown in cattle (6). Using embryonic chick intestine, Corradino (1) has demonstrated that vitamin D_3 and also its metabolites 25-hydroxycholecalciferol (25-HCC) and 1,25-dihydroxycholecalciferol (1,25-DHCC) stimulate the synthesis of intestinal calcium-binding protein and enhance the uptake of labelled calcium. Similarly, using both intact and anephric rats, it has been shown that the synthetic analogue of 1,25-DHCC, 1α-hydroxy vitamin D_3 (1α-HCC), increases intestinal calcium transport with a potency similar to 1,25-DHCC on a weight basis (4).

The similarity in intestinal action between the water soluble factor in SM leaves and 1,25-DHCC has been unequivocally demonstrated by Wasserman (8) and Corradino and Wasserman (2) and receives further support in the results described below.

Department of Animal Physiology and Nutrition, University of Leeds.

148

METHODS

A Thiry-Vella loop of jejunum, 180 cm long, the proximal end of which was 180 cm from the pylorus, was isolated *in situ* in each of eight Large-White pigs (4—12 months old) as chronic preparations. These loops were perfused with a nutrient solution as described (7) except that ^{51}Cr-EDTA was used to measure net water transport and the concentration of calcium lactate used was reduced to 5 meq/l. Three of these pigs were parathyroidectomized (PTX), the effectiveness of which was confirmed by measurement of the plasma calcium concentration after an overnight fast. A comparison was made between the effects on water and electrolyte absorption of 25-HCC, 1,25-DHCC, 1α-HCC, and aqueous extracts of SM leaf. The appropriate amount of each vitamin D metabolite was taken up in one ml plasma and diluted in one litre perfusion fluid. An aqueous extract of SM leaf was made by shaking overnight, under nitrogen and in the dark, a known weight of leaf in a given volume of perfusion fluid. The composition of the fluid had been adjusted to take account of the electrolyte content of the SM leaf so as to keep the initial composition of the perfusion fluid approximately the same as the normally used. Control periods of perfusion were carried out before and after exposure of the SM extract (or vitamin D metabolite) to the loop mucosa. Perfusion periods were of one h duration; 5—7 of these were carried out each day.

RESULTS

25-hydroxycholecalciferol
The range of intra-luminal concentrations studied was 1.5—263 ng/ml in 7 experiments with one PTX and two intact pigs. In all experiments the addition of 25-HCC to the perfusing fluid for six hours caused a significant increase in the absorption rate of calcium with its peak usually on the day of the 25-HCC addition. In two of four experiments in which phosphate absorption rate was measured there was a significant increase in this rate on the day of 25-HCC addition. However, the peak effect on phosphate absorption usually followed that of calcium so that in all four experiments there was a significant increase in phosphate absorption rate during the day following the addition of 25-HCC to the perfusion medium. There was also a significant stimulation of net water absorption on the day of the metabolite addition in 4 of 7 experiments.

1α-hydroxycholecalciferol
In 6 experiments with one PTX and three intact pigs, 1α-HCC added to

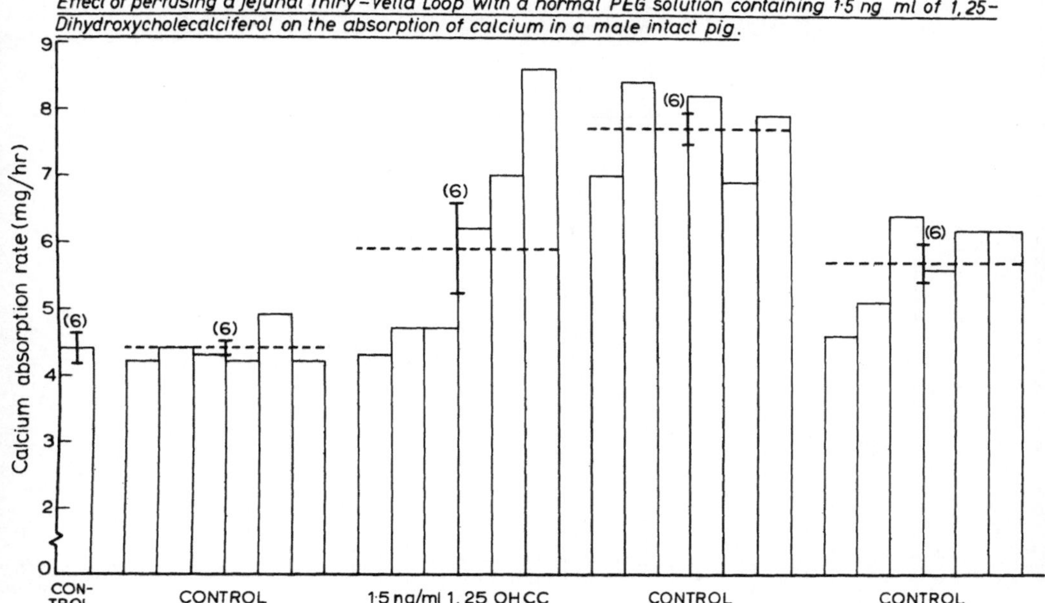

Figure 1. Changes in net calcium absorption from the fluid perfusing a healed Thiry-Vella loop of jejunum in a conscious pig associated with the addition to the fluid of 1,25-dihydroxycholecalciferol (1,25-DHCC) in a little plasma. After the first control period each vertical bar represents a perfusion period of one h. The calcium absorption during the day of 1,25-DHCC addition was significantly greater than that during the control period (P < 0.001). The figures in parentheses indicate the number of observations during each period.

the perfusion fluid (intra-luminal concentration 1.5—4.0 ng/ml) for six h caused a significant increase in calcium absorption on the day of the addition in all 6 experiments, but the peak effect was seen on the following day. As with 25-HCC, a significantly increased absorption of inorganic phosphate was only found on the day after the addition of the metabolite. In 3 experiments water absorption was increased during addition of the 1α-HCC.

1,25-dihydroxycholecalciferol

In all three experiments with two intact pigs, 1,25-DHCC (1.0—1.5 ng/ml), added intra-luminally for 6 h, caused a significant increase in calcium absorption on the day of addition of the metabolite and an even greater effect on the following day (Fig. 1). Although there was no effect on phosphate absorption rate on the day of 1,25-DHCC addition, there was a significant increase in its absorption on the following day, in line with the time course observed with the other metabolites tested. In only

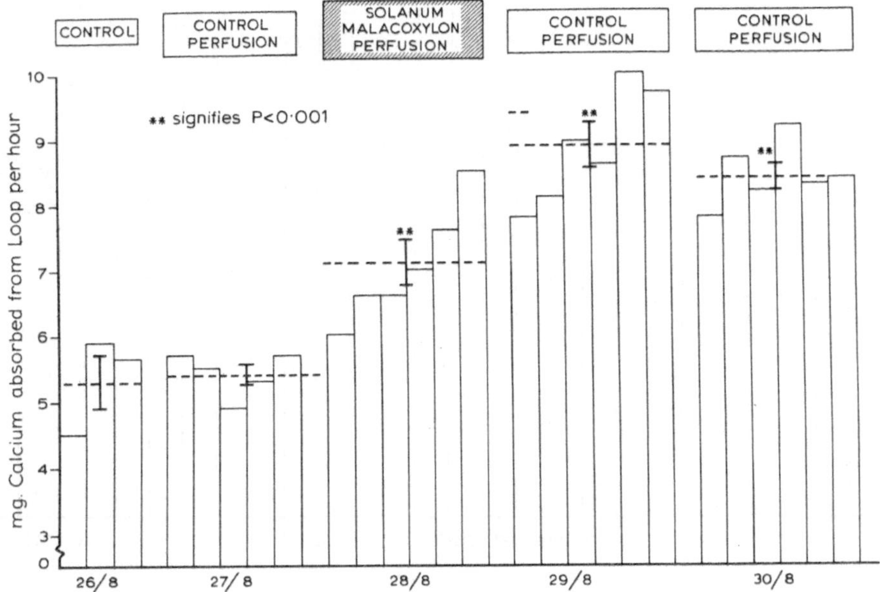

Figure 2. Changes in net calcium absorption from a perfused jejunal Thiry-Vella loop in a conscious pig associated with the use of a perfusion fluid containing a 1% aqueous extract of the dried leaves of Solanum malacoxylon. Each vertical bar represents a perfusion period of one h. The abscissa is in days.

one experiment was there a significant increase in water absorption.

Solanum malacoxylon

In nine experiments with two PTX pigs and four intact pigs, the intraluminal addition for six hours of an aqueous extract of SM leaves, to give a final concentration ranging from 0.5—2.5 per cent, stimulated calcium absorption from a jejunal loop (Fig. 2). In seven experiments the increase became significant during the day of addition and in the remainder it was delayed by one day. In all three experiments in which phosphate absorption rate was measured, a significant increase was found on the day of SM addition and in two of the three experiments it was maximal on this day. In five experiments an increase in water absorption was noted and accompanying rises in absorption of sodium and potassium were also observed.

DISCUSSION

There are several features common to the actions of SM leaves and vitamin D related sterols in this gut preparation. Calcium absorption was stimulated with a similar time course (Figs. 1 and 2), the rapidity of which suggests an initial action too quick to be accountable by the induced synthesis of mucosal calcium-binding protein (3). Often associated with this rapid increase in calcium absorption was an increase in the absorption of water, sodium and potassium. It seems possible that the rapid phase of the increased absorption of calcium may represent intercellular cation absorption, caused by the effect of the vitamin D analogues on the junctions between the intestinal epithelial cells rendering them "leaky" (9). The peak stimulatory effects on both calcium and phosphate absorption were usually observed on the day following the addition of the vitamin D analogue which would be in accord with an increase in intestinal calcium-binding protein production. The similarity in the timing of the response of calcium absorption to the active priciple in the SM leaves adds support to the evidence that this principle resembles 1,25-DHCC. However, there was one point of difference. Whereas stimulation of intestinal phosphate absorption by added vitamin D analogues was maximal on the next day and was seldom significant on the day of addition, the SM extract produced its effect on the same day in all three experiments.

There was no consistent effect on magnesium absorption caused by the addition of SM extract or any of the vitamin D analogues tested.

REFERENCES

1. Corradino, R.A.: Embryonic chick intestine in organ culture: response to vitamin D_3 and its metabolites. *Science* **179**, 402–405 (1973)
2. Corradino, R.A. & Wasserman, R.H.: 1,25-dihydroxycholecalciferol-like activity of Solanum malacoxylon extract on calcium transport. *Nature (Lond.)* **252**, 716–718 (1974)
3. Ebel, J.G., Taylor, A.N. & Wasserman, R.H.: Vitamin D-induced calcium-binding protein of intestinal mucosa. *Amer. J. clin. Nutr.* **22**, 431–436 (1969)
4. Holick, M.F., Semmler, E.J., Schnoes, H.K. & DeLuca, H.F.: 1α-hydroxy derivatives of vitamin D_3: a highly potent analogue of 1α,25-dihydroxy vitamin D_3. *Science* **180**, 190–191 (1973)
5. Mautalen, C.A.: Mechanism of action of Solanum malacoxylon upon calcium and phosphate metabolism in the rabbit. *Endocrinology* **90**, 563–567 (1972)
6. Sansom, B.F., Vagg, M.J. & Döbereiner, J.: The effects of Solanum malacoxylon on calcium metabolism in cattle. *Res. Vet. Sci.* **12**, 604–605 (1971)

7. Swaminathan, R., Ker, J. & Care, A.D.: Calcitonin and intestinal calcium absorption. *J. Endocr.* **61**, 83—94 (1974)
8. Wasserman, R.H.: Calcium absorption and calcium-binding protein synthesis: Solanum malacoxylon reverses strontium inhibition. *Science* **183**, 1092—1094 (1974)
9. Wasserman, R.H., Taylor, A.N. & Lippiello, L.: Effect of vitamin D_3 on lanthanum (La^{3+}) translocation: evidence for a shunt path. *Fed. Proc.* **32**, 3931 Abstr. (1973)
10. Worker, N.A. & Carrillo, B.J.: "Enteque Seco", calcification and wasting in grazing animals in the Argentine. *Nature (Lond.)* **215**, 72—74 (1967)

Current Concepts of the Chemical Conformation, Metabolism, and Interaction of the Steroid, Vitamin D, with the Endocrine System for Calcium Homeostasis

A.W. NORMAN, W.H. OKAMURA, E.J. FRIEDLANDER, HELEN L.HENRY, R.L. JOHNSON, M.N. MITRA, D.A. PROSCAL & W. WECKSLER

In recent years, there have been intensive efforts by several laboratories to elucidate various aspects of the parameters involved in the regulation of calcium and phosphorus metabolism. Three of the most important of these biological regulators are the seco steroid, calciferol (vitamin D), parathyroid hormone, and calcitonin. With each of these regulators, outstanding developments and advances have been made in recent years. It is the purpose of this article to briefly outline some of the developments that have occurred specifically with regard to our understanding of the shape or chemical conformation of vitamin D, its subsequent metabolism by the kidney in a regulated fashion, the interaction of the product steroid hormone (1,25-dihydroxycholecalciferol) with both the intestinal mucosal system where it mediates calcium absorption, and with the parathyroid gland. We have reviewed this subject in depth previously (2, 6).

A major advance has been made within the past year in terms of our understanding of the shape or conformation of vitamin D, its metabolites, and related analogs when they are in solution. In Fig. 1 is shown a summary of some of these important advances. Solution conformations of the A and seco B rings of vitamin D_3, 1α,25-dihydroxy-vitamin D_3, 1α-hydroxy-vitamin D_3, octa-nor-vitamin D_3 (a molecule with no side chain), and dihydrotachysterol have been determined via high resolution PMR spectroscopy (8, 9). The A ring of these steroids was found to be dynamically equilibrated between two chair conformers. A major conse-quence of this new conceptualization of the vitamin is the fact that

Departments of Biochemistry & Chemistry, University of California, Riverside, California.

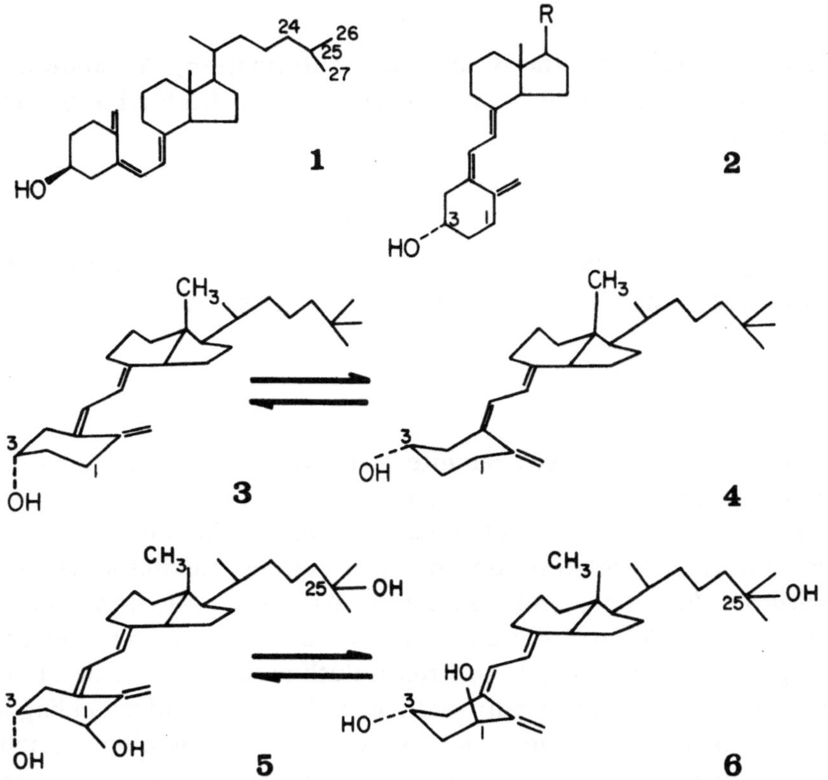

Figure 1. Evolution of conformational representations of vitamin D. Representation one resulted from the original chemical structural determination of vitamin D carried out in the 1930's. The first X-ray crystallographic analysis indicated the presence of a single A-ring chair conformation as shown in #3, but this normally simplified to that shown above in #2. Our recent report emphasizes that in solution there is a rapid equilibration between the two A-ring chair conformations as shown in 3 and 4. In structures 3 and 4, the 3-hydroxyl in both instances is geometrically β. Structures 5 and 6 are a similar pair of rapidly equilibrating conformers of $1\alpha,25\text{-(OH)-D}_3$.

different chair conformations produce different orientations of substituent groups in the A-ring. In one chair conformer for $1\alpha,25$-dihydroxyvitamin D, the 1α-hydroxyl is oriented equatorially, while in the other conformer this same hydroxyl is oriented axially (Fig. 1). We have proposed that the 1α-hydroxyl of $1\alpha,25$-dihydroxy-vitamin D_3 or its geometric equivalent[1] in analogs must occupy the equatorial, as opposed to the axial orientation for optimization of the steroid's biological activity (7). This proposal was based on an evaluation of existing published data concerning structure-function relationships and the steroid hormone model of action for the biologically active form of vitamin D.

Thus, while it is definitely a fact that all vitamin D-seco steroids undergo the dynamic equilibration between the two conformers, it remains to the future to delineate the implications of this phenomenon of vitamin D related molecules to their detailed biochemical mode of action. The concept of conformational optimization of biological activity of vitamin D-like compounds, is clearly one that bears further examination in the future.

The elucidation of the factors which govern the production by the kidney of $1\alpha,25$-dihydroxyvitamin D, and its subsequent interaction distally in the target intestine and bone, is critical to our understanding of calcium and phosphorus homeostasis. The mitochondrial enzyme of the kidney cortex which produces $1\alpha,25\text{-(OH)}_2\text{-D}_3$, the 25-OH-D_3-1-hydroxy-lase, is a classic cytochrome P–450 linked steroid hydroxylation system. The enzyme is sensitive to the presence of carbon monoxide and the wave length of light most effective in reversing this CO inhibition id 450 nm. The activity of the 1-hydroxylase is, in our view, altered by two basically different mechanisms: (a) the short term output (minute-to-minute) of $1\alpha,25\text{-(OH)}_2\text{-D}_3$ is affected by the ionic environment of the mitochondria, e.g. the presence of Ca^{2+}, 10^{-5} M, or phosphate, 10^{-3} M result in a 50% inhibition of the enzyme activity; the long term output (hour-to-hour) of $1\alpha,25\text{-(OH)}_2\text{-D}_3$, as determined by the steady state level of the 1-hydroxylase, is modulated directly by a parathyroid hormone status, and inversely by dietary vitamin D (4). These two modulators exert their effects by changing the rate of biosynthesis of this key enzyme. Together these two mechanisms provide an efficient means by which the organism can modulate the output of this highly potent steroid hormone.

The mechanism by which $1\alpha,25\text{-(OH)}_2\text{-D}_3$ stimulates intestinal absorption has been shown to involve a two step interaction of this seco-steroid with a specific cytosol-chromatin receptor system in this target organ (6). A key aspect of developing a molecular description of this system is to relate structural parameters of the effector steroid with

---

[1]) There are certain analogs of 1,25-dihydroxyvitamin D which are referred to as "pseudo-1" compounds. A pseudo-1α-hydroxyl group is defined as a hydroxyl that occupies geometrically a position equivalent to that of 1α-hydroxyl of $1\alpha,25$-$(OH)_2\text{-D}_3$. The α and β designation of positions on the A-ring of vitamin D-like seco-steroids is determined by reference to the original orientation on the A-ring in 7-dehydrocholesterol. Typical representatives of the class of "pseudo 1" compounds are dihydrotachysterol (DHT_3) and 5,6-trans-vitamin D_3.

its receptor system. A competitive binding assay for $1\alpha,25\text{-}(OH)_2\text{-}D_3$ has been developed using the intestinal cytosol-chromatin system as a receptor. A series of structural analogs for $1\alpha,25\text{-}(OH)_2\text{-}D_3$ [e.g. in the 5,6-cis series, $1\alpha\text{-}(OH)\text{-}D_3$, 3-deoxy-$1\alpha$, $25\text{-}(OH)_2\text{-}D_3$, 3-deoxy-$1\alpha$-OH-$D_3$, 25-OH-$D_3$, 24(R and S), $25\text{-}(OH)_2\text{-}D_3$, D_3 and in the 5,6-trans series, 25-OH-DHT$_3$, 25-OH-5,6-t-D_3, 5,6-t-D_3, DHV-IV] have all been evaluated for their relative ability to compete with $1\alpha,25\text{-}(OH)_2\text{-}D_3$. Results from these studies have defined the minimum structural requirements for interaction of a steroid in this receptor system. In this respect, the most biologically active analog of $1\alpha,25$-dihydroxy-vitamin D_3, which is also the most effective competitor in the steroid receptor assay, is the compound 3-deoxy-$1\alpha,25\text{-}(OH)_2\text{-}D_3$. It is also possible to utilize this receptor system as an assay to evaluate the circulating concentration of $1\alpha,25\text{-}(OH)_2\text{-}D_3$ in the plasma of normal man or in certain disease states.

A most interesting and exciting development in the vitamin D field concerns the demonstration that the parathyroid gland has a certain affinity to accumulate $1\alpha,25\text{-}(OH)_2\text{-}D_3$ (8). When ^{3}H-1,25-$(OH)_2$-D_3 was administered to vitamin D-deficient chicks, within two h the parathyroid glands were found to accumulate the steroid to a concentration four times that present in the blood, and equivalent to levels observed in the target intestine. If the animals were pre-treated with non-radioactive $1\alpha,25$-$(OH)_2$-D_3, then this competed or inhibited the localization of the subsequently administered radioactive $1\alpha,25\text{-}(OH)_2\text{-}D_3$, whereas the prior administration of $24,25\text{-}(OH)_2\text{-}D_3$ had no effect on this localization. It remains to be determined what the specific function of the $1\alpha,25\text{-}(OH)_2$-D_3 is in the parathyroid gland. One possibility is that it induces the synthesis of a calcium binding protein analogous to that produced in the intestinal response to the steroid; alternatively, it may modulate the secretion of parathyroid hormone. Unfortunately, at the present time, there are no assays capable of detecting immunoreactive PTH in avian species. It seems likely, however, that an important component of the complex endocrine regulatory system related to calcium and phosphorus homeostasis is the presence of the steroid, $1\alpha,25\text{-}(OH)_2\text{-}D_3$ in the parathyroid gland.

One point that should be obvious in the complex set of interactions and interrelationships discussed above, is that there is likely a multiplicity of possible sites and steps where the calciferol endocrine system may be disrupted. Thus, it is not surprising that there is quite an array of disease states which are known to be related in some fashion to vitamin D. These include chronic renal disease, hypophosphatemic VDRR, vitamin D-dependent rickets, sarcoidosis, hypoparathyroidism, drug induced metabolism of vitamin D as in the case of chronic administration of

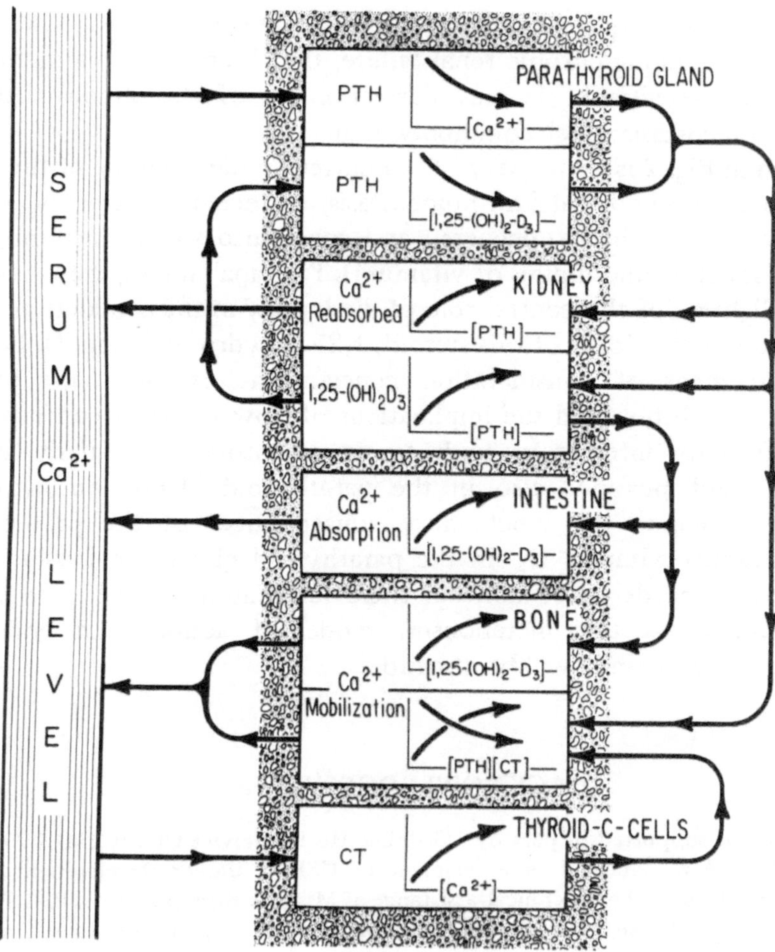

Figure 2. Calcium homeostasis-vintage 1975.

anti-convulsant drugs, and a host of bone diseases, e.g. osteomalacia, rickets, osteitis fibrosa cystica, renal osteodystrophy, and osteoporosis. Given the primary thesis of this article, that 1,25-dihydroxy-vitamin D_3 is the biologically active form of vitamin D, it is quite apparent why the kidney plays such an important and integral role in health and diseases which are known to be related to vitamin D. In a disease state which directly impairs the production of 1,25-dihydroxy by the kidney or which interferes with any of the multiplicity of steps involved in calcium homeostasis, an abnormal feedback signal to the kidney may result in an altered production of the hormone. Coburn, Norman, and coworkers have pioneered in an evaluation of the effects of both short term and chronic administration of $1\alpha,25$-dihydroxyvitamin D_3 in many of these disease states (1, 2). It is quite apparent that in several of these disease states, principally including chronic renal failure, that long term treatment with $1\alpha,25$-dihydroxyvitamin D may correct many of the abnormalities of calcium homeostasis which commonly occur.

Shown in Fig. 2 is a summary of our current understanding of the many complicated facets of calcium homeostasis. In recent years many striking relationships have been uncovered and come into focus concerning the metabolism and functioning of vitamin D. Principal among these has been the highlighting of the central role of the kidney in the production of the biologically active form of this steroid, 1,25-dihydroxyvitamin D. Equally important areas of consideration concern the precise shape of this molecule in solution and the implications of how shape or conformation may affect its interaction with target receptors, particularly in the intestine, and possibly also in the parathyroid gland. An intriguing question concerns the biochemical consequences of the presence of $1\alpha,25$-dihydroxyvitamin D_3 in the parathyroid gland. Obviously, much further work needs to be done in these fascinating areas before a clear understanding of the metabolism, mode of action, and endocrine regulation of vitamin D will be at hand.

ACKNOWLEDGEMENTS

This work was supported in part by US Public Health Service Grants AM–09012 and AM–14,750. A.W. Norman is a recipient of USPHS Career Development Award 1–KD–AM–13,654. The technical assistance of Ms. J.E. Bishop and Ms. P.A. Roberts and the secretarial assistance of Ms. S. Staton in these ongoing studies is gratefully acknowledged.

REFERENCES

1. Brickman, A.S., Sherrard, D.J., Jowsey, J., Singer, F.R., Baylink, D.R., Maloney, N., Massry, S.G., Norman, A.W. & Coburn, J.W.: 1,25-Dihydroxycholecalciferol. Effect on skeletal lesions and plasma parathyroid hormone levels in uremic osteodystrophy. *Arch. intern. Med.* **134**, 883–888 (1974)
2. Coburn, J.W., Hartenbower, D.L. & Norman, A.W.: Metabolism and action of the hormone, vitamin D: Its relation to diseases of calcium homeostasis. *West. Med.* **121**, 22–24 (1974)
3. Henry, H.L. & Norman, A.W.: Studies on calciferol metabolism IX. Characteristics of the renal 25-hydroxy-vitamin D_3-1-hydroxylase. *J. biol. Chem.* **249**, 7529–7535 (1974)
4. Henry, H.L., Midgett, R.J. & Norman, A.W.: Regulation of 25-hydroxyvitamin D_3-1-hydroxylase, *in vivo*. *J. biol. Chem.* **249**, 7584–7592 (1974)

5. Henry, H.L. & Norman, A.W.: Studies on the mechanism of action of calciferol VII: Localization of 1,25-dihydroxy-vitamin D_3 in chick parathyroid glands. *Biochem. Biophys. Res. Commun.* **62**, 781–788 (1975)
6. Norman, A.W. & Henry, H.: 1,25-dihydroxycholecalciferol-A hormonally active form of vitamin D_3. In: *Recent Progress in Hormone Research.* R.O. Greep (ed.). **30**, 431–480, 1974
7. Okamura, W.H., Norman, A.W. & Wing, R.M.: Vitamin D: Concerning the relationship between molecular topology and biological function. *Proc. nat. Acad. Sci.* (Wash.) **71**, 4194–4197 (1974)
8. Wing, R.M., Okamura, W.H., Pirio, M.R., Sine, S.M. & Norman, A.W.: Vitamin D_3 in Solution: Conformations of vitamin D, $1\alpha,25$-dihydroxyvitamin D_3, and dihydrotachysterol$_3$. *Science* **186**, 939–941 (1974)
9. Wing, R.M., Okamura, W.H., Rego, A., Pirio, M.R. & Norman, A.W.: Studies on vitamin D and its analogs. VII. Solution comformations of vitamin D_3 and $1\alpha,25$-dihydroxyvitamin D_3 by high resolution proton magnetic resonance spectroscopy. *J. Amer. chem. Soc.* **97**, 4980–4985 (1975)

Interaction of Parathyroid Hormone and 25-hydroxycholecalciferol on Renal Handling of Phosphate

A. JELONEK

INTRODUCTION

Since Harrison carried out his fundamental experiments on the anti-phosphaturic renal effect of vitamin D (5) a number of clinical and experimental observations have been made trying to elucidate the effect of vitamin D and/or the interaction with PTH on renal transport of phosphate. Results obtained by various authors have often been controversial (1, 6, 8, 9, 10). The difficulties encountered in interpreting the changes in urinary excretion of phosphate have been related to the following experimental conditions: 1) hypercalcaemic action of vitamin D, 2) experimental animals are volume expanded, 3) different effects of physiological and pharmacological doses of PTH and vitamin D or its metabolites, 4) experimental animals without vitamin D deficiency, 5) "acute" studies.

In the current experiments the majority of these potential difficulties have been obviated. The present study was undertaken to investigate the interaction of PTH and 25HCC on renal handling of phosphate in vitamin D-deficient thyroparathyroidectomized rats, using a prolonged perfusion technique without volume expansion.

MATERIALS AND METHODS

Young Holtzman female rats weighing ca 60 g were maintained 3—4 weeks on a special vitamin D-depleted test diet containing 0.400% calcium and 0.426% phosphorus. Objective criteria for vitamin D-deficiency have been established (12, 13). Groups of rats in which the mean serum calcium level did not fall below 6.5 mg% were excluded. Thyroparathyreoidectomy was

Institute of Pediatrics, Medical Academy, Kraków.

performed by electrocauthery under light anesthesia. The urinary bladder was exposed and a polyethylene catheter (No. 240) was placed through an incision for urine collections. The animals were then placed in a restraining cage similar to that described by Cotlove (2). Perfusion solution containing 10 mM calcium chloride, 20 mM sodium chloride, 2.5 mM potassium chloride, 5 mM magnesium chloride, 2.5 g inulin and 40 g glucose per 1 liter of distilled water was administered through a tail vein at the rate of 1.24 ml/h using a Harvard pump. Urine samples were collected at 2 h intervals with a fraction collector. Following a 16 h equilibration period either 25HCC or PTH or both were administered at the rate of 5 u/h respectively. Control animals received the 25HCC vehicle (propylene glycol). In each group 6–9 animals were sacrificed after 16 h of perfusion (equilibration period) and then after 4, 14 and 32 h beginning from the onset of infusion of the test substance. Blood samples were taken from the abdominal aorta immediately after the perfusion. Both blood serum and urine phosphate were simultaneously analyzed using a Technicon Autoanalyzer (7). Blood serum and urine inulin were determined according to Davidson and Sackner (3). The animals were grouped in the following way: *Control group*: animals receiving the vehicle (propylene glycol) at the rate of 0.031 ml/h for 4 h (8 rats), 14 h (7 rats), and 32 h (6 rats). *PTH group*: animals receiving PTH* at the rate of 5 u/h for 4 h (6 rats), 14 h (7 rats), and 32 h (6 rats). *25HCC group*: animals receiving 25HCC** at the rate of 5 u/h for 4 h (8 rats), 14 h (7 rats), and 32 h (6 rats). *PTH + 25HCC group*: animals receiving simultaneously PTH at the rate of 5 u/h and 25HCC at the rate of 5 u/h for 4 h (7 rats), 14 h (8 rats), and 32 h (7 rats).

RESULTS

Table I summarizes the effects of intravenous infusion of PTH and 25HCC (separately and simultaneously) on the fractional excretion of phosphate (C_p/C_{In}), serum phosphorus and serum calcium.

After the 4 h intravenous infusion of 25HCC separately or simultaneously with PTH, C_p/C_{In} was markedly lowered (Fig. 1) in comparison

---

Abbreviations used in this paper: PTH, parathyroid hormone; 25HCC, 25-hydroxycholecalciferol; 1,25DHCC, 1,25-dihydroxycholecalciferol; S_P, serum phosphorus; S_{Ca}, serum calcium; C_{In}, clearance of inulin; C_P, clearance of phosphate; TPTX, thyroparathyreoidectomy.

*   parathyroid hormone highly purified, supplied by Wilson Lab. Division, Chicago, Ill.

** 25 hydroxycholecalciferol hydrate. Research Lab. of the Upjohn Company, Kalamazoo, Michigan.

to the control group (respectively: $p < 0.005$, $p < 0.001$) whereas in the PTH group an increase was noted ($p < 0.005$).

After 14 h of infusion there were no significant differences in C_p/C_{In} between the 25HCC group, PTH + 25HCC group and the corresponding control group. At that time C_p/C_{In} in the PTH group was significantly higher ($p < 0.05$).

After the 32 h infusion of 25HCC no differences were noted in C_p/C_{In} in comparison to the control group, whereas the simultaneous infusion of PTH and 25HCC for that period of time resulted in a pronounced decline in C_P/C_{In} compared to the control group ($p < 0.02$). At the same time, in the PTH group C_P/C_{In} was significantly higher ($p < 0.001$). S_P (Table I) after the 4 h infusion appeared to be significantly higher only in the 25HCC group ($p < 0.02$). After 14 h of infusion there were no significant

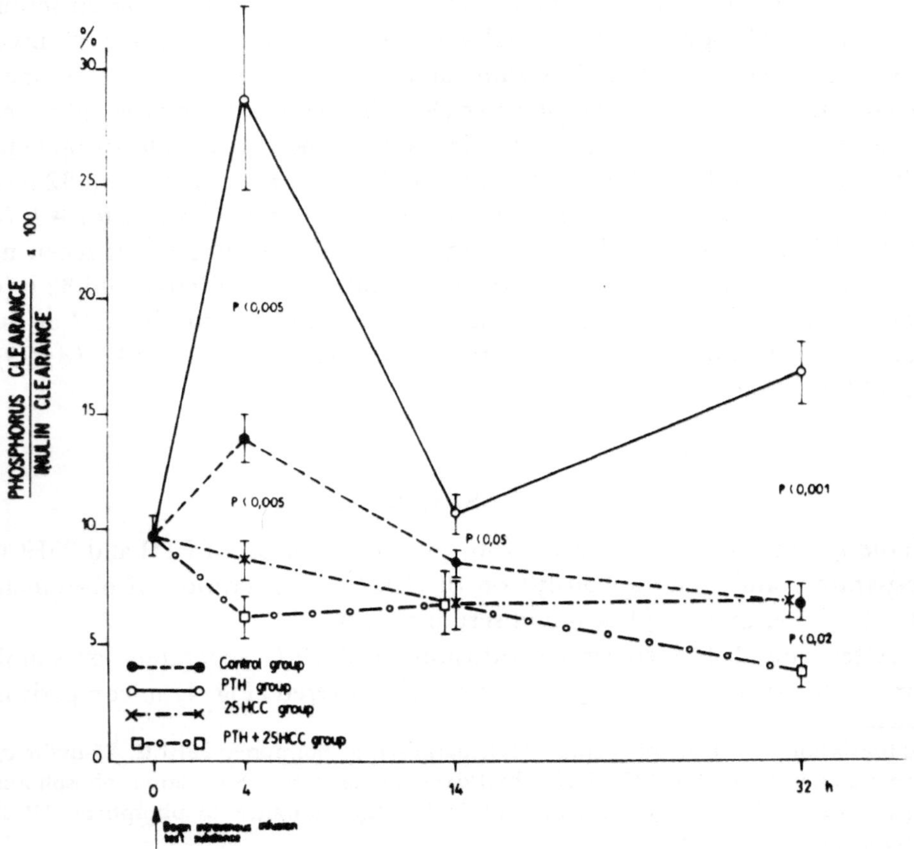

Figure 1. Effects of PTH, 5 u/h (open circles), 25HCC, 5 u/h (crosses) and both (open squares) on the fractional excretion of phosphate, compared to the control animals (closed circles).

differences in S_P between the experimental groups and the corresponding control group.

After the 32 h infusion of the test substances S_P was higher in both the 25HCC and PTH + 25HCC groups as compared with the control group (respectively: $p < 0.001$, $p < 0.05$).

After a 4 h infusion of the test substances (Table I) S_{Ca} was higher only in the PTH + 25HCC group as compared with the control group ($p < 0.05$). After a 14 h infusion no significant differences in S_{Ca} were noted in the experimental groups as compared with the control group. After 32 h infusion S_{Ca} was lowered only in the PTH group ($p < 0.005$).

There were no essential differences noted in inulin clearance (Table I) except for the 4 h infusion period (PTH group).

Group	Time[a]	n	C_{in}	$C_P/C_{in} \times 100$	S_P	S_{Ca}
	h		µl/min. $\bar{x} \pm SE$	%	mg%	mg%
	0[xx]	8	1.03±0.14	11.80±0.90	9.34±0.39	6.06±0.37
Control	4	8	1.12±0.08	13.99±1.17	7.77±0.36	6.26±0.25
	14	7	0.87 0.08	8.67 0.66	8.71 0.50	7.47 0.37
	32	6	0.91 0.12	6.85 0.84	9.30 0.59	10.22 0.33
PTH (Su/h)	4	6	2.21±0.30	28.70±4.06	7.33±0.19	6.28±0.46
	14	7	0.95 0.07	10.73 0.96	7.84 0.52	7.18 0.42
	32	6	1.11 0.21	16.87 1.36	8.20 0.53	8.48 0.28
25HCC (Su/h)	4	8	0.96±0.10	8.76±0.81	8.97±0.29	5.76±0.21
	14	7	0.88 0.13	6.82 1.24	9.21 0.33	8.30 0.40
	32	6	0.95 0.11	6.90 0.89	12.00 0.17	10.50 0.14
PTH+25HCC (Su+Su/h)	4	7	1.09±0.08	6.19±0.91	8.02±0.31	7.84±0.59
	14	8	0.90 0.07	6.86 1.46	9.75 0.29	8.43 0.41
	32	7	1.04 0.15	3.87 0.64	12.10 0.91	9.77 0.15

Table I. The Effects of PTH and 25HCC, Separately and Simultaneously on Fractional Excretion of Phosphate, Serum Phosphorus, and Calcium, Compared to the Control Animals.

DISCUSSION

The results of the present studies indicate that 25HCC in the applied dose slightly increases fractional tubular phosphate reabsorption. The presence of PTH definitely enhances and lengthens the renal effect of 25HCC. The observed changes in fractional excretion of phosphate were not accompanied by essential alterations in glomerular filtration rate. While attempting to explain the mechanisms of the antiphosphaturic effect of 25HCC and its augmentation in the presence of PTH the following hypothesis is set forth:

In the first period of infusion of 25HCC as well as in combination with PTH the occurrence of the antiphosphaturic effect was parallel to the increase of calcaemia. It is possible that the increase of Ca^{++} shift to the tubular cells is a sufficient factor enhancing renal phosphate conservation. However, the mechanism would have a limited maximum efficiency. The presence of PTH might intensify the synthesis of 1,25DHCC (4) and enhance in this way bone calcium mobilization (11). It does not explain, however, why PTH has augmented and prolonged the antiphosphaturic effect of 25HCC. It is possible that PTH in combination with 25HCC, beside its calcaemic effect, stimulated more efficient mechanisms saving the renal phosphate excretion.

ACKNOWLEDGEMENT

This study was carried out at the University of Pennsylvania Medical Service, VA Hospital, Philadelphia; and was included in the thesis of the author for the partial fulfillment of the requirement for a Ph.D. degree in Pediatrics at the Medical Academy in Krakow. The author is grateful to Professor J.B. Puschett for his indispensable guidance and help.

REFERENCES

1. Arnaud, C., Rasmussen, H. & Anast, C.: Further studies on the interrelationship between parathyroid hormone and vitamin D. *J. clin. Invest.* **45**, 1955–1964 (1966)

2. Cotlove, E.: Simple tail vein infusion method for renal clearance measurements in the rat. *J. Appl. Physiol.* **16**, 764–768 (1961)

3. Davidson, W.D. & Sackner, M.A.: Simplification of the anthrone method for the determination of inulin in clearance studies. *J. Lab. clin. Med.* **62**, 351–356 (1963)

4. Garabedian, M., Holick, M.F., DeLuca, H.F. & Boyle, I.T.: Control of 25-hydroxy-cholecalciferol metabolism by parathyroid glands. *Proc. nat. Acad. Sci. (Wash.)* **69**, 1973–1981 (1972)

5. Harrison, H.E. & Harrison, H.C.: The renal excretion of inorganic phosphate in relation to the action of vitamin D and parathyroid hormone. *J. clin. Invest.* **20**, 47—55 (1941)

6. Harrison, H.E. & Harrison, H.C.: The interaction of vitamin D and parathyroid hormone on calcium, phosphorus, and magnesium homeostasis. *Metabolism* **13**, 952—963 (1964)

7. Kessler, G. & Wolfman, M.: An automated procedure for simultaneous determination of calcium and phosphorus in serum and urine. *Clin. Chem.* **8**, 429—436 (1962)

8. Popovtzer, M.M., Robinette, J.B., DeLuca, H.F. & Holick, M.F.: The effect of 25-hydroxycholecalciferol on renal handling of phosphorus. *J. clin. Invest.* **53**, 913—921 (1974)

9. Puschett, J.B., Moranz, J. & Kurnick, W.S.: Evidence for a direct action of cholecalciferol and 25-hydroxycholecalciferol on the renal transport of phosphate, sodium and calcium. *J. clin. Invest.* **51**, 373—385 (1972)

10. Puschett, J.B., Beck Jr., W.S., Jelonek, A. & Fernandez, P.C.: Study of the renal tubular interactions of thyrocalcitonin, cyclic adenosine $3',5'$-monophosphate, 25-hydroxycholecalciferol and calcium ion. *J. clin. Invest.* **53**, 756—767 (1974)

11. Raisz, L.G., Trummel, C.L. & Simmons, H.: Induction of bone resorption in tissue culture: Prolonged response after brief exposure to parathyroid hormone or 25-hydroxycholecalciferol. *Endocrinology* **90**, 744—752 (1972)

12. Rasmussen, H., DeLuca, H., Arnaud, C., Hawker, C. & Stedingk, M.: The relationship between vitamin D and parathyroid hormone. *J. clin. Invest.* **42**, 1940—1946 (1963)

13. Steenbock, H. & Herting, D.C.: Vitamin D and growth. *J. Nutr.* **57**, 449—467 (1955)

Calcium Deficiency Osteoporosis and the Role of the Parathyroids for the Adaptation to a Low Calcium Intake

S.-E. LARSSON, O. AHLGREN & R. LORENTZON

Previous studies of the adaptation to a low Ca intake in one-year-old rats have demonstrated the occurrence of osteoporosis (2) and parathyroid hyperplasia (4). The conversion of vitamin D into 1,25-dihydroxycholecalciferol (1,25-DHCC) is essential for the adaptation by increasing the intestinal Ca absorption. The regulation of this conversion is not clear but the parathyroids have been considered as a major regulator of Ca absorption by influencing the synthesis of 1,25-DHCC in response to Ca intake and plasma Ca level. However, earlier studies designed to answer the question to what extent endogenous parathyroid hormone has an influence on Ca absorption have not completely resolved this problem. Recently, the plasma level of inorganic P has also been ascribed a regulatory role for the production of 1,25-DHCC besides the plasma Ca level and parathyroid hormone.

The present investigation was performed to examine the effects of selective parathyroidectomy (PTX) upon the factors involved during the adaptation to a low Ca intake; *i.e.* intestinal absorption and urinary excretion of Ca and inorganic P, the mobilization of minerals from the skeleton and the synthesis of 1,25-DHCC.

MATERIAL

The effects of PTX upon bone and the metabolism of Ca and P were studied in a total of 137 one-year-old male Sprague-Dawley rats with a body weight of 515—645 g, 95 of which underwent PTX. Intact and PTX animals were then kept for 2 and 6 months on a normal (1,2% Ca, 0.70% P) and a low (0.04% Ca, 0.70% P) Ca diet containing 600 I.U. of vitamin D_3 per 100 g and deionized water *ad lib.* (3).

Department of Orthopaedic Surgery, University of Umeå, S—901 85, Umeå.

The effect of PTX upon the metabolism of vitamin D was studied in 50 one-year-old rats which were kept on the normal calcium diet with no vitamin D added for 1½ month. Then, 33 animals underwent PTX (3) and intact and PTX animals were given the vitamin D-deficient normal and low-calcium diet for 4½ months.

METHODS

Standardized microradiographs of the distal femur metaphysis and the femoral mid-shaft were studied quantimetrically (3). Measurements of calcium, magnesium and inorganic phosphate concentrations in plasma, faeces and urine were performed as described previously (1). Calcium metabolism was further studied after administration of 20 μCi ^{45}Ca (Amersham) by stomach tube (1). The metabolism of vitamin D was studied by i.v. injection of 400 pmoles/100 g body weight of ^{3}H-25-hydroxycholecalciferol (Amersham). Subsequently, chloroform extracts of plasma, liver, kidneys and small intestine mucosa were chromatographed on Sephadex LH20 columns which were eluted with 35% Skellysolve B in 65% chloroform.

RESULTS

The PTX animals responded with reduced plasma Ca (4.1 mEq./l or below) in 86 per cent at a normal and in 100 per cent at a low-Ca intake. At a normal Ca intake, 17 per cent normalized their plasma Ca within 2 months and another 40 per cent within 2 to 6 months of observation. At a low-Ca intake, only 10 per cent showed a normalization after the long-term period of observation. This was brought about by increased mobilization of skeletal Ca with resulting osteoporosis similar to that found in Ca-deficient intact animals (Fig. 1).

PTX animals with persistently reduced plasma Ca showed a significantly increased bone mass on the normal level of dietary Ca (Fig. 1) which was paradoxycal in view of their inability of adaptation. On the low level of dietary Ca a normal bone resorptive activity was maintained despite PTX, possibly through the action of increased levels of 1,25-DHCC provoked by the profoundly reduced plasma Ca. This was, however, insufficient for adaptation and no osteoporosis developed. For adaptation to a reduced Ca intake, skeletal Ca reserves had to become mobilized through the action of the parathyroids with resulting osteoporosis. The described changes were progressive in character. The factor of bone growth was not involved in these old rats.

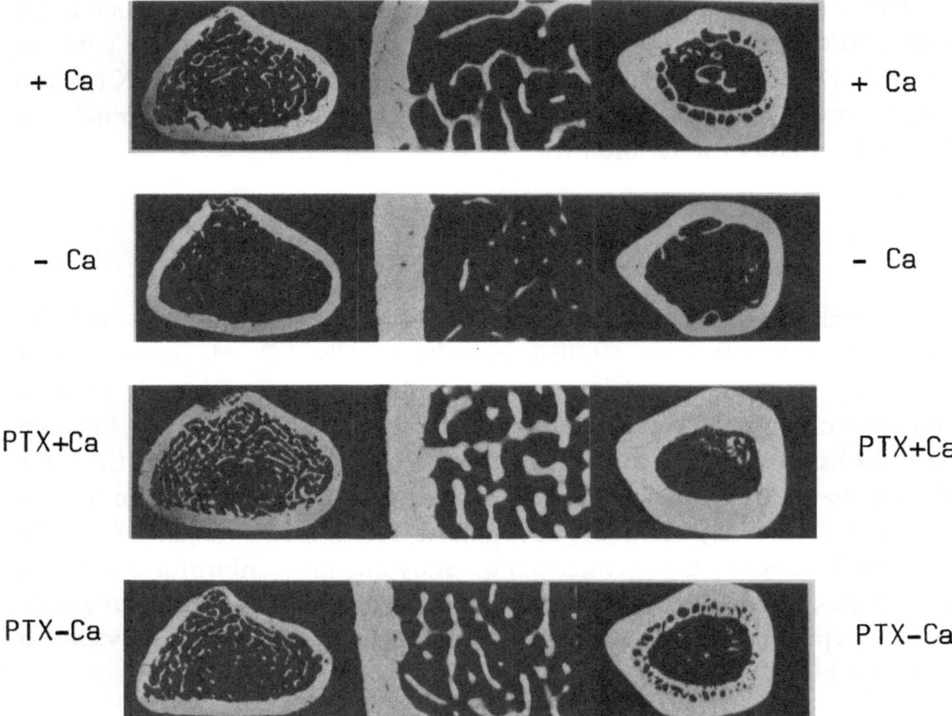

Figure 1. Standardized microradiographs of undecalcified sections of the distal femur metaphysis at low and higher magnification (to the left) and of the mid-shaft of the femur (to the right). Note the evident osteoporosis in the intact Ca-deficient animals which are absent in the Ca-deficient PTX animals. The PTX animals kept on the normal Ca intake show a paradoxical increase in bone mass despite their consistently reduced plasma Ca level.

Data regarding the metabolism of Ca, Mg and inorganic P showed that adaptation, viz. normalization of plasma Ca, was accomplished by the action of the parathyroids, and in the adapting PTX animals by renewed parathyroid activity, with mobilization of skeletal Ca which gave rise to osteoporosis. Intestinal net absorption of Ca showed no significant difference between normocalcaemic normophosphatemic intact animals and hypocalcaemic hyperphosphatemic parathyroidectomized animals at the respective level of dietary Ca. Thus, PTX animals with persistently reduced plasma Ca showed a normal adaptory increase in intestinal Ca absorption upon chronically restricted Ca intake indicating that the metabolism of vitamin D is not regulated by the parathyroids.

With regard to the metabolism of vitamin D the following percentage distribution was obtained for the eluted radioactivity corresponding to the various vitamin D metabolites of the kidneys:

Group	Peak I	Peak II	25-HCC	24,-25-DHCC	1,25-DHCC
+Ca/-D	1.6	1.7	84.5	-	12.2
-Ca/-D	3.3	2.1	74.4	-	20.3
PTX + Ca/-D	3.4	1.1	74.0	-	21.5
PTX - Ca/-D	1.3	0.9	77.2	-	20.7
PTX_N* + Ca/-D	2.6	1.1	77.9	-	18.5

*) PTX_N: PTX animals showing a normalization of their plasma calcium level after an initial reduction.

Thus, the values for 1,25-DHCC showed an increase from 12 per cent in the intact normal Ca animals to 20 per cent in the intact Ca-deficient animals, both groups of animals showing normocalcaemia or slight hypocalcaemia and normophosphatemia. No further increase was obtained in the hypocalcaemic, hyperphosphatemic PTX animals kept on the normal Ca intake as well as in the profoundly hypocalcaemic and hyperphosphatemic PTX animals with the low Ca intake. The production of 1,25-DHCC was still increased in the group of PTX animals which were normalizing their levels of plasma Ca and inorganic P.

DISCUSSION

The results of the present investigation clearly indicate that the regulation of intestinal Ca absorption and the synthesis of 1,25-DHCC are not primarily regulated by the parathyroids. Both adaptory mechanisms were stimulated by the chronically restricted dietary Ca intake even in the absence of the parathyroids. A slightly subnormal or significantly reduced plasma Ca level was the primary stimulator of the synthesis of 1,25-DHCC, as strongly suggested by our data. The level of plasma inorganic P had no primary influence. However, it is possible that the hyperphosphatemia of the PTX animals might have some inhibitory effect upon the synthesis of 1,25-DHCC by the kidneys, stimulated primarily by the hypocalcaemia, thus preventing an optimal conversion of 25-HCC into 1,25-DHCC. In fact, we found a maximum level for the peak corresponding to 1,25-DHCC above which there was no further production of this active metabolite of vitamin D despite a profoundly reduced plasma Ca level. In comparison to our previous studies of the vitamin D metabolism in intact animals of the same age and strain fed the same diet but with

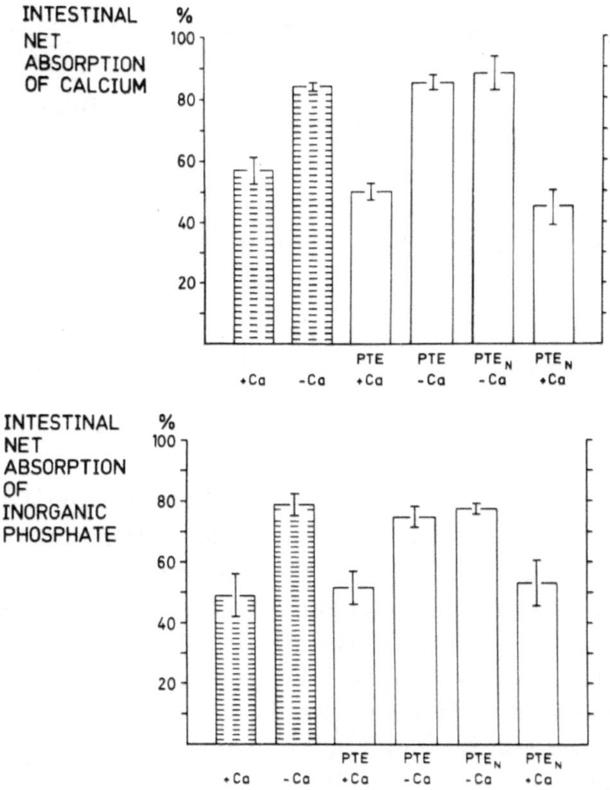

Figure 2. The intestinal net absorption of Ca and inorganic P in intact and PTX animals at normal and low Ca intake. Note, that there is no disturbance of the increase in intestinal Ca absorption adaptory to the reduced Ca intake after PTX. The absorption of inorganic P follows closely that of Ca.

vitamin D added, there was a 2—3 fold increase in the synthesis of 1,25-DHCC induced by vitamin D deficiency alone and a 4—5 fold increase induced by the low Ca, low vitamin D intake in the slightly hypocalcaemic normophosphatemic intact animals. The same level of 1,25-DHCC was found in the profoundly hypocalcaemic and hyperphosphatemic PTX animals and also in those PTX animals which were normalizing their plasma Ca by renewed parathyroid activity.

SUMMARY

Chronically restricted Ca intake in one-year-old rats resulted in osteoporosis in intact animals induced by increased parathyroid activity, and did not occur in PTX animals. Intestinal Ca absorption and the synthesis of the active vitamin D metabolite,

1,25-DHCC, was stimulated primarily by a reduced blood Ca level. The level of inorganic P might have a secondary influence. The parathyroids had no primary regulatory role in the metabolism of vitamin D. A maximum 4—5 fold increase was found in the production of 1,25-DHCC. This maximum level of 1,25-DHCC was not sufficient for the adaptation to a low Ca intake which then necessitated stimulated parathyroid activity.

ACKNOWLEDGEMENTS

This investigation was supported by a grant No. B76—17X—04063—04 from the Swedish Medical Research Council, the Faculty of Medicine at the University of Umeå and King Gustaf the Fifth's Eightieth Birthday Fund.

REFERENCES

1. Ahlgren, O. & Larsson, S.-E.: The role of the parathyroids for the adaptation to a low calcium intake. II. The short-term effect of parathyroidectomy on the adaptation to a low calcium intake in adult rats with special reference to calcium metabolism. *Acta path. microbiol. scand. Sect. A* 83, 13—24 (1975)
2. Larsson, S.-E.: On the development of osteoporosis. Experimental studies in the adult rat. *Acta orthop. scand. Suppl. No. 120* (1969)
3. Larsson, S.-E. & Ahlgren, O.: The role of the parathyrosis for the adaptation to a low calcium intake. I. The short-term effect of parathyroidectomy on the adaptation to a low calcium intake in adult rats with special reference to plasma calcium, bone tissue and adrenal glands. *Acta path. microbiol. scand. Sect. A* 83, 1—12 (1975)
4. Sevastikoglou, J.A. & Larsson, S.-E.: Osteoporosis and parathyroid glands. I. The effect of prolonged calcium deficiency on the parathyroids of the adult rat. *Clin. orthop. scand.* 85, 163—170 (1972)

Characteristics of the Vitamin D Binding Protein in Different Species

R. BOUILLON, P. VAN KERKHOVE & P. DE MOOR

Since 25-hydroxyvitamin D_3 (25-OH-D_3) is bound to serum and cytosol proteins, its metabolism could be largely dependent on the characteristics of these binding proteins (9, 7). Moreover, as these binding proteins are used for the measurement of 25-OH-D_3 (1, 8, 2) optimalisation of this assay could result from better knowledge of this binding.

1. *The affinity and capacity* of the 25-OH-D_3 binding proteins have been studied by competitive protein binding between very small concentrations (9.10^{-11} mol/l) of ^{3}H-25-OH-D_3 (specific activity 9.5 Ci/mmol, NEN) and variable amounts of stable 25-OH-D_3 (Philips Duphar, The Netherlands). The results were analysed as a Scatchard plot (14) according to Rosenthal (13) by subtracting a (constant) amount of non-specific bound 25-OH-D_3.

a) In six normal *human sera* and five sera from rachitic or osteomalacic patients very high association constants (K_A) at $0°C$ were found (Table I), but to our surprise, the capacity of these sera for 25-OH-D_3 was also very high.

This makes this protein unique among other hormone or vitamin binding proteins because it combines high capacity and high affinity. As the circulating levels of 25-OH-D_3 are about $3.75 \cdot 10^{-8}$ mol/l in our controls (2), the saturation of this circulating binding protein must be extremely low (1.5%). Pathologic variations in capacity or affinity of this 25-OH-D_3 binding protein have not yet been studied in much detail. Vitamin D-deficiency, however, resulted in a 42% increase in the total serum binding capacity.

b) The affinity and capacity of serum and tissue cytosols (105.000 G supernatant of liver, kidney and intestinal mucosa) of *rachitic rats* was

Departement voor Ontwikkelingsbiologie, Laboratorium voor Experimentele Geneeskunde, Katholieke Universiteit Leuven.

Table I

Binding of 25 OHD to Human Serum

0 °C	K_A (l/mol)	Cap (mol/l)
normal human serum (n=6)	$1.9 \pm 0.1 \times 10^{10}$	$2.6 \pm 0.3 \times 10^{-6}$
rachitic human serum (n=5)	$1.5 \pm 0.4 \times 10^{10}$	$3.7 \pm 0.6 \times 10^{-6}$
student t-test : p	$\leqslant 0.05$	< 0.005

Binding of 25 OHD to Rachitic Rat Proteins

		SERUM	CYTOSOL		
			LIVER	KIDNEY	INTESTINE
0 °C	K_A (l/mol) 10^{10}	2.2	2.6	1.7	0.7
	Cap (mol/g prot) 10^{-9}	32.5	3.6	7.4	2.5
37 °C	K_A (l/mol) 10^{9}	4.2	1.9	4.2	2.3
	Cap (mol/g prot) 10^{-9}	29.0	5.0	16.9	2.7

assessed both at $0°C$ and $37°C$ (Table I). Again a very high association constant was found for all preparations with a slightly lower value for the intestinal cytosol. Even at physiologic temperature ($37°C$) the affinity was still very high. The capacity of the rat serum binding protein was similar to that of the human one. In cytosol preparations, however, lower concentrations were present but, nevertheless, these capacities are more than 100 times higher than those of classic cytosol receptors for other steroid hormones (11). Diluted rat or human serum thus proved to be superior to kidney cytosol as binding protein for the assay of 25-OH-D$_3$ since serum has higher binding capacity, is easier to obtain, and has at least a similar affinity for 25-OH-D$_3$.

c) *Chick* serum has an association constant of $6.2 \ 10^{-8}$ l/mol at $0°C$ or more than 30 times lower than rat or human serum, whereas the capacity is comparable ($1.8 \ 10^{-6}$ mol/l). Somewhat higher association constants were found in kidney (3.6 10^9) and liver cytosol (2.8 10^9) of rachitic chicks.

The results of our measurement of association constant in the chick serum is comparable to those obtained by Edelstein (3). But our values for the K_A of cytosol preparations of rachitic rat kidneys, liver and intestine are higher than those obtained by Haddad (7, 10) or Edelstein (4), possibly because they used inappropriately high tracer concentrations. The great sensitivity (about 10—20 pg) of the 25-OH-D$_3$ assay (5, 12) is a further argument for a high K_A (5).

2. *The molecular weight* of the binding proteins for 25-OH-D$_3$ from rachitic rats was measured by gel filtration on a calibrated Sephadex G200 column. A value of 52.000 for serum, 80.800 for intestinal cytosol, 98.000 for liver cytosol and 109.000 for kidney cytosol was found. The same sequence of sedimentation coefficients was found on sucrose gradient ultracentrifugation.

3. The dissociation rate constant of 25-OH-D$_3$ in diluted rat serum was measured at different temperatures (37°, 24°, 12° and 4°C) by precipitation, at 0°C, of free 25-OH-D$_3$ with dextran-coated charcoal in samples taken out of a common stirred mixture. The dissociation was started by adding an excess (at least 500 times) of unlabelled 25-OH-D$_3$ after a preincubation of the labelled and the diluted rat serum for 30

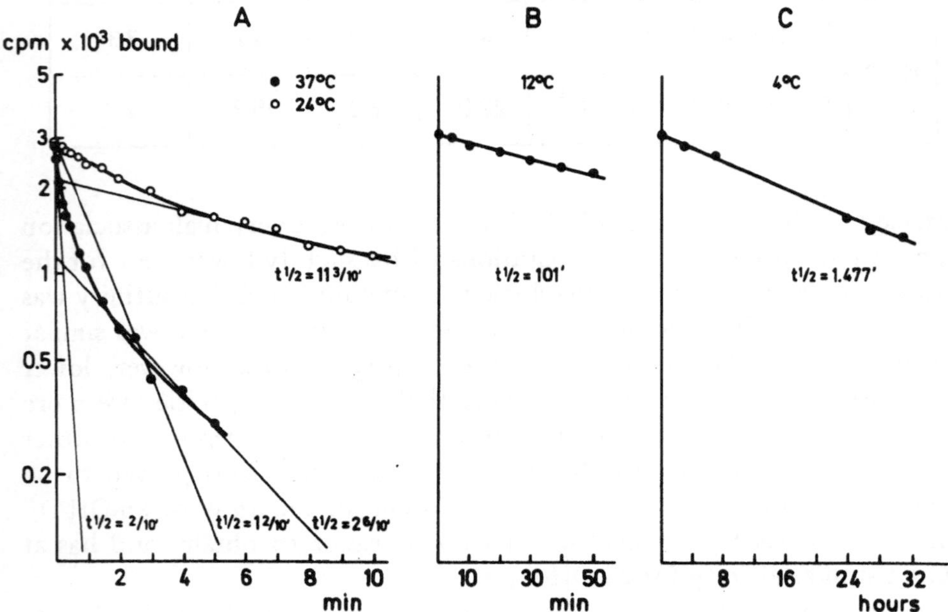

Figur 1. Dissociation of labelled 25-OH-D$_3$ from diluted rat serum at different temperatures. The indicated t 1/2 (dissociation half-time) is given in minutes.

minutes (60 minutes for the experiments at 12° and 4°C). The dissociation was very slow (T½ more than 24 h) at 0°C (Fig. 1), but biphasic and much more rapid at higher temperatures (24° and 37°C). Using the steepest part of the dissociation at these temperatures, an Arrhenius plot was made (r = 0.95) between the log of the dissociation rate constant versus the reciprocal of the absolute temperature. An activation energy of 47.6 kcal/mol was calculated. There remains, however, some reservation about this method of calculation because the dissociation did not represent a simple first order equation.

The association rate was only measured at 0°C with diluted rat serum and labelled 25-OH-D$_3$ (1.34 10^{-9} mol/l). The half-association time was about one minute, with a calculated (6) association rate constant (between 0 to 8 minutes) of 2.6 10^7 1 mol^{-1} min^{-1}. The association constant (K$_A$) calculated from the association rate constant (at 0°C) and dissociation rate constant (4°C) was 2.6 10^{10} l/mol a value comparable to the one obtained by a direct measurement.

In view of this rapid association and slow dissociation at low temperatures short incubation times in a competitive protein binding assay can be used but the labelled and unlabelled steroid must be added simultaneously to the binding protein to avoid unequal non-equilibrium concitions.

SUMMARY

1. The serum binding protein for 25-OH-D$_3$ from the rat and the human, and to a lesser extent also the chich, represents a high affinity-high capacity protein. This also hold true for the cytosol binding proteins from the rat and the chick (liver, kidney, intestine and muscle). These proteins, therefore, probably have a physiological role rather as a buffer than as a regulating system in the overall metabolism of vitamin D.

2. The binding proteins from rat serum and cytosol preparations have apparently different molecular weights.

3. The dissociation of 25-OH-D$_3$ from rat serum is highly temperature dependent, being rapid but biphasic at high temperatures and very slow at 4°C. The association is, however, extremely rapid even at low temperatures.

REFERENCES

1. Belsey, R., DeLuca, H.F. & Potts, J.T. Jr.: Competitive binding assay for vitamin D and 25-OH-Vitamin D. *J. Clin. Endocr.* 33, 554–557 (1971)
2. Bouillon, R., Van Kerkhove, P. & De Moor, P.: The measurement of 25-hydroxy-vitamin D in serum. (1976). Clin. Chem. In Press.

3. Edelstein, S., Lawson, D.E.M. & Kodicek, E.: The transporting proteins of cholecalciferol and 25 hydroxycholecalciferol in serum of chicks and other species. *Biochem. J.* **135**, 417–426 (1973)

4. Edelstein, S.: Vitamin D binding proteins. *Vitam. and Horm.* **32**, 407–428 (1974)

5. Ekins, R.P., Newman, G.B. & O'Riordan, J.L.H.: Theoretical aspects of "saturation" and radioimmunoassay. In: *Radioisotopes in Medicine: In vitro studies.* Hayes, R.L., Goswitz, F.A., Murphy, B.E.P. (eds.). U.S.Atomic Energy Commission, pp. 59–100 (1968)

6. Glasstone, S.: In: *Textbook of Physical Chemistry.* MacMillan, pp. 1054–1057 (1962)

7. Haddad, J.G. & Birge, S.J.: 25-hydroxycholecalciferol: Specific binding by rachitic tissue extracts. *Bioch. Biophys. Res. Commun.* **45**, 829–834 (1971)

8. Haddad, J.G. & Chyu, K.J.: Competitive protein-binding radioassay for 25-hydroxycholecalciferol. *J. clin. Endocr.* **33**, 992–995 (1971)

9. Haddad, J.G. & Chyu, K.J.: 25-hydroxycholecalciferol-binding globulin in human plasma. *Biochim. Biophys. Acta (Amst.)* **248**, 471–481 (1971)

10. Haddad, J.G., Hahn, T.J. & Birge, S.F.: Vitamin D metabolites: Specific binding by rat intestinal cytosol. *Biochim. Biophys. Acta* **329**, 93–97 (1973)

11. McGuire, W.L.: Quantitation of estrogen receptor in mammary carcinoma. In: *Methods in Enzymology.* O'Malley, B.W.O. and Hardman, J.G. (eds.). Academic Press. **36A**, 248–254 (1975)

12. Preece, M.A., O'Riordan, J.L.H., Lawson, D.E.M. & Kodicek, E.: A competitive protein-binding assay for 25-hydroxycholecalciferol and 25-hydroxyergocalciferol in serum. *Clin. Chim. Acta* **54**, 235–242 (1974)

13. Rosenthal, H.E.: A graphical method for the determination and presentation of binding parameters in a complex system. *Analyt. Biochem.* **20**, 525–532 (1967)

14. Scatchard, G.: The attractions of proteins for small molecules and ions. *Ann. N.Y. Acad. Sci.* **51**, 660–672 (1949)

Maturation of Chick Bone Collagen and Quantification of Its Structural Crosslinks: Vitamin D Status and Cohesiveness of the Collagen Macromolecular Matrix

G. L. MECHANIC

INTRODUCTION

The intermolecular covalent crosslinks of collagen are directly responsible for the stabilization of the macromolecular matrices of supporting tissues. Crosslink maturational changes occur relative to the age of the animal (8, 1, 2, 15, 19), and from tissue to tissue in the prenatal and postnatal animal (10, 9). A profound age-related event occurs in $[^3H]NaBH_4$-reduced bone collagen; the ratio of dihydroxylysinonorleucine (DHLNL) to hydroxylysinonorleucine (HLNL) decreases from 5.2 in very young bone to the much lower ratio of 1.0 in mature bone (9). It has been suggested that post-translational phenomena responsible for these differences in crosslink chemistry in collagenous tissues are under fine biochemical control (19) and may involve vitamin D-dependent structure-function relationships in bone. We have also recently reported an elevation in lysyl oxidase activity in extracts of rachitic bone (7). The results suggested that vitamin D could affect the quantitative relationship between maturational crosslinks in bone collagen by regulating the amount of activity of lysyl oxidase or perhaps a specific hydroxylysyl oxidase, and hence alter aldehydic crosslink precursor formation.

Toole *et al.* (20) indicate that only a small percentage of bone collagen is soluble from 3-weeks-old lathyritic rachitic chicks. Deshmukh *et al.* (6), in addition, report that only one third of the lathyritic soluble bone collagen is recovered from the CM cellulose solumns and caution that interpretation of results obtained using a small portion of the total bone collagen may be erroneous. In this study we have used the total diaphyseal bone collagen to obtain our data.

Dental Research Center and the Department of Biochemistry and Nutrition, University of North Carolina, Chapel Hill, North Carolina.

We have reported earlier (11) that $[^3H]NaBH_4$-reduced bone collagen from 4-week-old rachitic chicks had higher ratios of DHLNL/HLNL than normal chicks, indicating that vitamin D might perform a role in the biochemical regulation involved in the maturation of bone collagen. To extend this original study, times earlier than 4 weeks were used to ascertain if the quantitative change in the collagen crosslink ratio and extent of lysine hydroxylation could be observed before appreciable decreases in circulating calcium level and body growth occur. Another objective was to determine if a high, but non-toxic, daily intake of vitamin D affects the maturation. In this study we also investigated whether or not vitamin D affects chick skin collagen crosslinks.

We also report here vitamin D status and its effect upon the quantitative amounts of the $[^3H]NaBH_4$-reducible Schiff base and stable keto-imine crosslinks in bone collagen.

METHODS

Three groups of chicks fed *ad libitum* diets containing, 1) no vitamin D, 2) 1.4U cholecalciferol, and 3) 70U of cholecalciferol and all containing 1.34% Ca, 1.04% P and 0.23% Mg. High D diet (70U) had been established earlier as non-toxic (14).

Bones were taken from 5 chicks in each group after 1, 2, 3 and 4 weeks of age.

The diaphysis of the right tibia from two or three chicks in each group was treated as described previously for analysis of $[^3H]NaBH_4$-reduced collagen crosslinks (8, 1, 2, 15, 19, 10, 9). Analytical chromatography for reduced crosslinks was performed according to methods described elsewhere (12, 13). The extent of lysine hydroxylation was determined by automatic amino acid analyses. Defatted skin samples were also analyzed for $[^3H]NaBH_4$-reduced crosslinks by the same techniques. The quantitative amounts of $[^3H]NaBH_4$-reducible Schiff base and stable keto-imine crosslinks were determined directly by methods described elsewhere (14).

RESULTS

Histological sections of vitamin D-deficient diaphyseal bone only revealed evidence of rickets as early as 2 weeks of age, while there were no differences between the Control-D and High-D chicks at any time.

In Table I are presented the data obtained and confirm our earlier results (11). It is quite evident that there was a progressive decrease in the

Table I.

Effects of Vitamin D on Diaphyseal Bone Collagen Crosslinks
(Dihydroxylysinonorleucine and Hydroxylysinonorleucine)
Lysine Hydroxylation, Serum Calcium, and Body Weight in the Chick.

Group	Age (weeks)	DHLNL/ HLNL	Lysine Hydroxylation %	Serum Calcium (mg/100 ml)	Body Weight g
D-deficient	1	4.2	29	9.75	77
	2	3.9	28	8.40	140
	3	6.6	30	6.47	168
	4	7.6	29	7.10	176
Control-D*	1	3.6	29	10.50	69
	2	3.0	22	10.67	138
	3	2.2	22	10.33	258
	4	1.8	23	9.55	320
High-D**	1	3.2	30	10.90	77
	2	2.2	24	10.15	147
	3	1.8	22	10.63	237
	4	1.8	23	9.50	238

*) 1.4 I.U. of Vitamin D_3/g diet DHLNL — dihydroxylysinonorleucine

**) 70 I.U. of Vitamin D_3/g diet HLNL — hydroxylysinonorleucine

ratio with age in the Control-D group, suggesting the same normal maturational change previously observed with bovine bone (9). This was also noted in the High-D group for weeks 1 through 3. At 2 weeks, with increasing intake of vitamin D, there was a successive decrease in the ratios indicating that vitamin D has a dose-related effect on the maturational phenomena related to bone collagen structural crosslinks.

The collagen crosslink elution profile ratios obtained from samples of skin from Control-D chicks was similar to that previously obtained in mammalian species (10). We have found that DHLNL does exist in the skin of young chicks (G.L. Mechanic, unpublished results). Calculation of the DHLNL/HLNL ratios in skin samples from 1 or 2 chicks in each group at each time revealed no consistent change with time in any group.

Table I shows that the extent of hydroxylation of the bone collagen was elevated in all D-deficient chicks and in 1-week-old Control-D and High-D chicks compared to the older Control-D and High D-chicks.

Quantification of the $[^3H]NaBH_4$-reducible crosslinks and the direct determination of the stable keto-imine forms of the Schiff bases

Table II.

Quantitative Determination of Reducible Labile and Stable
Keto-Imine Crosslinks in Control and Rachitic Chick Bone Collagen.
3-weeks-old Chicks.

Collagen	$\dfrac{\text{Mole DHLNL}}{\text{Mole Collagen}}$	% DHLNL as Keto-Imine	$\dfrac{\text{Mole HLNL}}{\text{Mole Collagen}}$	% HLNL as Keto-Imine
Control	1.001		0.747	
		46%		63%
Control (Ac$_2$O/AcOH)[1]	0.464		0.47	
Rachitic	0.993		0.373	
		45%		14%
Rachitic (Ac$_2$O/AcOH)	0.447		0.051	

1) Method used reported in reference 14.
DHLNL — dihydroxylysinonorleucine
HLNL   — hydroxylysinonorleucine
Keto-Imine of DHLNL — 5-keto-5'-hydroxylysinonorleucine
Keto-Imine of HLNL   — 5-keto-lysinonorleucine

Δ^6-dehydrodihydroxylysinonorleucine and Δ^6-dehydrohydroxylysinonor-
leucine of 3-weeks-old normal and rachitic chick bone collagen of another
group are presented in Table II. It can be seen that the total amount of
DHLNL remains the same in the normal and rachitic bone as well as the
amounts of keto-imine, 5-keto-5'-hydroxylysinonorleucine. However,
there is a reduced quantity of HLNL in the latter as well as diminished
amount of stable keto-imine.

DISCUSSION

The data reported here confirm our earlier observation that there is a
quantitative alteration in the ratio of the structural crosslinks of
[^{3}H]NaBH$_4$-reduced chick rachitic bone collagen (11). This ratio shift
(increase of DHLNL/HLNL) may be observed as early as 1 week after
newly-hatched chicks are placed on a vitamin D-deficient diet. This
evidence is one of the earliest (18) and most sensitive indications yet
described for a bone disturbance caused by vitamin D deficiency.
Crosslink analysis may, therefore, be useful for investigating other diseases
involving bone, as previously described (16).

The data from the Control- and High-D groups (Table I) distinctly indicate that a progressive decrease in the ratios occurs with age; this is normal crosslink maturational change (2, 15, 19, 10, 9). However, this decrease in ratios occurs at a more rapid rate in the High-D group, reaching the 4-weeks Control-D level at 3 weeks. This response to vitamin D and vitamin D-deficiency suggests that cholecalciferol plays an important biochemical role in the maturational changes in the crosslinking of the organic matrix of bone. This seems to be specific for bone, because there were no significant differences in the crosslink ratios of skin between any of the groups.

Although the crosslink ratios in the D-deficient group at 1 and 2 weeks were higher than those in the Control- and High-D groups, the fact that the ratios remained stable, or were slightly higher at 1 week than at 2 weeks in the D-deficient group, probably reflects the presence of vitamin D metabolite stores carried from the egg. However, the subsequent rise in ratio in this group at 3 weeks is presumably due to the disappearance of these stores. The present data do not support the possibility that the increased ratios were due to reduced general protein synthesis due to rickets since there was no inhibition of growth in the 1- and 2-weeks-old animals (Table I).

In correlating circulating Ca with crosslink ratios in the D-deficient groups, it should be noted that although the greatest increase in crosslink ratio from 2 to 3 weeks corresponded to the greatest drop in serum calcium, the decrease in serum calcium from 1 to 2 weeks was not accompanied by an increase in crosslink ratio. The ratio continued to rise at 4 weeks, but the serum calcium showed no corresponding further decrease compared to the 3-weeks value. The circulating Ca remained the same at all ages in the Control- and High-D groups (Table I) although the ratios of DHLNL/HLNL progressively decreased in each group. It is also clear that this decrease was accelerated in the High-D group. These data suggest that the quantitative relationship between the crosslinks is independent of the circulating calcium level and may be due to some specific effect of vitamin D.

The percentage of lysine hydroxylation in the total diaphyseal bone collagen (Table I) demonstrates that the DHLNL/HLNL ratios do not necessarily reflect the degree of hydroxylation of the lysine residues, since the D-deficient chicks maintained a constant degree of lysine hydroxylation throughout the 4-weeks period, which was the same as for the 1-week-old chicks in the Control- and High-D groups. Furthermore, the extent of hydroxylation of the Control-D and High-D groups at 2, 3 and 4 weeks remained constant while the crosslink ratio decreased. The numbers in Table I corroborate the data reported by others (20, 4): that an increase of lysine hydroxylation does occur in vitamin D-deficient bone collagen, although the values correspond more closely to those reported by Toole *et al.* (20) and Miller *et al.* (17) for 3 weeks lathrytic chicks. Higher lysine hydroxylation results recently reported in calcium deficient chicks receiving physiological doses of vitamin D (5), were interpreted to mean that vitamin D deficiency is probably not responsible for hyperhydroxylation of lysine residues in bone collagen but that it is an effect caused by the ensuing hypocalcemia. We cannot support their suggestion, since we obtained higher hydroxylation values at 1 week in all the groups and at all ages in the vitamin D-deficient group (Table I) even though no apparent hypocalcemia occurred in any of the 1-week animals. It is suggested here that the circulating calcium level does not affect the extent of lysine hydroxylation, although our data do not exclude that possibility; nor, as mentioned earlier, does it affect the ratios of the structural crosslinks of bone collagen. Furthermore, no relationship between the overall lysine hydroxylation and the ratio of DHLNL/HLNL can be seen. The only obvious relationship shown in this report is between the ratios of the bone collagen crosslinks and the level of the vitamin D fed to the animals.

Direct analyses for reduced Schiff base and stable keto-imine crosslinks indicate there is no change in DHLNL while total HLNL is lowered in rachitic bone collagen. Significantly too, there was a much lesser amount of stable HLNL-keto-imine crosslink in the collagen. The decreased amount of HLNL may occur from lack of formation of sufficient amounts of aldehydic crosslink precursor and, therefore, result in less HLNL. It was stated earlier that there was an elevation of lysyl oxidase activity in extracts of rachitic bone which seems inconcordant with the amounts of the crosslinks found. It is, therefore, suggested that the increase in lysyl oxidase activity is a compensatory response caused by insult or trauma to the tissue in vitamin D-deficiency and has no relation to the biochemical phenomena involved in the formation of structural crosslinks in bone collagen.

It was also known that in addition to a lesser quantity of HLNL being present there was a significant decrease in the amount of stable form of Δ^6-dehydrohydroxylysinonorleucine, 5-keto-lysinonorleucine (keto-imine). These data indicate that the collagen macromolecular matrix in rachitic bone is a less cohesive matrix and this may result in causing aberrations in the organization, and packing, and the structure of the matrix is insufficient to ensure normal bone mineralization and turnover. It is suggested that vitamin D is essential and plays some regulatory role for the normal maturation of bone and its stabilization.

SUMMARY

The quantitative relationships were determined between the structural crosslinks, dihydroxylysinonorleucine (DHLNL) and hydroxylysinonorleucine (HLNL) in [^{3}H]NaBH$_4$-reduced diaphyseal bone collagen from 1-, 2-, 3- and 4-weeks-old chicks fed either a vitamin D-deficient diet, a normal-vitamin D diet or a high-, but non-toxic, vitamin D diet from time of hatching. Chicks fed the normal diet showed a progressive decrease in the ratio of DHLNL/HLNL with age. This decrease was accelerated in chicks receiving the High-D diet. In the D-deficient group, the ratio was higher than controls at 1 and 2 weeks and increased further at 3 and 4 weeks. Similar changes in DHLNL/HLNL did not occur in skin collagen. Compared to Control D animals, the increased crosslink ratios in the D-deficient bone collagen occurred prior to changes in growth rate and could not be correlated with lysine hydroxylation or the hypocalcemia seen in this group. These results suggest that the type of crosslink analysis used in this study provides one of the earliest and most sensitive indications of a bone disturbance due to vitamin D deficiency and that vitamin D specifically acts to increase the rate of maturation of bone collagen.

Direct quantitative determination of the reducible labile (imminium) and stable (keto-imine) crosslinks of 3-weeks-old bone collagen indicated no change in the DHLNL content while there was a reduction of HLNL in the rachitic bone. Data are also reported that indicates no alteration in the amount of 5-keto-5'-hydroxylysinonorleucine but a large reduction of 5-keto-lysinonorleucine. These data indicate that the rachitic bone reported previously by us, might be due solely to a compensatory response resulting from insult or trauma to the bone due to lack of vitamin D$_3$ metabolites.

ACKNOWLEDGEMENT

I thank Marshall Perry and Betty Hilliard for their technical assistance and Jean Cochran for typing the manuscript. This investigation was supported by N.I.H. research grant number DE 02668 from the National Institute of Dental Research and by N.I.H. grant number RR 05333 from the Division of Research Facilities and Resources.

REFERENCES

1. Bailey, A.J., Peach, C.M. & Fowler, L.J.: Chemistry of Collagen Cross-Links. "Isolation and Characterization of Two Intermediate Intermolecular Cross-Links in Collagen". *Biochem. J.* 117, 819—832 (1970)

2. Bailey, A.J. & Shimokamaki, M.S.: Age Related Changes in the Reducible Cross-Links of Collagen. *FEBS Lett.* 16, 86—88 (1970)

3. Barnes, M.J.: Bone Collagens. In: *Hard Tissue Growth, Repair and Remineralization*, Ciba Found. Associate Scientific Press, Amsterdam. Symposium II (new series), pp. 247—261 (1973)

4. Barnes, M.J., Constable, B.J., Morton, L.F. & Kodicek, E.: Short Communications: Bone Collagen Metabolism in Vitamin D Deficiency. *Biochem. J.* 132, 113—115 (1973)

5. Barnes, M.J., Constable, B.J., Morton, L.F. & Kodicek, E.: The Influence of

Dietary Calcium Deficiency and Parathyreoidectomy on Bone Collagen Structure. *Biochim. Biophys. Acta* **328**, 373–382 (1973)

6. Deshmukh, K., Just, M. & Nimni, M.E.: A Defect in the intramolecular and intermolecular Cross-Linking of Collagen Caused by Penicillamine. III. Accumulation of Acid Soluble Collagen with a High Hydroxylysine Content in Bone. *Clin. Orthop.* **91**, 186–196 (1973)

7. Gonnerman, W.A., Mechanic, G.L., Ramp, W.K. & Toverud, S.U.: Effect of Vitamin D on Lysyl Oxidase Activity in Chick Bone Metaphyses. *Fed. Proc.* **34**, 2714 (1975)

8. Mechanic, G.L. & Tanzer, M.L.: Biochemistry of Collagen Crosslinking: Isolation of a New Crosslink, Hydroxylysinohydroxynorleucine and Its Reduced Precursor, Dihydroxynorleucine, from Bovine Tendon. *Biochem. Biophys. Res. Commun.* **41**, 1597–1604 (1970)

9. Mechanic, G.L., Gallop, P.M. & Tanzer, M.L.: The Nature of Crosslinking in Collagens from Mineralized Tissues. *Biochem. Biophys. Res. Commun.* **45**, 644–653 (1971)

10. Mechanic, G.L.: The Intermolecular Crosslink Precursors in Collagen as Related to Function. Identification of δ-Hydroxy-α-amino-adipic-δ-semialdehyde in Tendon Collagen. *Israel J. Med. Sci.* **1**, 458–462 (1971)

11. Mechanic, G.L., Toverud, S.U. & Ramp, W.K.: Quantitative Changes of Bone Collagen Crosslinks and Precursors in Vitamin D Deficiency. *Biochem. Biophys. Res. Commun.* **47**, 760–765 (1972)

12. Mechanic, G.L.: An Automated Scintillation Counting System with High Efficiency for Continuous Analysis: Crosslinks of [^{3}H]NaBH$_4$ Reduced Collagen. *Anal. Biochem.* **61**, 349–354 (1974)

13. Mechanic, G.L.: A Two Column System for Complete Resolution of NaBH$_4$-Reduced Crosslinks from Collagen. *Anal. Biochem.* **61**, 355–361 (1974)

14. Mechanic, G.L., Kuboki, Y., Shimokawa, H., Nakamoto, K., Sasaki, S. & Kawanishi, Y.: Direct Quantitative Determination of Stable Structural Crosslinks in Bone and Dentin Collagens. *Biochem. Biophys. Res. Commun.* **60**, 756–764 (1974)

15. Mechanic, G.L.: Collagen Structure, Organization and Packing Reflected by Crosslinking Studies. In: *International Colloquium on the Physical Chemistry and Crystallography of Apatites of Biological Interest*, G. Montel (ed.), CNRS, Paris, pp. 212–213 (1975)

16. Mechanic, G.L. & Bullough, P.: The Qualitative and Quantitative Crosslink Structure of Human Normal and Osteogenesis Imperfecta Bone Collagens. In: *International Colloquium on the Physical Chemistry and Crystallography of Apatites of Biological Interest*, G. Montel (ed.), CNRS, Paris, pp. 226–230 (1975)

17. Miller, E.J., Martin, G.R., Piez, K.A. & Powers, M.J.: Characterization of Chick Bone Collagen and Compositional Changes Associated with Maturation. *J. Biol. Chem.* **242**, 5481–5489 (1967)

18. Ramp, W.K., Toverud, S.U. & Gonnerman, W.A.: *J. Nutr.* **104**, 803–809 (1974)

19. Robins, S.P., Shimokamaki, M. & Bailey, A.J.: The Chemistry of the Collagen Cross-Links: Age-Related Changes in the Reducible Components of Intact Bovine Collagen Fibres. *Biochem. J.* **131**, 771–780 (1973)

20. Toole, B.P., Kang, A.H., Trelstad, R.L. & Gross, J.: Collagen Heterogeneity within Different Growth Regions of Long Bones of Rachitic and Non-Rachitic Chicks. *Biochem. J.* **127**, 715–720 (1972)

CHAPTER IV
Vitamin D. Clinical Problems

Parathyroid Function in Chronic Vitamin D Deficiency in Man: A Model for Comparison with Chronic Renal Failure

S.W. STANBURY & G.A. LUMB

SUMMARY

It has become obvious, from the elegant *in vitro* studies of Cohn *et al.* (3) and of others, that calcium may act at various steps in the biosynthesis, storage, degradation and secretory release of parathyroid hormone. It would be premature to attempt to interpret our simple clinical observations in terms of these still incompletely defined mechanisms. These clinical studies have, however, identified several different components in the overall pattern of parathyroid response in chronic vitamin D deficiency. In significantly hypocalcaemic patients there was an inverse proportional relationship between serum calcium and serum iPTH, both in the steady-state (9) and during calcium infusion (Fig. 2A); a similar pattern was seen in primary hyperparathyroidism (Fig. 2B). One is tempted to regard this relationship as reflecting an action of calcium on hormonal biosynthesis. There is the further suggestion that profound hypocalcaemia in vitamin D deficiency might limit the capacity of the parathyroid gland to release its hormone (Figs. 3, 4). In the majority of cases of chronic vitamin D deficiency, the serum calcium is maintained close to normal but varying between 8.1 and 11.1 mg/dl in different individual patients of the present series; and in all these cases there is very obvious evidence of increased parathyroid activity. In such patients, it appears that the serum iPTH represents the maximum rate of secretion that the parathyroid gland can currently produce; and the level of serum calcium reflects the extent to which that parathyroid secretion compensates for lack of vitamin D action on gut, bone and kidney. If this equilibrium state is disturbed by an induced, barely detectable, increment in serum calcium there appears to be an immediate inhibition of parathyroid secretion (Fig. 5). There is nothing in the data to suggest that a direct action of vitamin D on the parathyroid glands is required for any of these patterns of parathyroid response, since all were observed in patients with clinical vitamin D deficiency. Only the very slow decay in the steady-state concentration of serum iPTH after correction of vitamin D deficiency in the two hypercalcaemic patients (Fig. 6) suggests a possible direct action of vitamin D on the parathyroid glands. The very slowness of this decay must imply a process of structural involution. Is it possible that vitamin D is involved in some way in coupling the calcium dependent processes of hormonal synthesis and secretion with a calcium dependent mechanism controlling cell

Department of Medicine, The Royal Infirmary, Manchester.

division? At present one can do little more than ask the question and wonder it it is a disturbance of such coupling that permits the development of hypercalcaemic secondary hyperparathyroidism.

Although there have been considerable advances in the understanding and treatment of azotaemic osteodystrophy, a variety of problems remain. The osteomalacic component in renal osteodystrophy is reasonably well understood but the recent independent demonstration, by workers in Heidelberg (10) and in Padua (1), of osteomalacia at the earliest stages of renal disease poses a challenging new problem. As yet there are no data, either on mineral metabolism or vitamin D metabolism, in these patients and the significance of the findings is obscure; but this is a phenomenon warranting further detailed study. It is, however, the hyperparathyroid component in azotaemic osteodystrophy which continues to pose most problems and to produce the most devastating clinical effects. It has been possible to reverse severe secondary hyperparathyroidism in renal failure by parathyroidectomy, by treatment with vitamin D preparations or dihydrotachysterol, by dialysis against high concentrations of calcium and by massive oral supplements of calcium salts. But is has remained difficult to monitor the accompanying changes in parathyroid function, either in longitudinal studies in the course of treatment or in cross sectional studies of the reactivity of the parathyroid glands at different stages in the evolution of the disease.

In view of these difficulties, and believing that advanced renal failure is a state of 'effective vitamin D deficiency', a study has been undertaken of parathyroid responses in chronic simple vitamin D deficiency rather than in renal failure. This option has two distinct advantages. Firstly, since renal function is normal in the patients studied, the Bricker-Slatopolsky factor of phosphate retention as an indirect stimulus to secondary hyperparathyroidism is effectively eliminated. Secondly, one avoids the problem of the renal retention of parathyroid hormonal fragments, which is generally agreed to hinder the interpretation of measurements of serum iPTH in renal failure. It has also been demonstrated that the severity of the secondary hyperparathyroidism in chronic simple vitamin D deficiency may approach that seen in chronic renal failure, whether such severity is assessed radiologically, by the bone histology or in terms of serum iPTH (9). In studying these patients, several different patterns of parathyroid secretory response have been recognized. The reasons for and the mechanisms underlying these different responses are as yet far from clear; but it is suggested that the study of these responses in privational vitamin D deficiency might throw some light on the relationships between 'effective vitamin D deficiency' and parathyroid function in chronic renal failure.

METHODS

The patients studied were adolescents or adults from among the immigrant Asian population in Britain, whose diet may be very low in vitamin D and who also get little exposure to sunshine. All had bone disease due to vitamin D deficiency and were suffering from the symptoms of rickets, osteomalacia or tetany. These clinical effects are corrected by as little as 450 I.U. (11 μg) per day of vitamin D given orally but they often relapse if this supplement is withdrawn or the patient defaults from treatment. Thus, individual patients may sometimes have had a relative insufficiency of vitamin D for many years — and in this respect they resemble the chronic conditioned deficiency of vitamin D in chronic renal failure.

Immunoreactive parathyroid hormone was measured by a radio-immunoassay believed to recognize the whole molecule and capable of detecting rapid changes in parathyroid secretion, as described elsewhere (9, 11). Following parathyroidectomy or after arrest of secretion by induced hypercalcaemia, the serum immunoreactive hormone (iPTH) measured by the antiserum used (AS 211/32) decays with a $t^{1/2}$ of 5—15 min. Serum iPTH has been related to the prevailing levels of serum calcium in the steady state and also as the serum calcium was raised progressively by calcium infusion. Belatedly, it was appreciated that phenomena of significant interest also occurred during the period immediately following cessation of the calcium infusion, and a less detailed study has been made of the parathyroid response at this time.

The clinico-pathological classification of chronic vitamin D deficiency
As was found to be both possible and necessary in analyzing the features of azotaemic osteodystrophy (17), the clinical and biochemical features of chronic vitamin D deficiency in adolescents and adults can be classified into several distinctive clinical syndromes (9, 18). A somewhat similar classification into 'stages of deficiency' was proposed by Fraser, Kooh and Scriver (6) for infants with vitamin D deficiency; but many of their patients were unstable and were observed to evolve from one to another 'stage' within a short period of time. In the present study, the patients were much older, their vitamin D deficiency was probably of long duration, and the clinical syndrome was apparently stable in the individual until the vitamin D deficiency was corrected.

188

In the most common syndrome (group I), rickets or osteomalacia is associated with a moderately depressed or normal serum calcium, hypophosphataemia and degrees of radiographic and histological secondary hyperparathyroidism which is sometimes extremely severe. Serum iPTH in these patients is always high, up to 10 times the highest normal value.

A second group of patients (group II) is profoundly hypocalcaemic (serum calcium < 7 mg/dl and as low as 5—6 mg/dl); radiographic and histological secondary hyperparathyroidism and osteomalacia are less severe; and the serum iPTH, although raised, is lower than in group I. None of the patients in this group was significantly hypomagnesaemic. Some these hypocalcaemic patients are also hypophosphataemic (IIc) but in an important sub-group (IIa,b) the serum phosphorus is normal or actually raised. This relative or absolute hyperphosphataemia is unresponsive to the administration of exogenous parathyroid extract. These cases thus represent an acquired form of apparent pseudohypoparathyroidism; but, although the hyperphosphataemia is uninfluenced by parathyroid extract, this may none the less raise the serum calcium. An identical pattern of serum biochemistry is a commonplace in chronic renal failure, and it is of interest that a combination of profound hypocalcaemia *plus* hyperphosphataemia in these vitamin D deficient patients has

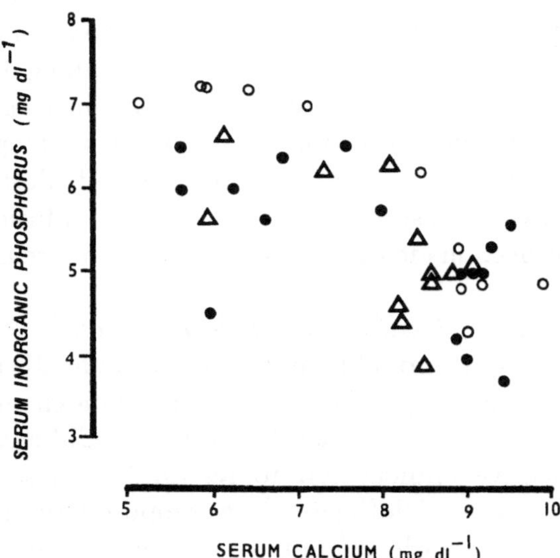

Figure 1. The relationship between the serum phosphorus (P$_i$) and serum calcium in three adolescent Asian boys with vitamin D deficiency, hypocalcaemia and hyperphosphataemia.

As the serum calcium increased during treatment with vitamin D there was a reciprocal reduction in serum P$_i$.

provoked only a modest degree of secondary hyperparathyroidism. Yet, when these patients are treated with small doses of vitamin D (11 μg/day) the serum phosphorus falls reciprocally with the induced rise in serum calcium (Fig. 1) and these changes are accompanied by an actual reduction in serum iPTH. This hyperphosphataemic group poses two immediate questions. Firstly, as to why there is so small an apparent increase of parathyroid secretion, despite the presence of conditions optimal for parathyroid stimulation. Secondly, when renal responsiveness to parathyroid hormone is restored during treatment with vitamin D, is this induced change due to the rise in serum calcium, to some action of vitamin D on the kidney, or to a combination of such effects? These problems can best be discussed after description of the parathyroid secretory responses in these patients.

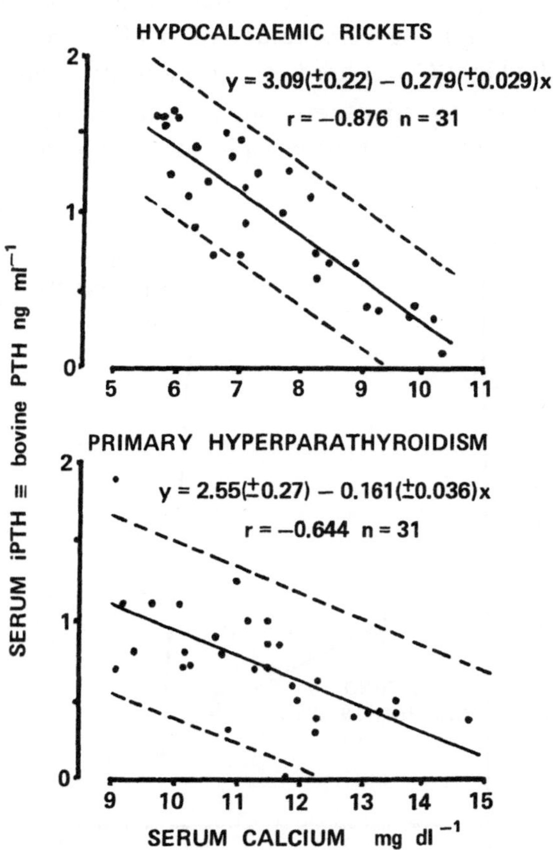

Figure 2. The relationship between the serum calcium and serum iPTH, A (above), in hypocalcaemic patients with vitamin D deficiency, before and during calcium infusion; B (below), in hypercalcaemic primary hyperparathyroidism, before and during infusion of calcium (3 patients) or EDTA (5 patients).

A third clinical syndrome includes only two individual examples (group III); chronic vitamin D deficiency, severe osteomalacia, very severe osteitis fibrosa and high serum iPTH and a sustained, minor but definite *hyper*calcaemia. It is open to argument that these were cases of primary hyperparathyroidism complicated by vitamin D deficiency; but the case we put is that this is hypercalcaemic secondary hyperparathyroidism, analogous to the same phenomenon seen in chronic renal failure but developing without the stimulus of phosphate retention.

Parathyroid responses in hypocalcaemic patients (group II)
Among these patients with vitamin D deficiency, the simplest pattern of parathyroid response to calcium infusion is seen in those with profound hypocalcaemia. As the serum calcium is raised from 5–6 mg/dl to 10 mg/dl or above there is a proportionate reduction in serum iPTH, regression analysis implying zero values of iPTH at a serum calcium of about 11 mg/dl (Fig. 2A). This is an apparently appropriate response, completely analogous to the observations made during infusion of EDTA or calcium in the cow (13) but differing in that the patients were clinically deficient of vitamin D.

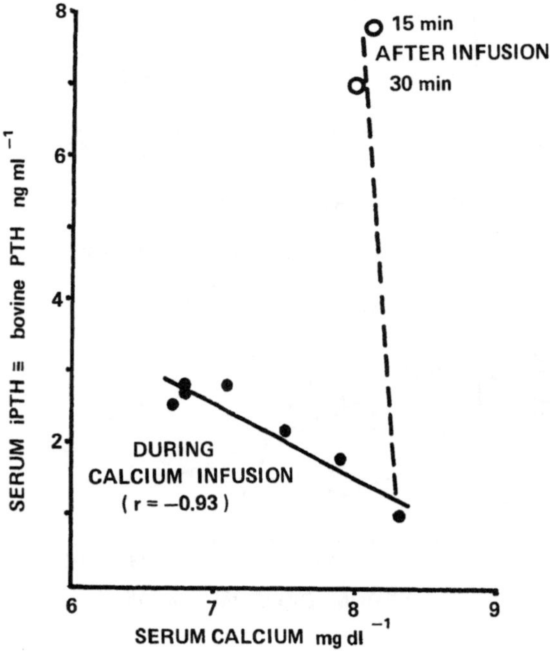

Figure 3. The relation between the serum calcium and serum iPTH in a single case of hypocalcaemic vitamin D deficiency during calcium infusion and during the 30 min after stopping the infusion.

Observations of this kind are usually terminated with the calcium infusion; but, in casual and often single, blood specimens obtained after the infusion it was not noted that the serum iPTH was higher than in any earlier specimen (Fig. 3). When this 'recovery' phase of the procedure was examined in detail, it was again found that there was a proportionate relation between serum calcium and serum iPTH; but there was a highly significant difference in the slope of the relationship between the two variables (Fig. 4). Following infusion of the calcium load, there appears to have been a temporary increase in the reactivity of the parathyroid glands, whereby a unit change of serum calcium is associated with a greater proportionate change of serum iPTH. This change is described as temporary since, when the serum calcium has restabilized at the level prevailing before the infusion, the serum iPTH has also reverted to its original level; but the data used to construct the post-infusional regression in Fig. 4 were obtained during a period of 3 h following each calcium infusion.

It is tentatively suggested that this phenomenon may be analogous to the effects of profound hypocalcaemia in inhibiting the release of insulin in

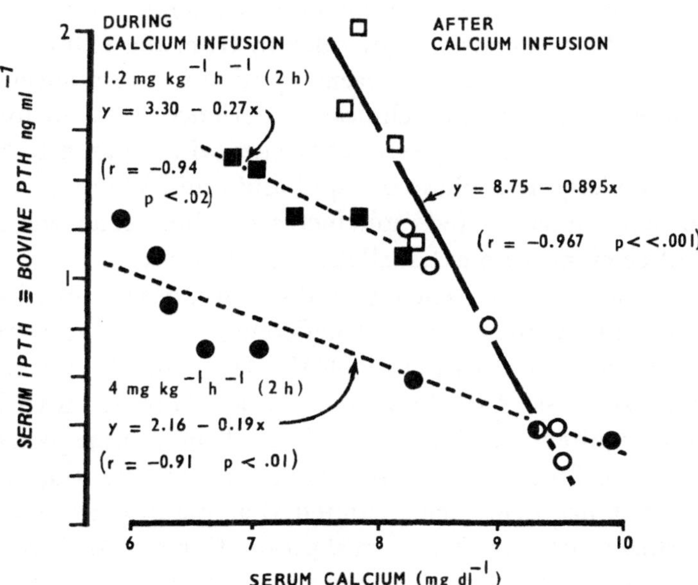

Figure 4. The relation between the serum calcium and serum iPTH in another hypocalcaemic patient, during and following calcium infusion. Two separate calcium infusions are shown (●, ■); the slope of the two regressions during infusion (interrupted lines) is not significantly different. Data obtained during the three hours following each infusion (○, □) are treated as a single regression (solid line), the slope of which is significantly different from the regressions during each infusion.

cows with milk fever (8). It seems possible that some cellular pool, perhaps intimately associated with the cell membrane, must be appropriately filled with calcium if the parathyroid cell is to secrete optimally in response to a change in circumambient calcium concentration. If a continued secretion of synthesized hormone is the factor that determines in some way the process of glandular hyperplasia (12), incomplete hormonal release might effectively prevent a hyperplastic response; and this, in turn, might explain why parathyroid hormone production in the profoundly hypocalcaemic patients fails to increase sufficiently to correct the hypocalcaemia. It is appreciated that there could be alternative explanations for the phenomenon described. The calcium infusion might activate the enzymic production of hormonal fragments recognized by the antiserum as intact hormone. This seems unlikely in view of the maintained proportionality between the serum calcium and immunoreactive PTH after the infusion (Fig. 4); and the same phenomenon was documented by an alternative antiserum (AS 367) known to recognize the hormonal fragments retained in renal failure.

Irrespective of these hypothetical considerations, it seems necessary to seek the primary cause of the hypocalcaemia outside the parathyroid glands. In those patients who are hyperphosphataemic the raised serum P_i will tend to facilitate deposition of calcium from the extracellular fluid into the bones; but some hypocalcaemic patients have reduced levels of serum phosphate. It is our clinical experience that many of our profoundly hypocalcaemic adolescent patients with vitamin D deficiency have recently undergone a phase of rapid and conspicuous growth; and it is suggested that the associated rapid increase in bone mass will also cause a deviation of calcium from extracellular fluid to bone.

It is unfortunate that we encountered our hypocalcaemic, hyperphosphataemic patients with vitamin D deficiency before we were able to measure the urinary output of cyclic AMP. Consequently, it is not known if the unresponsiveness of the hyperphosphataemia to parathyroid extract — that is, the acquired pseudohypoparathyroidism in these patients — was associated with a failure of cAMP production by the kidneys. This is unlikely, since it has been demonstrated that parathyroid hormone will induce a normal or greater than normal production of cAMP by renal cells of the vitamin D deficient rat (7). The recent delineation of the so-called type II idiopathic pseudohypoparathyroidism suggests a functional analogy with the hypocalcaemic hyperphosphataemic cases of vitamin D deficiency. Drezner *et al.* (5) described a child with pseudohypoparathyroidism in whom the administration of parathyroid hormone increased the urinary output of cAMP but failed to produce phosphaturia. Rodriguez *et al.* (14) describe an adult with a similar functional lesion in whom a normal phosphaturic response to administered parathyroid hormone was restored by infusions of calcium sufficient to raise the serum calcium to normal. It is not known if such artificial elevation of the serum calcium in hyperphosphataemic vitamin D deficiency would similarly restore renal responsiveness to exogenous hormone; but it is evident that responsiveness to endogenous hormone is restored as the serum calcium rises during treatment with vitamin D (Fig. 1). Thus, the combination of profound hypocalcaemia *and vitamin D deficiency* can produce a reversible abnormality of the renal handling of phosphate that simulates the lesion of type II pseudohypoparathyroidism. The profound hypocalcaemia of these vitamin D deficient individuals cannot alone explain the renal tubular abnormality, since hypocalcaemia of similar degree does not inhibit the phosphaturic response to parathyroid hormone in patients with simple hypoparathyroidism. It seems possible that 'vitamin D' may have some direct effect on the renal cellular distribution of calcium (7), possibly related to the formation of the renal 'calcium binding protein' (19); and that when the serum calcium is very low this effect of vitamin D is required for a normal phosphaturic response to parathyroid hormone.

These various phenomena of hypocalcaemic vitamin D deficiency in man require further detailed study but it seems self evident that what is learned from such studies may have potential application in the 'conditioned vitamin D deficiency' of renal failure. In essence, we are suggesting that mineral deposition in bone in subjects with impaired intestinal absorption of calcium may effectively starve the extracellular fluid of calcium, with consequential reduced secretory capacity of the parathyroid gland and reduced phosphaturic responsiveness of the kidney. In our adolescent patients with simple vitamin D deficiency, bodily growth and also hyperphosphataemia may cause deviation of calcium to the bone. In renal failure, the same two factors may be operating; and, as suggested long ago (16), the cancellous hyperostosis in azotaemic osteomalacia may have the same functional effect as skeletal growth.

Parathyroid responses in chronic vitamin D deficiency with normal or near-normal levels of serum calcium (groups I and III)

A qualitatively completely different parathyroid response to calcium infusion is seen in the commonest syndrome of chronic vitamin D deficiency (group I). These are patients in whom the serum calcium is normal or, if reduced, is no lower than about 8 mg/dl. The mean serum iPTH (2.25 ± 2.4 (SD) ng/ml; normal 0.35 ± 0.3 ng/ml) is higher than in the previously discussed group II (1.46 ± 0.6 ng/ml). The severity of the bony destruction, whether assessed radiographically or histologically, suggests that the parathyroid response in these patients maintained the higher serum calcium by its action on bone — despite the presumed lack of the 'permissive' action of vitamin D (or 1,25-dihydroxycholecalciferol). Thus, as we envisage the relationship between bone and extracellular fluid, the situation is diametrically in contrast with group II.

It was shown previously in these patients (9) that a trivial increase of serum calcium, of less than 1 mg/dl and probably considerably less than this, causes an apparently immediate arrest of hormonal secretion; the serum iPTH decays with a $t^{1/2}$ of 5—15 min, to reach values 20% or less of the initial value. In Fig. 5, the decrement of serum iPTH in six individuals is expressed as a percentage of the initial value and is related to the increment of serum calcium; an increase of 0.5 mg/dl in serum calcium

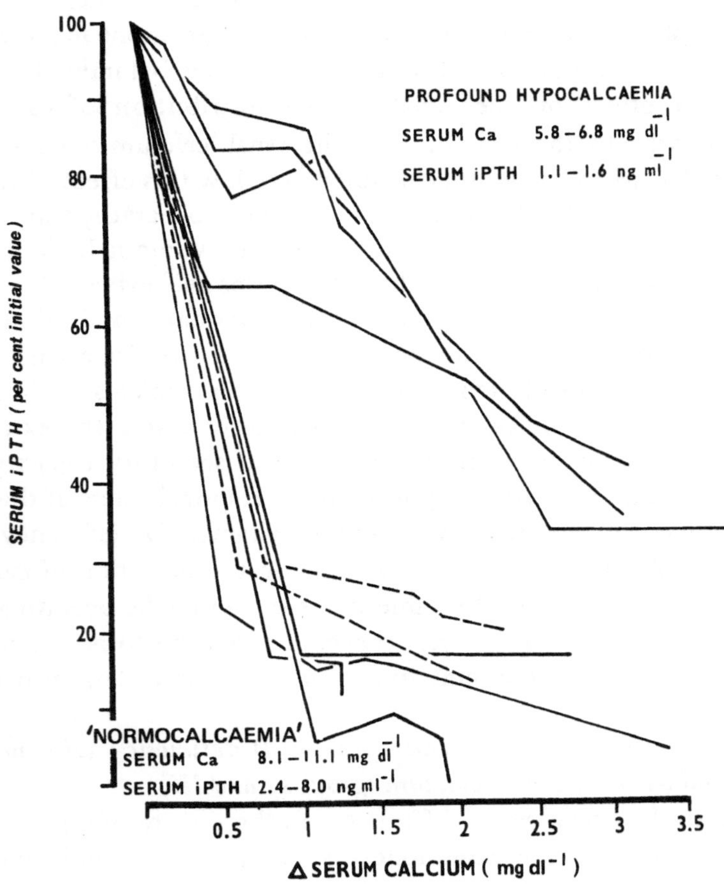

Figure 5. The relationship between the decrement in serum iPTH and increment in serum calcium, during calcium infusion, in six patients of groups I and III (left) compared with four patients of group II (right).

Serum iPTH is expressed as a percentage of control values before the infusion. The two hypercalcaemic cases of group III are shown by interrupted lines.

was associated with an average fall in serum iPTH of about 50%. This relationship implies, as was actually observed in some cases, that this major inhibition of parathyroid secretion could occur when the serum calcium was still actually subnormal. Thus, the parathyroid glands in these patients appear to respond sensitively or normally as if to maintain the prevailing concentration of serum calcium. The fact that the equilibrium concentration of serum calcium might be below the conventional normal cannot be ascribed to an abnormal parathyroid cell function.

Two of the six patients in Fig. 5 belonged to group III rather than group I; that is, both were hypercalcaemic with respective stable serum calcium concentrations of 10.6 and 11.1 mg/dl before the calcium infusion. As in the patients of group I, a barely detectable further rise in serum calcium appeared to arrest parathyroid secretion and the serum iPTH fell very rapidly to levels one fourth or one fifth of its initial value. But, since the initial values in these two patients were very high ($\sim$ 8 ng/ml, or more than 20 times the average normal value with this antiserum), the serum iPTH remained elevated above normal even after an 80% reduction; whereas the same proportionate reduction in the mean value of 2.25 ng/ml in the patients of group I brought the serum iPTH into the normal range.

There were literally no detectable differences between the two cases of group III and those of group I, other than the severity of the secondary hyperparathyroidism and the hypercalcaemia. Thus, under these conditions of testing by calcium infusion, there was an apparently identical pattern of parathyroid reactivity in patients with stable values of serum calcium between 8.1 and 11.1 mg/dl. In these patients, the serum calcium is presumably maintained at normal or 'near to normal' levels by the maximum secretory activity of their parathyroid glands; and the actual serum calcium attained may be a function of the mass of parathyroid tissue in the individual patient, or the degree of parathyroid hyperplasia. This, of course, is an oversimplification which would involve the assumption that all these patients were equally and absolutely deficient of vitamin D; and, although all had rickets or osteomalacia, such an assumption is not strictly justified. None the less, it is considered that the moderate hypercalcaemia in the two patients of group III was determined by massive hyperplasia. This contention is supported by the fact that the serum iPTH remained above normal when the serum calcium was raised to 13—14 mg/dl; and by the clinical evidence that these two middle-aged vegetarians may have had degrees of vitamin D insufficiency for most of their lives (9).

It is suggested that these two patients are analogous to that fraction of patients with chronic renal failure who also develop hypercalcaemia as a consequence of secondary hyperparathyroidism; but in whom the heterogeneity of immunoreactive hormonal species in the blood makes it much more difficult to document the reactivity of the parathyroid glands. In such cases of renal failure Slatopolsky *et al.* (15) have assessed parathyroid reactivity indirectly be measuring the effects of elevating the serum calcium on the renal tubular reabsorption of phosphate (TRP). Induced hypercalcaemia produced no effect on TRP; but, after removal of 3 to 3½ hyperplastic parathyroid glands, the residual parathyroid tissue appeared to respond normally to hypercalcaemia. Thus, the initial *apparent* unresponsiveness was a function of the mass of parathyroid tissue. The present experience with hypercalcaemic osteomalacia in simple vitamin D deficiency (group III) suggests that parathyroid secretion in Slatopolsky's patients probably responded similarly before and after sub-total parathyroidectomy; but that the pre-operative mass of tissue, even when suppressed by hypercalcaemia, still produced sufficient hormone to sustain renal phosphaturia.

Hypercalcaemic secondary hyperparathyroidism versus primary hyper-parathyroidism with vitamin D deficiency

Before discussing the clinical evolution of the two hypercalcaemic patients of group III, one must anticipate the challenge that they may have been cases of primary hyperparathyroidism with vitamin D deficiency or of 'tertiary' hyperparathyroidism — if there is such a thing (Clinicopatho-logical Conference (2)). The radioimmunoassay used in these studies is capable of detecting changes of serum iPTH, in patients with primary hyperparathyroidism, in response to induced changes of serum calcium. Generally we have tested parathyroid responsiveness in primary hyper-parathyroidism by reducing the serum calcium by EDTA infusions and have not felt ethically justified in inducing a further increase in serum calcium except in a very few cases. In those few patients who received a calcium infusion, the serum iPTH has fallen with increase in the serum calcium but without showing the apparently immediate arrest of secretion demonstrated in the secondary hyperparathyroidism of groups I and III (Fig. 5). If one aggregates the data from several patients, three having had calcium infusions and five EDTA infusions, they show a proportionate relationship between serum iPTH and serum calcium (Fig. 2B) entirely similar to that in the hypocalcaemic patients with vitamin D deficiency (group II, Fig. 2A) but differing, of course, in the range of calcium concentrations through which this pattern of response occurs. Thus, if these two hypercalcaemic cases of vitamin D deficiency (group III) are considered to have primary hyperparathyroidism, they demonstrate secretory phenomena that we have not so far detected in primary hyperparathyroidism by the use of the same techniques. The same two patients differ from hypercalcaemic cases of primary hyperparathyroidism with low vitamin D status in another way, the reasons for which are not yet understood. When given a pulse dose of radioactive cholecalciferol, patients with primary hyperparathyroidism and low vitamin D nutrition produce both radioactive 1,25-dihydroxycholecalciferol and radioactive 24,25-dihydroxycholecalciferol (11); the two patients of group III produced only radioactive 1,25-dihydroxycholecalciferol, like other normocalcaemic or hypocalcaemic patients with simple vitamin D deficiency.

Involution of the hyperparathyroidism in the patients of group III
If it is considered that these two patients have developed hypercalcaemia because of the degree of hyperplasia of their parathyroid glands, the analogy of hypercalcaemic secondary hyperparathyroidism in renal failure suggests that involution of the hyperplasia after correction of its primary cause could be extremely slow. Slatopolsky and his collaborators (15) could detect no evidence of parathyroid suppression after 2 months or more of induced hypercalcaemia in renal failure; hypercalcaemia due to persistent hyperparathyroidism may persist for 2 to 3 years after successful renal transplantation (4).

The clinical course in one of the cases of group III, during the first year after correction of vitamin D deficiency, has been documented in a previous publication (9); the extended course up to 2 years in both patients is shown in Fig. 6. The treatment given in hospital produced in both patients relief of pain and muscular weakness, healing of Looser's nodes and reversal of the radiographic signs of hyperparathyroidism; these changes were accompanied by a slow exponential fall of serum alkaline phosphatase into the normal range. All these changes were progressive in one patient, who continued her treatment and maintained adequate levels of serum 25-hydroxycalciferol; the other patient defaulted from taking her vitamin D supplement and relapsed after a year, when the serum 25-hydroxycalciferol was 4 ng/ml.

From the commencement of treatment in both patients there was an exponential fall in serum iPTH in parallel with the fall in serum alkaline phosphatase. Initially, this reduction in serum iPTH occurred without significant increase of serum calcium above pre-treatment values. There was a delayed rise of serum calcium after about a year of treatment, which is considered to be related to healing of the bone disease (9); but throughout the two years of observation, although there was a general inverse correlation between serum calcium and serum iPTH, this was not statistically significant.

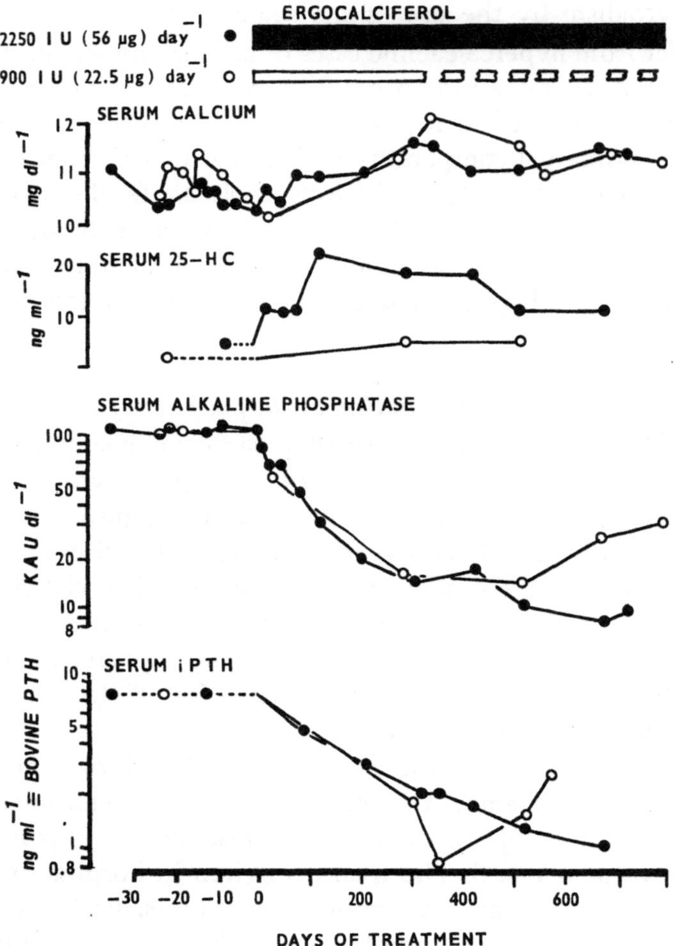

Figure 6. The involution of hyperparathyroidism in the two female hypercalcaemic patients (group III) during treatment with vitamin D.

Note: no measurements of serum 25-hydroxycalciferol were made when the second patient (O) was receiving vitamin D treatment in hospital. It is assumed that this treatment was sufficient to initiate the healing of her condition and that there was probably a rise in serum 25-hydroxycalciferol, as in the other patient (●); pathologically low values were again measured a year later when the patient defaulted from treatment.

During the period of 2 years, the steady-state concentration of serum iPTH had fallen very slowly, with a half-time of about 180 days, to reach levels of approximately 15% of the initial values; but these low concentrations were still 3 times the average normal value and the patients remained hypercalcaemic. There was thus a slow involution of the hyperparathyroidism and the long persistence of a lesser degree of hypercalcaemic hyperparathyroidism; a situation closely simulating the sequence of events encountered after renal transplantation (4). The low concentrations of serum iPTH achieved after prolonged treatment with vitamin D were of the same order as the concentrations produced acutely before treatment by a trivial rise in serum calcium (Fig. 5); under these conditions, the serum iPTH decayed with a half-time of 10—15 min as compared with six months. Thus, the same end result, in terms of serum iPTH, was produced by two totally different mechanisms.

The long-term suppression of parathyroid secretion was accompanied by healing of the bone disease, in both its osteomalacic and hyperparathyroid components; but, as is quite common with the osteodystrophy of vitamin D deficiency (18), histological healing was far from complete after a year of treatment. Since the patients were hypercalcaemic throughout the course of observation and since the correlation between serum iPTH and serum calcium was not statistically significant, it was suggested that the observed parathyroid suppression may have been produced by a direct action of one or other vitamin D sterol on the parathyroid glands themselves (9). The possibility cannot be excluded that the observed rise in serum calcium, and perhaps also undetected post-prandial increases of serum calcium after correction of the vitamin D deficiency, was responsible for parathyroid involution. But if the serum calcium is the only factor influencing the parathyroid glands, there is no reason why the parathyroid response should result in a mass of tissue in excess of physiological requirements and the production of actual hypercalcaemia. This, and the analogous situation in chronic renal failure, seems to imply an uncoupling of the mechanisms by which the serum calcium influences the synthesis and secretion of parathyroid hormone, and the mechanisms by which calcium influences the rate of mitosis in parathyroid cells.

ACKNOWLEDGEMENT

These studies were supported by a Programme Grant from the British Medical Research Council and the Department of Health and Social Security.

The antiserum 211/32 was kindly provided by the M.R.C. Division of Biological Standards.

200

REFERENCES

1. Bonucci, E. & Maschio, G.: Morphological aspects of bone tissue in chronic renal disease. In: *Vitamin D and Problems Related to Uremic Bone Disease.* Norman, A.W., Schaefer, K., Grigoleit, H.G., van Herrath, D. & Ritz, E. (eds.), de Gruyter, Berlin, New York, pp. 523–530, 1975

2. Clinicopathologic Conference. Tertiary hyperparathyroidism – does it exist? *Amer. J. Med.* **52**, 254–266 (1972)

3. Cohn, D.V., MacGregor, R.R., Chu, L.L.H., Huang, D.W.Y., Anast, C.S. & Hamilton, J.W.: Biosynthesis of proparathyroid hormone and parathyroid hormone. Chemistry, physiology and role of calcium in regulation. *Amer. J. Med.* **56**, 767–773 (1974)

4. David, D.S., Sakai, S., Brennan, B.L., Riggio, R.A., Cheigh, J., Stenzel, K.H., Rubin, A.L. & Sherwood, L.M.: Hypercalcaemia after renal transplantation. Long-term follow-up data. *New Engl. J. Med.* **289**, 398–401 (1973)

5. Drezner, M., Neelon, F.A. & Lebovitz, H.E.: Pseudohypoparathyroidism type II: a possible defect in the reception of the cyclic AMP signal. *New Engl. J. Med.* **289**, 1056–1060 (1973)

6. Fraser, D., Kooh, S.W. & Scriver, C.R.: Hyperparathyroidism as the cause of the hyperaminoaciduria and phosphaturia in human vitamin D deficiency. *Pediat. Res.* **1**, 425–435 (1967)

7. Kimmich, G.A. & Rasmussen, H.: Regulation of pyruvate carboxylase activity by calcium in intact rat liver mitochondria. *J. biol. Chem.* **244**, 190–199 (1969)

8. Littledike, E.T., Witzel, D.A. & Whipp, S.C.: Insulin: evidence for inhibition of insulin release in spontaneous hypocalcaemia. *Proc. Soc. exp. Biol. (N.Y.)* **129**, 135–139 (1968)

9. Lumb, G.A. & Stanbury, S.W.: Parathyroid function in human vitamin D deficiency and vitamin D deficiency in primary hyperparathyroidism. *Amer. J. Med.* **56**, 833–839 (1974)

10. Malluce, H.H., Ritz, E., Kutschera, J., Krause, G., Werner, E., Gati, A., Seiffert, U. & Lange, H.P.: Calcium metabolism and impaired mineralisation in various degrees of renal insufficiency. In: *Vitamin D and Problems Related to Uremic Bone Disease.* Norman, A.W., Schaefer, K., Grigoleit, H.G., von Herrath, D. and Ritz, E. (eds.), Gruyter, Berlin, New York, pp. 513–522, 1975

11. Mawer, E.B., Backhouse, J., Hill, L.F., Lumb, G.A., de Silva, P., Taylor, C.M. & Stanbury, S.W.: Vitamin D metabolism and parathyroid function in man. *Clin. Sci. Molec. Med.* **48**, 349–365 (1975)

12. Parfitt, A.M.: Relation between parathyroid cell mass and plasma calcium contration in normal and uremic subjects. *Arch. intern. Med.* **124**, 269–273 (1969).

13. Ramberg, C.F. Jr., Mayer, G.P., Kronfield, D.S., Aurbach, G.D., Sherwood, L.M. & Potts, J.T. Jr.: Plasma calcium and PTH response to EDTA infusion in the cow. *Amer. J. Physiol.* **213**, 878 (1967)

14. Rodriguez, H.J., Villarreal, H., Klahr, S. & Slatopolsky, E.: Pseudohypoparathyroidism type II; restoration of renal responsiveness to parathyroid hormone by calcium administration. *J. clin. Endocr.* **39**, 693–701 (1974)

15. Slatopolsky, E., Rutherford, W.E., Hoffsten, P.E., Elkan, I.O., Butcher, H.R. & Bricker, N.S.: Non-suppressible secondary hyperparathyroidism in chronic progressive renal disease. *Kidney Internat.* **1**, 38–46 (1972)

16. Stanbury, S.W. & Lumb, G.A.: Metabolic studies of renal osteodystrophy. I. Calcium, phosphorus and nitrogen metabolism in rickets, osteomalacia and hyperparathyroidism complicating chronic uraemia and in the osteomalacia of the Fanconi syndrome. *Medicine (Baltimore)* **41**, 1–31 (1962)
17. Stanbury, S.W. & Lumb, G.A.: Parathyroid function in chronic renal failure. *Quart. J. Med.* **35**, 1–23 (1966)
18. Stanbury, S.W., Torkington, P., Lumb, G.A., Adams, P.H., de Silva, P. & Taylor, C.M.: Asian rickets and osteomalacia: patterns of parathyroid response in vitamin D deficiency. *Proc. Nutr. Soc.* **34**, 111–117 (1975)
19. Taylor, A.N. & Wasserman, R.H.: Vitamin D-induced calcium-binding protein: comparative aspects in kidney and intestine. *Amer. J. Physiol.* **223**, 110–114 (1972)

Calcium Metabolism in Patients with Chronic Non-dialytic Renal Disease

S. Pors Nielsen, O. Helmer Sørensen, B. Lund, O. Bärenholdt, O. Munck & K. Pedersen

Recent research has disclosed that the kidney has a hitherto unrecognized role in the regulation of calcium metabolism. The normal kidney is able to convert 25-hydroxycholecalciferol (25-OH-D_3) to 1,25-dihydroxy-cholecalciferol (1,25-(OH)$_2$-D_3) (4), which is more active in promoting intestinal calcium absorption than the native form of the vitamin or any other known metabolite. The diseased kidney has more or less lost this ability, causing in some cases a metabolic derangement with symptoms of vitamin D deficiency (osteomalacia). The osteomalacia and the hyper-phosphataemia apparently are major causes of hypocalcaemia in chronic renal disease, and therefore indirectly responsible for the secondary hyperparathyroidism.

Many sophisticated methods are available for diagnosing calcium-metabolic disturbances, and great efforts have been made to treat these disturbances in patients with chronic renal disease. We have recently demonstrated that the synthetic vitamin D analogue 1α-hydroxy-cholecalciferol (1α-OH-D_3) can reverse the biochemical manifestations of vitamin D deficiency in dialysis patients with renal osteodystrophy, and ameliorate the radiological bone lesions of such patients (8).

Little is known about the severity and incidence of calcium-metabolic derangement in patients with chronic non-dialytic renal disease. Knowledge hereof is important to achieve in order to clarify the need of those patients for treatment with 1,25-(OH)$_2$-D_3, 1α-OH-D_3 or other active vitamin D analogues. The aim of this work was to define the incidence and gravity of calcium-metabolic derangement in a population of non-dialyzed out-patients with chronic renal disease of varying degrees. We have, in particular, been interested in the metabolism of vitamin D in these

Department of Clinical Physiology & Department of Medicine E, Frederiksberg Hospital, Copenhagen and Department of Clinical Physiology & Department of Medicine B, Glostrup Hospital, Glostrup.

patients. It has been suggested (10) that the hepatic 25-hycroxylation is impaired in chronic non-dialytic renal disease, and that this should be the explanation of the low plasma values of 25-OH-D$_3$ sometimes found. We have studied this, and shall in the following show that the mean plasma 25-OH-D$_3$ concentration was elevated in our patients, and present evidence that if low 25-OH-D$_3$ plasma concentrations are encountered in patients with chronic non-dialytic renal disease the explanation of this phenomenon is unlikely to be an impaired hepatic 25-hydroxylation of vitamin D.

MATERIAL AND METHODS

Patients

A total number of 81 adults with non-dialytic chronic renal disease were studied, all out-patients of the nephrology department of Glostrup Hospital.

The following parameters were studied: 1) Bone density (photon absorptiometry), 2) concentrations in serum of 25-OH-D$_3$, creatinine, Ca, total protein, alkaline phosphatase and inorganic phosphatate (P$_i$).

Bone density

The bone density (bone mineral content, BMC) was determined by photon absorptiometry of both forearms (1).

Biochemical measurements

All blood samples were drawn without stasis of the forearm. Serum was separated immediately. 25-OH-D$_3$ was measured by competitive protein binding assay (modification of the method, described by Haddad & Chyu (5)).

The recovery of 25-OH-D$_3$ after extraction and chromatography was 79,1% (SD = 4.2%) (n = 65). The recovery of added 25-OH-D$_3$ was 94% (SD = 19%) (n = 10) after correction for procedural losses. Standard deviation of the difference of paired samples was 1.6 ng (n = 48). The coefficient of variation of repeated measurements at the level of 15 ng was 13,5% (n = 30). The sensitivity in our routine assay was 1,5 ng/ml.

Alkaline phosphatase was measured by the Merckotest® for alkaline phosphatase, P$_i$ was determined by the phosphomolybdate method, and creatinine by the picrate method. Refractometry was used for protein measurement and atomic absorption spectrophotometry for Ca measurement (Perkin-Elmer Atomic Absorption Spectrophotometer 403).

Calculations

The creatinine clearance of the patients was estimated from the serum creatinine level, age and body weight by the nomogram of Kampmann and co-workers (6).

RESULTS

The alkaline phosphatase concentration of the patients, like the normals, had a Gaussian-type distribution (log. scale). The mean concentration of alkaline phosphatase was, although significantly higher than in normals (p < 0.001) only moderately elevated and not as high as in overt osteomalacia (Fig. 1a). The highest levels of alkaline phosphatase were encountered at low creatinine clearances.

The patients, like the normals, had a Gaussian-type distribution of BMC with about the same variance as the normals (Fig. 1b). It can be seen that the BMC values of the patients were significantly lower than in the normals (p < 0.01). In addition, we found that patients with very low bone densities had higher serum alkaline phosphatase levels than patients with high BMC-values (Fig. 1c). BMC tended to be low at low creatinine clearance, but there was no clear-cut correlation between the two parameters.

The mean serum Ca concentration of the patients and its variance did not differ much from that found in normals. There was no correlation between the serum Ca concentration of the patients and their serum

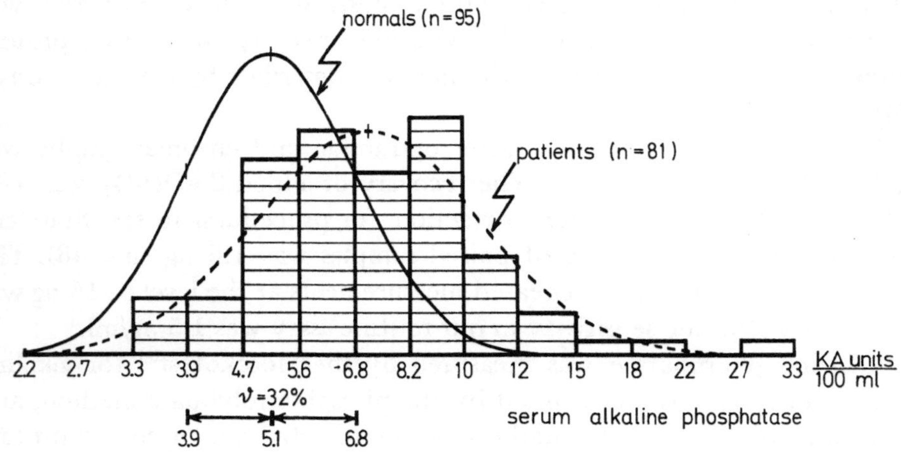

Figure 1a. Distribution of serum alkaline phosphatase concentration (log. scale, 20% intervals) in normals (1) and in the patients; ϑ = coefficient of variation of the normal subjects.

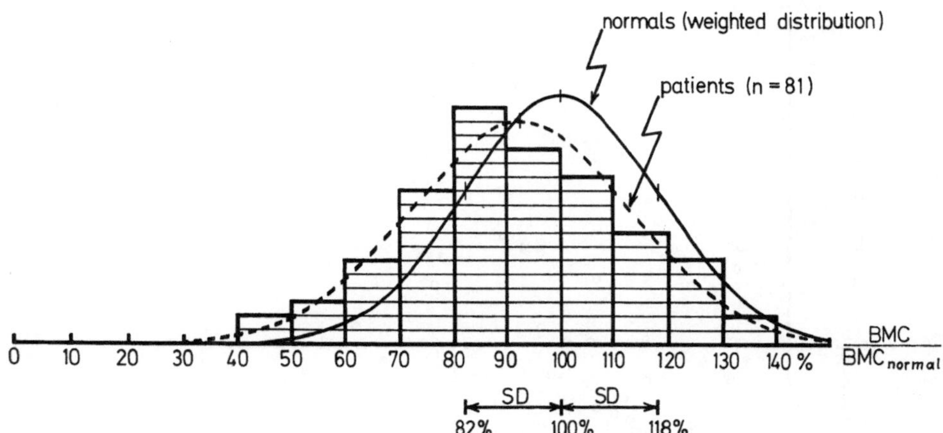

Figure 1b. Distribution of bone density of the forearm (BMC), expressed as measured BMC values divided by the normal BMC value (normalized values in per cent) (1); SD = standard deviation of normals.

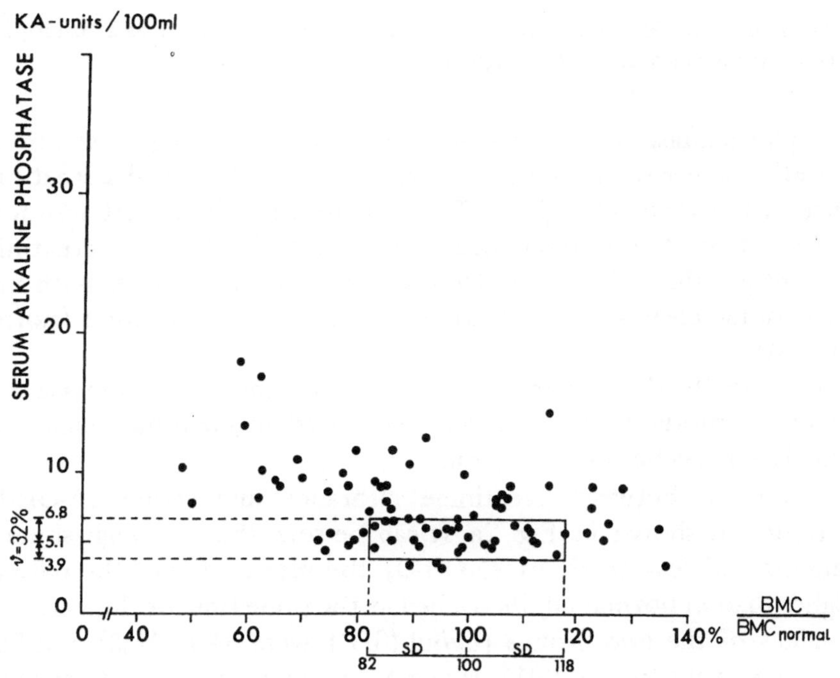

Figure 1c. Serum alkaline phosphatase concentration and normalized BMC values in the patients. Rectangular area denotes normal values (see legends of Fig. 1a and Fig. 1b).

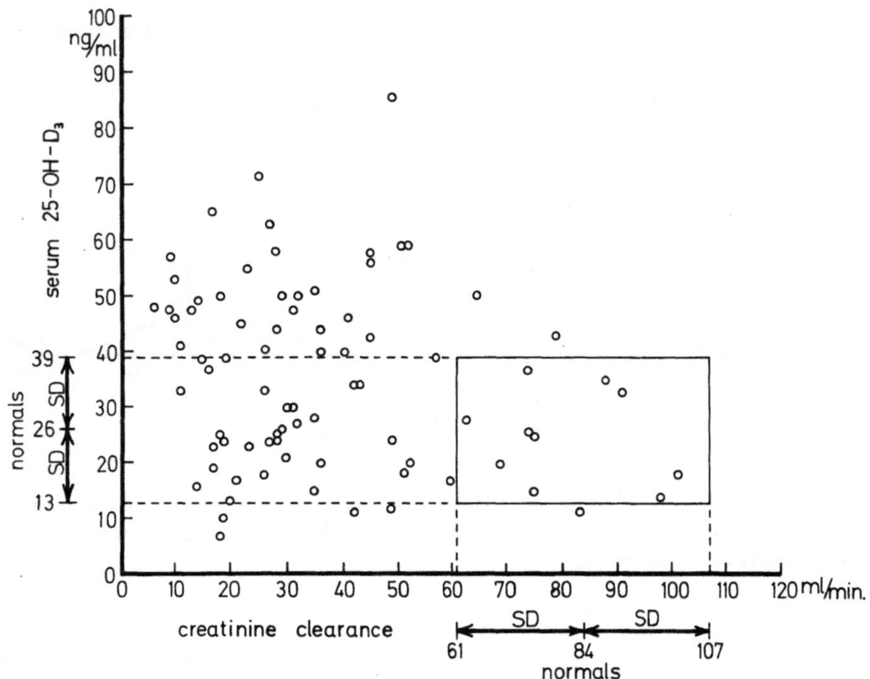

Figure 2a. Serum 25-OH-vitamin D₃ concentrations of the patients as a function of estimated creatinine clearance. Rectangular area denotes normal values.

alkaline phosphatase concentration. The patients having a serum Ca concentration lower than the normal mean minus 1 SD all had a creatinine clearance of less than 53 ml/min. The patients with the lowest serum Ca values (less than the normal mean minus 2 SD) all had a creatinine clearance of less than 33 ml/min. However, most of the patients with such a low creatinine clearance had a normal serum Ca concentration (normal mean ± 1 SD).

In our patients the serum P_i concentration and the serum Ca x P_i concentration product was inversely correlated to creatinine clearance, when the latter was below 40 ml/min.

The relationship between creatinine clearance and serum 25-OH-D₃ concentration is shown in Fig. 2a. It can be seen that although some of the patients had low levels of 25-OH-D₃ the mean value of the patients was higher than in normal adults studied at the same time of the year. The mean values of the two groups (ng/ml (SD)) were 34,1 (16,3) and 25,8 (13,3), respectively (p < 0.001). However, some of the patients had very low 25-OH-D₃ concentrations. Such low concentrations could be due to nutritional osteomalacia or to an impaired conversion of vitamin D₃ to 25-OH-D₃ (hepatic 25-hydroxylation). In order to find out which were

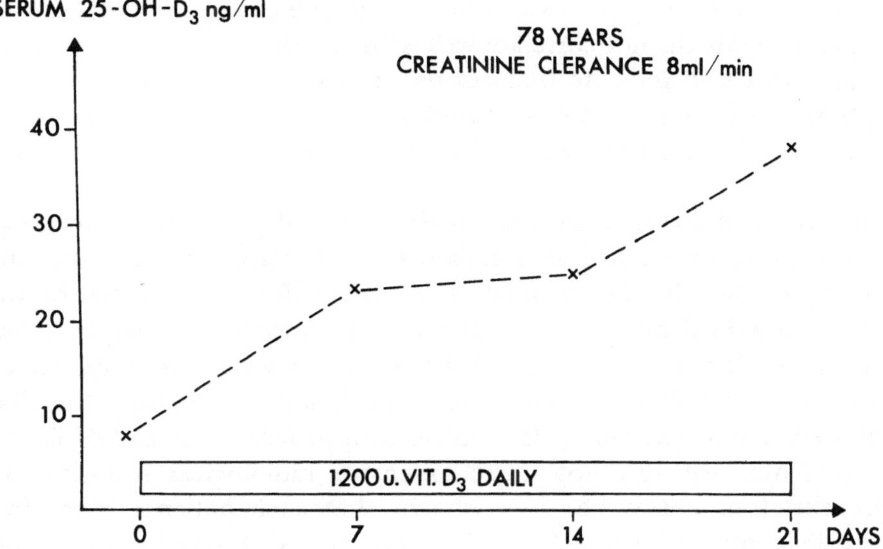

Figure 2b. Effect of oral administration of vitamin D_3 on serum concentrations of 25-OH-vitamin D_3 in one patient.

true we investigated one patient with chronic non-dialytic renal disease (GFR = 8 ml/min) and low plasma concentration of 25-OH-D_3. The patient was given 1200 U of vitamin D_3 daily by mouth for 3 weeks. Serum 25-OH-D_3 during that treatment progressively rose from 8 to 38 ng/ml, suggesting that the ability of the patient to 25-hydroxylate vitamin D_3 was not impaired (Fig. 2b).

DISCUSSION

The fact that serum alkaline phosphatase levels of our patients were only moderately elevated, even if the creatinine clearance was low, would suggest that our patients with chronic non-dialytic renal disease did not suffer from any severe degree of uraemic osteodystrophy. This is corroborated by the fact that BMC was only moderately decreased. It can be seen from our data that the BMC values of the patients to a great extent overlap the BMC values of the normals. Normalizing for skeletal size does not significantly improve the separation of the groups. The phenomenon probably mirrors not only varying degrees of uraemic osteodystrophy, but also a variability of diet, treatment and duration of uraemia. As shown by S.H. Cohn and co-workers elsewhere in this book (2) in a population of renal patients there is a significant correlation between BMC and total body Ca (TBCa), although changes of BMC in the individual patients do not correlate well with changes in TBCa. Therefore, our data might well indicate that our patients as a population do in fact have diminished TBCa. This is, however, not necessarily suggestive of osteomalacia of any greater extent but can be explained by diet and lack of exercise.

Comparison of mean values for 25-OH-D$_3$ is not particularly meaningful unless the group studied are standardized with respect to sun exposure and diet (vitamin D intake included). Low serum 25-OH-D$_3$ concentration have been described by others in patients with chronic non-dialytic renal disease (7, 9). It has been suggested that such low values were due to an impaired hepatic 25-hydroxylation (10). Our data do not confirm this. We did, however, find extremely low serum-concentrations of 25-OH-D$_3$ in some patients. We did not make extensive radiological examination of the skeleton, but it has been shown that such patients more frequently than others have radiographic signs of osteomalacia (3). It was not possible to group our patients according to their diet, but the fact that a substantial number of the patients took vitamin D supplements could account for the relatively high 25-OH-D$_3$ levels found. The fact that the serum 25-OH-D$_3$ levels progressively rose from subnormal to high normal values in one patient during treatment with vitamin D$_3$ would suggest that the hepatic 25-hydroxylation is unimpaired. It is possible, but difficult to prove, that the high mean concentration of 25-OH-D$_3$ found in the patients might represent an adaptation to increased needs in a situation of reduced production of 1,25-(OH)$_2$D$_3$ and subclinical osteomalacia.

It is still an open question whether patients with chronic non-dialytic renal disease should be treated with 1,25-(OH)$_2$D$_3$, 1α-OH-D$_3$ or other analogues, or whether such treatment should be given to patients with overt osteomalacia only. The present data reveal no greater need for treatment with active vitamin D in patients with chronic non-dialytic bone disease, but it is possible that a long-term profylactic, carefully controlled treatment with e.g. 1α-OH-D$_3$ would be advisable.

ABSTRACT

Calcium metabolism was evaluated in 81 patients with chronic non-dialytic renal disease. Bone mineral content measured by photon absorptiometry was significantly lower than in normals, and serum alkaline phosphatase levels higher than in normals. However, the differences were small indicating that the calcium-metabolic disturbance was moderate. None of the patients suffered from overt osteomalacia.

The mean serum level of 25-hydroxy-cholecalciferol (25-OH-D$_3$) was slightly, but significantly higher than in normals studied at the same time of the year, but some patients had subnormal 25-OH-D_3 levels. One such patient was given a daily dose of 1200 U of vitamin D_3 by mouth for 3 weeks. Hereafter, the serum 25-OH-D_3 concentration rose progressively to high normal values. These two observations would suggest that the hepatic 25-hydroxylation in such patients is unimpaired.

The results are in accordance with the existence of a subclinical vitamin D deficiency in patients with non-dialytic renal disease, due to an impaired production of 1,25-$(OH)_2D_3$. It is an open question whether such patients should be given (prophylactic) treatment with 1α-hydroxycholecalciferol.

ACKNOWLEDGEMENT

This work was supported by a grant from "Den lægevidenskabelige forskningsfond for Storkøbenhavn, Færøerne og Grønland".

REFERENCES

1. Christiansen, C., Rødbro, P. & Lund, M.: Incidence of anticonvulsant osteomalacia and effect of vitamin D: controlled therapeutic trial. *Brit. med. J.* IV, 695–701 (1973)
2. Cohn, S.H., Ellis, K.J., Caselnova, R.C., Asad, S.N. & Letteri, J.M.: Loss of calcium from axial and appendicular skeleton in patients with chronic renal failure. *Proc. XIth Europ. Symp. Calc. Tiss.*
3. Fairney, A., Bowling, K., Varghese, A., Moorhead, J.F. & Wills, M.R.W.: Serum 25-hydroxycholecalciferol (25 HCC) concentration in chronic renal failure. In: *Vitamin D and Problems related to Uremic Bone Disease*, Norman, A.W., Schaefer, K., Grigoleit, H.-G., von Herrath, D. and Ritz, E. (eds.), de Gruyter, Berlin, New York, pp. 319–323, 1975
4. Fraser, D.R. & Kodicek, E.: Unique biosynthesis by kidney of a biologically active vitamin D metabolite. *Nature (Lond.)* 228, 764–766 (1970)
5. Haddad, J.G. & Chyu, K.J.: Competitive protein-binding radioassay for 25-hydroxycholecalciferol. *J. Clin. Endocr.* 33, 992–995 (1971)
6. Kampmann, J.P., Siersbæk-Nielsen, K., Kristensen, K. & Mølholm Hansen, J.: Aldersbetingede variationer i urinkreatinin og endogen kreatininclearance. *Ugeskr. Læg. (Copenhagen)* 133/48, 2369–2372 (1971)
7. Letteri, J., Roginsky, M., Moo, F., Scipione, R., Ellis, K. & Cohn, S.: The relationship between calcified tissue mass and plasma 25-hydroxycholecalciferol in chronic renal failure. In: *Vitamin D and Problems related to Uremic Bone Disease*, Norman, A.W., Schaefer, K., Grigoleit, H.-G., von Herrath, D. and Ritz, E. (eds.), de Gruyter, Berlin, New York, pp. 303–309, 1975
8. Nielsen, S. Pors, Binderup, E., Godtfredsen, W.O., Jensen, H. & Ladefoged, J.: Long-term treatment of uraemic osteodystrophy with 1α-hydroxycholecalciferol. In: *Vitamin D and Problems Related to Uraemic Bone Disease*, Norman, A.W., Schaefer, K., Grigoleit, H.-G., von Herrath, D. & Ritz, E. (eds.), de Gruyter, Berlin, New York, pp. 623–628, 1975
9. Offermann, D., von Herrath, D. & Schaefer, K.: Serum 25-hydroxycholecalciferol in uremia. *Nephron* 13, 269–277, (1974)
10. Wake, C.J. & Maddocks, J.L.: Vitamin-D metabolism in chronic renal failure. *Lancet* (1975) I, 516

Intestinal Calcium Absorption and Whole-body Calcium Retention in Various Stages of Renal Insufficiency

E. WERNER, H. H. MALLUCHE, J. KUTSCHERA, M. HODGSON & W. SCHOEPPE

Pathogenetic mechanisms and clinical signs of bone disease in *terminal* renal failure are well described in literature (1, 8). More recent results show that already in *early* renal failure peripheral parathyroid hormone resistance (6) and changes in bone histology (5) will be found.

It was the purpose of the study reported in this paper to investigate whether changes of intestinal absorption of calcium and/or whole body retention of calcium (^{47}Ca) are also demonstrable in these early phases of renal insufficiency. In addition the influence of treatment with 5,6 *trans* 25 Hydroxycholecalciferol (5,6-*trans*-25-OHCC), a synthetic active vit. D metabolite, on these parameters was studied.

MATERIAL AND METHODS

Fractional intestinal absorption of calcium (A_{Ca}) and whole body retention of ^{47}Ca i.v. (R_{Ca}) were measured with a whole body counter (WBC) in 103 patients (42 males, mean age 43 ± 13 years, range 19—63 years; 61 females, mean age 43 ± 12 years, range 19—73 years) in various stages of renal insufficiency. The same measurements were performed in 37 healthy volunteers (24 males, mean age 33 ± 10 years, range 24—68 years; 13 females, mean age 31 ± 8 years, range 21—45 years). All persons investigated lived on a normal diet, none of the patients received vit. D, vit. D analogous sterols, or calcium and aluminium hydroxyde respectively. Each patient was fasted from the night before until noon on the day of the test. At 9 a. m. the patient was by mouth given 18 mg calcium (as calcium gluconate) in 100 ml of deionized water which contained 0.5—1 μCi ^{47}Ca. The natural body radioactivity was counted by the WBC before

Ges. f. Strahlen- und Umweltforschung, Frankfurt/Main.

administration of radiocalcium. To control the intake of the test dose, we measured the patient immediately after the administration. The ^{47}Ca activity retained 7 days after the test dose was measured again with the WBC. To correct for the excretion of absorbed ^{47}Ca within this period a tracer dose of ^{47}Ca (4 μCi, spec. activity $>$ 100 μCi ^{47}Ca/mg Ca $\hat{=}$ i.v. Ca-application $<$ 0.003 meq. $\hat{=}$ $<$ 0.5 meq./l inj. sol.) was given intravenously. Two to four hours after the injection the initial counting rate for whole body calcium retention (after correction of rest activity from the oral dose) was measured by WBC. The fractional intestinal absorption of Ca was calculated by:

$$A_{Ca} = \frac{Z_{7,oral} \times a_{inj.}}{Z_{7,inj.} \times a_{oral}}$$

A_{Ca} : fractional intestinal absorption of calcium
a_{oral} : ^{47}Ca-activity oral
$a_{inj.}$: ^{47}Ca-activity i.v.
$Z_{7,oral}$: net count rate of ^{47}Ca 7 days after oral application
$Z_{7,inj.}$: net count rate of ^{47}Ca 7 days after i.v. application.

A_{Ca} represents the actual fractional amount of Ca absorbed (unidirectional influx).

To determine whole body retention of ^{47}Ca two additional measurements two and four weeks after i.v. ^{47}Ca were performed respectively. All count rates were corrected for radioactive decay and normalized to the first counting value after injection. The following additional parameters were measured: total and ionized calcium, serum alkaline phosphatase, serum phosphate, arterial serum pH, total serum protein, glomerular filtration rate (measured as endogenous creatinine clearance), urinary hydroxyproline, and 24 h urinary excretion of calcium and phosphate. Serum parathyroid hormone concentration was determined by radioimmunoassay.

Iliac crest biopsy was performed in 57 patients (undecalcified sections after embedding in methylmetacrylate) and evaluated by micromorphometry after the method of Schenk and Merz (7).

Immediately after the measurement of A_{Ca} and R_{Ca} in 23 patients treatment with 2000 I.U. 5,6-*trans*-25-OHCC/day was started. 3 to 9 months later a control measurement was performed under the same conditions as before.

RESULTS AND DISCUSSION

To get high absorption values in normal controls we used a low calcium content of the test dose. Thus, we were able to detect decreases in A_{Ca} due to reduction of active intestinal calcium transport earlier than with higher amounts of calcium in test doses. The mean value of A_{Ca} in healthy volunteers was 0.72 ± 0.10 ($\bar{x} \pm 1$ SD). There was no significant difference between male and female controls. The mean normal value of R_{Ca} of ^{47}Ca four weeks after injection was 0.41 ± 0.06. There was a significant difference between men (mean value 0.43 ± 0.05) and women (mean value 0.37 ± 0.06) ($p < 0.005$). R_{Ca} varied with age. Measurements in children (various metabolic diseases) showed significantly higher values than in adults. We found — in good agreement to the bone remodelling data of Frost (4) — the highest rates of R_{Ca} in newborn, decreasing values in adolescents, and in adults rather constant rates.

Fractional intestinal absorption of calcium was significantly diminished in patients with renal insufficiency ($p < 0.001$). There was a good correlation between A_{Ca} and glomerular filtration rate (GFR) ($r = 0.37$) (Fig. 1a). The first decreased values for A_{Ca} were seen at a GFR of about 80 ml/min $\times$ 1.73 m^2. It is noteworthy that, on the other hand,

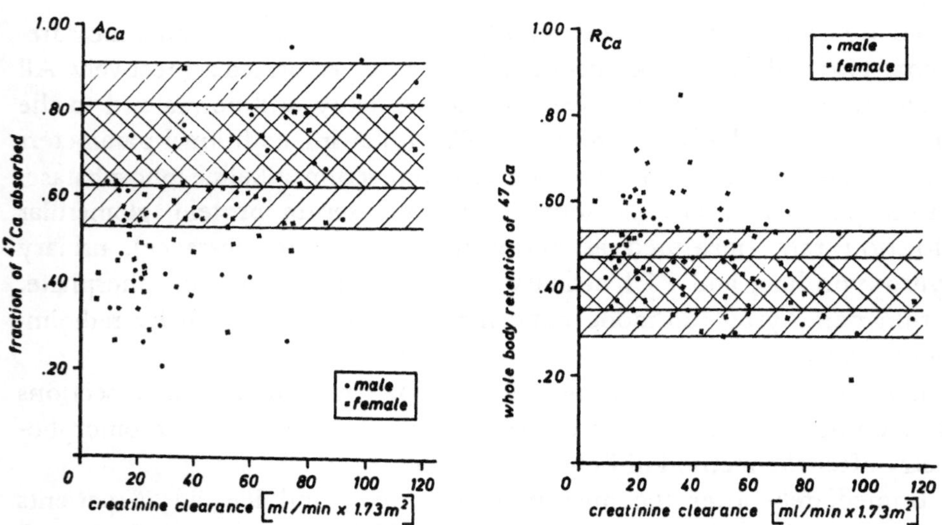

Figure 1. Variations of intestinal absorption (A_{Ca}) and whole body retention of calcium (R_{Ca}) in various stages of renal insufficiency.
a) *Left:* relation between GFR and A_{Ca};
b) *Right:* relation between GFR and R_{Ca} (hatched areas depict normal range (1 SD/2 SD)).

some patients even with a GFR lower than 20 ml/min $\times$ 1.73 m² showed normal values of A_{Ca}. This is in good agreement with the findings of Brickman (2). Additionally there were good correlations between A_{Ca} and urinary Ca-excretion (r = 0.34) and serum alkaline phosphatase (r = 0.40). Since serum alkaline phosphatase reflects skeletal turnover one may conclude that patients with higher degrees of metabolic osteopathy show a lower intestinal calcium absorption. All the other measured serum and urinary parameters showed no significant correlations with A_{Ca}. Complementarily, there were significant correlations between histomorphometrically defined changes of bone histology and A_{Ca}. A_{Ca} correlates with neutral trabecular surface (NS) (r = 0.36), inversely with osteoid volume (VO) (r = − 0.35) as well as with osteoid seam thickness (ȿ) (r = − 0.35) (Fig. 2a), and with the degree of endosteal fibrosis (EOFS) (r = − 0.31). The correlations between the skeletal histology and A_{Ca} are possibly coincidental since changes in skeletal histology increase and A_{Ca} decreases with decreasing GFR. The correlations between A_{Ca} and alkaline phosphatase as well as micromorphometric parameters of osteoid (osteoid seam thickness, osteoid volume) may suggest that even in these early stages of renal insufficiency a disturbance of vit. D metabolism (reduced conversion of 25-hydroxycholecalciferol (25-OHCC) to 1,25-dihydroxycholecalciferol (1,25-DHCC) or a peripheral 1,25-DHCC resistance) exists. These results are in parallel with the experimental data of Massry (6) who found even at a GFR of about 80 ml/min $\times$ 1.73 m² a peripheral resistance of endogenous and exogenous parathyroid hormone. Investigations with the recently described radioimmunoassay for 1,25-DHCC (3) will be able to clarify this problem.

Whole body retention of [47]Ca increased with decreasing GFR (r = − 0.47) (Fig. 1b) and was directly proportional to PTH concentrations in serum (r = 0.49). This latter correlation could be again coincidental due to the analogous increase of PTH concentrations and R_{Ca} with falling GFR. On the other hand, these correlations could be explained by the fact that in hyperparathyroidism with consecutively elevated bone turnover avidity of skeleton for calcium increases and consequently R_{Ca} increases, too. This assumption is confirmed by the negative correlations between R_{Ca} and serum as well as urinary calcium (r = − 0.48 rsp. r = − 0.42). Similar to the correlation between R_{Ca} and PTH concentrations we found correlations between R_{Ca} and PTH mediated parameters such as phosphate clearance (r = − 0.32), fractional phosphate clearance (r = 0.38), alkaline phosphatase (r = 0.42), bone phosphatase (r = 0.43), and arterial serum pH (r = 0.42). No correlations were seen between R_{Ca} and total protein, urinary hydroxyproline and 24 h phosphate excretion in the urine.

Pathological changes of bone histology were in parallel with variations of

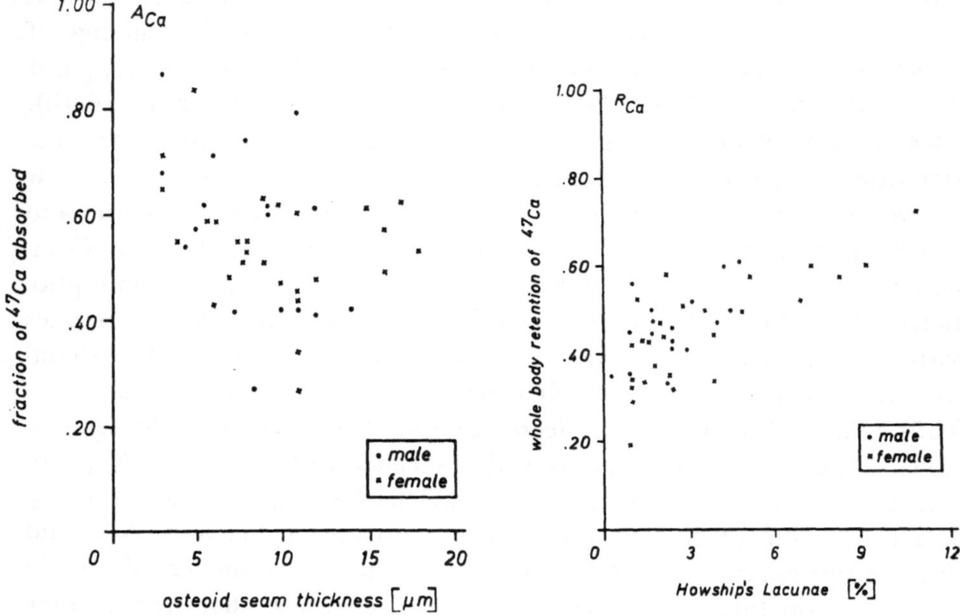

Figure 2. Relationship between fractional absorption (A_{Ca}), whole body retention of calcium (R_{Ca}) and histological parameters of bone apposition and bone resorption.
a) *Left*:   relation between A_{Ca} and osteoid seam thickness ($\bar{s}$);
b) *Right*:   relation between R_{Ca} and Howship's Lacunae filled with osteoclasts.

R_{Ca}. The closest correlation was seen between R_{Ca} and endosteal fibrosis (SVEOF/R_{Ca}: r = 0.58). Nearly the same correlation was seen between R_{Ca} and neutral trabecular surface (NS) but logically it was inverse (r = − 0.57). Parameters of osteoid (VO/R_{Ca}: r = 0.56; $\bar{s}$/R_{Ca}: r = 0.55; OS/R_{Ca}: r = 0.53) correlated with R_{Ca}, too. In addition not only osteoid parameters but also parameters of bone resorption (HO/R_{Ca}: r = 0.50; OI/R_{Ca}: r = 0.48) correlated to R_{Ca} (Fig. 2b). These close correlations between R_{Ca} and histological parameters of bone apposition and resorption demonstrate that R_{Ca} represents a simple and reliable method for assessing bone turnover, preferably bone accretion rate.

After treatment with 5,6-*trans*-25-OHCC for 3 to 9 months we did not find a significant increase of A_{Ca} although there was a considerable number of patients showing an increase of A_{Ca} (preferably patients with higher levels of GFR). In our opinion a daily dose of 2000 I.U. 5,6-*trans*-25-OHCC is not sufficient to improve the decreased A_{Ca} due to uremia. Preliminary results with higher doses confirm this explanation. These findings correspond with the experiences of Brickman with 1,25-DHCC (2). R_{Ca} also showed a tendency to increase during treatment with 5,6-*trans*-25-OHCC, but with the dose of 2000 I.U. daily this was not statistically significant.

The presented data show that A_{Ca} and R_{Ca} changed in parallel to our histological findings (5) already in early stages of renal insufficiency. They are useful to judge azotemic osteopathy in the earliest clinically asymptomatic stages.

ACKNOWLEDGEMENT

We thank Dr. W.H.L. Hackeng, Rotterdam, for the determination of PTH concentrations.

REFERENCES

1. Bricker, N.S., Slatopolsky, E., Reiss, E. & Avioli, L.V.: Calcium, Phosphorous and Bone in Renal Disease and Transplantation. *Arch. intern. Med.* 123, 543—553 (1969)
2. Brickman, A.S., Massry, S.G., Norman, A.W. & Coburn, J.W.: On the Mechanism and Nature of the Defect in Intestinal Absorption of Calcium in Uremia. *Kidney Internat.* 7, Suppl. 2, 113—117 (1975)
3. Broumbaugh, P.F., Haussler, D.H., Bressler, R. & Haussler, M.R.: Radioceptor-Assay for 1,25 Dihydroxyvitamin D_3. *Science* 183, 1089—1091 (1974)
4. Frost, H.M.: *Bone Remodelling Dynamics.* Thomas, pp. 17—41, 1963
5. Malluche, H.H., Ritz, E., Hodgson, M., Kutschera, J., Krause, G., Seiffert, U., Gati, A. & Lange, H.P.: Skeletal Lesions and Calcium Metabolism in Early Renal Failure. *Proc. Eur. Dial. Transp. Ass. XI*, Tel Aviv (1974). In Press
6. Massry, S.G., Llach, F., Coburn, J.W., Singer, F.R., Kurokawa, K. & Kaye, J.H.: Homeostasis and Action of Parathyroid Hormone in Normal Men and in Patients with Mild Renal Failure. *Proc. Eur. Dial. Transp. Ass. XI*, Tel Aviv (1974). In Press
7. Merz, W.A. & Schenk, R.K.: Quantitative Structural Analysis of Human Cancellous Bone. *Acta Anat. (Basel)* 75, 54—66 (1970)
8. Ritz, E., Malluche, H.H., Bommer, J., Mehls, O. & Krempien, B.: Metabolic Bone Disease in Patients on Maintenance Haemodialysis. *Nephron* 12, 393—404 (1974)

Loss of Calcium from Axial and Appendicular Skeleton in Patients with Chronic Renal Failure

S. H. COHN, K. J. ELLIS, A. N. MARTINO*, S. N. ASAD*, & J. M. LETTERI*

The widespread prevalance of bone disease in chronic renal failure both prior to and during hemodialysis is an important aspect of uremia. Loss of bone mineral of skeleton in renal disease can be measured directly by total-body neutron activation analysis (TBNAA). The absorptiometric technique, using monochromatic photons from ^{125}I, applied to the appendicular skeleton (radius) also reflects the loss of bone mineral content (BMC) in renal disease. In the present study the results of these two techniques are compared in 25 patients with renal insufficiency, 53 with end stage renal failure on dialysis, and 24 normal control subjects.

The range in absolute levels of total-body calcium in the renal patients is very large (466—1307 g) (see Table I). This variability renders an average TBCa value for the group meaningless. However, normalization of the data for sex, age and skeletal size greatly reduce this variability. In order to measure the relative deficit in TBCa in individual patients from the absolute Ca measurement, it is necessary to normalize the data for sex, age and skeletal size. For this purpose an empirically derived relationship was used to predict the normal skeletal Ca in each subject, based on weight, height, sex and age (1, 2). The measured TBCa divided by the predicted TBCa is referred to as the calcium ratio. This ratio is shown to be useful in expressing the relative deficit of Ca in individual renal patients (see Table I).

The mean values for the calcium ratios for males and females of Group I were 1.033 ± 15.4% and 0.901 ± 12.5%, respectively. For males and females of Group II, the corresponding mean ratios were 1.015 ± 15.6% and 0.970 ± 16.1% as compared to 0.997 ± 5.6% and 0.987 ± 4.6% for normal male and female subjects. The corresponding BMC ratios for Group I were 1.044 ± 15.2% and 0.848 ± 14.4% as compared with the

Medical Research Center, Brookhaven National Laboratory, Upton, New York and *) Division of Renal Diseases, Nassau County Medical Center, East Meadow, New York.

Table I.

Total-body Calcium and Radial Bone Mineral Content of Patients with Renal Disease.

Patient Category	No.	Sex	Total-body Calcium		Bone Mineral Content			
			$\dfrac{TBCa}{g}$	$\dfrac{TBCa}{Ca_p}$	$\dfrac{BMC}{g/cm}$	$\dfrac{Width}{cm}$	$\dfrac{BMC/W}{g/cm^2}$	$\dfrac{BMC}{BMC_p}$
I Renal (non-dialysis)	15	M	1018 ±14.1*	1.033 ±15.4	1.154 ±14.7	1.491 ± 8.0	0.766 ±11.4	1.044 ±15.2
	10	F	728 ±13.7	0.901 ±12.5	0.753 ±15.9	1.171 ±10.2	0.652 ±20.0	0.848 ±14.4
II Renal (dialysis)	29	M	1048 ±16.7	1.015 ±15.6	1.159 ±16.1	1.525 ±11.5	0.760 ±12.5	1.003 ±16.6
	24	F	753 ±15.9	0.970 ±16.1	0.812 ±19.7	1.249 ± 9.4	0.650 ±17.1	0.955 ±19.5
Normal Contrast	12	M	1077 ±15.7	0.997 ± 5.6	1.192 ±15.8	1.469 ± 9.7	0.813 ±13.2	0.994 ± 6.4
	12	F	795 ±12.2	0.987 ± 4.6	0.893 ±11.0	1.204 ±12.7	0.749 ± 9.2	1.011 ± 5.5

*	= coefficient of variation (per cent)
TBCa	= total body calcium
Ca_p	= predicted total-body calcium
BMC	= bone mineral content of radius
BMC/W	= bone mineral content of radius/width of radius
BMC_p	= predicted bone mineral content of radius

normal ratios of $0.994 \pm 6.4\%$ and $1.011 \pm 5.5\%$.

The calcium ratio is plotted against the BMC ratio for each individual in Groups I and II in Fig. 1. The distribution of Ca ratios in Groups I and II, indicate that about 50% of the patients fall within 2 SD of the normal mean (1.014 ± 0.082). About 20% of the patients in Groups I and II have Ca ratios greater than 2 SD of the mean, while 32% from Group I and 25% from Group II have Ca ratios below 2 SD of the mean. Thus, about half of the patients were in negative or positive balance for sufficient time to alter the total-body calcium. This wide range in Ca ratio reflects the diversity with regard to the extent of osteodystrophy and degree of soft tissue calcification. The heterogeneity with regard to skeletal mass in uremic patients results from variability in dietary calcium intake, duration and extent of uremia, course of treatment, and type of disease.

The bone mineral content, a measure of the linear density of the bone scanned (density per unit length of bone, g/cm) also varies widely, from

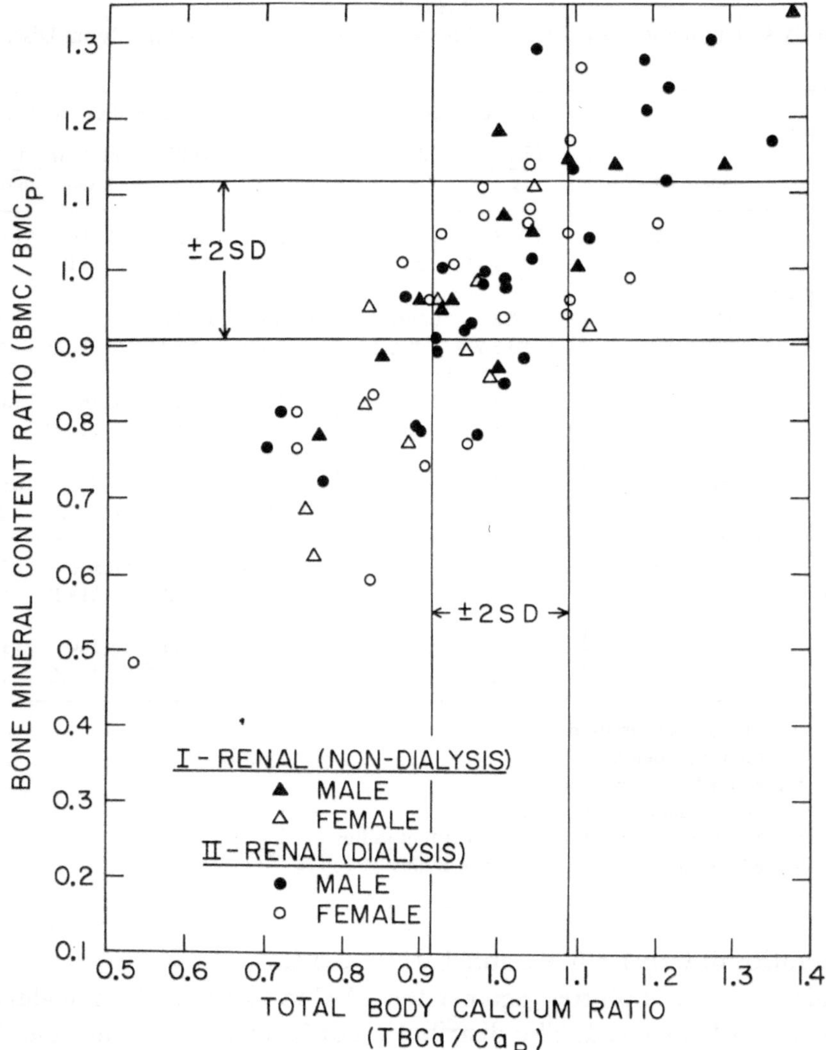

0.465 to 1.568 g/cm in the renal patients. The large variation again reflects sex, age, and size of the individual. For example, even in a large normal population, the coefficient of variation in BMC (at the 8 cm site) in various age groups ranged from 6 to 24% (4). To compare BMC in individuals of different sizes, an index of size and age is required (2). A small degree of normalization is effected by dividing the radius BMC by the radius width, (see Table I), but the correlation of BMC/W with the Ca ratio is poor, and reflects the inability of the width to normalize the data effectively. This result is not surprising since there is poor correlation between radius width and the parameters of height, age and skeletal size.

For example, despite cortical thickening, radius width does not change markedly with age (6).

The mean values of the BMC ratios for males in both renal Groups I and II do not differ significantly from the values of the ratios for normal males. The females of Group I, however, have a mean value lower than that of the controls. In both renal groups, the females have lower mean values of BMC and Ca ratios than those of males. The greater loss of Ca from the body and the radius in uremic females is associated with amenorrhea and depression of ovulatory function. The disturbances of the endocrine factors which contribute to the anovulatory state in uremia may contribute to the more extensive loss of calcified tissue mass in the female uremic patient. Katz (4) and Ritz (5) also reported a significant decrease in BMC associated with increased immunoreactive parathyroid hormone blood levels, as well as a higher prevalence of radiographic abnormalities in the female renal patients on maintenance hemodialysis.

The data indicate a significant correlation between TBCa and BMC in all groups studied. The correlation was the highest (0.944) in the normal contrast group, as would be expected. There is no disturbance in the calcium metabolism in these subjects, and all parts of the skeleton show the same constituent proportions. The correlation coefficient between BMC and TBCa in patients with renal disease (Group I), was 0.919 which is not as high as that for the normal subjects, but equally significant. Even in patients on dialysis, (Group II), with more extensive osteodystrophy, the correlation was still highly significant, although lower than that for the other two groups: 0.892.

The findings suggest that there is a differential rate of loss of Ca from the different parts of the skeleton in renal patients. It is well known that cortical bone and trabecular bone have differing turnover rates. While the 8 cm site on the radius, of course, consists primarily of cortical bone, the total body contains both cortical and trabecular bone. In addition, patients with renal disease have a potential abnormal pool of calcium in sites of soft tissue calcification.

Finally, the question arises on the relative utility of quantitating changes in skeletal Ca by both TBNAA and BMC measurements. The correlation coefficient relating the relative change in TBCa ($\Delta\%$) and the change in BMC ($\Delta\%$) in 16 patients following 9–12 months of dialysis was very poor: 0.25.

It is clear from the results that changes in BMC in individual patients do not necessarily relate to changes in TBCa. In like manner, the TBCa measurement alone does not define the distribution of total-body Ca between the skeleton and soft tissue in renal patients. However, taken together, the BMC measure along with that of TBCa does suggest possible alterations in the skeletal Ca distribution associated with renal disease.

ACKNOWLEDGEMENT

Research supported by the U.S. Energy Research and Development Administration and the Meadowbrook Foundation and Research Foundation, and the Kidney Foundation of New York, Inc.

REFERENCES

1. Cohn, S.H., Ellis, K.J., Zanzi, I., Letteri, J. & Aloia, J.: Correlation of radius bone mineral content with total-body calcium in various metabolic disorders. Int. Conf. on Bone Measurement, Chicago, 1973
2. Cohn, S.H., Shukla, K.K. & Ellis, K.J.: A multivariate predictor of total-body calcium. *Int. J. Nucl. Med. Biol.* 1, 131–134 (1974)
3. Johnston, C.C., Smith, D.M., Yu, P.L. & Deiss, W.P.: *In vivo* measurement of bone mass in the radius. *Metabolism* 17, 1140–1149 (1968)
4. Katz, A.J., Hampers, C.J. & Merrill, J.P.: Secondary hyperpara-thyroidism and renal osteodystrophy in chronic renal failure. *Medicine (Baltimore)* 48, 333–348 (1969)
5. Ritz, E., Krempien, B., Mehls, O. & Malluche, H.: Skeletal abnormalities in chronic renal insufficiency before and during maintenance hemodialysis. *Kidney International* 4, 116–127 (1973)
6. Smith, D.M., Johnston, C.C. & Yu, P,L.: *In vivo* measurement of bone mass. *J. Amer. Med. Ass.* 219, 325–329 (1972)

Comparative Effects of 1α-hydroxycholecalciferol in Children and Adults with Renal Glomerular Osteodystrophy

R.G. Henderson, J.A. Kanis, J.G.G. Ledingham, D.O. Oliver, R.G.G. Russel, R. Smith & R.J. Walton

INTRODUCTION

Recent short-term studies have shown that microgram doses of both 1,25-dihydroxycholecalciferol ($1,25\text{-}(OH)_2\text{-}D_3$), the hormonally active metabolite of vitamin D, and 1α-hydroxycholecalciferol ($1\alpha\text{-}OH\text{-}D_3$), a synthetic analogue can increase intestinal absorption of calcium in patients with chronic renal failure in whom large (mg) doses of vitamin D are ineffective (1, 4, 5, 6). In patients with renal osteodystrophy, these compounds may reduce plasma parathyroid hormone concentration and osteoclastic bone resorption (2) and increase the calcium content of the bones of the hands (3). There is, however, little published work on the long-term effects of $1,25\text{-}(OH)_2\text{-}D_3$ or $1\alpha\text{-}OH\text{-}D_3$ in patients with renal osteodystrophy. The present paper describes the effects of $1\alpha\text{-}OH\text{-}D_3$ in four children and three adults studied for up to 14 months.

PATIENTS AND METHODS (TABLE I)

The adults received haemodialysis for an average of three 6-hour periods each week, using Meltec Kiil dialysers in a single pass system. The final dialysate was made from 34 parts softened tap water and one part concentrated dialysate solution. Two patients (D.M^{C}G. and E.S.) were dialysed against a final dialysate which contained 1.5 mmol litre^{-1} (6.0 mg%) of calcium while B.M. was dialysed against a calcium dialysate of 1.75 mmol litre^{-1}.

The biochemical and radiological methods have been previously described (5).

Renal Unit, Churchill Hospital, Oxford and Metabolic Unit, Nuffield Departments of Orthopaedic Surgery and Clinical Medicine, University of Oxford, Oxford.

RESULTS

Children (Table Ia)

In two patients (A.P. and G.T.), X-ray appearances improved and the alkaline phosphatase and total hydroxyproline excretion (in A.P.) fell. In the other two children, treatment was either of short duration (S.T.) or intermittent due to illness (D.D.). The alkaline phosphatase rose when treatment was stopped and fell again when it was restarted. Hypercalcaemia was not seen in the children.

Adults (Table Ib)

The radiological improvement which was seen in two adults occurred more slowly than in the children. In all three patients, hypercalcaemia occurred during treatment and disappeared within a few days when this was stopped, but rapidly recurred when the 1α-OH-D$_3$ was started again, even in reduced doses. In one patient (D.M^{C}G.), periarticular metastatic calcification appeared during treatment but subsequently improved when 1α-OH-D$_3$ was stopped. In this patient, plasma alkaline phosphatase and non-protein-bound hydroxyproline both increased when treatment was stopped, and parathyroidectomy was performed subsequently (Fig. 1).

Table Ia.

The Use of 1α-OH-D$_3$ in Renal Glomerular Osteodystrophy in Non-dialysed Children.

Patient	Age Sex	Renal disease	Skeletal symptoms	1α-OH-D$_3$ daily dose duration	Biochemistry	Radiology	Improvement of symptoms
						Outcome	
A.P.	14 F	Interstitial nephritis	Knock knees	1.5 μg 14 months	Alk.P'ase↓↓ T.H.P.↓↓ Ca	Erosions healed; mineralisation	++
D.D.	14 F	Obstructive Uropathy	Knock knees	1.5 μg 11 months	Alk.P'ase↓ (Ca)→	Unchanged	±
G.T.	16 M	Obstructive Uropathy	Dwarfism; gross deformity	1.5 μg 5 months	Alk.P'ase↓ (Ca)→	Erosions improved mineralisation	±
S.T.	16 M	Membranous glomerulonephritis	None	2.0 μg 2 months	Alk.P'ase↓ (Ca)→	Unchanged	−

Table Ib.
The Use of 1α-OH-D₃ in Renal Glomerular Osteodystrophy
in Adults Receiving Haemodialysis.

Patient	Age Sex	Renal disease	Skeletal symptoms	1α-OH-D₃ daily dose duration	Bio-chemistry	Outcome	
						Radiology	Improve-ment in symptoms
D.MᶜG.	33 F	Pyelonephritis; bilateral nephrectomy	Bone pain	2.0 2.0 μg 7 months	Alk.P'ase→ P.H.P.→ (Ca)↑	Erosions improved; metastatic calcification	± pain from metastatic calcification
B.M.	43 M	Glomerulo-nephritis	None	1.0 μg 8 months	Alk.P'ase→ P.H.P.→ (Ca)↑	Unchanged	–
E.S.	53 M	Unknown	Bone pain	2.0 μg 9 months	Alk.P'ase↓ P.H.P.↓ (Ca)↑	Erosions improved	+

Tables Ia and b. Arrows indicate increased (↑)
           unchanged (→)   or reduced (↓)
           Alk.P'ase   = Alkaline Phosphatase
           THP   = 24 h total urinary hydroxyproline
           PHP   = Plasma non-protein-bound hydroxyproline
           (Ca)   = Total Plasma Calcium

           Symptoms: ++ markedly improved
                      ± equivocal
                      – unchanged

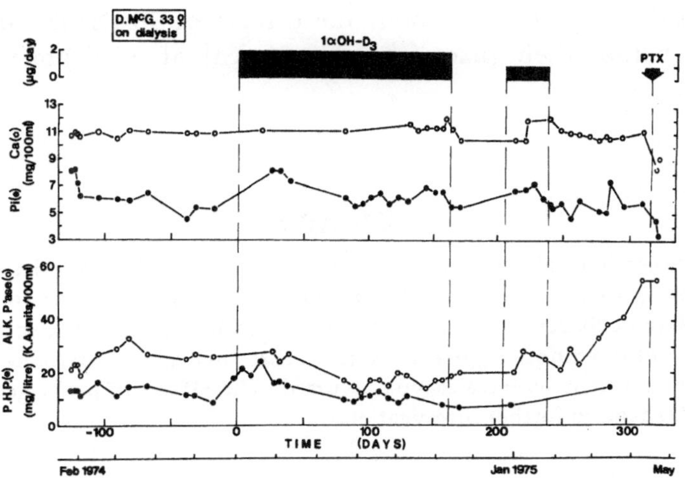

Figure 1. Patient D.MᶜG. Biochemical changes and the effects of 1α-OH-D₃. PTX — parathyroidectomy.

The plasma alkaline phosphatase and hydroxyproline did not alter significantly in the two other patients. In two adults, bone pain diminished within 1—2 weeks of 1α-OH-D$_3$ administration but recurred when treatment was stopped (Table I).

DISCUSSION

This limited experience suggests that renal osteodystrophy may be reversed by 1α-OH-D$_3$ more effectively in children than in adults. Puberty and slowing of skeletal growth could contribute to the radiological and biochemical improvement in the children, but are unlikely to be major factors since this improvement ceases and bone disease progresses when 1α-OH-D$_3$ is stopped. It is not known how 1α-OH-D$_3$ reverses parathyroid induced bone resorption. Of several possibilities, a slight increase in plasma ionised calcium, or a direct effect on the parathyroid gland seem most obvious. 1α-OH-D$_3$ may offer an effective treatment for renal osteodystrophy in non-dialysed children, without the danger of hypercalcaemia, and may allow the surgical correction of severe deformities.

In contrast, in dialysed adults with renal bone disease, bone pain is rapidly relieved by 1α-OH-D$_3$ but hypercalcaemia is a problem and may lead to metastatic calcification. The reasons for the different responses of children and adults to 1α-OH-D$_3$ are obscure, although more rapid formation of bone in children may be a factor. Nor is it clear why a second course of 1α-OH-D$_3$ should produce hypercalcaemia more rapidly than the first, since the compound is unlikely to accumulate in the body.

More needs to be known about the effects of 1α-OH-D$_3$ and related substances before their place in the treatment of renal bone disease is established.

SUMMARY

1α-hydroxycholecalciferol (1α-OH-D$_3$) has been given in daily oral microgram doses for periods of up to 14 months to patients with renal glomerular osteodystrophy. In four non-dialysed children there was radiological and biochemical evidence of healing within 3—6 months, without hypercalcaemia. In three dialysed adults, bone healing was less rapid and recurrent hypercalcaemia occurred. The effects of prolonged treatment with 1α-OH-D$_3$ require further investigation.

ACKNOWLEDGEMENT

We are grateful to the nursing and technical staff who made this study possible. This work was supported by the Wellcome Trust and National Fund for Research into Crippling Diseases. J.A.K. holds an MRC Fellowship and R.J.W. the Goodger Scholarship. R.G.H. was a recipient of the Drummond Fellowship.

REFERENCES

1. Brickman, A.S., Coburn, J.W. & Norman, A.W.: *New Engl. J. Med.* 287, 891 (1972)
2. Brickman, A.S., Coburn, J.W., Massry, S.G.: *Ann. Int. Med.* 80, 161 (1974)
   Massry, S.G., Norman, A.W. & Coburn, J.W.: *Ann. Int. Med.* 80, 161 (1974)
3. Catto, G.R.D., MacLeod, M., Pelc, B. & Kodicek, E.: *Brit. Med. J.* I, 12 (1975)
4. Chalmers, T.M., Davie, M.W., Hunter, J.O., Szaz, K.F., Pelc, B. & Kodicek, E.: *Lancet* (1973) II, 696
5. Henderson, R.G., Russell, R.G.G., Ledingham, J.G.G., Smith, R., Oliver, D.O., Walton, R.J., Small, D.G., Preston, C., Warner, G.T. & Norman, A.W.: *Lancet* (1974) I, 379
6. Peacock, M., Gallagher, J.C. & Nordin, B.E.C.: *Lancet* (1974) I, 385

1α-hydroxycholecalciferol and 25-hydroxycholecalciferol in Renal Bone Disease

A. E. FOURNIER[1], P. J. BORDIER[2], J. GUERIS[2], J. CHANARD[3], P. MARIE[2], C. FERRIERE[2], M. OSARIO[4], J. BEDROSSIAN[4] & H. F. DE LUCA[5]

INTRODUCTION

The better control of calcium-phosphate homeostasis in uremic and chronically hemodialysed patients by the use only of phosphate binders, oral calcium supplement, high dialysate calcium, and/or more frequent dialysis, has considerably improved the course of renal osteodystrophy. Thus, progressive decrease in serum parathyroid hormone and prevention or reversal of radiographic evidence of sub-periosteal bone resorption have been reported (14). However, bone biopsies still remained quite abnormal with increased resorption and decreased calcification front (although this latter may occasionally be seen normal even in anephric patients) (14, 2, 4, 5). It may, therefore, be anticipated that the above mentioned therapeutic measures will not completely prevent fragility of the skeleton on a long-term basis, stressing the importance of additional therapeutic approaches in these patients who have now a longer life expectancy. Though it was found that large doses of vitamin D could produce marked improvement of renal osteodystrophy, the dangers of vitamin D toxicity are considerable (17). Therefore, the discovery that the kidneys were the unique site of 1α-hydroxylation of 25-OH-cholecalciferol and that $1,25\text{-}(OH)_2$-cholecalciferol ($1,25\text{-}(OH)_2 D_3$) the most rapid and the most potent vitamin D metabolite to promote calcium transport in the intestine, was undetectable in uremic patients (10) suggested that lack of $1,25\text{-}(OH)_2 D_3$, was the basic explanation for the vitamin D-resistant state and led to extensive evaluation of this compound in uremic patients. It was found that at a daily dose of 0.5 to 1.5 μg, $1,25\text{-}(OH)_2 D_3$ was able to

1) Clinique Médicale B. — C.H.U. Amiens.
2) André Lichwitz Research Unit — INSERM, Hôpital Lariboisiere Paris.
3) Service d'Etudes Métaboliques, Département de Néphrologie. Hôpital Necker Paris.
4) Clinique Médicale Néphrologique. Hôpital Broussais Paris.
5) College of Agricultural & Life Sciences, Madison, Wisconsin.

increase calcium absorption, decrease the high plasma levels of iPTH and even to improve the bone lesions as assessed on X-rays and on biopsies (6). However, $1,25\text{-}(OH)_2 D_3$ is very expensive to synthetize and seems to have little effect on osteoid mineralisation. Therefore, $1\alpha\text{-}OH\text{-}D_3$ which is much less expensive to synthetize and which is approximately one half as active as $1,25\text{-}(OH)_2 D_3$ in the cure of rachitic lesions in the rat, seems to be the drug of choice for extensive clinical trial. Short-term trials (6—15 days) on limited series (3 and 6 patients) have already shown that 10 μg i.v. (8) or 25 μg orally (15) can increase calcium absorption as well as plasma levels of calcium of uremic patients and decrease their elevated serum alkaline phosphatase. A long-term trial (9—10 weeks) in 3 patients at the dose of 2 μg daily has recently shown that not only calcium absorption increased but also that the calcium content of the bone as measured by a neutron activation technique, was increased (7). However, no data have yet been published on the effects of $1\alpha\text{-}OH\text{-}D_3$ on parathyroid secretion and bone histology of uremic patients. Therefore, we have studied the effect of $1\alpha\text{-}OH\text{-}D_3$ given orally in 10 uremic patients not only on calcium absorption and plasma levels of calcium, phosphate and alkaline phosphatase but also on the plasma levels of immunoreactive parathyroid hormone (iPTH) and on bone biopsy abnormalities as assessed by histomorphometry.

To this study was added 3 other patients maintained on chronic hemodialysis and treated with $25\text{-}OH\text{-}D_3$. As a matter of fact, the usefulness of this drug in the treatment of renal osteodystrophy has been rather disregarded since it has been proved that this compound had to be first 1α-hydroxylated by the kidney to become active on calcium transport in the intestine. However, Bordier *et al.* have stressed that normal mineralisation can take place in the absence of kidneys i.e. in the absence of 1α-hydroxylated metabolites of vitamin D_3 (2) and have shown in another study that at a dose of 30—200 μg daily, it could significantly increase the calcification front in uremic patients not yet on dialysis (5). Furthermore, it has been reported in uremic patients, that low plasma levels of $25\text{-}OH\text{-}D_3$ were associated with radiographic evidence of osteomalacia and that this could be related to an increase of hepatic microsomal turnover of vitamin D into inactive metabolites. Therefore, the evaluation of $25\text{-}OH\text{-}D_3$ in the treatment of osteodystrophy of patients on chronic hemodialysis appeared warranted (12).

PATIENTS AND METHODS

Patients and treatment protocol

Thirteen uremic patients (12 men, 1 woman) aged from 27 to 58 years were studied. 25-OH-D$_3$ was given orally 3 times a week at a dose, 200, 200 and 300 μg (700 μg per week) for 4—8 weeks in 3 patients on chronic were studied.

25-OH-D$_3$ was given orally 3 times a week at a dose of 200, 200 and 300 μg (700 μg per week) for 4—8 weeks in 3 patients on chronic hemodialysis. On the figures, they are represented by a circle: one patient was anephric and is represented by a solid circle; the 2 other patients are represented by hatched circles. One of these two patients had his dialysis duration reduced from 6 hours 3 times a week to 3 and a half hours 3 times a week after the cuprophan membrane was replaced by a polyacrylonitril membrane. This patient is represented by a circle in a box.

1α-OH-D$_3$ was given orally 3 times a week at a dose of 4, 4 and 6 μg (14 μg/week) for 1—12 weeks to 10 patients: 2 were not yet on dialysis and had a creatinine clearance of 7 ml/mn which decreased at the end of the treatment period to 5 ml/mn. They are represented on the figures by an open triangle in the circle; the 8 other patients, represented by black triangles, had been on chronic hemodialysis for 2—48 months; 5 of them had been anephric for 6—48 months. For 2 of them, the dialysis duration was reduced: they are represented by solid triangles in a box.

In all patients, dialysis was performed with a dialysate calcium of 7 mg/100 ml. The diet contained approximately 1 g of calcium and 1.3 g of phosphorus. Al-(OH)$_3$ (3 g per day) was given to all patients. In 4 patients, however, its dosage was increased between the control period and the period of treatment with 1α-OH-D$_3$: these patients are represented by a triangle within a circle.

METHODS

In all patients, blood samples were taken 3 times a week (before each dialysis for those on maintenance dialysis) 2 weeks before and throughout the therapeutic period. Plasma calcium and plasma phosphate were measured by autoanalyzer on each sample. Plasma iPTH was determined every 2 weeks by Gueris (13) using a modification of the method of Arnaud *et al.* (1) using the GP$_1$M antibody (kindly provided by C. Arnaud) which recognizes the C. terminal region of the molecule. It is expressed in ng of protein/ml of parathyroid adenoma-culture medium (normal range 4—8).

In the table are given the mean values of these biological data for the group treated with 1α-OH-D₃ and for the group treated with 25 OH-D₃. The means were calculated from the means of individual patients calculated with the data recorded during the 2 weeks before vitamin D therapies and during the last 2 weeks of treatment.

In 12 patients, 2 transiliac bone biopsies were performed, the first one before treatment and the other at the end of the 4—12 weeks therapy. Serial sections of 5—6 μg thickness were obtained and stained with Toluidine blue. The following parameters were quantitated: osteoclast count (OC) expressed in number per mm^2 of bone section; active resorption surface (ARS) i.e. cancellous bone surface showing Howship's lucanae with osteoclasts, expressed in per cent of the total cancellous bone surface; active formation surface (AFS) i.e. the osteoid surface covered with an active osteoblast layer, expressed in per cent of the total cancellous bone surface; osteoid volume expressed in per cent of the cancellous bone tissue, i.e. calcified + osteoid bone tissue; osteoid surface, expressed in per cent of the total cancellous bone surface; calcification front labelled by methylchlortetracycline, expressed in per cent of the osteoid surface. The normal value for all ages and sex are given in the table.

In 6 patients, fractional absorption of calcium was measured according to the method of Chanard *et al.* (9) before and at the end of 1—10 weeks treatment with 1α-OH-D₃.

RESULTS AND INTERPRETATION

The table summarizes all the biological and histomorphometric data before and at the end of the treatment period with either 1α-OH-D₃ or 25-OH-D₃.

Plasma minerals and iPTH

Plasma calcium P.Ca increased in the ten patients treated with 1α-OH-D₃ so that its mean increased significantly from 8.9 to 10.1 mg/100 ml. 25-OH-D₃ augmented plasma calcium in only 2 patients so that the mean was increased (not significantly) from 9.4 to 9.9 mg/100 ml.

Plasma phosphate P.Po₄ increased in 4 patients, remained unchanged in one and decreased in 5 patients treated with 1α-OH-D₃ so that the mean value slightly decreased from 6.1 to 5.6 mg/100 ml. However, 4 out of the 5 in whom plasma phosphate decreased, had increased their Al-(OH)₃ intake and 2 out of the 4 in whom plasma phosphate increased have had their dialysis duration reduced, so that the effect of 1α-OH-D₃ alone on

Table I. Mean Biological and Histomorphometric data.

		NORMALS Mean ± SD	n	1α-OH-D₃ Mean + SD control	treated	paired t test p.	n	25-OH-D₃ Mean ± SD control	treated	paired t test p.
P Calcium mg/100 ml		10.0 ± 0.2	10	8.9 ± 1.3	10.1 ± 1.5	<0.01	3	9.4 ± 0.8	9.9 ± 0.6	NS
P PO₄	all patients	3.2 ± 0.3	10	6.1 ± 1.1	5.6 ± 1.3	NS	3	5.2 ± 1.1	6.5 ± 1.8	<0.1
mg/100 ml	subgroup without interfering factor on PO₄		4	5.5 ± 1.3	6.3 ± 0.8	NS	2	4.8 ± 1.1	5.5 ± 0.8	NS
P Ca x	all patients	32 ± 3	10	53.1 ± 7.9	55.8 ± 16	NS	3	49.6 ± 14	64.9 ± 17	<0.05
PPO₄	subgroup without interfering factor on PO₄		4	53.3 ± 12.3	65.0 ± 9.0	<0.05	2	44.2 ± 15	56.1 ± 12	NS
PiPTH ng of protein/ml		6 ± 1	9	47.1 ± 15.4	36.4 ± 11.8	<0.01	3	31.5 ± 7.9	23.7 ± 2.1	NS
S. Alk. Phosphatase K.A.U.		8 ± 2.5	10	21 ± 13.5	14.8 ± 6.8	<0.05	3	14.7 ± 2.3	23.7 ± 4.9	<0.05
Osteoclast count N/mm² of bone section (0C)		37 ± 0.7	9	1.2 ± 0.6	0.95 ± 0.87	NS	3	0.84 ± 0.24	0.86 ± 0.75	NS
Active resorption surface % of total cancellous bone surface (ARS)		59 ± 0.17	9	2.6 ± 1.4	2.3 ± 1.8	NS	3	2.3 ± 0.6	2.0 ± 1.5	NS
Active formation surface % of total cancellous bone surface (AFS)		4.9 ± 1.4	9	13.2 ± 6.8	8.7 ± 2.9	<0.01	3	8.2 ± 6.4	11.4 ± 7.2	<0.05
Osteoid volume % of (0V) cancellous bone tissue		4.1 ± 1.1	8	16.6 ± 12.9	12.2 ± 11.9	<0.05	3	3.5 ± 1.9	3.7 ± 1.4	NS
Osteoid surface % of total cancellous bone surface (OS)		14.8 ± 3	9	49.5 ± 21.2	43.4 ± 18	NS	3	44.1 ± 25.4	42.9 ± 23.4	NS
Calcification front % of osteoid surface (CF)		63 ± 4.5	9	17.6 ± 7.6	27.4 ± 9.6	<0.01	3	11.3 ± 1.0	27.9 ± 7.7	<0.05
Fractional Calcium Absorption % + SEM		56.8 ± 1.8	6	32.8 ± 3.5	72.5 ± 6.9	<0.01				

plasma phosphate could be evaluated only in a subgroup of 4 patients. In this subgroup, a slight increase of 0.8 mg/100 ml was observed. In the 25-OH-D₃ treated group, there was also a slight increase of plasma phosphate (+ 0.75 mg/100 ml) when the patient with decrease of dialysis duration was excluded. The same subgrouping was made for the evaluation of the changes of the calcium phosphate product; in the subgroups without independent factors interfering with plasma phosphate, there was an increase, both with 1α-OH-D_3 and 25-OH-D_3.

Plasma iPTH (P.iPTH) decreased in 8 out of the 9 patients treated with 1α-OH-D_3 in whom it was measured so that the mean value went down from 47 to 36 (upper normal limit = 8), 2 out of 3 patients had their iPTH levels decreased with 25-OH-D_3.

There was no correlation between the changes of plasma iPTH and the changes of plasma calcium (r = -0.47: NS), nor with the changes of plasma phosphate (r = $+0.44$: NS).

Fractional calcium absorption (FCa A%)

FCa A% was depressed in all patients before treatment and went up into the normal range in 3 and above the normal range in the 3 others. The mean increase from 32.8 to 72.5% was highly significant; this confirms the study of Catto *et al.* (7) in 3 patients showing that at low doses (2 μg/day) 1α-OH-D_3 is still able to increase calcium absorption and that it is not necessary, therefore, to use higher doses as used by Chalmers (10 μg I.V.) or Peacock (25 μg orally/day) (8, 15).

Bone resorption and P.iPTH

There was no significant change in osteoclast count and activity when the 2 groups (the one treated with 1α-OH-D_3 and the other treated with 25-OH-D_3) were considered. However, there was a significant positive correlation between the changes of iPTH and the changes of osteoclast count (p $<$ 0.01) and active resorption surface (p $<$ 0.05). Therefore, individual data correlating PiPTH and OC or ARS were plotted (Fig. 1) before and after treatment. This group could then be distinguished into: (1) a group of 6 patients in which there was a parallel decrease in plasma iPTH, osteoclast count and active resorption surface; in this group were 6 patients on dialysis, 4 being treated with 1α-OH-D_3 and 2 with 25-OH-D_3; (2) a group with the 2 patients not yet on dialysis (represented by an open triangle in a circle) in whom plasma iPTH decreased slightly and OC and ARS did not change significantly; (3) a group of 4 patients in whom plasma iPTH did not change significantly whereas OC and ARS increased significantly. This latter group was further characterized by a significant difference as regards the changes in plasma phosphate: thus, in this group there was an increase in plasma phosphate (+ 1.5 mg/100 ml) (because 3 of them had their dialysis duration reduced) whereas in the 2 other groups plasma phosphate decrease (-0.85 mg/100 ml).

These data suggest that the increase in plasma phosphate induced by the decrease of dialysis duration in spite of the use of polyacrylonitril

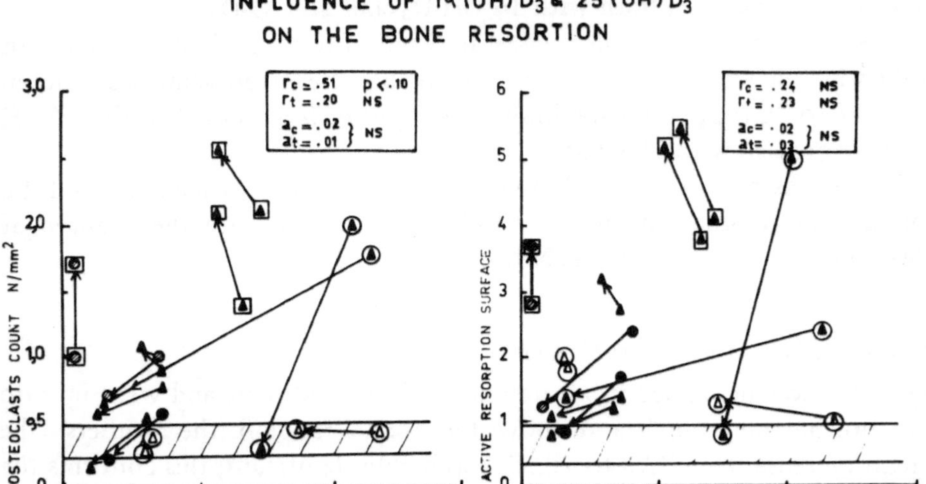

membrane prevent the suppression of iPTH by the administration of 25-OH-D$_3$ or 1α-OH-D$_3$. As iPTH levels are kept constant, the vitamin D metabolites seems to potentiate the action of PTH on the differentiation of progenitor cells to osteoclasts, and this results in an increase of osteoclasts number and activity.

From a therapeutical point of view, decrease of dialysis duration appears thus particularly hazardous when vitamin D metabolites are given. Therefore, the subsequent increase of plasma phosphate should be carefully prevented by increasing the dosage of phosphate binders.

Bone formation and serum alkaline phosphatase

There was a striking difference between 1α-OH-D$_3$ and 25-OH-D$_3$ as regards their effect on active formation surface and serum alkaline phosphatase: whereas 1α-OH-D$_3$ induced a significant decrease of both AFS and serum alkaline phosphatase, 25-OH-D$_3$ induced a significant increase of these 2 parameters. A significant positive correlation was found between the changes of AFS and the changes of serum alkaline phosphatase. These results have to be brought together with those found by Bordier *et al.* (3) in nutritional osteomalacia: in this situation, during the first two weeks following an i.v. pulse, AFS did not change or increase moderately with 1α-OH-D$_3$ and markedly with 25-OH-D$_3$. When the second biopsy was performed after 3 weeks of therapy, the AFS was, however, no longer different from controls. However, since the duration of treatment was shorter with 25-OH-D$_3$ than with 1α-OH-D$_3$ the differende between these 2 drugs as regards AFS and alkaline phosphatase, can not yet be ascribed to the nature of the drug.

Osteoid volume was also significantly decreased in the group treated with 1α-OH-D_3 whereas it increased in 2 out of 3 patients treated with 25-OH-D_3. The changes of osteoid volume were positively correlated with the changes of AFS but were not negatively correlated with the variations of the calcification front. This demonstrates that the increase of osteoid volume in a few patients was due to an increased formation of new osteoid and not to a decreased mineralisation.

Calcification front and plasma calcium phosphate product
The table shows that there was a significant increase of calcification front in the group treated with 1α-OH-D_3 (in 2 cases, it increased by more than 100% and in 6 cases by more than 30%) as well as in the group treated with 25-OH-D_3 (more than 100% in 2 cases and more than 30% in 1 case). This increase in calcification front was not associated with a decrease of the osteoid surface. Since CF is expressed in per cent of the osteoid surface, it is thus excluded that the increase of CF was actually due to a decrease of the osteoid surface. When the whole group (α-OH-D_3 and 25-OH-D_3 patients) was considered, there was no close correlation between the changes of calcification front and the changes of either plasma calcium, plasma phosphate or the plasma Ca x PO_4 product. This lack of correlation suggests a direct effect of 25-OH-D_3 and 1α-OH-D_3 on the bone mineralisation as it has been already proposed and demonstrated by Eastwood *et al.* (11).

CONCLUSIONS

1. Both 25-OH-D_3 and 1α-OH-D_3 significantly increased plasma calcium and decreased plasma iPTH. They also moderately increased plasma phosphate.
2. The decrease of iPTH is not significantly correlated either with the changes of plasma calcium nor with the changes of plasma phosphate.
3. 1α-OH-D_3 increased significantly fractional intestinal calcium absorption in the 6 patients in whom it was measured.
4. There was a parallel decrease in osteoclast count and activity with the decrease in iPTH, except in 4 patients characterized by an increase in plasma phosphate and non-significant decrease in iPTH.
5. There was a striking difference between 1α-OH-D_3 and 25-OH-D_3 as regards bone formation: 1α-OH-D_3 *decreased* alkaline phosphatase and

active formation surface, whereas 25-OH-D$_3$ *increased* it. However, since the duration of treatment was shorter with 25-OH-D$_3$ this difference cannot be ascribed with certainty to the nature of the drug.

6. The changes of osteoid volume was positively correlated with the changes of active formation surface, but not negatively correlated with the changes of calcification front.

7. Calcification front increased significantly in the 3 patients (one anephric) treated with 25-OH-D$_3$ and in 8/9 of the patients treated with 1α-OH-D$_3$.

8. There was no correlation between the changes of calcification front and the changes of either plasma Ca, phosphate or the plasma calcium x phosphate product, suggesting a direct effect of 25-OH-D$_3$ and 1α-OH-D$_3$ on bone mineralisation.

9. Thus, both 1α-OH-D$_3$ and 25-OH-D$_3$ (this latter, however, at a dose 50 times higher) appear to be valuable additional therapeutic agents in uremic osteodystrophy since, on a medium-term basis, they tend to reverse hyperparathyroidism and to improve the mineralisation defect. However, in no case did plasma iPTH nor the histomorphometric parameters return to normal. This warrants further comparative studies of these 2 drugs on a long-term basis.

SUMMARY

Four μg of 1α-OH-D$_3$ were given orally every other day to 10 uremic patients (8 on chronic hemodialysis) for 1—12 weeks and 200 μg of 25-OH-D$_3$ to 3 patients on chronic hemodialysis for 4—8 weeks. Before and at the end of therapy a transilial bone biopsy was analysed. Serial evaluations of serum immunoreactive PTH (P.iPTH), calcium, phosphate and alkaline phosphatase were made.

It is concluded that both 1α-OH-D$_3$ and 25-OH-D$_3$ (this latter, however, at a dose 50 times higher) are able to depress hyperparathyroidism in 2/3 of the cases and to improve consistently the defect of mineralisation. Neither PiPTH nor the bone histomorphometric parameters returned to normal in any case, so that long-term evaluation of these 2 drugs is warranted.

ACKNOWLEDGEMENTS

The authors express their thanks

1. To the nurse staff of the dialysis centers of Hôpital Broussais and of the C.H.U. of Amiens for their collaboration in collecting blood samples.

2. To Dr. Mathieu Des Fossey of Roussel Laboratories for the supply of 25-OH-D$_3$.

REFERENCES

1. Arnaud, C., Tsao, H.S. & Little, D.T.: Radioimmunoassay of human parathyroid hormone in serum. *J. clin. Invest.* 50, 21—34 (1971)

2. Bordier, P.J., Tun Chot, S., Eastwood, J.B., Fournier, A.E. & De Wardener, H.E.: Lack of histological evidence of vitamin D abnormality in the bones of anephric patients. *Clin. Sci.* **44**, 33–41 (1973)

3. Bordier, P.J., Pechet, M., Hesse, R., Marie, P. & Rasmussen, H.: Response of adult patients with osteomalacia to treatment with cristalline 1α-hydroxyvitamin D_3. *New Engl. J. Med.* **291**, 866–871 (1974)

4. Bordier, P.J., Marie, P., Arnaud, C.D., Gueris, J., Ferriere, C. & Norman, A.W.: Early effects of vitamin D_3 and some analogues 25-OH-D_3, 1,25-$(OH)_2D_3$, 1α-OH-D_3 upon resorption and mineralization of bone in nutrional or malabsorption osteomalacia. In: *Vitamin D and Problems related to Uremic Bone Disease*, Norman, A.W., Schaefer, K., Grigoleit, H.-G., von Heaarath, D. and Ritz, E. (eds.), de Gruyter, Berlin, New York, pp. 133–141, 1975

5. Bordier, P.J., Marie, P.J. & Arnaud, C.D.: Evolution of renal osteodystrophy: Correlation of bone histomorphometry and serum mineral and immunoreactive parathyroid hormone values before and after treatment with calcium carbonate and 25-OH-calciferol. *Kidney Internat.* **7**, (Suppl. 2) 102–112 (1975)

6. Brickman, A.S., Sherrard, D.J., Massry, S.G., Norman, A.W. & Coburn, J.W.: 1,25-$(OH)_2D_3$ — Effect on skeletal lesions and parathyroid hormone levels in uremic osteodystrophy. *Arch. intern. Med.* **134**, 883–888 (1974)

7. Catto, G.R.D., Mac Leod, M., Pelc, B. & Kodicek, E.: 1α-OH-D_3: a treatment for renal bone disease. *Brit. med. J.* **1**, 12–14 (1975)

8. Chalmers, T.M., Davie, M.W., Hunter, J.O., Szazk, F., Pelc, B. & Kodicek, E.: 1α-OH-D_3 as a substitute for the kidney hormone 1,25-$(OH)_2D_3$ in chronic renal failure. *Lancet* (1973) ii, 696–699

9. Chanard, J., Assailly, J., Bader, L. & Funck Brentano, J.L.: A rapid method for measurement of fractional intestinal absorption of calcium. *J. nucl. Med.* **15**, 588–592 (1974)

10. De Luca, H.F.: The kidney as an endocrine organ involved in the function of vitamin D. *Amer. J. Med.* **58**, 39–47 (1975)

11. Eastwood, J.B., Bordier, P.J., Clarkson, E.M., Tun Chot, S. & De Wardener, H.E.: The contrasting effects on bone histology of vitamin D and of calcium carbonate in the osteomalacia of chronic renal failure. *Clin. Sci. Molec. Med.* **47**, 23–42 (1974)

12. Eastwood, J.B. & De Wardener, H.E.: Vitamin D metabolism in chronic renal failure. *Lancet* (1975) i, 981

13. Gueris, J.: Radioimmunological determination of plasmatic parathyroid hormone in man. In: *Radioimmunoassay and Related Procedures in Medicine*, Edited by the International agency of Nuclear Power, Vienna, Vol. I, pp. 337–352, 1974

14. Johnson, W.J., Goldsmith, R.S., Beabout, J.W., Jowsey, J., Kelly, P.J. & Arnaud, C.D.: Prevention and reversal of progressive secondary hyperparathyroidism in patients maintained by hemodialysis. *Amer. J. Med.* **56**, 827–839 (1974)

15. Peacock, M., Gallagher, J.C. & Nordin, B.E.C.: Action of 1 hydroxyvitamin D_3 on calcium absorption and bone resorption in man. *Lancet* (1974) i, 385–389

16. Rasmussen, H. & Bordier, P.J.: *The Physiological and Cellular Basis of Metabolic Bone Disease.* Williams & Williams, 1974

17. Stanbury, S.W.: Azotemic renal osteodystrophy. *Clin. Endocr. Metab.* **1**, 267–304 (1972)

The Clinical Use of Synthetic 1,25-Dihydroxycholecalciferol

I.M.A. EVANS, M. BOULTON-JONES, F.H. DOYLE, G.F. JOPLIN, M. LOCKWOOD, E.W. MATTHEWS & I. MACINTYRE

INTRODUCTION

1,25-dihydroxycholecalciferol ($1,25\text{-}(OH)_2 D_3$) is the most active metabolite of vitamin D_3 yet known (5, 4) and the chemically synthesized material has recently become available for clinical use. After first confirming its identity using high speed liquid chromatography, we studied the effects of oral administration in two main clinical settings: hypoparathyroidism and chronic renal failure.

1,25-Dihydroxycholecalciferol

Chemically synthesized $1,25\text{-}(OH)_2 D_3$ was provided by Professor A.W. Norman and Hoffmann-La-Roche Inc. in a diluent solvent of ethanol: 1,2-propanediol 1:1. This material was characterised by means of high speed liquid chromatography (HSLC) on a Dupont 830 liquid chromatograph using an ODS-Permaphase column and monitoring absorption at 254 nm as previously described (3). A linear solvent gradient was used from methanol to water 3:7 v/v to methanol to water 8:2 v/v, increasing the gradient by 2% v/v per minute, and a homogeneous peak of $1,25\text{-}(OH)_2 D_3$ was seen together with a few minor later-eluting peaks which had their origin in the diluent solvent, as shown when the solvent was chromatographed alone. Using HSLC the $1,25\text{-}(OH)_2 D_3$ was shown to be stable on storage in the diluent solvent and no break-down peaks were observed.

For clinical use the $1,25\text{-}(OH)_2 D_3$ was made up in propylene glycol to give a stock solution containing 0.5 μg per 200 μl, this was kept in 4 ml aliquots in dark glass bottles at $-20°$C. These aliquots were dispensed for immediate use and subsequently stored at $4°$C. The $1,25\text{-}(OH)_2 D_3$ was

Endocrine Unit, Royal Postgraduate Medical School, Ducane Road, London.

administered orally on a sugar lump in the doses described elsewhere in the text.

PATIENTS AND METHODS

1) Hypoparathyroidism

One male and one female patient, aged 56 and 53 years respectively, were studied; both were hypoparathyroid due to previous surgery. These patients had required vitamin D preparations to maintain normocalcaemia but such treatment had been discontinued more than two months prior to administration of $1,25\text{-}(OH)_2 D_3$.

2) Chronic Renal Failure

a) Absent or Minimal Renal Bone Disease

One female patient aged 62 years with polycystic kidneys had initial parathyroid hormone levels within the normal range and no evidence of bone disease radiologically. One male patient aged 34 years with chronic pyelonephritis secondary to spina bifida and neurogenic bladder also had initially normal parathyroid hormone levels and no radiological evidence of bone disease.

b) Moderate or Severe Renal Bone Disease

One female patient aged 8 years with chronic glomerulonephritis had high parathyroid hormone levels (above 2 ng/ml) and radiologically a combination of rickets and hyperparathyroid bone disease. One female patient aged 54 years with chronic glomerulonephritis had parathyroid hormone levels of 1.1 ng/ml and moderate hyperparathyroid bone disease as manifest by phalangeal subperiosteal resorption.

3) Biochemical Methods

Serum calcium and phosphate were measured by a Technicon Auto-Analyzer. Plasma parathyroid hormone (PTH) was measured by immunoassay using antiserum 211/32 (Burroughs Wellcome) giving a detection limit of 100 pg/ml and an upper limit of normal of 730 pg/ml.

RESULTS

Hypoparathyroidism (Fig. 1)

$1,25\text{-}(OH)_2 D_3$ treatment was initiated in both patients at a dose of 1.0 μg

238

daily. After 10–14 days the serum calcium had risen to over 12 mg/100 ml and 1,25-(OH)$_2$D$_3$ was discontinued. The serum calcium took between one and two weeks to fall within the normal range and 1,25-(OH)$_2$D$_3$ was reinstituted at a dose of 0.25 μg daily three to four weeks after discontinuation. This dose has satisfactorily maintained the serum calcium within the normal range in both patients. Serum phosphate did not change in one patient and rose by an average of 1.0 mg/100 ml in the other (comparing the 0.25 μg/day level with pre-treatment values).

Chronic Renal Failure
a) Absent or Minimal Renal Bone Disease (Fig. 2)
In both patients PTH levels were suppressed. The patient with polycystic kidneys and average pre-treatment levels of PTH of 320 pg/ml had consistently undetectable levels after two weeks treatment with 0.5 μg 1,25-(OH)$_2$D$_3$ daily. Serum calcium rose by an average of 0.5 mg/100 ml and serum phosphate did not change. The second patient, with pyelonephritis, had pre-treatment PTH levels ranging from undetectable values to 320 pg/ml which rose to 1.0 ng/ml after six weeks treatment with 0.5 μg

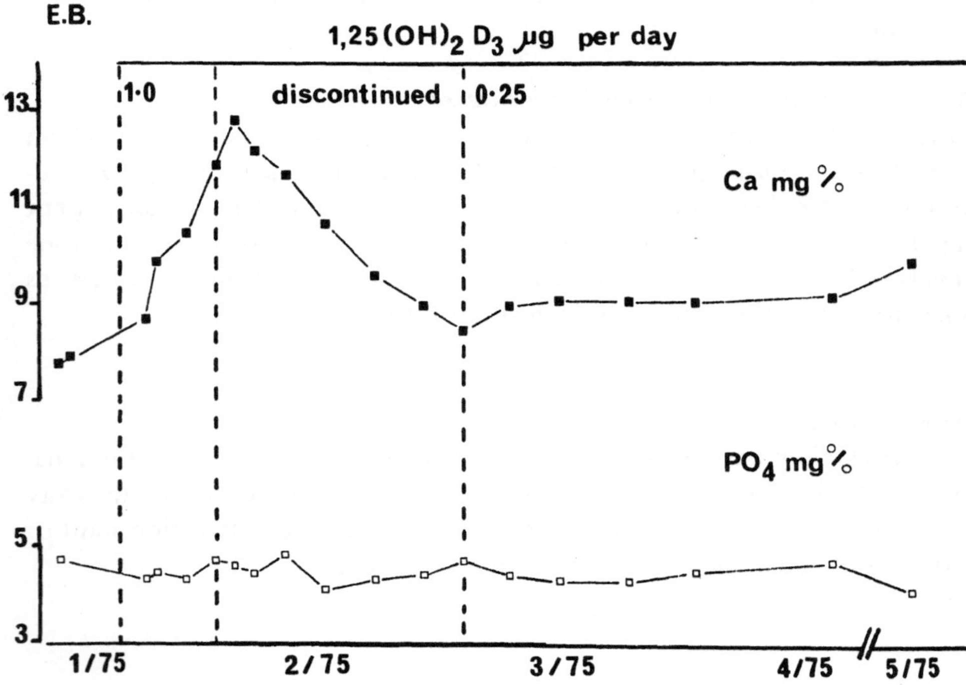

Figure 1. The effect of oral 1,25-(OH)$_2$D$_3$ on serum calcium ■ and phosphate □ (mg/100 ml) in a 56 year old man with hypoparathyroidism.

1,25-(OH)₂ D₃ daily. At this time the dose of 1,25-(OH)₂ D₃ was increased to 1.0 μg daily and after ten days the PTH levels become undetectable and have remained so. Serum calcium rose by an average of 3 mg/100 ml over pre-treatment values and serum phosphate did not change significantly.

b) Moderate or Severe Renal Bone Disease

PTH was not suppressed in either patient on a dose of 0.5 μg 1,25-(OH)₂ D₃ daily, after four months in one (aged 8 years) and after two months in the other (aged 54 years). The adult patient, however, noted a dramatic improvement in bone pain one week after initiating therapy and radiologically some healing was noted in the bone lesions after two months. At this time the serum calcium had risen by 2.0 mg/100 ml to 12.5 mg/100 ml and parathyroidectomy was performed. The dose of 1,25-(OH)₂ D₃ was then increased to 1.0 μg daily and the radiological improvement continued. In the other patient (aged 8 years) some healing was also noted radiologically of both rachitic and hyperparathyroid changes after four months treatment.

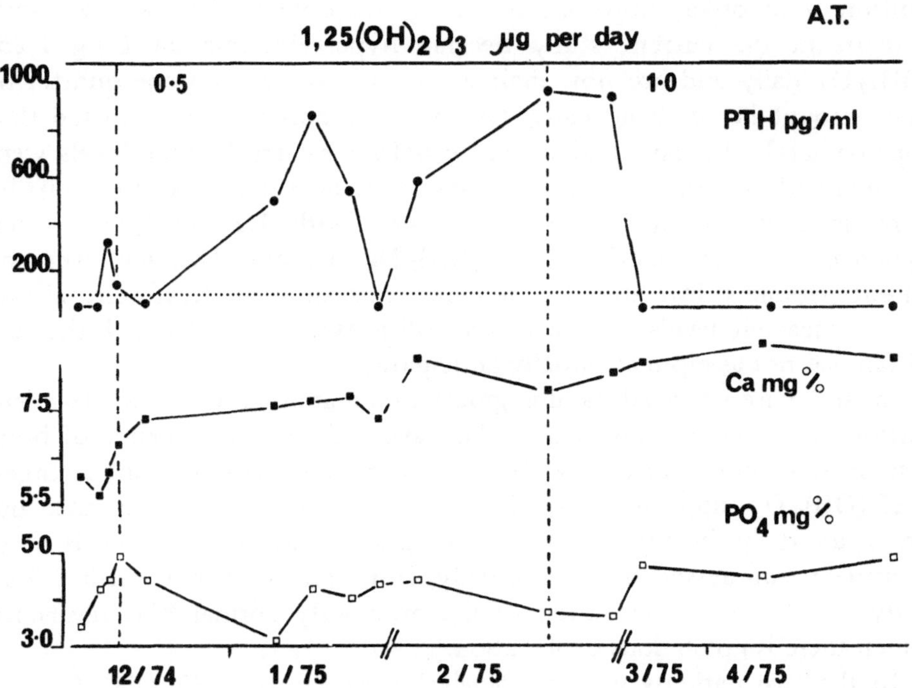

Figure 2. The effect of oral 1,25-(OH)₂ D₃ on plasma immunoreactive parathyroid hormone ● (pg/ml), serum calcium ■ and phosphate □ (mg/100 ml) in a 34 year old man with chronic renal failure. .... indicates the detection limit of the PTH immunoassay used (100 pg/ml).

DISCUSSION

Biosynthetic $1,25\text{-}(OH)_2 D_3$ has previously been reported to be an effective form of therapy in hypoparathyroidism (6) and chronic renal failure (2, 1). In hypoparathyroidism, 0.68 to 2.5 μg daily rapidly corrected the low plasma calcium and this was subsequently maintained within the normal range (6). In chronic renal failure amounts ranging from 0.35 to 2.7 μg $1,25\text{-}(OH)_2 D_3$ daily were shown to increase calcium absorption over 4 to 8 days while a longer period of treatment led to progressive healing of radiological rickets and secondary hyperparathyroidism in one case (2). In another study, a dose of 0.68 μg daily was shown to elevate serum calcium whilst avoiding the serious hypercalcaemia and evidence of enhanced bone resorption seen with the 2.7 μg dose (1).

In the present study chemically synthesized $1,25\text{-}(OH)_2 D_3$ has been shown to be effective in small amounts in hypoparathyroidism. A word of caution is in order, however, regarding the dose to be used, as both patients in the current study became hypercalcaemic on 1 μg $1,25\text{-}(OH)_2 D_3$ daily and are now maintained satisfactorily on one quarter of that amount i.e. 0.25 μg daily. On the 1 μg dose it is to be noted that approximately 10 days had to elapse before hypercalcaemic levels were reached and this relatively slow response as regards serum calcium must be born in mind when therapy is initiated with $1,25\text{-}(OH)_2 D_3$. When hypercalcaemia is manifest, $1,25\text{-}(OH)_2 D_3$ has an advantage over the parent vitamins D_2 and D_3 in that cessation of treatment allows normocalcaemic levels to be achieved within two weeks, although the rate of fall was not as rapid as initially anticipated.

In the chronic renal failure group there appears to be a clear cut distinction between those cases who have radiological evidence of bone disease and those who do not. In the two patients without bone changes, $1,25\text{-}(OH)_2 D_3$ suppressed the PTH levels, on 0.5 μg daily in one case and on 1 μg daily in the other. In the patient whose PTH levels were eventually suppressed on 1.0 μg daily, initiation of therapy with 0.5 μg daily resulted in an elevation of the previously normal PTH levels for which there is no obvious explanation.

In the two patients with radiological bone disease $1,25\text{-}(OH)_2 D_3$ in a dose of 0.5 μg daily did not suppress the PTH levels. Despite this, there was radiological evidence of healing of hyperparathyroid bone changes in one patient and some healing of both rachitic and hyperparathyroid features in the other.

Based on these preliminary findings, chemically synthesized 1,25-(OH)$_2$D$_3$ appears to be an effective agent in maintaining normal serum calcium levels in hypoparathyroidism and it could well prove useful in the management of some of the calcium metabolic abnormalities seen in chronic renal failure.

SUMMARY

The effects of chemically synthesized 1,25-dihydroxycholecalciferol (1,25-(OH)$_2$D$_3$) were evaluated in 2 clinical settings: hypoparathyroidism and chronic renal failure. The material was administered orally in all cases. In hypoparathyroidism 0.25 μg of 1,25-(OH)$_2$D$_3$ was effective in maintaining normal serum calcium levels. In chronic renal failure small amounts (0.5 − 1.0 μg) of 1,25-(OH)$_2$D$_3$ suppressed parathyroid hormone levels in patients without radiological evidence of bone disease but failed to do so when bone disease was present, although some radiological healing did occur.

ACKNOWLEDGEMENTS

This work was supported in part by the Wellcome Trust. I.M.A.E. is in receipt of a MRC Clinical Research Fellowship. Our thanks are due to Dr. L. Galante, Dr. S. Girgis, Mrs. C. Hillyard and Miss L. Heywood for performing the parathyroid hormone assays.

REFERENCES

1. Brickman, A.S., Coburn, J.W., Massry, S.G. & Norman, A.W.: 1,25-Dihydroxy-Vitamin D$_3$ in normal man and patients with renal failure. *Ann. Intern. Med.* **80**, 161−168 (1974)
2. Henderson, R.G., Russell, R.G.G., Ledingham, J.G.G., Smith, R., Oliver, D.O., Walton, R.J., Small, D.G., Preston, C., Warner, G.T. & Norman, A.W.: Effects of 1,25-Dihydroxycholecalciferol on calcium absorption, muscle weakness and bone disease in chronic renal failure. *Lancet* (1974), i, 379−384
3. Matthews, E.W., Byfield, P.G.H., Colston, K.W., Evans, I.M.A., Galante, L.S. & MacIntyre, I.: Separation of hydroxylated derivatives of vitamin D$_3$ by high speed liquid chromatography (HSLC). *FEBS Letters* **48**, 122−125 (1974)
4. Norman, A.W. & Wong, R.G.: The biological activity of the vitamin D metabolite, 1,25-dihydroxycholecalciferol in chickens and rats. *J. Nutr.* **102**, 1709−1718 (1972)
5. Omdahl, J., Holick, M., Suda, T., Tanaka, Y. & DeLuca, H.F.: Biological activity of 1,25-dihydroxycholecalciferol. *Biochemistry* **10**, 2935−2940 (1971)
6. Russell, R.G.G., Smith, R., Walton, R.J., Preston, C., Basson, R., Henderson, R.G. & Norman, A.W.: 1,25-Dihydroxycholecalciferol and 1α-Hydroxycholecalciferol in hypoparathyroidism. *Lancet* (1974), ii, 14−17

The Value of Bone Density Measurements in Predicting the Risk of Developing Avascular Necrosis following Renal Transplantation

R. LINDSAY, S.G. McPHERSON, J.B. ANDERSON & D.A. SMITH

INTRODUCTION

The effect of renal transplantation is, in most instances, to restore a functioning kidney which can respond to the metabolic status of the patient. The serum phosphate returns to normal, the serum calcium rises and the hyperparathyroid state, during maintenance dialysis, resolves in many cases. In a variable proportion of those patients who have received a renal allograft, structural bone failure may occur, within the next few years (2). This condition, aseptic necrosis, is a major influence in determining how quickly a recipient of a successful graft can return to a completely normal way of life. Its aetiology is still obscure; of importance may be continued hyperparathyroidism, steroid therapy or both (2).

We have studied peripheral bone density after renal transplantation following the changes occurring during the first post-transplant year when circulating parathyroid hormone activity is likely to be diminishing and when a high-dosage steroid regime is required to support the allograft.

PATIENTS & METHODS

From May 1969 to November 1973, 52 patients received renal allografts. Of these, 34 patients whose operations occurred during 1971—1973 were studied in depth. At transplantation, 1/g Prednisolone was given intravenously and the patients commenced on Prednisolone 200 mg orally daily, reducing by 10 mg per day to 50 mg with further reductions to a maintenance dose of 10—15 mg at 3 months. The patients were given, in addition, Imuran (Azathioprine) 3 mg/kg/day orally; this initial dose being varied to maintain the total white cell count between 5,000 and 7,000 cells/mm^3. Episodes of acute rejection of the graft were treated with

University Departments of Medicine & Surgery, Western Infirmary, Glasgow.

intravenous Prednisolone (1/g) with no alteration of the oral immunosuppresive therapy.

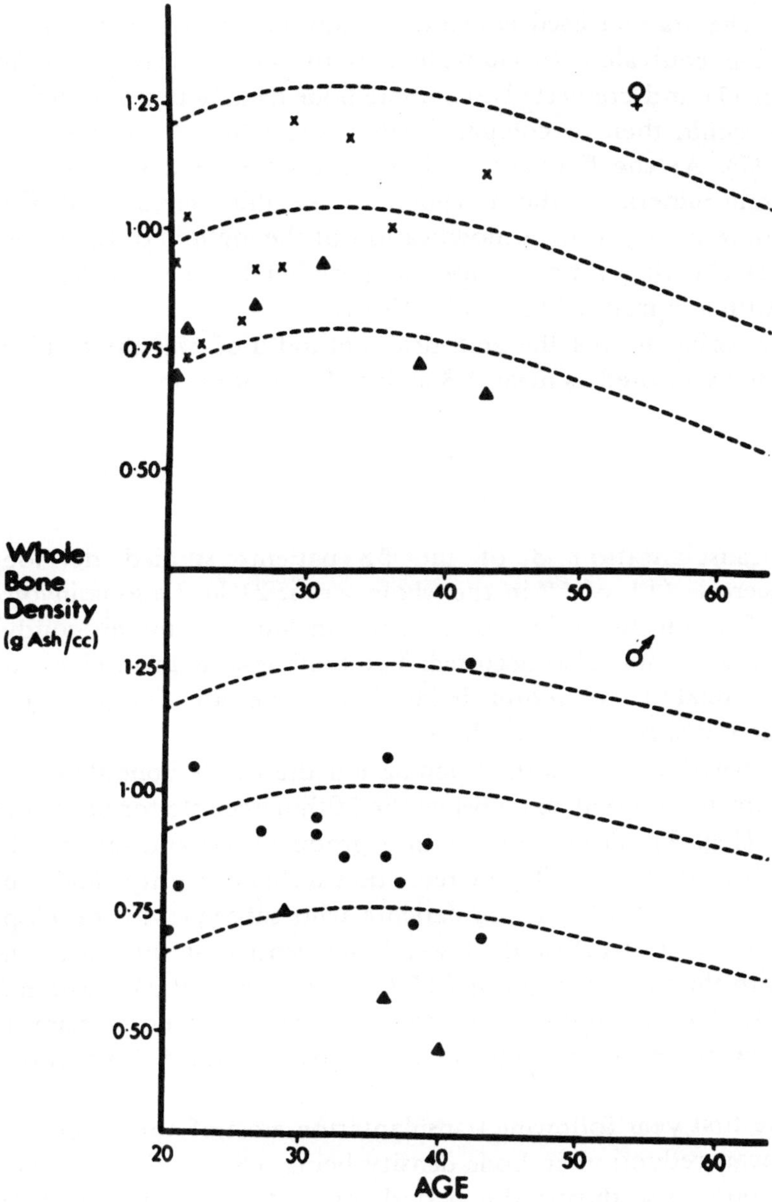

Figure 1. Whole bone density (g ash per cm^3) of patients who had received a renal graft six weeks previously. The normal distributions (50th %ile with 5 and 95% confidence limits) for females (upper) and males (lower) are shown by the dotted lines. The patients represented by the filled triangles are those who subsequently developed clinical avascular necrosis.

Peripheral bone mineral content was estimated initially 6 weeks after transplantation. At this visit, a standard radiograph of the right hand was obtained for densitometric and morphological measurements of the third metacarpal. The standardised aluminium equivalent was calculated from the absorption equivalent in aluminium of the bone, measured using a densitometer (1) and converted to a whole bone density in g ash/cm^3 (4). The patients could then be compared with a previously described normal population (7). At the first visit and at each subsequent visit measurements of bone mineral at the mid-point of the third metacarpal of the right hand were made, using a modification of the method of Cameron & Sorenson (3) allowing serial changes in peripheral bone density to be monitored with a reproducibility of 2–3% (5).

When comparing means the statistical method used was the Student's t-test. Results are quoted as mean ± Standard Error of mean.

RESULTS

Following transplantation 8 of the 52 patients studied developed ischaemic necrosis (11 of 52 in the whole series; 21%). Lesions involved mostly the femoral head but other sites including femoral condyle, humeral head and talus also occurred. The bone lesions themselves were not directly related to serum lipids levels, steroid dosage, age at transplantation or survival of the allograft.

When compared to the normal population the whole bone density of the graft recipients tended to lie below the 50th percentile for both males and females (Fig. 1). Comparison with a group of age and sex-matched controls confirmed statistically the reduction in bone density (Males $p <$ 0.005; Females $p < 0.01$). Those patients who subsequently developed bone necrosis had a significantly lower bone density at this visit when compared with the remaining group (0.78 ± 0.66; 0.95 ± 0.03 g ash/cm^3 p < 0.005). As far as could be determined there was no relationship between whole bone density and the period of time on dialysis prior to transplantation.

During the first year following transplantation most of our patients lost bone; the mean reduction in bone density being 6.8 ± 1.8% ($p < 0.01$). The rate of bone loss during this period was not significantly different between those who developed bone necrosis and those who did not (7.3 ± 2.0% and 6.6 ± 1.3% respectively). However, there was a significant relationship between the rate at which each patient lost bone and the setting of the level of phosphate excretion in the transplanted kidney (Fig. 2) measured approximately 6 months following operation. This relation-

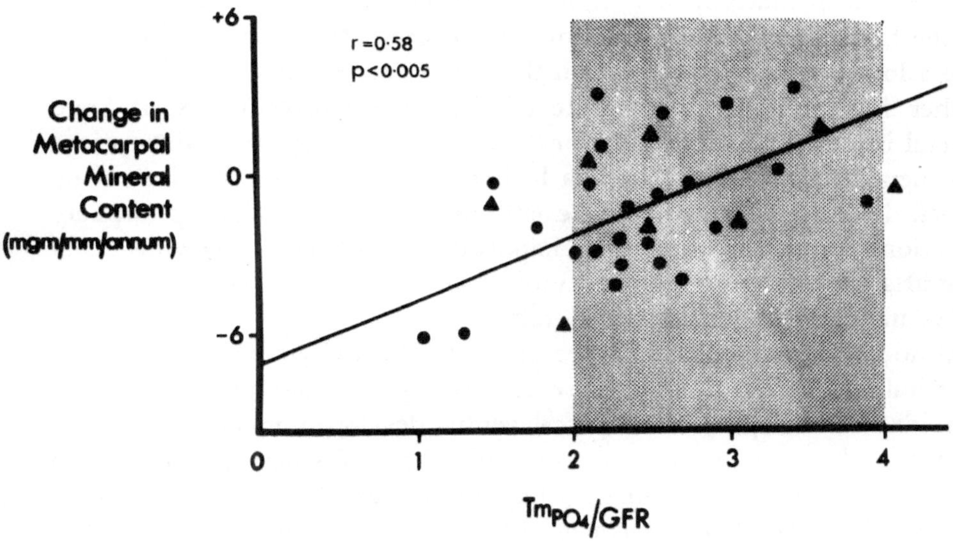

Figure 2. The relationship between the rate of bone loss measured during the first post-transplant year and the setting of the phosphate excretion (tubular maximum for phosphate divided by GFR). The stippled area represents the normal range for $TmPo_4$/GFR. Again the patients represented by the filled triangles are those who developed avascular necrosis.

ship held good for both groups of patients. There was no relationship between the rate of bone loss and the total steroid dose during the first post-operative year, or the number of intravenous 1/g doses required by each patient.

DISCUSSION

The problem of ischaemic necrosis developing after renal transplantation is one which will require solution before the full beneficial effects of the transplanted kidney will be evident. Structural bone collapse may be severe enough not only to cause symptoms but also to require major surgical intervention. Crude analysis of the factors involved have revealed no single definitive precipitating cause. Our more detailed study has confirmed a higher risk among patients with low peripheral bone density at transplantation. As there is good correlation between different sites using this technique, then the bone in other situations of those patients with lower mineral content may be less likely to withstand structural failure (6). However, there is some overlap between the two groups of patients and it is clear that some further insult must occur before final structural collapse.

After transplantation there is, in most instances, continued loss of bone for the first year. We would assume that those patients who at operation have a low mineral content within their bone are less likely to withstand further loss of mineral. The rate of this post-transplant loss of bone mineral is related to the setting for the control of phosphate reabsorption. In general those patients with a higher phosphate excretion lose bone fastest. It is well known that several factors can influence phosphate excretion. Of these the most important in patients receiving renal allografts are likely to be parathyroid hormone activity, steroid therapy and vitamin D metabolism by the graft.

Although we have no direct measure of PTH activity in plasma, we have little indirect evidence to support the concept of continued hyperparathyroidism. Post-operative hypercalcaemia developed in only 7 of our patients and tended to be transitory, only 2 patients having raised serum calcium at the end of the first post-operative year. In one instance only, were any fresh radiological changes of hyperparathyroidism noted and this patient has not developed ischaemic necrosis.

Although steroid therapy is known to reset the level of phosphate excretion, there was no direct link between the total dose of steroid and the amount of bone lost, as one might have expected, if the steroid therapy were the principal cause of bone resorption. As yet we have no evidence on how our transplanted kidneys metabolise Vitamin D although none of our patients have received Vitamin D supplements postoperatively.

REFERENCES

1. Anderson, J.B., Shimmins, J. & Smith, D.A.: A new technique for the measurement of metacarpal density. *Brit. J. Radiol.* 39, 443—450 (1966)
2. Briggs, W.A., Hampers, C.L., Merrill, J.P., Hager, E.B., Wilson, R.E., Birtch, A.G. & Murray, J.E.: Aseptic necrosis in the femur after renal transplantation. *Ann. Surg.* 175, 282—289 (1971)
3. Cameron, J.R. & Sorenson, J.: Measurement of bone mineral content *in vivo*: An improved method. *Science* 142, 35—44 (1963)
4. Shimmins, J.: *Ph.D. Thesis*, Glasgow University, 1972
5. Shimmins, J., Smith, D.A., Aitken, J.M., Anderson, J.B. & Gillespie, F.C.: The accuracy and reproducibility of bone mineral measurements *in vivo* (b) Methods using sealed isotope sources. *Clin. Radiol.* 23, 47—51 (1972)
6. Smith, C.B., Horton, P.W., Aitken, J.M. & Smith, D.A.: The estimation of bone mineral content at selected skeletal sites by gamma-ray absorption. *Brit. J. Radiol.* 47, 314—318 (1974)
7. Smith, D.A., Anderson, J.B., Shimmins, J., Speirs, C.F. & Barnett, E.: Changes in metacarpal mineral content and density in normal male and female subjects with age. *Clin. Radiol.* 20, 23—32 (1969)

A Quantitative Analysis of Bone Changes Following Anticonvulsant Therapy

F. MELSEN & L. MOSEKILDE

INTRODUCTION

The bone manifestations in patients on anticonvulsants have been described as analogous to osteomalacia (5, 6, 8, 12), but no information on the frequency and degree has been published. The present paper is a study of the bone characteristics in 60 outpatients treated with phenytoin for more than 10 years. It comprises morphometric analysis of undecalcified and decalcified bone from a standard area of the right iliac crest.

MATERIALS AND METHODS

The investigation comprised 60 epileptics between 18 and 50 years of age: 30 men (mean 27 years o.a.) and 30 women (mean 26 years o.a.). All were out-patients having been treated with phenytoin, in some cases combined with other anticonvulsants, for more than 10 years. None of the patients was disabled or had any endocrinological or known liver disease. The serum creatinine was below 1.2 mg per cent in all subjects and none was receiving corticosteroids or other potent medication other than the anticonvulsants.

The control group comprised 16 persons: 8 men and 8 women (with the same age and sex distribution) who had died suddenly from accident or coronary occlusion.

After double-labelling with chlortetracycline, biopsy was performed and two bone cylinders were obtained from the iliac crest (2). In the controls, specimens were obtained from the same area. The best specimen

University Institute of Pathology, Århus Amtssygehus, Århus and Department of Neurology, Århus Kommunehospital, Århus.

was embedded undecalcified in methylmethacrylate (13). 5–8 μ thick sections were produced using a Jung K microtome, and stained with toluidine. The other specimen was decalcified in EDTA–K and 4 μ thick sections were stained with haematoxylin-eosin.

The following parameters were measured in the cancellous bone: *absolute volume of trabecular bone (AVTB)* (2); *per cent osteoid surface (OS)* (14); *relative osteoid volume (OV)* (3); *index of osteoid (OI)* calculated as $\frac{OV \times 100}{OS}$ (16) and *trabecular osteoclastic surface resorption (RS)* (3). On the decalcified sections the *mean volume of periosteocytic lacunae (POL)* (microns^2) was measured (15).

In the specimens from the epileptics the *calcification rate (CR)* was

Table I.

	Variables	N	$\bar{X}$	S.E.	D.max	N	$\bar{X}$	S.E.	D.max	P
			Males				Females			
Epileptic	AVTB	29	23.41	0.90	.14	30	23.55	0.87	.13	N.S.
	OS	29	16.55	1.43	.15	30	13.28	1.50	.18	<0.05
	OV	29	4.61	0.38	.12	29	3.50	0.34	.12	<0.05
	OI	29	29.03	1.85	.15	29	27.85	1.48	.12	N.S.
	SR	29	5.31	0.32	.15	29	4.67	0.28	.14	N.S.
	POL	24	72.33	1.67	.07	24	70.83	1.66	.07	N.S.
	CR	21	0.72	0.03	.10	18	0.74	0.02	.14	N.S.
Normal	AVTB	7	21.86	1.06	.19	8	23.88	1.81	.15	
	OS	7	6.50	1.38	.12	8	11.89	2.32	.18	
	OV	7	1.26	0.30	.15	8	1.75	0.54	.19	
	OI	7	18.43	1.27	.09	8	13.39	2.67	.15	
	SR	7	3.36	0.13	.19	8	3.14	0.26	.11	
	POL	6	52.67	1.93	.22	7	51.29	0.81	.23	
	CR									

Statistical description of the variables.

Variables	Difference epilep/normal	
	F	M
AVTB	N.S.	N.S.
OS	N.S.	P<0.01
OV	P<0.05	P<0.01
OI	P<0.01	P<0.01
SR	P<0.01	P<0.01
POL	P<0.01	P<0.01
CR		

Comparison of the two groups by means of a t-test.

calculated after measurement in the fluorescence microscope on 20 μ thick undecalcified and unstained sections (7).

A statistical description of the variables was carried out including evaluation of departures from normality by means of a Kolmogorov-Smirnov test. Applying a Student's t-test, comparison of the various parameters was performed between sexes in the normal as well as in the epileptic group. Following this, the parameters obtained in the epileptic group were compared with those from the normal group (Table I).

RESULTS

There was no significant difference in the AVTB between the group of patients and the controls and between the two sexes. In the epileptics the OS was significantly increased in the males but not in the females (Table I). The OV was increased in both sexes, as was the mean thickness of osteoid expressed as OI. In both sexes the CR was decreased compared with the mean value recorded by Frost (7).

A significant increase in both RS and POL was found in both sexes.

Table II.

	AVTB	OS	OV	OI	RS	POL
Males	7	17	25	20	23	22
Females	1	2	7	12	13	23

The number of patients representing increased parameters ($> \bar{x} \pm 2$ SD) can be seen in Table II. Decreased values were measured only in AVTB and only seen in 2 males and 1 female. In none of the patients all the parameters were normal.

DISCUSSION

The bone changes found in the present work suggest a mineralization defect analogous to osteomalacia. This result is corroborated by the decreased serum calcium and increased serum alkaline phosphatase (19, 11, 4) and reduced serum 25-OH-cholecalciferol (9, 20, 1) found in previous studies.

Osteomalacia expressed as an increase of osteoid (OV) was found in 25 men and 7 women. None of the patients had changes as marked as described in casuistics from USA, Britain and India (5, 6, 8, 12, 22, 21).

250

The explanation may be a relatively high vitamin D content in the food of our patients. According to Hansen (10) the average daily vitamin D intake from food in the Danish population is about 200 I.U. Based on a careful anamnesis, the daily vitamin D intake for the patients in this study was between 35 and 980 I.U. ($\bar{x}$ 294 I.U.) (17).

The values of RS and POL are compatible with increased serum PTH (8). The increased resorption activity or the hyperparathyroidism is considered to be secondary to reduced calcium as serum PTH decrease following infusion of calcium (8). In 2 men and 1 woman RS and POL increased, but normal osteoid values were found. Hypocalcaemia and secondary hyperparathyroidism which are able to increase the renal transformation of 25-OH-cholecalciferol to $1,25\text{-}(OH)_2$-cholecalciferol (18) could explain the normal calcification in these cases.

SUMMARY AND CONCLUSIONS

Biopsies were obtained from a standardized area of the iliac crest of 60 epileptics, all having been treated with phenytoin for more than 10 years, and 16 controls of the same age and sex distribution. In toluidine-stained sections of methacrylate-embedded bone the following parameters were measured in trabecular bone: absolute volume of trabecular bone, per cent osteoid-covered surfaces, relative osteoid volume, trabecular osteoclastic resorption surfaces and mean thickness of osteoid, calculated as an osteoid index. In the tetracycline-labelled patients the calcification rate was estimated and on decalcified sections the mean volume of the periosteocytic lacunae was measured.

The results revealed an increased amount of unmineralized bone as an expression of osteomalacia. The changes were moderate. The calcification rate was low compared to what is considered normal elsewhere. The trabecular osteoclastic resorption surfaces and mean volume of periosteocytic lacunae were increased, suggesting a secondary hyperparathyroidism.

REFERENCES

1. Belsey, R.E., DeLuca, H.F. & Potts, J.T.: A rapid assay for 25-OH-vitamin D_3 without preparative chromatography. *J. clin. Endocr.* **38**, 1046–1051 (1974)
2. Bordier, P., Matrajt, H., Miravet, L. & Hioco, D.: Mesure histologique de la masse et de la résorption des travées osseuses. *Path. et Biol.* **12**, 1238–1243 (1964)
3. Courpron, P.: *Données histologiques quantitatives sur le vieillissement osseux humain.* Thesis, Lyon 1972
4. DeLuca, K., Masotti, R.E. & Partington, M.W.: Altered calcium metabolism due to anticonvulsant drugs. *Develop. med. Child. Neurol.* **14**, 318–321 (1972)
5. Dent, C.E., Richens, A., Rowe, D.J.F. & Stamp, T.C.E.: Osteomalacia with long-term anticonvulsant therapy in epilepsy. *Brit. med. J.* **4**, 69–72 (1970)

6. Flury, W.: Osteomalazie nach langdauernder antiepileptischer Behandlung. *Schweiz. med. Wschr.* 102, 1333—1338 (1972)

7. Frost, H.M.: Tetracycline-based Histological Analysis of Bone Remodelling. *Calc. Tiss. Res.* 3, 211—237 (1969)

8. Genuth, S.M., Kelin, L., Rabinovich, S. & King, K.C.: Osteomalacia accompanying chronic anticonvulsant therapy. *J. clin. Endocr.* 35, 378—386 (1972)

9. Hahn, T.J., Hendin, B.A., Scharp, C.R. & Haddad, J.G.: Effect of chronic anticonvulsant therapy on serum 25-hydroxycalciferol levels in adults. *New Engl. J. Med.* 287, 900—904 (1972)

10. Hansen, K.M.: *Husholdningsrådets tekniske meddelelser* 2, 16 (1973)

11. Hunter, J., Maxwell, J.D., Stewart, D.A., Parson, V. & Williams, R.: Altered calcium metabolism in epileptic children on anticonvulsants. *Brit. med. J.* 4, 202—204 (1971)

12. Marsden, C.D., Reynolds, E.H., Parsons, V., Harris, R. & Duchen, I.: Myopathy associated with anticonvulsant osteomalacia. *Brit. med. J.* 4, 526—527 (1973)

13. Melsen, B.: *The Cranial Base. The postnatal development of the cranial base studied histologically on human autopsy material.* Thesis. *Acta odontol. scand.* 32, suppl. 62 (1974)

14. Meunier, P.: *La dynamique du remaniement osseux etudiée par lecture quantitative de la biopsie osseuse.* These, Lyon 1967

15. Meunier, P., Bernard, J. & Vignon, G.: The Measurement of Periosteocytic Enlargement in Primary and Secondary Hyperparathyroidism. *Israel J. Med. Sci.* 3, 482—485 (1971)

16. Meunier, P., Edouard, C., Courpron, P. & Toussaint, F.: Morphometric Analysis of Osteoid in Iliac Trabecular Bone. Methodology. Dynamical Significance of the Osteoid Parameters. (Unpublished (1974))

17. Mosekilde, L. & Melsen, F.: Etiological factors in osteomalacia accompanying chronic anticonvulsant therapy and recorded by a quantitative analysis of bone changes. (To be published)

18. Rasmussen, H., Wong, M., Bikle, D. & Goodman, D.B.P.: Hormonal control of the renal conversion of 25-hydroxy-cholecalciferol to 1.25-dihydroxycholecalciferol. *J. clin. Invest.* 51, 2502—2504 (1972)

19. Richens, A.C. & Rowe, D.J.F.: Disturbance of calcium metabolism by anticonvulsant drugs. *Brit. med. J.* 4, 73—76 (1970)

20. Stamp, T.C.B., Round, J.M., Rowe, D.J.F. & Haddad, J.G.: Plasma levels and therapeutic effect of 25-hydroxycholecalciferol in epileptic patients taking anticonvulsant drugs. *Brit. med. J.* 4, 9—12 (1972)

21. Teotia, M. & Teotia, S.P.S.: Rickets precipitated by anticonvulsant drugs. *Am. J. Dis. Child.* 125,850—852 (1973)

22. Varkey, K., Raman, P.T., Bhaktaviziam, A. & Taori, G.M.: Osteomalacia due to phenytoin sodium. *J. Neurol. Sci.* 19, 287—295 (1973)

Treatment of Anticonvulsant Osteomalacia with Vitamin D

C. CHRISTIANSEN & P. RØDBRO

INTRODUCTION

It is well established that there is a disturbance in vitamin D metabolism in epileptics treated with anticonvulsant drugs. This so-called "Anticonvulsant Osteomalacia" includes features as: Hypocalcaemia, elevated serum alkaline phosphatase, a subnormal bone mineral content, and a subnormal serum concentration of 25-OH-D_3 (1, 3, 5, 6).

It is believed that the condition is due to an induction of liver enzymes and also that 25-OH-D_3 should be the drug of choice for treatment (4, 7).

The purpose of this study was to evaluate the optimal treatment of anticonvulsant osteomalacia from bone mineral content and serum calcium measurements with respect to dose and type of vitamin D.

EXPERIMENT I

226 epileptic patients who attended the epilepsy clinic at Glostrup Hospital at regular intervals consented to the study. They all fulfilled the following criteria: They were between 21 and 70 years of age; they had no symptoms of digestive or renal diseases; and they were all cooperative. According to the anticonvulsant treatment the 226 patients were divided into three groups:

Group A comprised 151 patients treated with phenytoin.

Group B comprised 16 patients treated with phenobarbitone *or* primidone (since the action of primidone is due to the metabolic conversion to phenobarbitone, these patients were taken as one group).

Group C comprised 59 patients treated with phenytoin plus phenobarbitone *or* phenytoin plus primidone.

Departments of Clinical Chemistry and Clinical Physiology, Glostrup Hospital, and Department of Clinical Physiology, Aalborg Sygehus.

The patients in groups A, B and C were allocated to "vitamin D_2 treatment groups" or "placebo treatment groups".

The bone mineral content was determined by direct photon absorptiometry on both forearms (1). Blood samples were drawn from all patients without stasis, and serum calcium was determined.

26 out of 226 epileptics (12%) had serum calcium levels lower than the normal range, and the mean value in the epileptic patients was significantly lower than normal at the 0.1% level.

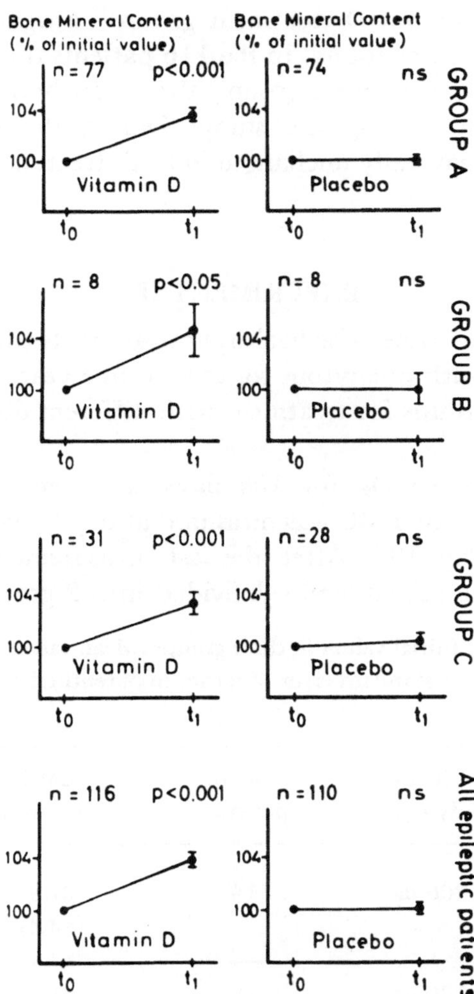

Figure 1. Effect on bone mineral content (BMC) of three months' treatment with vitamin D_2 plus calcium (left) and placebo plus calcium (right). BMC expressed as 100% at time t_0 and as percentage of initial value ($\pm$ 1 SEM) three months later, at time t_1.

ns = not significant.

20% of the epileptic patients had BMC values lower than the mean value minus two standard deviations of their corresponding normal age and sex group, and 80% of the patients had values lower than the mean normal value of their corresponding age and sex group.

After the initial bone mineral content measurement the "vitamin D group" was treated with calcium lactate (390 mg Ca daily) and vitamin D (50 µg daily) and the "placebo group" was treated with calcium lactate (390 mg Ca daily) and placebo.

To evaluate the effect of treatment, each patient's initial value was termed 100%. In all the "vitamin D treatment groups" a significant increase in bone mineral content was found. In the 116 patients the increase averaged 4%. In the "placebo treatment group" the mean bone mineral content remained unchanged during the study. During the study the serum calcium levels were virtually unchanged in both treatment groups (Fig. 1).

EXPERIMENT II

Fourty epileptic (but otherwise healthy) out-patients took part. They had been in treatment with phenytoin for at least one year. The patients were allocated to three groups for treatment with different doses of vitamin D_2 (Table I).

All patients were given D_2 for 105 days, supplemented with 500 mg calcium orally each day. BMC was measured at day 0, day 15, day 30, day 45, day 75, and day 105. After the last measurement the patients in groups I and II were mixed, and subdivided into 2 groups: One group of

Table I. Initial values in three groups. Mean values given (± 1 standard error of mean, in parenthesis).

Group	Vit. D_2 daily dose	No. of patients	BMC % of normal	Serum calcium (mg/l)
I	200 µg	14	87.9 (±4.4)	94.0 (±1.2)
II	100 µg	13	88.5 (±4.2)	96.2 (±1.1)
III	50 µg	13	88.2 (±2.3)	94.5 (±0.9)

13 patients received 25 μg vitamin D_2 daily for 150 days, another group of 14 patients received 5 μg vitamin D_2 daily for 150 days. Hereafter BMC and serum calcium were measured again. The results are given as per cent of initial values (Tables II and III). Normal values for serum calcium were: 99.0 ± 0.4 mg/l (mean ± 1 standard error of mean).

The initial values showed that the patients as a group had a subnormal BMC and hypocalcaemia (Table I). During initial treatment with vitamin D_2 BMC increased in all groups, mostly in group II, in which the value at day 105 was not significantly different from normal mean (Table II). During maintenance vitamin D_2 therapy BMC was maintained at the same level after 5 months as at the end of initial treatment, when the patients were given 25 μg vitamin D_2 daily, while BMC decreased to initial levels

Table II. Mean BMC ± 1 standard error of mean (in per cent of initial value at time 0) as a function of time in three groups before and during treatment with calcium and vitamin D_2.

	Day 0	Day 15	Day 30	Day 45	Day 75	Day 105
Group I	100	101.4 ± 0.6	103.4 ± 1.2	104.7 ± 1.0	105.6 ± 0.9	105.7 ± 1.5
Group II	100	103.0 ± 1.4	106.2 ± 2.1	106.7 ± 1.9	107.8 ± 1.9	108.9* ± 2.1
Group III	100	101.2 ± 0.4	102.4 ± 0.6	103.1 ± 0.7	102.7 ± 0.6	103.0 ± 0.9

*) Not significantly different from the normal mean (113.4 ± 1.7%).

Table III. BMC values at start of study, after initial dose for 3½ months (200 or 100 μg vitamin D_2 daily) and after a further 5 months on two different maintenance doses (25 or 5 μg vitamin D_2 daily) given to 13 and 14 epileptics, respectively. Mean values in per cent of initial values ± 1 standard error of mean.

Daily maintenance dose	Start of study	After 3½ months	After 8½ months
25 μg	100	105.5 ± 2.3	105.2 ± 1.8
5 μg	100	108.9 ± 2.7	100.9 ± 2.0

256

when only 5 μg was given (Table III). Mean values of serum calcium were unchanged through all 8½ months.

EXPERIMENT III

Fiftyfour epileptics attending an epilepsy clinic took part in the investigation. They were out-patients who had been on anticonvulsant therapy for at least one year. The patients were divided into six subgroups for treatment with three different doses of D_3 and 25-OH-D_3. The allocation was done so that initial mean BMC-values were alike in each subgroup. The daily doses were 200, 100, and 50 μg D_3 daily and 40, 20, and 10 μg 25-OH-D_3 daily. The treatment was given for 84 days.

BMC and serum calcium were measured at time t_0 (beginning of the study), time t_1 (day 21), time t_2 (day 42), time t_3 (day 63) and time t_4 (day 84).

The mean BMC-values in the six subgroups (in per cent of initial values) are given as a function of time during treatment in Fig. 2, which also

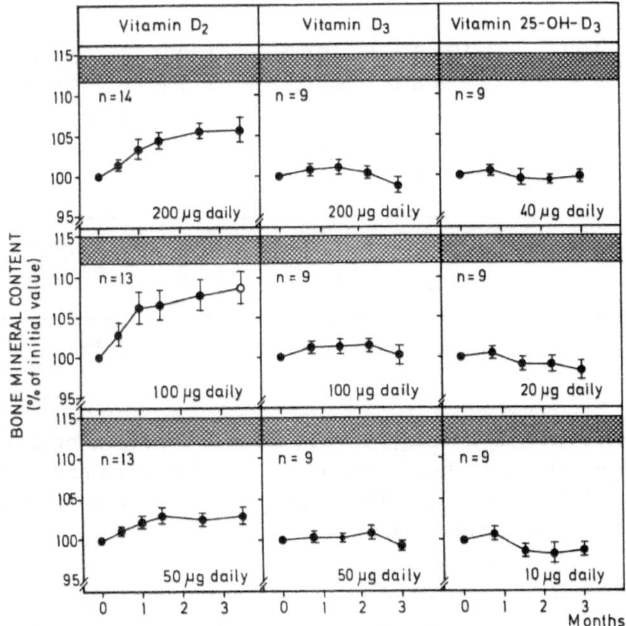

Figure 2. (BMC) Bone mineral content as a function of time during treatment with three different doses of vitamin D_3 (200, 100, and 50 μg daily) and vitamin 25-OH-D_3 (40, 20, and 10 μg daily) in 54 epileptics, divided into 6 subgroups.

At the left are given values in 40 epileptics treated with vitamin D_2 (200, 100, and 50 μg daily).

contains the values from 40 epileptics treated with vitamin D_2 (experiment II). There was no significant change in BMC during treatment with vitamin D_3 and 25-OH-D_3, opposed to the findings with vitamin D_2.

The mean serum calcium in the six subgroups (mg/l) are given as a function of time during treatment in Fig. 3, which also contains the values from 40 epileptics treated with vitamin D_2 (experiment II). A significant increase in serum calcium was seen during treatment with D_3 and 25-OH-D_3, opposed to the findings with vitamin D_2.

DISCUSSION

It is a cost-benefit problem whether epileptic patients should have prophylactic vitamin D treatment or not. The clinical significance of the condition remains to be clarified. Some reports indicate that fractures are more common in epileptics than in controls (8), and that vitamin D may reduce the frequency of epileptic seizures (2). If all epileptic patients should be treated prophylactically with vitamin D, an initial dose of 100

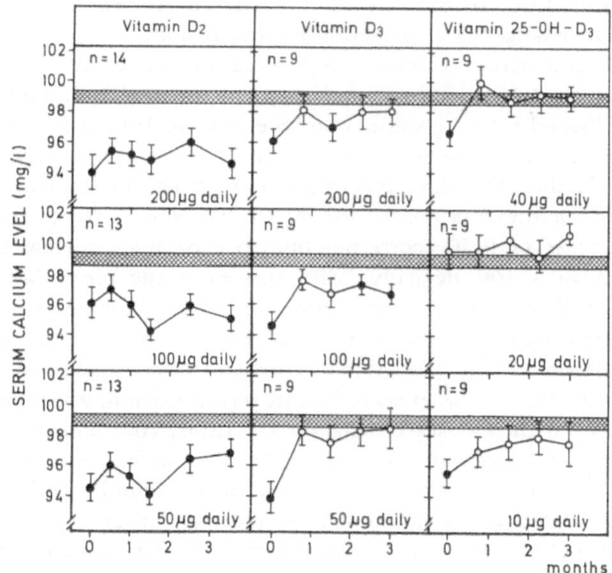

Figure 3. Serum calcium as a function of time during treatment with 3 different doses of vitamin D_3 (200, 100, and 50 µg daily), and vitamin 25-OH-D_3 (40, 20, and 10 µg daily) in 54 epileptics, divided into 6 subgroups.

At the left are given values in 40 epileptics treated with vitamin D_2 (200, 100, and 50 µg daily).

Unfilled circles: Values not significantly different from normal mean (hatched area: normal mean ± 1 SEM).

258

μg vitamin D_2 daily for 4–5 months, followed by a maintenance dose of 25 μg daily would seem appropriate.

The present findings have demonstrated (when the effect of vitamin D on anticonvulsant osteomalacia is evaluated from BMC and serum calcium) that D_3 and 25-OH-D_3 act similarly, whereas vitamin D_2 differs in its action from these vitamins. The findings suggest that induction of liver enzymes alone cannot explain the mechanism of anticonvulsant osteomalacia. If this were the case 25-OH-D_3 should have had an optimal effect. The results suggest that the metabolism of D_2 and D_3 are affected in different manners by anticonvulsant drugs.

ABSTRACT

Evaluated from two indices of osteomalacia: Hypocalcaemia, and lowered bone mineral content (BMC), the occurrence and incidence of the iatrogenic state of anticonvulsant osteomalacia is well established.

In a controlled therapeutic trial comprising 226 epileptic patients we have demonstrated that the mean BMC value for all epileptics was 87% of normal. During treatment with 50 μg vitamin D_2 daily for three months an average BMC increase of 4% was found. Normal levels of BMC were not reached by this relatively small vitamin D supplement, and no change was observed in serum calcium.

Hereafter, BMC and serum calcium level were measured in 40 epileptics on long-term phenytoin treatment, *before* and *during* treatment with vitamin D_2. Initially the patients were divided into 3 groups, who received 50, 100, and 200 μg daily for 105 days. Hereafter, the 27 patients on the two higher doses were subdivided into two groups, who for a further 150 days received either 25 μg or 5 μg daily. During the whole study serum calcium was unaffected by the vitamin D treatment. With all 3 doses an initial increase in BMC (corresponding to a positive calcium balance) was observed – highest with 100 μg daily. With this dose the mean BMC value was normalized. In the following maintenance dose period an estimated calcium balance close to zero was seen with 25 μg daily, whereas the patients on 5 μg daily showed a negative calcium balance, as their BMC values returned to the initial figures.

Finally, in 54 epileptic out-patients, BMC and serum calcium was measured before and during treatment with 3 doses of vitamin D_3 (200, 100, and 50 μg daily) and 25-OH-D_3 (40, 20, and 10 μg daily) for 12 weeks. The patients who received D_3 or 25-OH-D_3 showed a similar pattern, opposite to what was found with D_2: An increase in serum calcium but unchanged values of bone mineral content. The results suggest that liver enzyme induction cannot alone explain anticonvulsant osteomalacia.

REFERENCES

1. Christiansen, C., Rødbro, P. & Lund, M.: Incidence of anticonvulsant osteomalacia and effect of vitamin D: Controlled therapeutic trial. *Brit. Med. J.* 4, 695–701 (1973)

2. Christiansen, C., Rødbro, P. & Sjö, O.: "Anticonvulsant action" of vitamin D in epileptic patients? A controlled pilot study. *Brit. Med. J.* 2, 258–259 (1974)
3. Dent, C.C., Richens, A., Rowe, D.J.F. & Stamp, T.C.B.: Osteomalacia with long-term anticonvulsant therapy in epilepsy. *Brit. Med. J.* 4, 69–72 (1970)
4. Hahn, T.J., Hendin, B.A., Scharp, C.R. & Haddad, J.G.: Effect of chronic anticonvulsant therapy on serum 25-hydroxycalciferol levels in adults. *New Engl. J. Med.* 287, 900–904 (1972)
5. Hunter, J., Maxwell, J.D., Stewart, D.A., Parsons, V. & Williams, R.: Altered calcium metabolism in epileptic children on anticonvulsant therapy. *Brit. Med. J.* 4, 202–204 (1971)
6. Richens, A. & Rowe, D.J.F.: Disturbance of calcium metabolism by anticonvulsant drugs. *Brit. Med. J.* 4, 73–76 (1970)
7. Stamp, T.C.B., Round, J.M., Rowe, D.J.F. & Haddad, J.G., Plasma levels and therapeutic effect of 25-hydroxycholecalciferol in epileptic patients taking anticonvulsant drugs. *Brit. Med. J.* 4, 9–12 (1972)
8. Vasconcelos, D.: Compression fractures of the vertebrae during major epileptic seizures. *Epilepsia (Amst.)* 14, 323–328 (1973)

CHAPTER V
Parathyroid Hormone

Effects of Acute and Chronic PTH Stimulation on Osteoblasts and the Under-lying Bone Matrix

B. KREMPIEN, G. GEIGER & E. RITZ

Parathyroid hormone has a profound effect on the synthesis and removal of collagen by the action of bone cells. A peculiar biphasic response of bone collagen synthesis to the administration of PTH can be observed: In acute experiments a decrease of radioproline incorporation into bone collagen is found. After 2—5 days, the rate of collagen synthesis is back to normal and strikingly increased later on.

The regular three-dimensional structure of bone matrix results from the fact that collagen is organized along hierarchically ordered levels of organization: Suprahelix, subfibril, fibril, fiber and finally osteonic lamellae. For want of adequate methods, the interaction of collagen texture and bone cell function has not been investigated to any large extent. For this reason it is poorly understood, why the texture of collagen matrix is altered in hyperparathyroidism. We, therefore, used the methods of scanning electron microscopy (SEM) and transmission electron microscopy (TEM) to study this problem.

MATERIAL AND METHODS

100 g male Wistar rats were treated with 100 i.u. PTH (Lilly) daily for 1, 2 or 3 days and with 50 i.u. daily for a period of 3 weeks. Animals were perfused with glutaraldehyde. Specimens from the outer surface of the skull and from the tibial metaphysis were prepared by critical point drying for scanning electron microscopy (JEOL JSM S1) of osteoblasts. For the evaluation of bone mineral, organic non-mineralized matrix was digested with sodium hypochlorite before coating the bone surface with carbon gold. By this method the mineralized structures underneath the osteo-

Department of Pathology and Internal Medicine, University of Heidelberg.

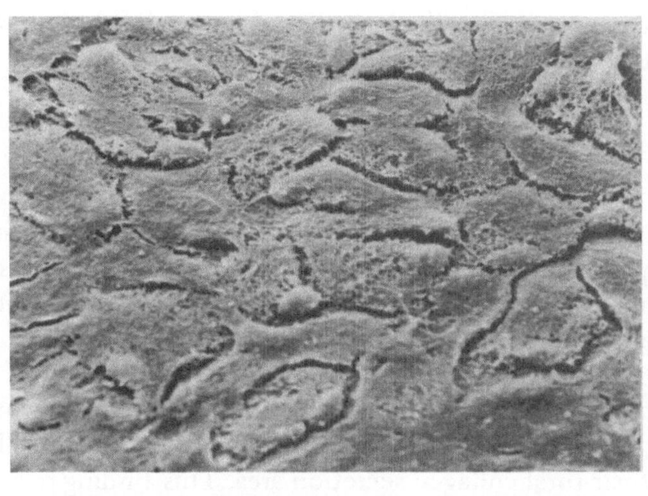

a) enl. 1:1000

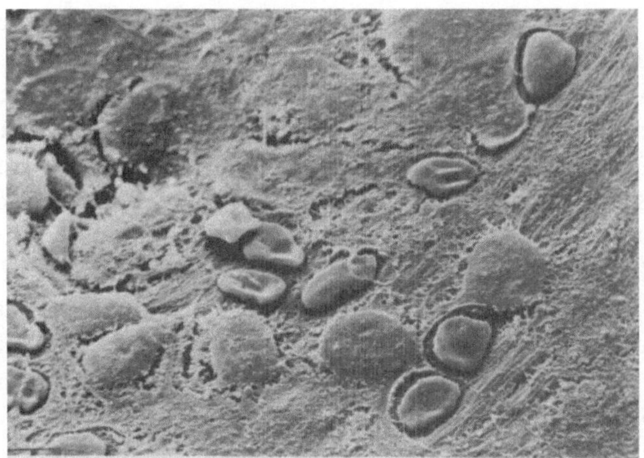

b) enl. 1:2000

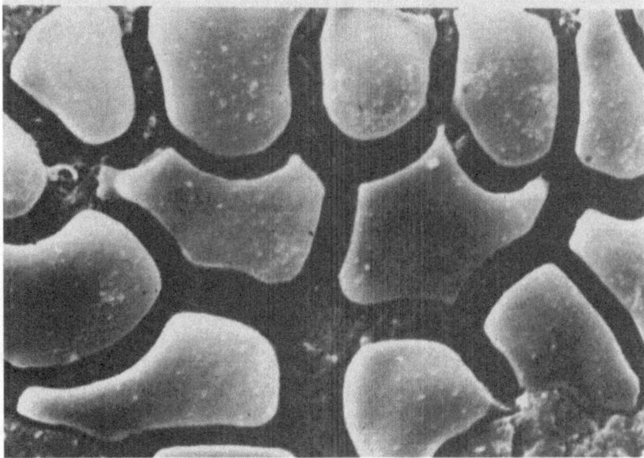

c) enl. 1:5000

Figure 1 a-c. SEM photographs of rat calvaria.

262

blasts are exposed. Ultrathin sections of mineralized bone were studied for TEM with a Siemens Elmiscope 7.

RESULTS AND DISCUSSION

a) *SEM findings in osteoblasts:* In normal animals osteoblasts form pseudoepithelial layers on the bone apposition surface as seen in Fig. 1 a. Intercellular spaces are narrow. Cells are joined by numerous inter-digitations. In rats, treated with 100 i.u. PTH for a period of one to three days, single cells overlaying the bony surface appear shrunken and detached from the matrix at their outer circumference where they appear to have been attached previously (Fig. 1 b). Thus, the cells do no longer occupy their total collagen secretion area. This finding may explain, why a decrease of collagen synthesis is found acutely after the administration of PTH. Shrinking was not seen in all cells simultaneously to an equal extent (Fig. 1 b). This may result from the fact that bone surface cell activity is not synchronized and in phase. In control animals bone surface cells seem to be "anchored" to the collagen underneath by ropelike cellular processes that apparently protrude from the cellular surface. Cells of PTH treated animals have lost these processes and appear rather smooth (Fig. 1 b) and c). In animals, treated with PTH for a period of 3 weeks, these findings are encountered no longer only in single osteoblasts but in whole groups of osteoblasts, perhaps cell clones (Fig. 1 c). The findings in metaphyseal trabeculae of the tibia were identical with those obtained in the endosteal surface of the skull.

We can only speculate, why cell shape changes under the influence of PTH: Osmotic effects may be involved or contraction of microfibrils of the cytosol under the influence of raised intracellular calcium concentrations.

b) *SEM findings in the mineralization front:* When organic matrix of bone is removed by digestion with sodium hypochlorite, apatite crystals ("calcospherites") become visible that are deposited along collagen fibers. With this technique "domain formation" can be visualized in bone forming surfaces (1). In domains collagen fibers are arranged in one preferential direction resulting from the action of a group of osteoblasts which synthesize collagen in phase. This synchronization of collagen deposition by osteoblasts is evidently lost after prolonged administration of PTH. This is clearly visible in Fig. 2 b, where the anarchy of mineralized collagen fibers texture is readily apparent when compared with a normal control animal (Fig. 2 a). These findings in experimental animals are analogous to findings in patients suffering from hyperpara-

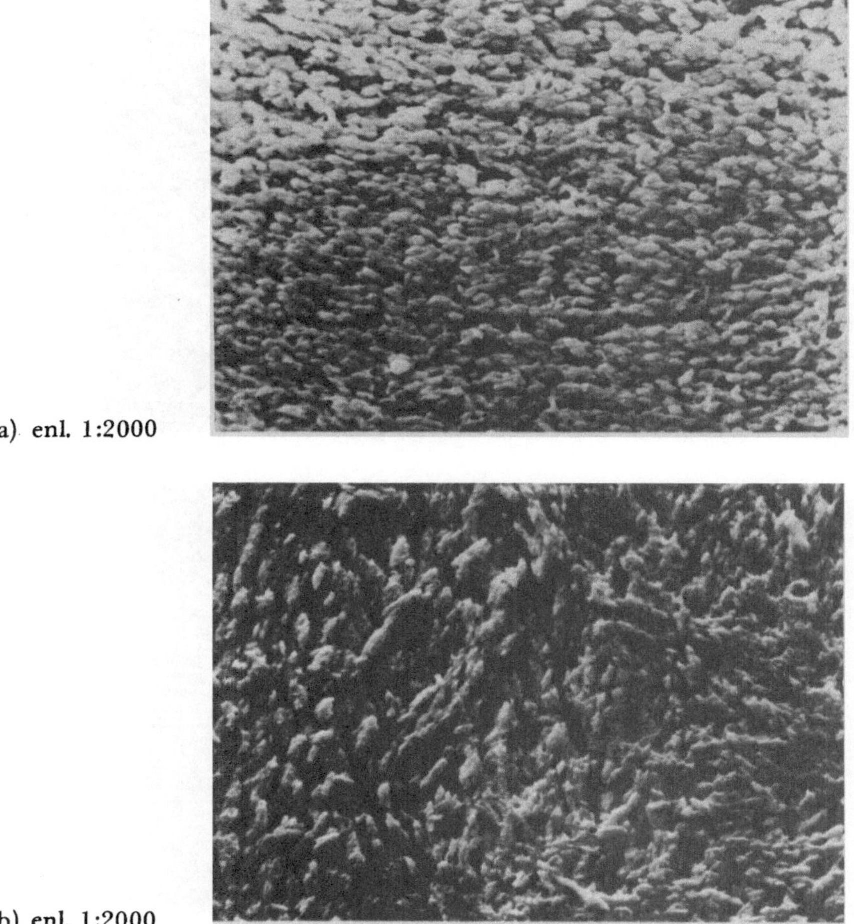

a) enl. 1:2000

b) enl. 1:2000

Figure 2 a and b. SEM photographs of rat calvaria after digestion with sodium hypochlorite.

thyroidism. In normal bone, areas of bone formation are characterized by regular domains (Fig. 3 a). Mineral deposits are of uniform size and almost cylindrical. In secondary hyperparathyroidism (Fig. 3 b), domain formation is no longer visible. Calcospherites are highly irregular and of non-uniform size and shape. A long and short diameter is no longer distinguishable.

c) *TEM findings in osteoblasts:* A disturbance of collagen synthesis and deposition by osteoblasts was also noticed in studies with transmission electron microscopy. Normal osteoblasts (Fig. 4 a) are characterized by numerous closely packed cellular processes, narrow intercellular spaces, close contact of osteoblasts with the collagen surface and finally

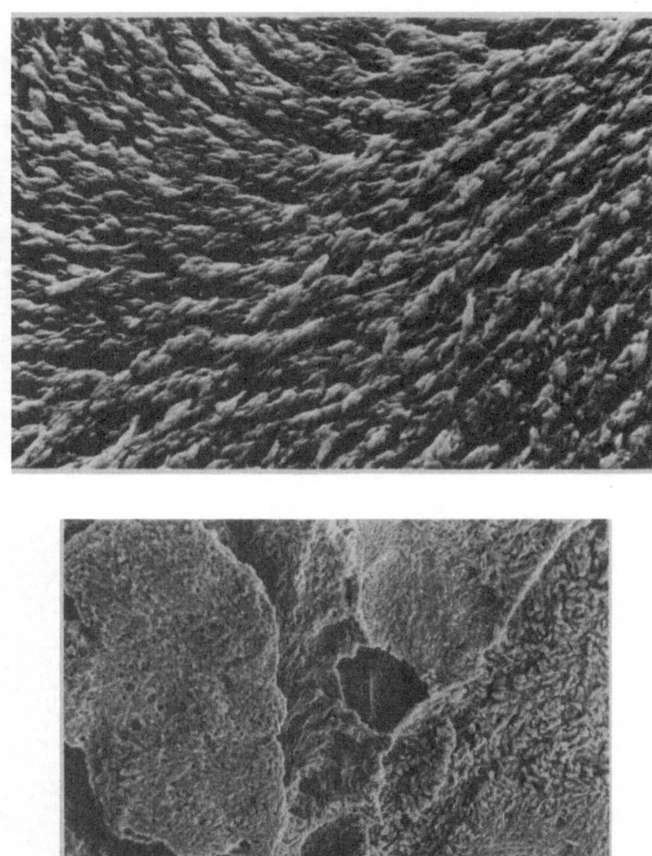

a) enl. 1:2000

b) enl. 1: 300

Figure 3 a and b. SEM photographs, endosteal human femur.

endoplasmic reticulum with closely spaced membranes within the cells. Collagen exhibits a plaited texture, the direction of collagen fibers alternating in regular intervals.

In contrast, when PTH is administered for a prolonged period (Fig. 4 b), cellular processes are shortened and reduced in number. Osteoblasts seem to have retracted from the collagen surface. The ergastoplasma is no longer composed of tightly packed membranes but appears to be a loose sacculated structure, pointing to a reduction of protein synthesis by osteoblasts. The alternating texture of orderly arranged collagen fibers is lost. The intercellular spaces between osteoblasts are markedly opened.

In the skeleton of individuals with hyperparathyroidism, loss of synchronized collagen deposition is reflected by the loss of lamellar bone

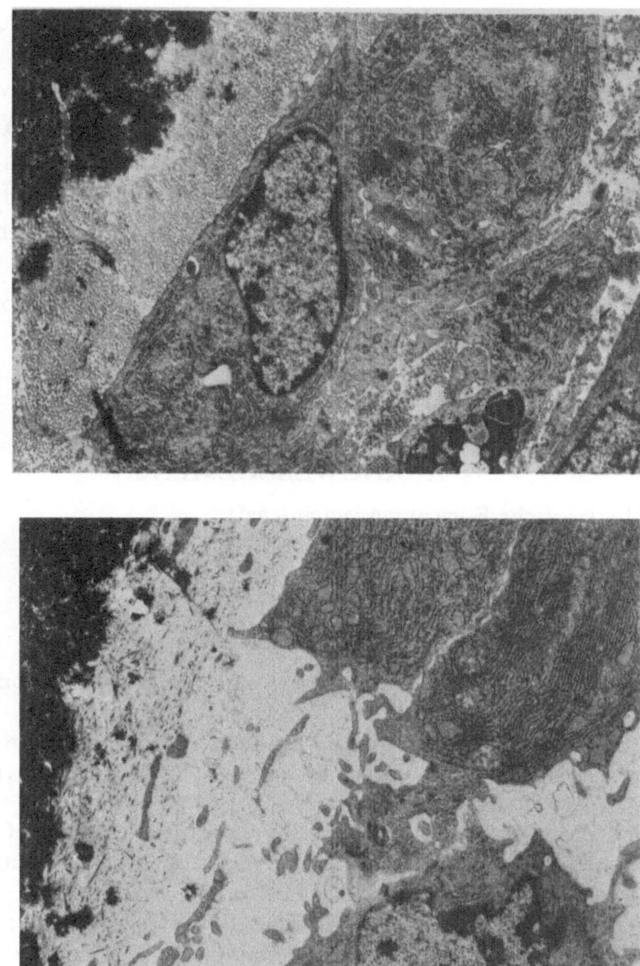

Figure 4 a and b. TEM photographs of rat calvaria.
enl. 1:10.200

structure and by its replacement with woven bone (3, 4). Some pecularities of macroscopic bone structure in hyperparathyroid bone disease are the direct consequences of this derangement of microscopical bone structure. Since the biomechanical quality of woven bone is inferior to the one of lamellar bone (2, 3, 4), increased amounts of woven bone must be deposited, when lamellar bone is replaced by woven bone. This causes "osteosclerosis" of woven bone. Since osteocytes in woven bone can no longer respond to mechanical strain like osteocytes in lamellar bone, the functional (i.e. trajectorial) structure of bone tissue is gradually lost. We conclude from our studies that bone cells, stimulated by large doses of PTH, escape from control mechanisms that normally regulate bone remodelling. As a consequence, collagen, that is laid down by osteoblasts after stimulation with PTH, looses its regular supramolecular structure. Our experimental findings explain the skeletal consequences of excessive secretion of PTH leading to osteitis fibrosa. Yet, we do not know, whether they really contribute to our understanding of what PTH does in the skeleton under physiological circumstances. It may well be that in our experiments we are dealing with pharmacological artefacts. In studies with continuous administration of physiological doses of PTH, parathyroid hormone has been shown to be an anabolic hormone without demonstrable effects on bone resorption (5).

ACKNOWLEDGEMENTS

Supported by Deutsche Forschungsgemeinschaft.

REFERENCES

1. Boyde, A. & Jones, S.J.: Osteoblasts and bone collagen orientation. Intern. Workshop Calc. Tiss., Gesher Haziv, Israel 1974
2. Krempien, B., Geiger, G. & Ritz, E.: Hardness of bone in various ages and diseases. *Calc. Tiss. 1972*, Proc. 9th Europ. Symp. Calc. Tiss., Czitober, H. and Eschberger, Y. (eds.), Facta Publication, Wien, p 95—99 1973
3. Krempien, B., Geiger, G. & Ritz, E.: Structural changes of cortical bone in secondary hyperparathyroidism. Replacement of lamellar bone by woven bone. *Virchows Arch. path. Anat.* **366**, 249 (1975)
4. Krempien, B., Geiger, G. & Ritz, E.: Alteration of bone tissue structure in secondary hyperparathyroidism. A scanning electron microschopical study. In: *Vitamin D and problems related to uremic bone disease.* A.W. Norman *et al.* (eds.), Gruyter, p 157—165, 1975
5. Parsons (see chapter IX, Reeve, J., Tregear, G.W. & Parsons, J.A.).

An *in vivo* Experimental Model for the Study of Resistance to the Renal Action of Parathyroid Hormone in Man

S. TOMLINSON, G.N. HENDY & J.L.H. O'RIORDAN

We have previously shown that intravenous administration of highly purified bovine parathyroid hormone (BPTH) to normal subjects results in rapid changes in plasma concentration and urinary excretion of adenosine $3',5'$-cyclic monophosphate (cyclic AMP). These changes appear to be due to the effects of the hormone on the kidney (3), presumably in stimulating adenylate cyclase of the renal cortex (2). We have also shown that resistance to the action of the hormone can occur; patients with severe chronic renal failure (creatinine clearance < 10 ml/min) had higher concentrations of plasma cyclic AMP (>25 pmol/ml) and lower excretion rates than normal (< 1.8 nmol/min). No change in plasma cyclic AMP and minimal increase in urinary excretion of the nucleotide occurred when 200 units of BPTH were administered as an intravenous injection to these patients. In some patients with primary hyperparathyroidism and normal renal function, impairment in the response to BPTH was observed. This appeared to be associated with high concentrations of endogenous parathyroid hormone. However, this resistance to BPTH was reversible since marked improvement in the response to exogenous hormone was seen after removal of the parathyroid adenoma, when the concentration of endogonous hormone had fallen (4).

The changes in plasma cyclic AMP are highly reproducible in any normal individual; when 200 units of BPTH were injected into a normal subject on three occasions at weekly intervals the increase in plasma cyclic AMP was remarkably similar. However, when 200 units of BPTH were injected into a normal subject at intervals of one hour, there was a progressive decline in the response to each injection (Fig. 1). The maximum concentration after the third injection was only one quarter of that following the first injection. Repeated intermittent stimulation of the

Department of Medicine, The Middlesex Hospital, London.

system, therefore resulted in a diminishing response. Furthermore, when 1000 units of BPTH were infused in a normal subject over two hours, after an initial increase, there was a progressive decline in plasma cyclic AMP concentration and in urinary cyclic AMP excretion during the

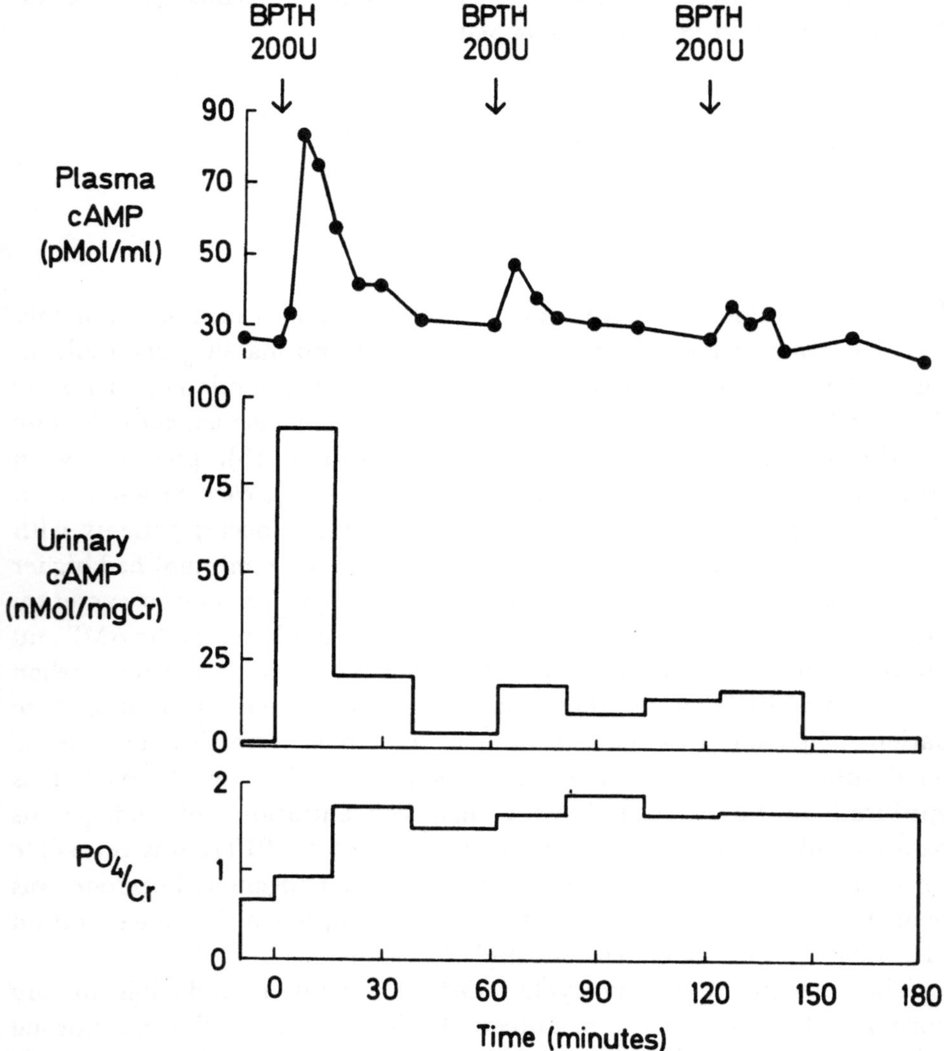

Figure 1. Shows the effects of injections of 200 units of BPTH given to a normal subject at hourly intervals. There was a progressive diminution in the increase in plasma cyclic AMP concentration and urinary cyclic AMP excretion. The phosphaturic response continued and was maintained throughout the experimental period. It should be noted, however, that the phosphaturic response, in this subject, after a single injection of 200 units of BPTH, continues for at least 165 minutes.

Time of injection indicated by ↓

(PO_4: phosphate; Cr: creatinine)

second hour of the infusion. This decline in plasma cyclic AMP concentration and urinary cyclic AMP excretion occurred despite the presence of high plasma concentrations of amino-terminal immunoreactive BPTH (1). In addition, when an injection of 200 units of BPTH was given at the end of such an infusion, the increase in plasma cyclic AMP was only one-twentieth of that observed following a control infusion of vehicle without BPTH (Fig. 2). This impairment in response to an injection of

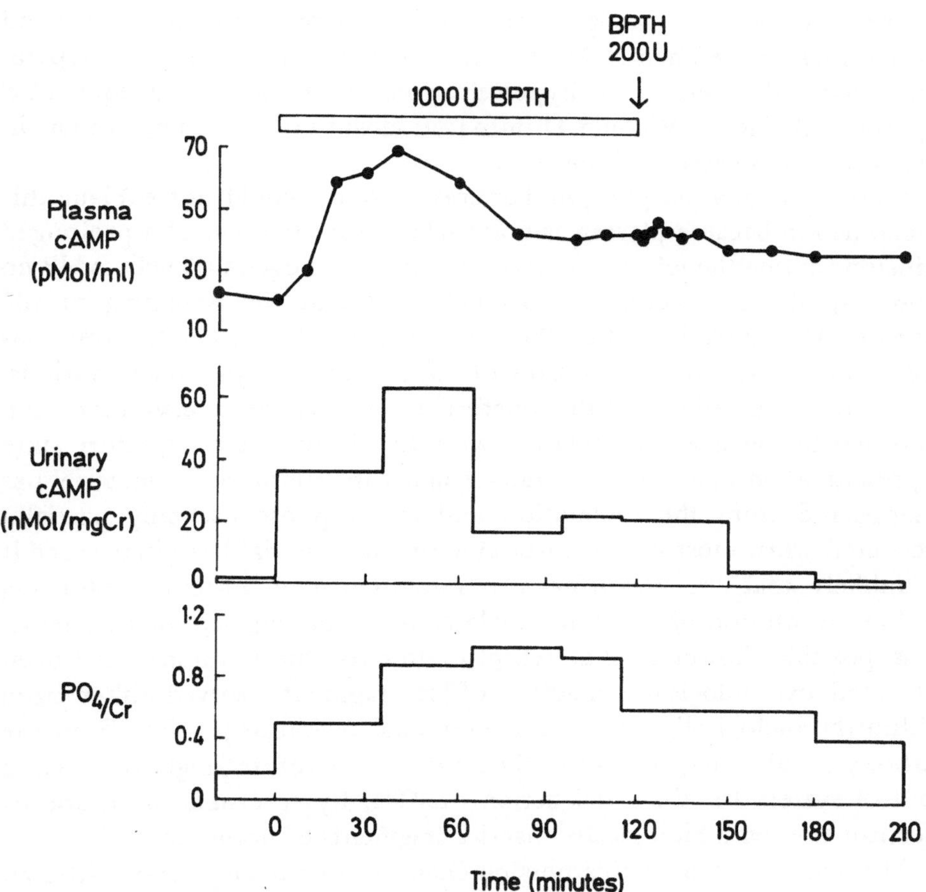

Figure 2. During a prolonged infusion of 1000 units of BPTH, plasma cyclic AMP concentration initially increased and then declined during the latter part of the infusion. At the termination of the infusion an injection of 200 units of hormone resulted in a markedly impaired response (cf. response to the first injection of hormone in Fig. 1). The changes in urinary cyclic AMP reflect those observed in the plasma. The phosphaturic response was maintained in spite of the declining cyclic AMP response, but there was no increase in phosphaturia following the injection of hormone.

(PO₄: phosphate; Cr: creatinine)

hormone was still present 30 minutes after the end of the infusion. When the injection of 200 units of BPTH was given four hours after the end of the infusion the response had improved, although it was still only one-third of that observed following a control infusion of vehicle alone. After 24 hours the normal increase in plasma cyclic AMP was observed following the injection of BPTH.

Thus not only repeated but also prolonged stimulation of the renal adenylate cyclase can induce a refractory state. This is not due to hypercalcaemia since no significant increase in plasma calcium occurred during these experiments. Moreover, a patient with surgical hypoparathyroidism who was made hypercalcaemic (total plasma calcium 11.8 mg/100 ml) due to Vitamin D intoxication did not show impairment in response to an injection of hormone.

Increase in plasma phosphodiesterase activity could not explain this phenomenon because plasma samples taken from the end of a prolonged infusion (during the refractory period) degraded exogenous cyclic AMP no more rapidly than samples taken before and at the beginning of the infusion. It is unlikely that all the available intracellular ATP substrate has been consumed since it is present in relatively high concentrations; however, it is possible that the kinetics of the adenylate cyclase have been modified by persistent stimulation with BPTH. Since the refractory state is present when most of the amino-terminal immunoreactive material has disappeared from the circulation and the response has only partially recovered when most of the carboxy-terminal material has disappeared it is unlikely that the impairment in response to an injection following prolonged infusion of hormone is related to circulating peptide fragments. It is possible, however, that receptor sites for the hormone have been saturated by biologically inactive BPTH fragments derived either from within the biologically active amino-terminal region or possibly from the carboxy-terminal region.[1] This phenomenon, therefore, might represent a control system for the renal action of PTH, by reversible saturation of receptor sites with biologically inactive fragments of hormone.

The reproduction of this phenomenon in normal subjects provides an *in vivo* experimental model for the study of resistance to this action of parathyroid hormone in man.

---

1) Footnote added in proof.
   Since this paper went to press, we have infused the synthetic 1—34 amino-terminal fragment of BPTH into a normal subject and have shown that resistance develops to this peptide. Clearly, therefore, carboxy-terminal immunoreactive material cannot be responsible for the development of this phenomenon.

REFERENCES

1. Barling, P.M., Hendy, G.N., Evans, M.C. & O'Riordan, J.L.H.: Region-specific immunoassays for parathyroid hormone. *J. Endocr.* **66**, 307–318 (1975)
2. Marcus, R. & Aurbach, G.D.: Bioassay of parathyroid hormone *in vitro* with a stable preparation of adenyl cyclase from rat kidney. *Endocrinology* **85**, 801–810 (1969)
3. Tomlinson, S., Barling, P.M., Albano, J.D.M., Brown, B.L. & O'Riordan, J.L.H.: The effects of exogenous parathyroid hormone on plasma and urinary adenosine $3',5'$-cyclic monophosphate in man. *Clin. Sci. Molec. Med.* **47**, 481–492 (1974)
4. Tomlinson, S., Hendy, G.N., Preece, M.A. & O'Riordan, J.L.H.: Acquired resistance to the effects of parathyroid hormone on plasma and urinary adenosine $3',5'$-cyclic monophosphate in man. *Clin. Sci. Molec. Med.* **48**, 11p-12p (1975)

Dietary Induction of *Osteitis Fibrosa Cystica* in the rat: Roles of Parathyroid Hormone and Calcitonin

I. CLARK & H.R. BRASHEAR

INTRODUCTION

It is well known that hyperparathyroidism in man may result in skeletal and renal lesions, although they are not invariable consequences (1). Experimental secondary hyperparathyroidism has been produced in a variety of species by feeding diets with low Ca/P ratios, i.e., either low calcium (6) or high phosphate levels in the diet (7, 8). Bone lesions varied from mild osteoporosis to severe osteitis fibrosa cystica depending on the experimental conditions. The stimulus for increased parathyroid secretion is the level of serum calcium (12) which is directly related to the logarithm of the dietary Ca/P ratio (13). The serum level of calcium is also influenced by other dietary factors, e.g. vitamin D and hormones. Thus, although the level of circulating parathyroid hormone may be the prime etiologic factor in producing bone lesions, its action may be modified by dietary means (3).

The present study was designed to determine the effects of alterations in parathyroid hormone and/or calcitonin secretion on the development of skeletal and renal lesions in rats fed diet with low Ca/P ratio (high phosphate diet).

MATERIALS AND METHODS

Adult male Holtzman rats weighing approximately 250 g were used: *group 1* — intact controls, *group 2* — rats whose parathyroid glands were removed surgically and implanted back into their original sites (parathyroid implants — PTI), *group 3* — rats whose parathyroid glands were removed surgically (parathyroid deficient — PD), *group 4* — rats whose

Department of Surgery, CMDNJ-Rutgers Medical School, Piscataway, New Jersey and Division of Orthopaedic Surgery, University of North Carolina, School of Medicine, Chapel Hill, North Carolina.

parathyroid glands had been transplanted into the strap muscle in the neck and whose thyroid glands had been removed (calcitonin deficient — CD) and *group 5* — rats whose parathyroid and thyroid glands had been removed (parathyroid and calcitonin deficient — PCD). All animals whose thyroid glands had been removed, received thyroxine replacement therapy. Half of the animals in each group were fed a semi-synthetic diet with a Ca/P ratio of 2 (0.8% calcium and 0.4% phosphorus) and the other half were fed a semi-synthetic diet with a Ca/P ratio of 1:8 (0.4% calcium and 3.2% phosphorus). The animals had free access to these diets for two months. After allowing at least 11 days for the animals to recover from the effects of surgery, serum levels of calcium were determined. If the levels of animals that had parathyroid implants (group PTI) or transplants (group CD) were not within the normal range, the rats were discarded. If the serum levels of calcium of the parathyroidectomized rats (group PD) and the thyroparathyroidectomized animals (group PCD) were not below 2 mM, they too were discarded.

RESULTS

When fed the control diet, animals with parathyroid deficiency gained considerably less weight than did the other groups which gained about the same (Table I). With the exception of the parathyroidectomized rats, none of the groups eating the high phosphate diet gained as much as did those eating the control diet. Only intact rats fed the high phosphate diet lost weight while the other groups gained similar amounts with the greatest gain by rats that were parathyroid and calcitonin deficient (PCD). It would appear that transplantation of the parathyroid gland, calcitonin deficiency, or parathyroid plus calcitonin deficiency reduces the toxicity of the high phosphate diet with respect to soft tissue growth.

The serum levels of calcium and phosphate of rats with reimplanted parathyroid glands (PTI) fed the control diet were within the normal range but those fed the high phosphate diet had serum levels of calcium higher and phosphate lower than the intact animals fed the same diet (Table I).

Parathyroid deficient rats (PD) fed the control diet had significantly lower levels of serum calcium and higher levels of phosphate than did the intact rats (Table I).

Chronic calcitonin deficiency (CD) lowered serum calcium but did not change serum phosphate of rats fed the control diet. Although the high phosphate diet lowered further serum levels of calcium and raised those of phosphate in these animals, the magnitude changes was not as large as in

Table I. Effect of dietary Ca/P ratios.

	Body wt. change (g)		Serum calcium (mM)		Serum phosphate (mM)		Kidney lesions		Bone lesions	
Ratio	2:1	1:8	2:1	1:8	2:1	1:8	2:1	1:8	2:1	1:8
Intact	+150	- 4	2.59	1.76**	2.48	6.28	0	++++	0	++++
PTI	+180	+ 74	2.56	1.94**	2.82	4.11**[a'']	0	++++	0	+
PD	+ 34	+ 32	1.93[a]	1.69	3.43[a']	4.95*	0	±	0	0
CD	+146	+ 78	2.47	1.95*	2.65	4.69**	0	+++	0	0
PCD	+150	+110	2.43	2.23[a'']	3.00	3.91[a'']	0	+++	0	0

** indicates the mean values are significantly different ($p < .01$) from those of animals fed the diet with Ca/P ratio of 2:1, * means $p < .05$, [a] indicates $p < .01$ and [a''] indicates $p < .05$.

intact rats (Table I).

Serum levels of calcium and phosphate of rats, which had their parathyroid and thyroid glands removed (PCD) and which were fed the control diet had returned to normal after two months (Table I). We have noted this recovery after thyroparathyroidectomy in other studies (to be published). Feeding these animals the high phosphate diet did not change their serum levels of calcium and phosphate anywhere near as much as it did when fed to intact rats. It would appear that given sufficient period of time to recover from surgery, animals lacking both parathyroid hormone and calcitonin establish a new calcium and phosphate homeostasis that is similar to animals secreting both hormones. However, under dietary conditions that cause a fall in serum calcium and a rise in serum phosphate in rats secreting both hormones, animals without the hormones are not as responsive.

The surgical manipulations had no apparent effect on skeletal or renal tissue of rats fed the control diet, but had varying effects when the high phosphate diet was fed (Table I). Nephrocalcinosis was observed in all groups except the parathyroid deficient rats. We had reported this previously, and felt that parathyroid hormone was essential for the production of these kidney lesions (4). However, animals lacking parathyroid hormone and calcitonin also had severe lesions, although not as great as did intact or parathyroid-implanted rats. Thus, it would appear that parathyroid hormone is not essential for the development of kidney lesions with this diet. Only the intact rats fed the high phosphate diet developed severe *osteitis fibrosa cystica*. The other animals were free of any bone damage except the animals with parathyroid implants which occasionally showed evidence of increased resorption cavities.

DISCUSSION

It is apparent that the diet with low Ca/P ratio (high phosphate content) has significantly different effects on calcium and phosphate metabolism and on kidney and skeletal pathology depending on the hormonal status of the animal. Intact animals fed the high phosphate diet had increased parathyroid activity (3) and increased thyroidal content (5) of calcitonin which implies decreased calcitonin secretion. The combination of these effects resulted in nephrocalcinosis and *osteitis fibrosa cystica*. The animals with parathyroid transplants (PTI) also had increased parathyroid hormone secretion and also, presumably decreased calcitonin secretion, but only developed renal lesions. However, in this instance, parathyroid hormone secretion may have been less than that by intact animals and although insufficient to cause skeletal lesions, it probably was enough to produce the renal lesions. The serum responses of calcium and phosphate suggest decreased parathyroid activity but definite proof will not be available until we determine the level of circulating immunoreactive parathyroid hormone by radioimmune assay. A certain degree of hypersecretion of parathyroid hormone appears to be essential for the development of *osteitis fibrosa cystica*. The parathyroidectomized rats (PD) fed the high phosphate diet developed neither skeletal nor renal lesions. Very possibly, endogenous secretion of calcitonin in the absence of parathyroid activity prevented the renal lesions. Others have shown that under certain conditions exogenous calcitonin prevents renal lesions induced by exogenous parathyroid hormone (10, 9). The calcitonin deficient animals (CD) also may secrete less parathyroid hormone as a consequence of transplantation and this could account for the lack of skeletal lesions and diminution of renal lesions. The animals lacking both hormones (PCD) also only had renal lesions, and inexplicably had more severe lesions than the rats lacking only calcitonin (CD).

One common denominator that is affected by changes in parathyroid activity, calcitonin levels, dietary levels of calcium or phosphate and serum levels of calcium and phosphate is the metabolism of vitamin D_3. It has been shown that parathyroid hormone increases the conversion of $25\text{-}OH(D_2)$ to $1,25\text{-}di(OH)D_2$, the metabolite most active in calcium mobilization from the skeleton while calcitonin inhibits its conversion (11). Low levels of serum calcium increase the production of the 1,25 derivative while high levels depress it (2). Low dietary phosphate also increases the production of the 1,25 derivative and in the thyroparathyroidectomized rat a significant correlation exists between the serum level of phosphate and the production of the 1,25 and 24,25 derivatives

(14). At low serum levels of phosphate the 1,25 derivative predominates and at high levels the 24,25 derivative. In addition to these known factors, there are undoubtedly others which may affect the relative amounts of these metabolites. We would like to suggest that the differences in renal and skeletal lesions in the animals fed the high phosphate diet may result from alterations in the production or levels of the various metabolites of vitamin D as a consequence of variations in hormonal secretion and/or the levels of calcium or phosphate in the serum.

SUMMARY

High phosphate diet which induces secondary hyperparathyroidism in adult intact rats cause severe nephrocalcinosis and *osteitis fibrosa cystica*. Removal of only the parathyroid gland, prevents the renal and skeletal damage. Transplantation of the parathyroid gland in the presence or absence of calcitonin prevents the bone lesions but not the kidney lesions. The data suggests that transplantation of the parathyroid gland impairs the ability of this gland to function normally under severe stimulation. Calcitonin has a protective action against nephrocalcinosis caused by a high phosphate diet only in the absence of endogenous parathyroid hormone. It is suggested that differences in skeletal and renal lesions may be the result of differences in the production or levels of the metabolites of vitamin D as consequence of variations in hormonal secretions and/or the circulating levels of calcium and phosphate.

REFERENCES

1. Albright, F. & Reifenstein, E.C. Jr.: *The Parathyroid Glands in Health and Disease*, Williams and Wilkins, Baltimore 1948
2. Boyle, I.T., Gray, R.W. & DeLuca, H.F.: Regulation by calcium of *in vivo* synthesis of 1,25-dihydroxycholecalciferol and 21,25-dihydroxycholecalciferol. *Proc. nat. Acad. Sci. (Wash.)* 68, 2131–2134 (1971)
3. Clark, I. & Cordero, F.: Effects of endogenous parathyroid hormone on calcium magnesium and phosphate metabolism in rats II. Alterations in dietary phosphate, *Endocrinology* 96, 360–369 (1974)
4. Clark, I. & Rivera-Cordero: The role of the parathyroid glands in phosphate-induced nephrocalcinosis. In: *Calcium, Parathyroid Hormone and the Calcitonins*, Talmage, R.W. & Munson, P. (eds.), Excerpta Medica, Amsterdam, pp 253–262 (1972)
5. Clark, I., Cooper, C. & Hirsch, P.: Variations in Dietary Ca/PO$_4$ Ratios and Thyrocalcitonin Content of the Thyroid Gland. *Fed. Proc.* 33, 242 (1974)
6. Jaffé, H.L., Bodansky, A. & Chandler, J.B.: Ammonium chloride decalcification, as modified by calcium intake: The relation between generalized osteoproosis and ostitis fibrosa, *J. exp. Med.* 56, 823–833 (1932)
7. Kinter, J.H. & Holt, R.L.: Equine osteomalacia, *Philipp. J. Sci.* 49, 1–89 (1932)
8. Krook, L., Barret, R.B., Usui, K. & Wolke, R.E.: Nutritional secondary hyperparathyroidism in the cat, *Cornell Vet.* 53, 224–240 (1963)

9. Peng, T.: Protective effect of thyrocalcitonin against nephrocalcinosis induced by parathyroid hormone in rats, *Molec. & Cell. Endocr.* **2**, 1–15 (1974)
10. Rasmussen, H. & Tenenhouse, A.: Thyrocalcitonin, osteoporosis and osteolysis, *Am. J. Med.* **43**, 711–726 (1967)
11. Rasmussen, H., Wong, M., Bickle, D. & Goodman, D.B.P.: *J. clin. Invest.* **51**, 2502 (1972)
12. Sherwood, L.M., Case, A.D., Mayer, C.P., Aurbach, G.D. & Potts, J.T. Jr.: Evaluation by radioimmunoassay of factors controlling the secretion of parathyroid hormone, *Nature* **209**, 52 (1966)
13. Stoerk, H.C. & Carnes, W.H.: The relation of dietary calcium: phosphate ratio to serum calcium and parathyroid volume, *J. Nutr.* **29**, 43 (1945)
14. Tanaka, Y. & DeLuca, H.F.: The role of inorganic phosphate in the control of vitamin D metabolism, *Arch. Biochem.* **154**, 566–574 (1973)

Bone Cells and Structure of Cancellous Bone in Primary Hyperparathyroidism — A Histomorphometric and Electron Microscopic Study

G. DELLING, R. ZIEGLER & A. SCHULZ

Primary hyperparathyroidism is more often found without any clinical important bone disease due to radioimmunological determination of serum parathyroid hormone and automatic screening of serum calcium (14). Morphological bone changes of these so-called mild forms of primary hyperparathyroidism were published by a few research groups only (1, 8, 10, 12). Periods of vitamin D defiency were discussed as a pathogenic factor for the development of parathyroid adenomas (16, 8). Increased parathyroid hormone secretion influences bone cells and vitamin D metabolism. In primary hyperparathyroidism, changes of bone formation and mineralization can be expected in addition to increased bone resorption.

Our own experience with 32 patients with primary hyperparathyroidism is reported below. From the morphological point of view the following questions are of special interest: 1. Changes of bone structure, bone cells and mineralization, 2. Existence of skeletal development stages, 3. Bone changes following removal of the parathyroid adenoma.

PATIENTS AND METHODS

Iliac crest biopsies of 32 patients with surgical proven hyperparathyroidism were undecalcified embedded in methylmetacrylate (2), cut and stained (Goldner, Kossa, Giemsa, 10 μ). The parameters of bone structure, bone formation and bone resorption were determined by an integration eyepiece using the point counting method (9) as well as partly by an image analysing computer system (Zeiss Mikrovideomat). 60 cases without bone disease at the same age served as controls (3, 4). All data were statistically evaluated by a computer program. In addition, in 10

Institute of Pathology, University of Hamburg and Centre of Internal Medicine and Pediatrics, Department of Endocrinology and Metabolism, University of Ulm.

cases the serum parathyroid hormone concentrations were measured. 6 biopsies were prepared for electron microscopic investigation.

RESULTS

1. Bone structure: The structure of trabecular bone remained intact (6). In 3 cases we found an osteitis fibrosa only, but these cases were not considered in the histomorphometry because they had fibrous bone (woven bone) and not regular lamellar bone. The volume density (bone mass) and the specific bone surface were within the normal range (Fig. 1). The trabeculae were sometimes slightly broadened, but not thinner than in normal cases. The surface density (total bone surface) was increased due to resorption lacunae and channels in a few cases only (Fig. 1).

2. Bone formation: The osteoid volume was increased in 18 out of 28 cases. Nearly all patients had increased osteoblasts/osteoid interfaces and an increase of the total extent of osteoid seams (surface density of osteoid seams). The explanation for the increase of the osteoid volume and the surface of the osteoid seams could either be a disturbance of the bone mineralization or an increased synthesis of bone matrix with a normal mineralization process. There was a significant correlation between the osteoid volume and the surface density of the osteoblast/osteoid interface (Figs. 1 and 2). Olah (11) described the same finding in his cases with primary hyperparathyroidism and concluded that no mineralization disturbance exists. In our material, however, a few cases showed an increase of the osteoid volume without the same increase of the osteoblasts (Fig. 2). This must be the result of a defective mineralization. In the ultrastructure, the osteoblasts are active cells with a large number of cell organels and they produce collagen fibres. Within the osteoid lie many membrane-bounded vesicles, which represent the first stages of mineral deposit into the bone tissue (13). In contrast, in vitamin D deficient rats and in osteomalacia, an obvious reduction of these membrane-bound vesicles occurs. Therefore, in most of the cases we have no morphological substrate for a severe disturbance of mineralization.

3. Bone resorption: Size and form of the osteoclasts can differ considerably in primary hyperparathyroidism. Very large, and in other cases flat osteoclasts can be observed. The parameters of bone resorption – surface density of the osteoclast/bone interface, the total extent of Howships lacunae and the index of osteoclasts showed a considerable variation in our studies (Fig. 1). The index of osteoclasts correlated with the serum parathyroid hormone concentration as reported by other authors. There was, however, no correlation between the interface osteoclasts/bone and osteoblasts/osteoid. The age of the patients might determine whether the

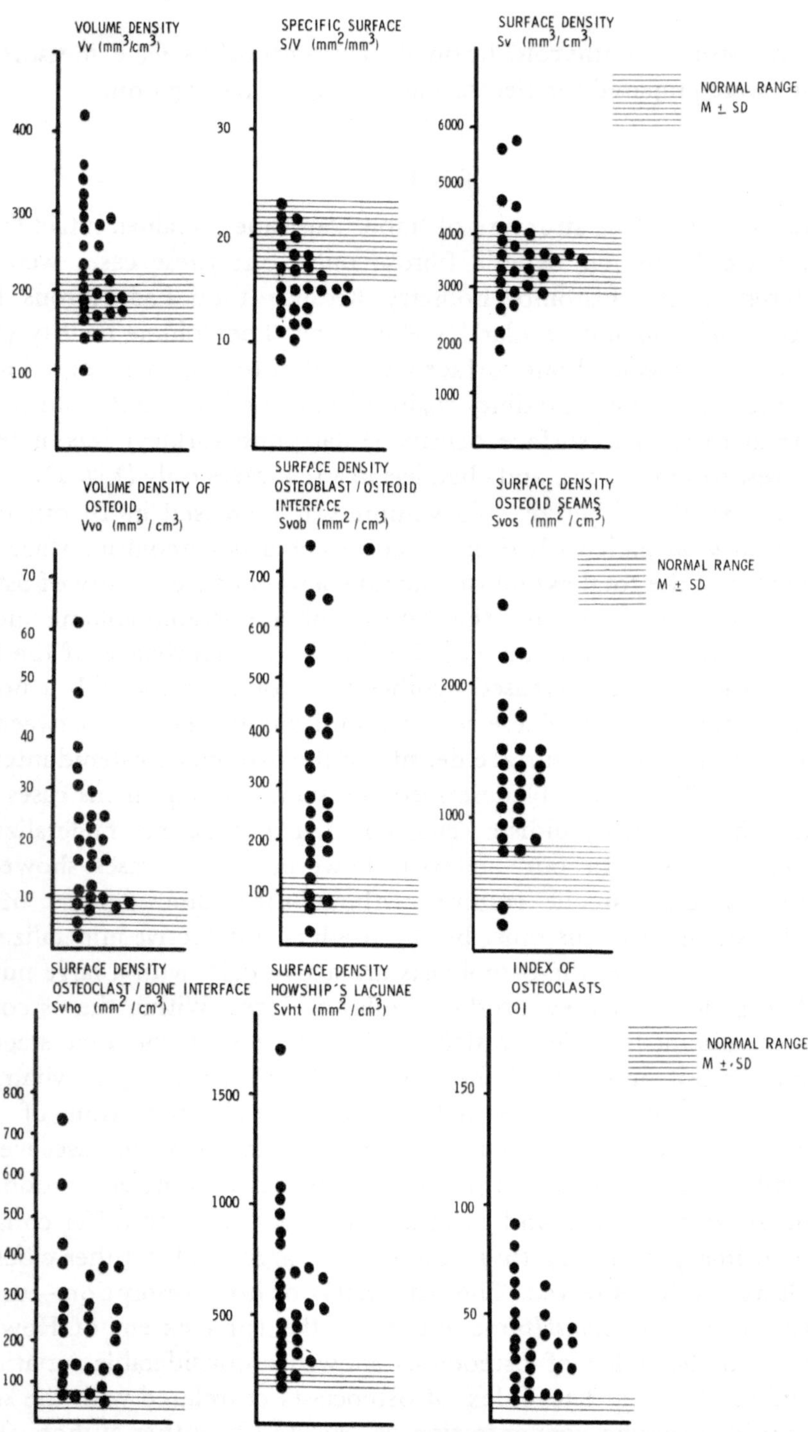

Figure 1. Bone structure, bone formation and bone resorption in 28 cases with primary hyperparathyroidism.

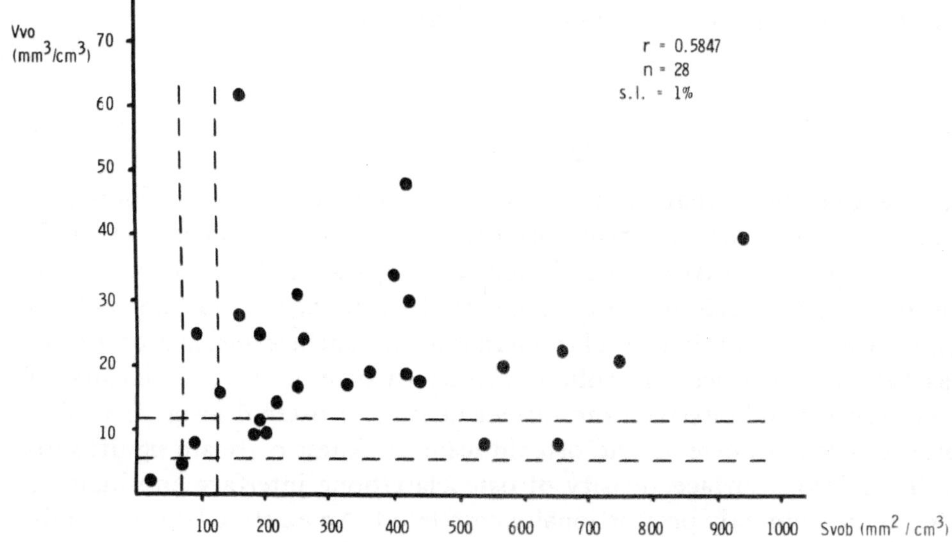

Figure 2. Correlation between volume density of osteoid (Vvo) and the surface density of osteoblast/osteoid interface (Svob) in 28 cases with primary hyperparahtyroidism.

osteoblasts or the osteoclasts are more sensitive to parathyroid hormone. The resorption "activity" of the osteoclasts is still under discussion in the case of many metabolic bone diseases. Electron microscopy might answer some of the related questions. In all osteoclasts we have seen under the electron microscope, the cytoplasm contains numerous lysosomes and an expanded endoplasmatic reticulum. The brush border is extended when compared to normal and the mineralized bone matrix on the adjacent bone shows a clear decalcification. "Naked" collagen fibres can be demonstrated.

The small degree of change in remodelling surfaces under increased parathyroid hormone secretion (15) needs an additional parameter for diagnostic purposes. This, as described by Jowsey (8), is the endosteal fibrosis, which can be seen in spite of a small increase of osteoclastic resorption. But in 20% of our cases an endosteal fibrosis is lacking. These findings lead us to classify bone changes in primary hyperparathyroidism in 4 stages. In stage I a normal bone tissue is present in spite of undecalcified preparation of the bone specimen and in spite of quantitative analysis. But we believe that these cases are rare and not typical of border-line cases. In stage II the bone remodelling surfaces are increased. In stage III an additional endosteal fibrosis of different degree occurs. Stage IV (to-day less and less frequent (7)) is the typical so-called osteitis fibrosa (own material: stage I 3%, stage II 19%, stage III 69%, stage IV 9%). By the application of these stages a better comparison of morphological and clinical parameters might be possible.

DISCUSSION

Our results show that in primary hyperparathyroidism an increased remodelling of trabecular bone occurs. Bone structure remains intact. It might be possible that osteitis fibrosa develops after decompensation of the intracellular calcium storage capacity. In general we were not able to demonstrate a disturbance of mineralization. Surface density of osteoblast/osteoid interface and volume density of osteoid, surface density of osteoblast/osteoid interface and surface density of osteoid seams as well as volume density of osteoid and osteoid seam thickness correlate significantly. In addition, surface density of osteoclast/bone interface and inactive osteoid are indirectly proportionally correlated. An existing latent vitamin D defiency could be compensated by parathyroid hormone, which increases the formation of 1,25 dihydroxycholecalciferol (5). After 3 weeks and 3 months following removal of a parathyroid adenoma, we observed an excessive osteomalacia, probably the result of a delayed mineralization due to a diminished formation of active vitamin D metabolites in the postoperativ hypoparathyroid stage.

ACKNOWLEDGEMENT

Supported by Deutsche Forschungsgemeinschaft, SFB 34 "Endokrinologie", Hamburg and SFB 87, Ulm.

REFERENCES

1. Byers, P.D. & Smith, R.: Quantitative histology of bone in hyperparathyroidism: its relation to clinical features, X-rays and biochemistry. *Quart. J. Med.* **40**, 471—486 (1971)
2. Delling, G.: Über eine vereinfachte Methacrylateinbettung für unentkalkte Knochenschnitte. *Beitr. path. Anat.* **145**, 100—105 (1972)
3. Delling, G.: Age related bone changes. *Current Topics in Pathology* **58**, 117—147 (1973)
4. Delling, G.: Endokrine Osteopathien. *Veröffentlichungen aus der Pathologie* 98 (Vol.), Fischer Stuttgart, 1975
5. DeLuca, H.F.: Regulation of vitamin D metabolism. In: *Proc. 4th Int. Symp. Endocrinology 1973*, MacIntyre, I. (ed.), William Heinemann, London, pp. 38—51, 1974

6. Genant, H.K., Horst, J.V., Lanzl, Z.H., Mall, J.L. & Doi, K.: Skeletal demineralization im primary hyperparathyroidism. In: *Int. Conf. Bone Mineral Meassurement 1973*, Mazess, R.Z. (ed.), U.S. Department of Health, Education, and Welfare (No. 75–683)

7. Haas, H.G.: *Knochenstoffwechsel- und Parathyreoidea-Erkrankungen*. Thieme, Stuttgart, 1966

8. Jowsey, J.: Bone histology and hyperparathyroidism. *Clin. Endocr. Metab.* 3, 267–303 (1974)

9. Merz, W.A.: Die Streckenmessung an gerichteten Strukturen im Mikroskop und ihre Anwendung zur Bestimmung von Oberflächen-Volumen-Relationen im Knochengewebe. *Mikroskopie* 22, 132–142 (1967)

10. Meunier, P., Vignon, G., Vauzelle, J.-L. & Zech, P.: Étude histologique quantitative de la résorption osteoclastique dans les hyperparathyroidies primitives et secondaires. *Path. et Biol.* 17, 927–983 (1969)

11. Olah, A.J.: Quantitative relations between osteoblasts and osteoid in primary hyperparathyroidism, intestinal malabsorption and renal osteodystrophy. *Virchows Arch. path. Anat. Abt. A* 358, 301–308 (1973)

12. Schenk, R., Olah, A.J. & Merz, W.A.: Bone cell counts. In: *Clinical aspects of metabolic bone disease*. Frame, B., Parfitt, A.M., Duncan, H. (eds.), Excerpta Medica (Amsterdam), pp. 103–113, 1973

13. Thyberg, J.: Electron microscopic studies on the initial phases of calcification in guniea pig epiphyseal cartilage. *J. Ultrastruct. Res.* 46, 206–218 (1974)

14. Watson, L.: Primary hyperparathyroidism. *Clin. Endocr. & Metab.* 3, 215–235 (1974)

15. Wilde, C.D., Jaworski, Z.F., Villanueva, A.R. & Frost, H.M.: Quantitative histological measurements of bone turnover in primary hyperparathyroidism. *Calc. Tiss. Res.* 12, 137–142 (1973)

16. Woodhouse, N.J.Y., Tun Chot, S., Bordier, P., Sigurdsson, G. & Joplin, G.F.: Vitamin D administration in primary hyperparathyroidism. *Proc. 9th Europ. Symp. Calc. Tiss.*, Wien/Baden 1972, Czitober, H. and Eschberger, J. (eds.), Facta Publication, Wien, p. 253, 1973

Parathyroid Hormone and Phosphaturia

T. Uchikawa, A.B. Borle & R.J. Midgett

In 1911, Greenwald reported that urinary phosphate was significantly decreased in parathyroidectomized dogs (3). He was the first to suggest a link between the observed drop in phosphate excretion and the removal of the parathyroid glands. Sixty-four years later, the mechanism by which parathyroid hormone (PTH) enhances phosphaturia is still largely speculative. In the last ten years many investigators have reported that PTH, in addition to its phosphaturic effect, increases sodium and bicarbonate excretion, raises the pH of the proximal tubule fluid and of the urine, produces a mild diuresis and a systemic metabolic acidosis. These results suggest that PTH may inhibit H^+ ion secretion in the renal tubules. Indeed, if one assumes that monovalent phosphate is more permeable or more readily transported than divalent phosphate, one could explain the phosphaturia as follows: a rise in intraluminal pH would depress phosphate reabsorption simply by reducing the monovalent to divalent phosphate ratio in the renal tubules. Two essential elements of this hypothesis need to be firmly established before it can be accepted: 1) monovalent phosphate is more readily transported than divalent phosphate; 2) PTH inhibits H^+ ion secretion in renal tubules. We undertook these studies to test the validity of both postulates (2, 3).

METHODS

Phosphate fluxes were determined by kinetic analyses of P_{32} movements in isolated kidney cells ($LLC-MK_2$) at different phosphate concentrations, and at pH ranging from 6.8 to 8.0. Hydrogen ion production was measured in isolated kidney cells and in rat kidney slices by the pH stat method.

Department of Physiology, University of Pittsburgh, School of Medicine, Pittsburgh, Pa.

RESULTS

a) *Phosphate transport.* Isolated kidney cells have two distinct cellular phosphate compartments. On every evidence available, it appears that the first compartment represents the intracellular inorganic phosphate pool, while the second compartment represents the organic fraction of the cell. The extracellular pH markedly influences phosphate uptake by the cells; phosphate transport is increased at low pH and decreased at high pH. Fig. 1 shows that the influx of phosphate into the inorganic phosphate pool is a function of the pH. The total phosphate being constant (10 mM), influx is directly and linearly proportional to the concentration of monovalent phosphate. The relation between the phosphate influx and the absolute concentration of monovalent phosphate can be used to calculate the permeability ratio of monovalent to divalent phosphate. The results reveal that monovalent phosphate is 12 times more permeable or more easily transported than divalent phosphate. Having calculated the absolute permeability for both monovalent and divalent phosphate, it is possible to predict cellular phosphate transport with less than 5% error, at any phosphate concentration between 1 and 20 mM and at any pH between 6.8 to 8.0. In these conditions, phosphate transport is still linearly related to the monovalent phosphate concentration. We conclude that the main ion crossing the cell membrane is monovalent phosphate and that it is transported by simple diffusion. Our calculations suggest that phos-

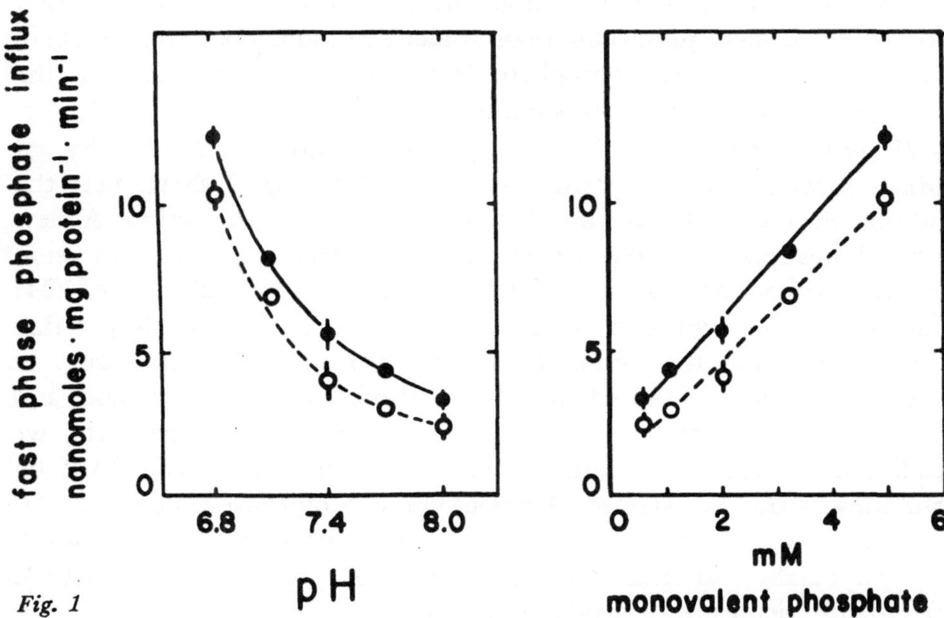

Fig. 1

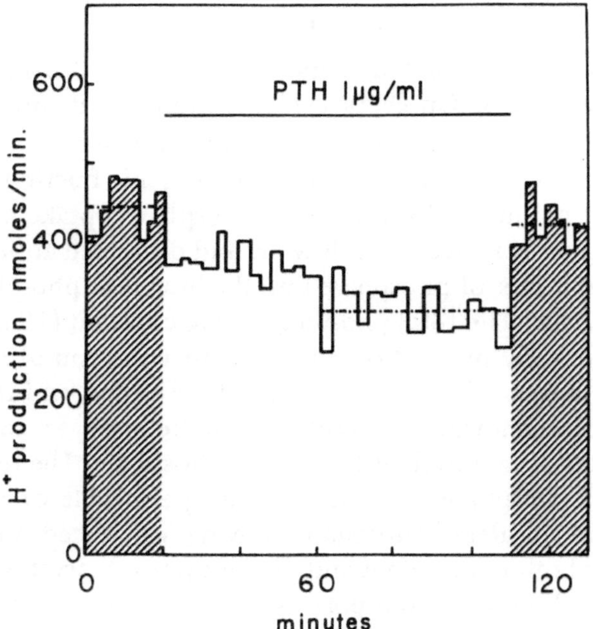

Fig. 2

phate is passively distributed across the cell membrane according to its Nernst equilibrium potential. Finally, the phosphate fluxes that we measured in isolated cells between the pH of 6.8 to 8.0 can account for the whole range of phosphate reabsorption measured in the mammalian kidney. We believe, therefore, that our first postulate is correct, that monovalent is more permeable than divalent phosphate and that a rise in pH can interfere with phosphate transport simply by altering the monovalent to divalent phosphate ratio.

b) *H^+ production.* H^+ production by isolated kidney cells and by rat kidney slices remains constant for at least 5 h. Fig. 2 shows that the addition of 1 µg PTH/ml immediately depresses H^+ production. A new reduced steady state is reached in about 1 h. The maximum depression obtained is about 35%. H^+ ion production is decreased by all types of PTH available: by the synthetic 34-peptide bovine PTH, by a highly purified PTH obtained from the Wilson Laboratories or from Dr. H. Rasmussen.

It is generally accepted that PTH stimulates adenyl cyclase and increases the intracellular concentration of cyclic AMP. Therefore, we studied the effects of aminophylline which increases cyclic AMP by inhibiting phosphodiesterase. Aminophylline progressively depresses H^+ production from 12 to 70% between the concentrations of 10^{-6} and 2×10^{-3} M. Finally, we studied the effect of dibutyryl cyclic AMP (DBC): 10^{-4} M DBC depresses H^+ production 35%.

DISCUSSION

We have calculated the magnitude of the pH change which should occur in the proximal tubule to account for the phosphaturia of hyperparathyroidism and for the decreased phosphate excretion observed in hypoparathyroidism: a rise in the intraluminal pH of the proximal tubule from 6.8 to 7.2 would totally account for a decreased tubular reabsorption of phosphate from 87 to 60%. A small depression of 0.1 pH unit, from 6.8 to 6.7 would fully explain a rise in tubular reabsorption of phosphate from 87 to 95%. Thus, a trivial modulation of the proximal tubule H^+ secretion, well within the physiological range can easily account for the fluctuations observed in renal phosphate reabsorption.

CONCLUSIONS

In conclusion, we have established that PTH inhibits H^+ ion production in isolated kidney cells and in rat kidney slices. We presented evidence that this effect is due to an increased cellular concentration of cyclic AMP. These data strongly support our hypothesis that the phosphaturic effect of PTH is due to a depression of H^+ secretion into the renal tubules which produces a rise in intraluminal pH. The rise in tubular fluid pH decreases the monovalent to divalent phosphate ratio. And since monovalent phosphate is at least 10 times more permeable than divalent phosphate, phosphate reabsorption is depressed and phosphate excretion increases.

ACKNOWLEDGEMENT

Supported by grant AM 07867 from the National Institutes of Health U.S.

REFERENCES

1. Greenwald, J.: The effect of parathyroidectomy upon metabolism. *Amer. J. Med.* **28**, 103–132 (1911)
2. Uchikawa, T. & Borle, A.B.: The effect of pH on phosphate transport in kidney cells. *Fed. Proc.* **33**, 241 (1974)
3. Uchikawa, T. & Borle, A.B.: The effect of parathyroid hormone (PTH) on hydrogen ion production. *Fed. Proc.* **34**, 312 (1975)

Parathyroid Activity in Hyperthyroidism

J. BOMMER, E. RITZ & E. GENGENBACH

Hyperthyroid patients are known to have higher serum P levels than euthyroid patients (1, 9). Hyperphosphatemia in hyperthyroid patients was thought to be the consequence of diminished parathyroid hormone (PTH) secretion. According to this concept, increased bone resorption, (known to be stimulated by thyroxine), would lead to a rise of ionised serum calcium levels. This would suppress PTH secretion and thus increase the reabsorption of P (phosphate). Hyperphosphatemia would ensue as a result of "secondary hypoparathyroidism".

Although increased tubular reabsorption of P in hyperthyroid patients was demonstrated by various authors (2, 6), "secondary hypoparathyroidism" has never unequivocally been documented. We, therefore, studied the long-term effect of thyroxine withdrawal (thyroidectomy) and of thyroxine excess (pharmacological doses of thyroxine) on serum P levels and renal handling of P in the rat, both in the presence and in the absence (parathyroidectomy) of endogenous parathyroid hormone.

MATERIAL AND METHODS

In 40 g male Wistar rats, hypothyroidism was induced by surgical thyroidectomy. Endogenous PTH secretion was preserved or abolished by concomitant auto-transplantation of the parathyroid glands or parathyroidectomy (PTX). Thyroidectomy (TX) caused a significant fall of PBI from 6.5 ± 1.2 μg/100 ml to 2.9 ± 0.8 μg/100 ml 3 weeks after the operation. The success of parathyroid auto-transplantation was controlled by measuring serum calcium (Ca) levels. In PTX animals, serum Ca dropped by more than 1.25 mEq/l after overnight fasting. Hyperthyroidism was induced in sham-operated and PTX animals by daily injections of 0.05 mg thyroxine i.p., starting on the 14th day after

Medizinische Universitätsklinik Heidelberg.

operation. Euthyroid control animals (sham-operated or PTX) received solvent injections.

Once weekly, the animals were placed in metabolic cages after overnight fasting; they were given 5% (w/v) glucose solution as drinking fluid. The measurements in metabolic cages were performed from the 4th to the 7th week. We determined serum levels and urinary excretion (6 h collecting period) of P (of Henry (7)), Ca, Mg (atomic absorption spectrophotometry, Perkin Elmer Type 290 B), Na (emission flame photometry, Eppendorf Co., Type 1101 M) and creatinine (of Folin (4)).

For the measurement of extracellular volume a 0.5 ml blood sample was taken 1, 30, 60, 90 and 120 min after an injection of 7 μCi ^{169}YbDTPA and replaced by equal amounts of saline. Serum radioactivity was counted in a Tracerlab Gamma Guard 150. Extracellular volume was calculated after Kuni (8).

RESULTS AND DISCUSSION

a) *Serum P levels:* In the presence of endogenous PTH secretion, serum P levels were significantly lower in TX animals than in euthyroid or hyperthyroid animals (Table I). A similar relationship was also found in the absence of endogenous PTH (PTX animals). In these animals TX resulted in a decrease and thyroxine excess in an increase of serum P levels.

These results show that thyroxine raises serum P in the presence as well as in the absence of endogenous PTH. In the absence of endogenous PTH, the increase of serum P was even more pronounced than in animals with intact parathyroid function. These findings clearly exclude "secondary hypoparathyroidism" as the cause of the increase of serum P under the influence of T_4.

The rise of serum P levels was even less in animals with intact parathyroid function. A decrease of fractional tubular reabsorption of P (TRP) under conditions of thyroxine excess was found only in parathyroid-intact, not, however, in PTX animals.

These findings point to an activation of PTH secretion in hyperthyroid animals with intact parathyroid function. Parathyroid overactivity would counteract the increase of serum P levels resulting from T_4 excess.

Changes of calcitonin secretion are not likely to have contributed to the hyperphosphatemia of hyperthyroid animals: PTX animals in whom thyroid hormone caused the greatest elevation of serum P levels are not likely to have had effective serum calcitonin concentrations in view of their persistent hypocalcemia.

Table I. Serum and Urinary P in Rats with Intact Parathyroid Function.

ANIMALS	SERUM P $mM\ l^{-1}$	$U_P \times V$ $\mu M\ min^{-1}\ kg^{-1}$	C_P $10^{-1}ml\ min^{-1}kg^{-1}$	TRP %	T_P/C_{Cr} $\mu M\ ml^{-1}$	C_{Cr} $ml\ min^{-1}\ kg^{-1}$	$U_{Cr} \times V$ $10^{-5}g\ min^{-1}\ kg^{-1}$
I HYPOTHYROID (n = 11)	2, 72 ± 0, 11 (2,3)	1, 66 ± 0, 42 (3)	6, 06 ± 1, 20 (3)	80, 1 ± 6, 9	2, 19 ± 0, 22 (2)	3, 07 ± 0, 51 (2,3)	2, 78 ± 0, 44 (2,3)
II EUTHYROID (n = 9)	2, 97 ± 0, 17	1, 98 ± 0, 42 (3)	6, 60 ± 1, 30 (3)	83, 5 ± 4, 6 (3)	2, 51 ± 0, 13	4, 06 ± 0, 38	3, 50 ± 0, 74
III HYPERTHYROID (n = 9)	3, 03 ± 0, 17	2, 93 ± 0, 34	9, 73 ± 1, 13	76, 7 ± 5, 1	2, 29 ± 0, 28	4, 36 ± 0, 46	3, 83 ± 0, 38

I   TX animals, solvent injection
II   sham-op, solvent injection
III sham-op, thyroxin injection
Wilcoxon test for random samples, significant ($p < 0.01$) difference with group II (2) resp. group III (3).

b) *Renal handling of P: Urinary excretion of P* and *clearance of P* increased in parallel with the effect of thyroxine both in the presence and in the absence of endogenous PTH. Since under steady state conditions at constant serum P levels urinary excretion of P must equal the balance between intestinal net absorption and P retention from growth, differences of urinary P excretion must result from differences of food uptake and/or intestinal absorption of P and/or growth rates.

The reabsorption of P per unit GFR (C_{cr}) increased in parallel with the level of thyroxine. Tubular reabsorption depends on filtered load of P and tubular reabsorption capacity. The filtered load of P in turn depends on serum P concentration and GFR. In the experimental animals, no correlation was found between TRP and GFR. This finding points to glomerulo-tubular balance with tubular reabsorption changing in proportion to changes of the glomerular filtration rate.

The fractional reabsorption of P was slightly but significantly decreased in hyperthyroid animals with intact parathyroid function. In contrast, in PTX animals TRP was unchanged in hyper- and hypothyroid animals.

However, it is not strictly permissible to compare directly the fractional rates of tubular P reabsorption in these three groups with widely different serum P levels. Under conditions of glomerulo-tubular

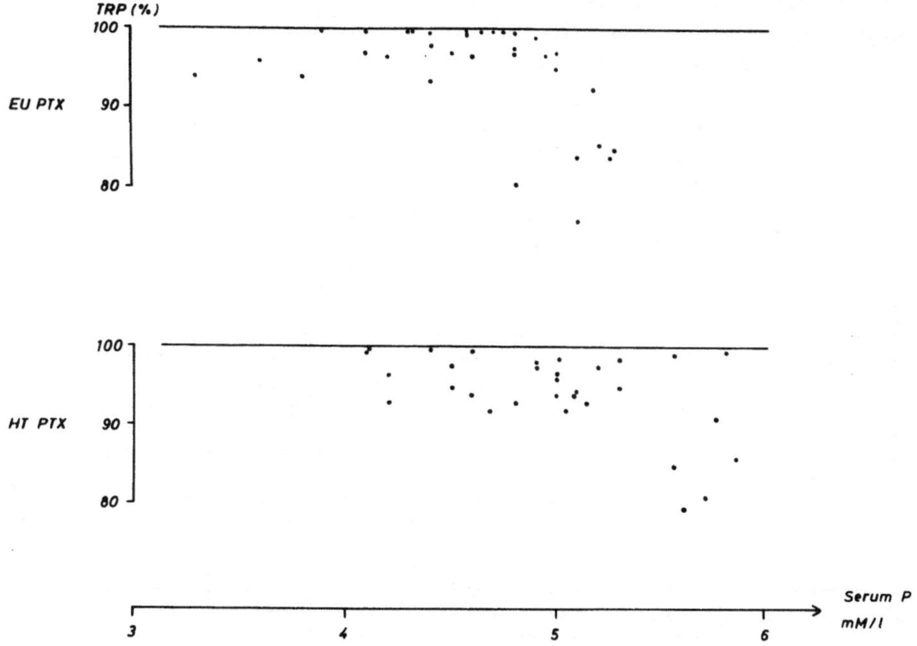

*Figure 1.*Correlation between the fractional tubular reabsorption of phosphate and serum phosphate levels of eu- and hyperthyroid PTX rats.

balance, fractional tubular reabsorption of P depends on serum P levels. Therefore, the relation between serum P levels and TRP was analysed. At high serum P levels, fractional tubular reabsorption was higher in hyperthyroid PTX animals than in euthyroid PTX animals (Fig. 1).

c) *Changes of extracellular volume:* Since thyroxine does not change tubular reabsorption of P by suppressing PTH secretion, there remain two mechanisms, by which thyroxine might alter tubular reabsorption of P: A direct metabolic effect on tubular cells or indirect effects mediated by changes of extracellular volume, changes of vitamin D metabolism etc. Proximal tubular reabsorption of P is known to respond to changes of extracellular fluid volume (5). In hyperthyroid PTX animals, the extracellular volume was significantly lower than in euthyroid PTX animals (hyperthyroid animals 26.1 ± 2.2 ml/100 g body weight; euthyroid: 29.2 ± 1.9 ml/100 g body weight; hypothyroid 32.4 ± 3.1 ml/100 g body weight) ($\bar{X}$ ± S.D.). In addition, urinary Na-excretion was lower in hyperthyroid animals (Table II). These results suggest that the effect of thyroxine on tubular reabsorption of P may result from changes of the extracellular space.

Table II. Serum Levels and Urinary excretion of Sodium, Calcium and Magnesium in Hypo-, Eu- and Hyper-thyroid Rats in the Presence or Absence of Endogenous PTH.

	Serum sodium meq/l	$U_{Na} \times V$ 10^{-4} mEq min^{-1} kg^{-1}	Serum calcium meq/l	$U_{Ca} \times V$ 10^{-4} mEq min^{-1} kg^{-1}	Serum magnesium meq/l	$U_{Mg} \times V$ 10^{-3} mEq min^{-1} kg^{-1}
① Hypothyroid n = 11	131,65 ± 6,45	7,63 ± 1,83	4,88 ±0,20	4,65 ± 1,59	1,70 ± 0,08	1,16 ±0,34
				p 1/3 < 0,01		p 1/3 < 0,01
② Euthyroid n = 9	131,27 ± 5,10	7,50 ± 2,14	4,83 ±0,14	4,23 ± 1,53	1,71 ±0,14	1,62 ± 0,62
				p 2/3 < 0,01		
③ Hyperthyroid n = 9	134,40 ± 3,71	6,65 ± 1,34	4,79 ± 0,23	1,98 ± 0,63	1,76 ± 0,06	1,62 ± 0,29
④ Hypothyroid n = 12	132,39 ± 2,43	9,40 ± 1,55	3,37[**] ± 0,28	4,43 ± 0,93	1,62 ± 0,14	1,22 ± 0,23
			p4/6 < 0,01	p 4/6 < 0,01	p4/6 < 0,01	p4/6 < 0,01
⑤ Euthyroid n = 11	132,28 ± 3,28	9,60 ± 2,60	3,30[**] ± 0,25	4,14 ±0,66	1,58 ± 0,13	1,58 ± 0,21
			p 5/6 < 0,01	p 5/6 < 0,01	p 5/6 < 0,01	
⑥ Hyperthyroid n = 10	132,48 ± 3,00	8,31 ± 2,22	3,06[**] ± 0,27	2,70 ± 0,69	1,74 ± 0,09	1,94 ± 0,69

Rows ①–③: with endogenous PTH. Rows ④–⑥: without endogenous PTH (PTX).

Significance of the effect PTX (Wilcoxon test for random samples.
**) p < 0.01; *) p < 0.05

CONCLUSION

Even in the absence of PTH, serum P levels rose in response to thyroxine. Thus, PTH secretion is not suppressed in hyperthyroidism as assumed hitherto. In contrast, the findings suggest an activation of endogenous PTH secretion in thyrotoxic animals. These results agree with the finding of increased serum PTH levels in thyrotoxic patients (radioimmuno-assay*), reported elsewhere (3). After parathyroidectomy TRP of hyper-thyroid animals was not different from that of euthyroid animals, although the serum P levels were elevated. At high serum P levels, TRP

*) Kindly performed by Dr. W.H.L. Hackeng, Rotterdam, Bergweg Ziekenhuis.

was higher in hyperthyroid PTX animals, pointing to a raised renal "threshold". In view of the contracted extracellular space in hyperthyroid animals, this effect of thyroxine might be mediated by changes of extracellular volume, although alternative mechanisms are not excluded.

ACKNOWLEDGEMENT

This work was supported by the Deutsche Forschungsgemeinschaft.

REFERENCES

1. Aub, J.C., Bauer, W. & Heath, C.: Studies of calcium and phosphorous metabolism: Effects of thyroid hormone and thyroid disease. *J. clin. Invest.* 7, 97–137 (1929)
2. Bijvoet, O.L.M.: Relation of plasma phosphate concentration to renal tubular reabsroption of phosphate. *Clin. Sci.* 37, 23–36 (1969)
3. Bommer, J., Ritz, E. & Schmidt-Gayk, H.: Hyperphosphatemia in Thyrotoxicosis: Hypo- or Hyperparathyroidism? *Proc. Int. Symp. Phosphate (Paris)*, 1975 (In Press)
4. Folin, O.: Bemerkungen zur Bestimmung von Kreatinin (und Kreatin) im Blut. *Hoppe-Seylers Z. physiol. Chem.* 228, 268–272 (1934)
5. Frick, A.: Reabsorption of inorganic phosphate in the rat kidney. *Pflügers Arch. ges. Physiol.* 304, 351–364 (1968)
6. Harden, R.McG., Harrison, M.T., Alexander, W.D. & Nordin, B.E.C.: Phosphate excretion and parathyroid function in thyrotoxicosis. *J. Endocr.* 28, 281–288 (1964)
7. Henry, R.: In: *Clinical Chemistry: Principles and Technics.* Harper & Row, New York, pp. 409–416 (1964)
8. Kuni, H. & Graul, E.H.: 51 CrEDTA Grundsätzliches zur Einführung einer neuen Substanz zur nuklearmedizinischen Bestimmung der glomerulären Clearance. *Atompraxis* 12, (1966) Suppl.
9. Malamos, R., Sfikakis, F. & Pandos, P.: The renal handling of phosphate in thyroid disease. *J. Endocr.* 45, 269–273 (1969)

CHAPTER VI
Calcitonin

Calcitonin and Blood Calcium Homeostasis during Intestinal Calcium Absorption in Man

G. COEN & B. PALAGI

The inhibitory effect of calcitonin on bone resorption is a well documented phenomenon (5, 7). However, its physiologic importance in man is still uncertain. From earlier studies (4, 6, 9, 11) counteraction against hypercalcemia through inhibition of bone calcium resorption appeared to be the most probable role. Since physiologic hypercalcemia may be provoked by intestinal absorption of dietary calcium, the role of calcitonin has been evaluated by administering oral calcium loads to athyroid animals (3, 8). Studies in man have also been carried out (2). In the present investigation on athyroid and control subjects, a larger oral calcium load than in the previous study was administered in order to further evaluate the role of endogenous calcitonin in the control of postprandial hypercalcemia. In addition the possibility that other factors beside calcitonin are playing a role in antagonizing hypercalcemia was taken into consideration and ruled out after comparison of the curves of blood stable calcium increment and of the cumulative fractional intestinal calcium absorption.

MATERIALS AND METHODS

The study was carried out in 7 normal controls aged 33—61 years and 7 patients aged 18—40 years who had in the past undergone total thyroidectomy for thyroid carcinoma and subsequent radioiodine treatment. None of the patients had evidence of metastases. The patients received full replacement doses of dessiccated thyroid. In addition they were euparathyroid and had never presented low blood calcium levels or tetanic symptoms.

Clinica Medica II, University of Rome.

Following an overnight fast, the subjects received at 9 A.M. 15 mg/kg b.w. calcium as the chloride salt dissolved in 100 ml distilled water containing 10–15 μci ^{45}Ca. Measurements of stable calcium and radioactivity were made on blood samples drawn in basal conditions and at intervals thereafter. The values of serum radioactivity were referred to a standard man of 70 kg b.w. Correction for body weight provides a statistically more significant correlation between intestinal per cent absorption of calcium and plasma radioactivity (10). In two normal subjects and three thyroidectomized patients blood samples after calcium ingestion were drawn at close intervals. In the same subjects 6–10 days after the oral load, the curve of dilution of the isotope following i.v. ^{45}Ca administration was obtained. Deconvolution of the function of the rate of entry of labelled calcium into the plasma was calculated according to the method of Birge *et al.* (1).

Serum radioactivity was determined by liquid scintillation counting. Blood Calcium was determined by the AutoAnalyzer according to the method described by Wills and Gray (12).

RESULTS

Mean values of serum calcium and phosphorus were 9.39 ± 0.13 and 3.67 ± 0.35 mg/dl respectively in the thyroidectomized patients and 9.79 ± 0.17 and 3.71 ± 0.19 mg/dl in normal subjects (mean ± SE). The difference between the mean values was not significant.

Fig. 1 shows the mean values (± SE) of the calcium increments following oral administration of calcium and of serum radioactivity as per cent of administered radioactive dose per litre of serum. The difference in mean values of the blood calcium increments was significant at 180 and 240 min (p < 0.025 and < 0.01, respectively), with higher average values in thyroidectomized patients than in normal controls (Fig. 1). The difference in the mean values of serum radioactivity was not significant.

The curves of cumulative intestinal absorption of calcium as per cent of the absorbed fraction was calculated in the five cases from the deconvoluted function of the rate of entry of labelled calcium. No substantial difference was found between controls and athyroid subjects. About 96% of the calcium load was absorbed within 3 h of the administration. It has already been suggested (10) that after an initial period the curve of cumulative intestinal calcium absorption approaches an inverted exponential, where $A_t = 1 - e^{-kt}$ (A_t is the absorption of calcium at time t). The assumption of a constant fractional rate k for the mean cumulative intestinal absorption curve calculated in the present

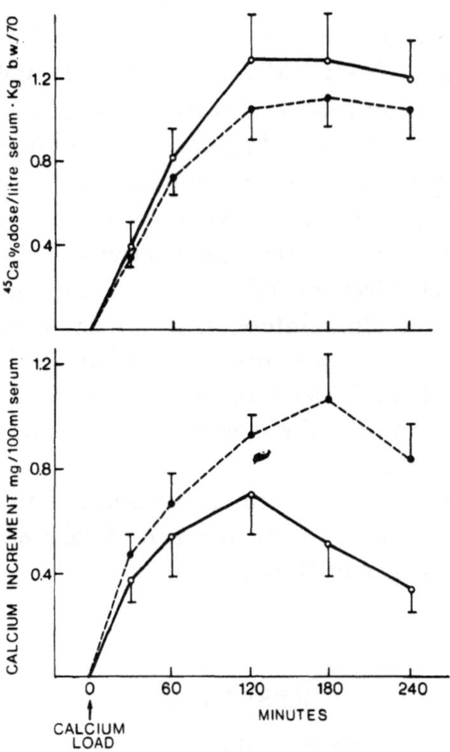

Figure 1. Time course of blood calcium increment and of plasma radiocalcium (mean ± SE) after oral calcium administration in controls (solid line) and thyroidectomized subjects (dotted line).

cases was confirmed by plotting $A_\infty - A_t$ versus t on semilogarithmic paper (Fig. 2) where a straight line was obtained. In the same figure an analogous plotting of the curve of blood calcium increments $(I_\infty - I_t)$ of thyroidectomized patients, unlike the controls, appears to coincide with the cumulative absorption line.

DISCUSSION

In a previous study (2) blood calcium increments following a 500 mg calcium load were higher in thyroidectomized subjects than in the controls. This finding was not due to differences in per cent calcium absorption or in clearance rate of absorbed calcium from the exchangeable compartment. Therefore, lack of calcitonin was considered to be responsible for impaired blood calcium homeostasis. Following the ingestion of the 15 mg/kg b.w. calcium load, as after the 500 mg load, the difference of the mean calcium increments of normal and pathologic

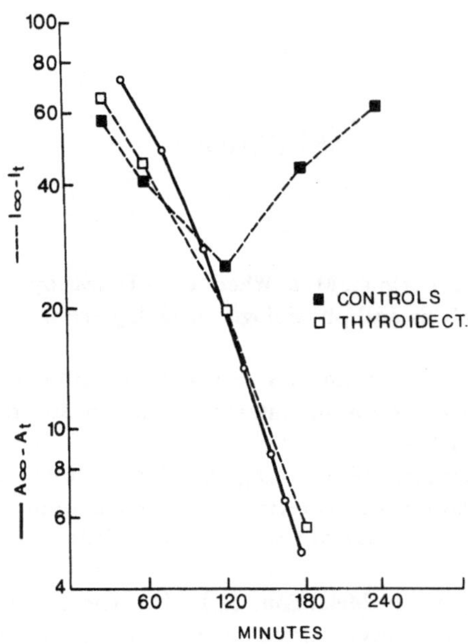

Figure 2. Cumulative per cent intestinal absorption of calcium (mean of 5 cases) expressed as $A_\infty - A_t$ versus t (solid line). Blood calcium increments of the thyroidectomized ($\square$) and control subjects ($\blacksquare$) are plotted as $I_\infty - I_t$ versus time.

subjects was higher at the third h. At this time the difference was 0.56 mg/dl (0.34 mg/dl after the 500 mg load), a value very close to the expected hypocalcemia in case of inhibition of the bone resorption process.

In thyroidectomized subjects, unlike the controls, the curve of blood calcium increments following the oral dose conforms to the cumulative calcium absorption curve (Fig. 2). A linear relationship appears to exist between calcium increments and the absorption process. Therefore, in the absence of calcitonin no humoral homeostatic factor appears to counteract absorption-induced hypercalcemia. Likewise, suppression of parathyroid glands does not seem to have a significant effect.

In conclusion: Calcitonin is able to counteract absorption hypercalcemia. However, transitory calcium increments are not abolished by the hormone. Considering that calcium absorption goes on for 3—4 h, it can be assumed that the action of calcitonin on the skeleton continues for several h with a resulting bone matrix and mineral sparing effect.

REFERENCES

1. Birge, S.J., Peck, W.A., Berman, M. & Whedon, G.D.: Study of calcium absorption in man: A kinetic analysis and physiologic model. *J. clin. Invest.* **48**, 1705–1713 (1969)
2. Coen, G., Mazzuoli, G.F., Antonozzi, I. & Scarda, A.: Role of the thyroid gland in blood calcium homeostasis during intestinal calcium absorption in man. *Metabolism* **23**, 709–714 (1974)
3. Cooper, C.W., Schwesinger, W.H., Mahgoub, A.M., Ontjes, D.A., Gray, T.K. & Munson, P.L.: Regulation of secretion of thyrocalcitonin. In: *Calcium, Parathyroid Hormone and the Calcitonins*, Excerpta Medica (Amst.), pp. 128–139 (1972)
4. DeLuca, H.F., Morii, H. & Melancon, M.J. Jr.: The interaction of vitamin D, parathyroid hormone and thyrocalcitonin. In: *Parathyroid Hormone and Thyrocalcitonin (Calcitonin)*, Excerpta Medica (Amst.), pp. 448–454 (1968)
5. Friedman, J. & Raisz, L.G.: Thyrocalcitonin: Inhibition of bone resorption in tissue culture. *Science* **150**, 1465–1467 (1965)
6. Hirsch, P.F. & Munson, P.L.: Importance of thyroid gland in the prevention of hypercalcemia in rats. *Endocrinology* **79**, 655–658 (1966)
7. Milhaud, G., Perault, A.M. & Moukhtar, M.S.: Étude du mechanisme de l'action hypocalcémiante de la thyrocalcitonine. *C.R. Acad. Sci. (Paris)* **261**, 813–816 (1965)
8. Munson, P.L. & Gray, T.K.: Function of thyrocalcitonin in normal physiology. *Fed. Proc.* **29**, 1206–1208 (1970)
9. Talmage, R.V., Neuenschwander, J. & Kraintz, L.: Evidence for the existence of thyrocalcitonin in the rat. *Endocrinology* **76**, 103–107 (1965)
10. Tothill, P., Dellipiani, A.N. & Calvert, J.: Plasma concentrations of radiocalcium after oral administration and their relationship to absorption. *Clin. Sci.* **38**, 27–39 (1970)
11. Sanderson, P.H., Marshall, F. & Wilson, R.E.: Calcium and phosphorus homeostasis in the parathyroidectomized dog: Evaluation by means of ethylenediaminetetracetate and calcium tolerance test. *J. clin. Invest.* **39**, 662–670 (1960)
12. Wills, M.R. & Gray, B.C.: Micromethod for the estimation of calcium by autoanalyzer. *J. clin. Path.* **17**, 687–689 (1964)

Gastrointestinal Effect of Calcitonin: Inhibition of Gastrin Secretion

I. HORNUM, J. FAHRENKRUG & J.F. REHFELD

The stimulatory effect of calcium on gastric acid secretion is well-known (6, 4), but the mechanism unknown. Gastrin secretion may be involved since calcium infusions increase the concentration of gastrin in serum (7).

In 1971 Hesch and co-workers (5) reported that calcitonin inhibited gastric acid secretion without concomittant changes in serum levels of calcium. In 1974 Becker and co-workers (1) found that calcitonin depressed the serum gastrin response to food in normals.

In the present study we have been interested in an effect of calcitonin on gastrin and the possible relations to calcium.

MATERIAL AND METHODS

Nine normal subjects without known gastrointestinal diseases and three patients with pernicious anaemia (p.a.) and hypergastrinaemia were studied.

All subjects were examined after overnight fasting:
1) the response to a large protein-rich meal was determined by blood samples taken at 15, 10, 5 min before and 2, 5, 10, 20, 30, 45, 60 and 90 min after onset of meal.
2) the food response was determined when synthetic salmon calcitonin (sCT) was given as an intravenous bolus injection with 0.7 MRC units per kg bodyweight at the start of the meal.
3) the response to sCT given intravenously to the patients was examined in the fasting state, using the same intervals for blood sampling as at the meal-studies.

The serum-gastrin concentration was determined by radioimmunoassay (8, 11), serumcalcium by fluorescent titration with EDTA using calcein as

Departments of Medicine (T) and Clinical Chemistry, Bispebjerg Hospital, Copenhagen.

indicator (2) and serumalbumin by the bromcresol-green method (10), and the results analysed for significance (p < 0.005) of differences between means using the Student's paired t-test.

RESULTS

In Fig. 1 the findings in the normal subjects are demonstrated.

The full-drawn line represents the meal-induced response in serum gastrin concentration, the broken line the response to meal plus calcitonin, and the punctuated line the effect of calcitonin in the fasting subjects. The vertical bars represent 1 SEM.

In the basal state the mean concentration of serum gastrin was 26 pmoles/1 ± 2 and the response to salmon calcitonin alone significantly decreased this level at 20, 30, 45, 60 and 90 min after injection.
The serum gastrin response to the meal was significantly reduced by sCT at 5, 10, 20 and 30 min after injection.

In the three patients with p.a. calcitonin reduced serum gastrin concentration as seen in Table I.

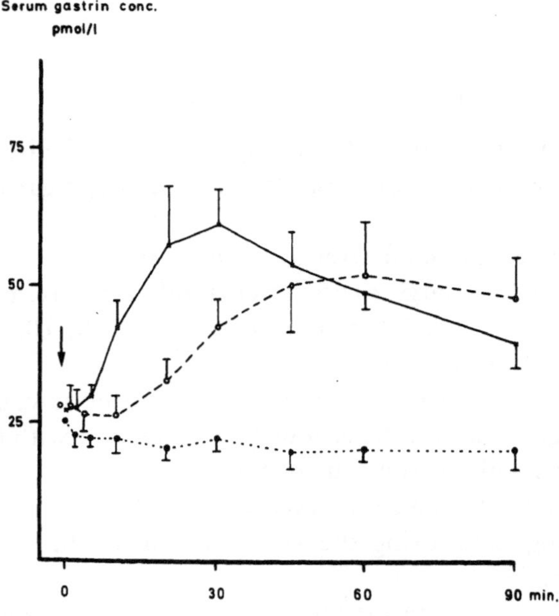

Figure 1. Effect of intraveneous injection of synthetic salmon calcitonin (sCT) on the serum-gastrin response to food and serum-gastrin concentrations in fasting normal subjects. Meal (x———x), meal + sCT (o———o), and sCT (o- - - - -o). Time of injection or start of the meal. Vertical bars represent 1 SEM.

Table 1. Total Serum Gastrin Concentration and Distribution of Components (I — IV) in 3 Patients with Pernicious Anaemia before and after Injection of Calcitonin.

Patients	Before calcitonin		After calcitonin	
	Serum gastrin (pmoles/l)	Components (fraction)	Serum gastrin (pmoles/l)	Components (fraction)
SM	272	0.08 0.48 0.28 0.16	99	0.07 0.73 0.13 0.07
MP	1262	0.13 0.64 0.21 0.02	194	0.13 0.74 0.13 0.00
KT	2324	0.20 0.62 0.15 0.03	1681	0.29 0.63 0.06 0.02
Mean	1286	0.14 0.58 0.21 0.07	629	0.16 0.70 0.11 0.03

As the serum gastrin concentration in normal subjects is so low that gel-filtration yields only minimal differences in the component pattern, we have chosen to analyse this pattern in the hypergastrinaemic p.a. patients. In Table I the numerical changes after calcitonin injection can be seen, and the same changes are illustrated in Fig. 2, where the pattern of components in one of the p.a. patients is shown before and after injection of sCT.

A decrease in the absolute concentrations of all four main components can be seen. The decrease, however, is most pronounced in the components III and IV, whereas the components I and II constitute relatively larger fractions of the total gastrin immunoreactivity after sCT than before.

The total serum calcium concentration was unaffected by sCT alone, meals, or meal plus calcitonin. The level of ultrafiltrable calcium calculated from the calcium and albumin contrations was similarily unchanged.

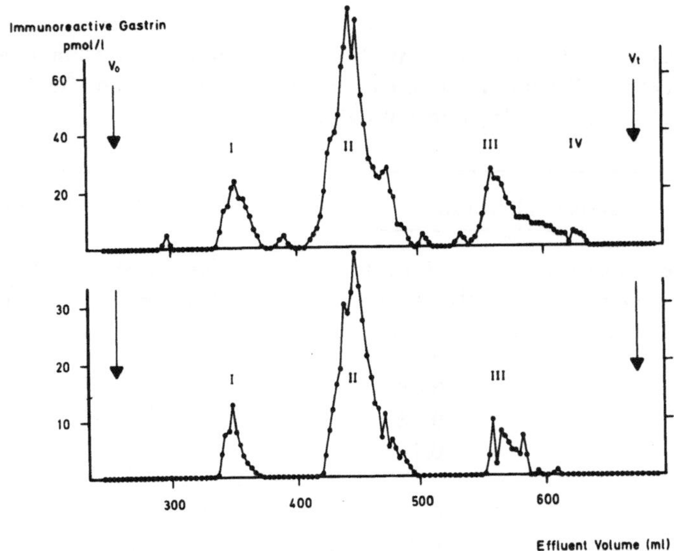

Figure 2. Diagram of gastrin components (I — IV) in serum from a patient with pernicious anaemia eluted on a Sephadex G-50 superfine column (27 x 2000 mm). Above: Gastrin components before injection of calcitonin. Below: Gastrin components after injection of calcitonin.

V_o: Void volume. V_t: Total mobile phase.

DISCUSSION

Our study shows that calcitonin lowers the gastrin-concentration in serum in the basal state, as well as the response to food without concommittant changes in the concentration of total and ultrafiltrable calcium in serum.

The discrepancy between our results and those of Becker *et al.* (1) and others (3) might be due to the way of calcitonin administration. While others have given prolonged infusion of calcitonin we have used the bolus-injection technique, recognizing that no matter which technique employed, the study still would be "unphysiological".

Immunoreactive gastrin in serum exists in four main components of different molecular size (9): One large and three smaller components.

Our finding of an effect of sCT primarily on the smaller components of gastrin may indicate a direct effect of calcitonin on the gastrin cells in the duodenum and antrum, as it is assumed that gastrointestinal tissue release immunoreactive gastrin mainly as component III (9).

REFERENCES

1. Becker, H.D., Reeder, D.D., Scurry, M.T. & Thompson, J.C.: Inhibition of gastrin release and gastric secretion by calcitonin in patients with peptic ulcer. *Amer. J. Surg.* 127, 71 (1974)
2. Bett, I.M. & Fraser, G.P.: A rapid micro-method for determining serum calcium. *Clin. chim. Acta* 4, 346—356 (1959)
3. Bieberdorf, F.A., Gray, T.K., Walsh, J.H. & Fordtran, J.S.: Effect of calcitonin on meal-stimulated gastric acid secretion and serum gastrin concentration. *Gastroenterology* 66, 343—346 (1974)
4. Donegan, W.L. & Spiro, H.M.: Parathyroids and gastric secretion. *Gastroenterology* 38, 750—759 (1960)
5. Hesch, R.D., Hüfner, M., Hasenhager, B. & Creutzfeldt, W.: Inhibition of gastric secretion by calcitonin in man. *Horm. Metab. Res.* 3, 140 (1971)
6. Murphy, D.L., Goldstein, H., Boyle, J.D. *et al.*: Hypercalcemia and gastric secretion in man. *J. Appl. Physiol.* 21, 1607—1610 (1966)
7. Reeder, D.D., Jackson, B.M., Ban, J. *et al.*: Influence of hypercalcemia on gastric secretion and serum gastrin concentrations in man. *Ann. Surg.* 172, 540—546 (1970)
8. Rehfeld, J.F., Stadil, F. & Vikelsøe, J.: Immunoreactive gastrin components in human serum. *Gut* 15, 102—111 (1974)
9. Rehfeld, J.F., Stadil, F., Malmström, J. & Miyata, M.: In: *Gastrointestinal Hormones*, Thompson, J.C. (ed.), University of Texas Press (1975), in press.
10. Schirardin, H. & Ney, J.: Eine vereinfachte Mikromethode zur Bestimmung von Serumalbumin mit Hilfe von Bromkresolgrün. *Z. klin. Chem. klin. Biochem.* 10, 338—344 (1972)
11. Stadil, F. & Rehfeld, J.F.: Determination of gastrin in serum. *Scand. J. gastroent.* 8, 101—112 (1973)

Immunoreactive Calcitonin in Non-thyroid Tumours

C.J. HILLYARD, R.C. COOMBES[1], P.B. GREENBERG[2] & I. MacINTYRE

INTRODUCTION

The production of calcitonin by malignant tissue is now a well recognised occurrence in medullary carcinoma of the thyroid (9, 8, 4, 2), carcinoid (10) and some cases of phaeochromocytoma (11).

At an early stage in the development of the radioimmunoassay for calcitonin in this laboratory, calcitonin was extracted from an oat cell lung carcinoma (16) and this finding prompted us to investigate the occurrence of calcitonin in other types of malignant disease. Preliminary findings indicated that plasma levels of calcitonin are raised in many patients with non-thyroid tumours (3) and we have now measured calcitonin in extracts of malignant tissue taken from such patients.

METHODS

Heparinised blood samples were obtained from fasting patients, kept in ice for up to 1 hour, separated and stored at $-18°C$. All samples were assayed within 2 weeks of collection.

Tissues for extraction were obtained at operation or autopsy, snap-frozen, lyophilized and thinly sliced. Excess fat was removed by ether, the tissues weighed and subsequently extracted 3 times with butanol: acetic acid: water (75:7.5:21) v/v. (12). The extracts were then lyophilized and dissolved in o.1 M formic acid for immunoassay and determination of biological activity in the rat.

Radioimmunoassay
Extracts were dissolved in 0.05 M phosphate buffer pH 7.4 and assayed

Endocrine Unit, Royal Postgraduate Medical School, Ducane Road, London.

(2). Blood samples were assayed using the assay previously reported (3). Samples were assayed in 6 serial dilutions. One extract was assayed before and after addition of 500 mg Spherosil XOA 400 (3).

Hypocalcaemic Activity of Tissue Extracts

The biological activity of the extracts was measured in 50 g Wistar rats (12). The activity of extracts of 2 breast cancers, a fibroadenoma and normal breast tissue was compared with the effect of the MRC Standard B pig thyroid calcitonin.

RESULTS

Radioimmunoassay

Elevated levels (> 0.1 ng/ml) of immunoreactive calcitonin were found in the plasma of four patients with carcinoma of the prostate, eight of the 11 patients with oat cell carcinoma of the lung and all eight patients with disseminated breast cancer. Elevated levels of circulating calcitonin were

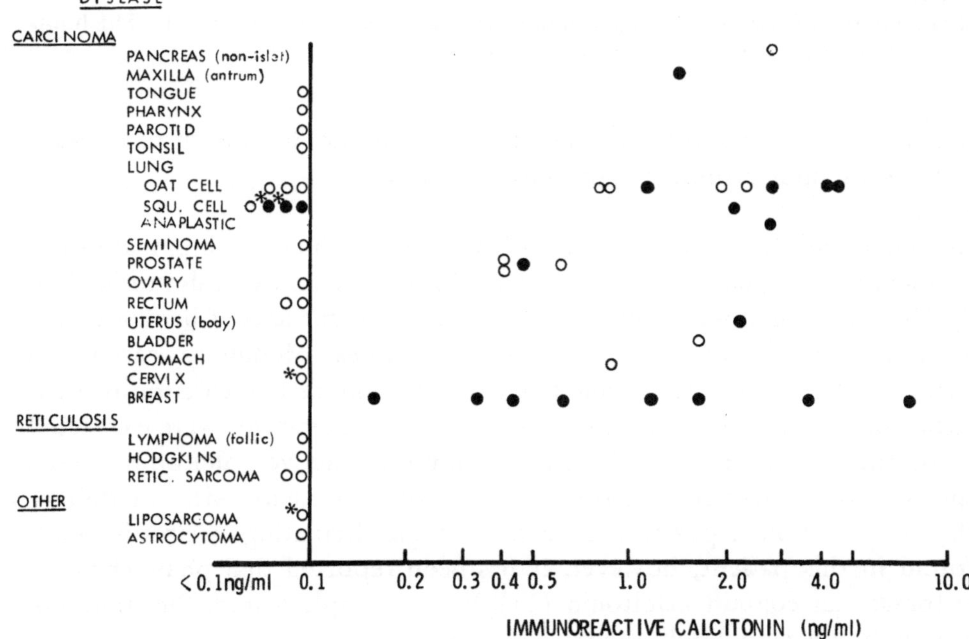

Figure 1. Plasma calcitonin levels in patients with cancers.
● with skeletal metastases
○ no skeletal metastases
∗ hypercalcaemic

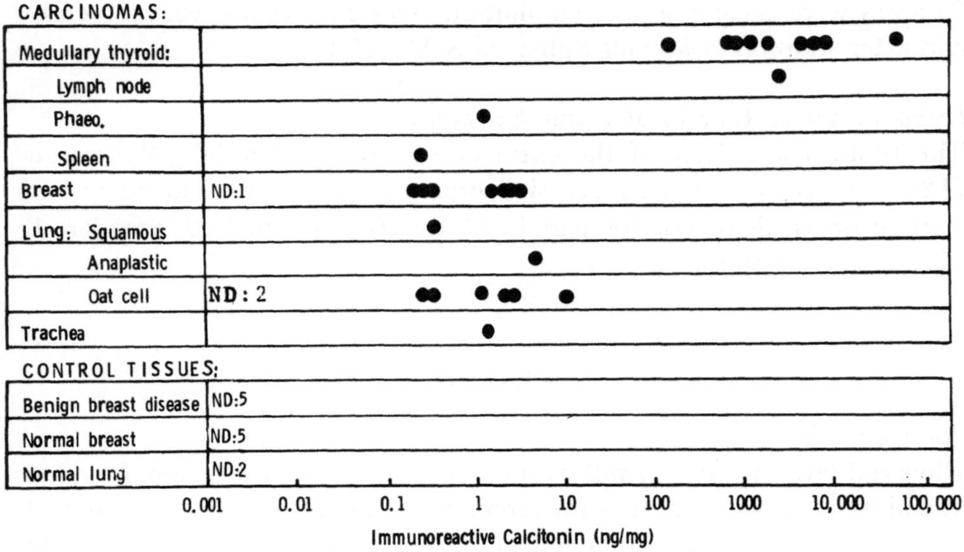

CARCINOMAS:

Medullary thyroid:	
Lymph node	
Phaeo.	
Spleen	
Breast	ND:1
Lung: Squamous	
Anaplastic	
Oat cell	ND : 2
Trachea	

CONTROL TISSUES:

Benign breast disease	ND:5
Normal breast	ND:5
Normal lung	ND:2

0.001 0.01 0.1 1 10 100 1000 10,000 100,000

Immunoreactive Calcitonin (ng/mg)

Figure 2. Calcitonin content of tissue extracts. The calcitonin level is plotted on the abscissa, on a log scale.

N.D.: number of tissues in each group in which calcitonin was undetectable. The limits of detection vary with the weight of tissue available.

also seen in patients with various other carcinomas, but not in those with reticuloses, liposarcoma or astrocytoma (Fig. 1).

All but three (one breast cancer and two oat cell lung carcinomas) of the extracts of malignant tumours contained immunoreactive calcitonin (Fig. 2). However, calcitonin could not be measured in extracts of benign breast lesions, normal lung or normal breast tissues. Medullary carcinoma extracts contained greater quantities of immunoreactive calcitonin, than either breast or lung cancers, however, significant amounts were extracted from these tumours. The finding of immunoreactive calcitonin in the spleen and phaeochromocytoma tissue of a patient with medullary thyroid carcinoma, possibly reflects the high circulating calcitonin levels found in this patient, however, it has been reported that phaeochromocytomas can contain calcitonin (11). In the sample tested, the immunoreactivity was completely removed by 'Spherosil'.

Hypocalcaemic Activity of Tissue Extracts
Both breast carcinoma extracts produced hypocalcaemia in rats and the hypocalcaemic effect was greater with increasing quantities of extract.

Extracts of fibroadenoma and normal breast, however, produced only slight, non-specific hypocalcaemia, even in high concentrations.

DISCUSSION

The observation that immunoreactive calcitonin was present in the plasma of patients with several non-thyroid cancers, prompted us to extract tumours taken from such patients at operation or autopsy.

The tumours extracted contained a material which possessed many of the characteristics of human calcitonin. The inhibition of binding of ^{125}I labelled calcitonin to antibody paralleled that produced by synthetic human calcitonin standard. Immunological activity was removed by 'Spherosil' which has previously been shown to extract iodinated calcitonin (3). Also, the two breast tumour extracts which were bio-assayed were capable of producing hypocalcaemia in the rat.

The finding that tumour extracts contained immunoreactive calcitonin might possibly have been explained by the "sponge-effect", taking up and storing calcitonin produced by the thyroid. However, further studies indicate that proliferating lung carcinoma cells produce calcitonin *in vitro* over an 18 month period (5) and that monolayers of breast cancer cells release this hormone. (To be published.)

Breast carcinoma has rarely been associated with ectopic hormone production and only two hormones, parathyroid hormone (6, 7) and human chorionic gonadotrophin (1) have been reported to be produced by this tumour. In contrast, lung cancers are known to be associated with a wide variety of hormones (14) and calcitonin can be produced by oat cell carcinoma of the lung (15).

The explanation for calcitonin production by tumours remains to be clarified. Although the 'APUD' cell characteristics of medullary thyroid carcinoma cells are shared by some other tumours from which calcitonin has been extracted (13), these have not yet been demonstrated in many of the cancers associated with high circulating levels of calcitonin (3). Breast cancer is a common malignancy for which there is currently no satisfactory marker to predict recurrence or to monitor treatment and it seems possible that the radioimmunoassay of calcitonin, at least in patients with advanced disease could be used for this purpose. We are currently working on a sensitive assay which we hope will permit the detection of disease at an operable stage, before dissemination has occurred, or the early detection of residual tumour after mastectomy.

CONCLUSIONS

Calcitonin, which is both immunologically and biologically active, can be produced by tumour tissues.

Measurements of plasma calcitonin may be of value in the management of patients with malignant diseases other than medullary thyorid carcinoma, particularly breast cancer.

SUMMARY

Immunoreactive calcitonin was found in extracts of seven out of eight consecutive breast carcinomas and in eight lung carcinomas, but not in extracts of benign breast lesions or normal tissues.

These results suggest that the high plasma calcitonin levels observed in patients with a wide variety of cancers may reflect ectopic production of calcitonin by cancer tissue.

The radioimmunoassay for calcitonin could prove to be of value in the management of malignant disease, especially breast cancer.

ACKNOWLEDGEMENTS

We thank the following surgeons for providing tissues:

Mr. J.L. Boak, Mr. N. O'Higgins and Mr. R. Corey-Pearce (Hammersmith Hospital) and Mr. J. Bradbeer and Mr. B. Tanner (Mayday Hospital).

We are grateful to Dr. L. Galante, Dr. P.G.H. Byfield and Mr. E.W. Matthews for performing the bioassays.

Synthetic human calcitonin was supplied by Ciba-Geigy (Basle) Ltd.

This work was supported in part by the Medical Research Council and the Wellcome Trust. R.C.C. is in receipt of an MRC Clinical Research Fellowship and P.B.G. was in receipt of an overseas scholarship from the Royal Australasian College of Physicians.

REFERENCES

1. Braunstein, G.D., Vaitukaitis, J.L., Carbone, P.P. & Ross, G.T.: Ectopic production of human chorionic gonadotrophin by neoplasms. *An. Intern. Med.* **78**, 39–45 (1973)
2. Clark, M.B., Boyd, G.W., Byfield, P.G.H. & Foster, G.V.: A radioimmunoassay for human calcitonin M. *Lancet* (1969), **ii**, 74–76
3. Coombes, R.C., Hillyard, C.J., Greenberg, P.B. & MacIntyre, I.: Plasma immunoreactive calcitonin in patients with non-thyroid tumours. *Lancet* (1974), **i**, 1080–1083

4. Cunliffe, W.J., Black, M.M., Hall, R., Johnstone, I.D.A., Hodson, P., Shuster, S., Gudmundsson, T.V., Joplin, G.F., Williams, E.D., Woodhouse, N.J.Y., Galante, L. & MacIntyre, I.: A calcitonin-secreting thyroid carcinoma. *Lancet* (1968), **ii,** 63–66

5. Ellison, M., Woodhouse, D., Hillyard, C.J., Dowsett, M., Coombes, R.C., Gilby, E.D., Greenberg, P.B. & Neville, A.M.: Immunoreactive calcitonin production by human lung carcinoma cells in culture. *Brit. J. Cancer* **32,** 373–379 (1975)

6. Mavligit, G.M., Cohen, J.L. & Sherwood, L.M.: Ectopic production of parathyroid hormone by carcinoma of the breast. *New. Engl. J. Med.* **285,** 154–156 (1971)

7. Melick, R.A., Martin, T.J. & Hicks, J.D.: Parathyroid hormone production and malignancy. *Brit. Med. J.* **2,** 204–205 (1972)

8. Melvin, K.E.W. & Tashjian, A.H. Jr.: The syndrome of excessive thyrocalcitonin produced by medullary carcinoma of the thyroid. *Proc. nat. Acad. Sci. (Wash.)* **59,** 1216–1222 (1968)

9. Milhaud, G., Tubiana, M., Parmentier, C. & Coutris, G.: Epithelioma de la thyroide secretant de la thyrocalcitonine. *C.R. Acad. Sci. (D) (Paris)* **266,** 608–610 (1968)

10. Milhaud, G., Calmettes, C., Raymond, J.P., Bignon, J. & Moukhtar, M.S.: Carcinoide secretant de la thyrocalcitonine. *C.R. Acad. Sci. (D) (Paris)* **270,** 2195–2198 (1970)

11. Milhaud, G., Calmettes, C., Jullienne, A., Tharaud, D., Bloch-Michel, H., Cavaillon, J.P., Colin, R. & Moukhtar, M.S.: A new chapter in human pathology: calcitonin disorders and therapeutic uses. In: *Calcium, Parathyroid Hormone and the Calcitonins.* R.V. Talmage & P.L. Munson. (Eds.), Excerpta Medica, Amsterdam. 56–70 (1972)

12. Moseley, J.M., Matthews, E.W., Breed, R.H., Galante, L., Tse, A. & MacIntyre, I.: The ultimobranchial origin of calcitonin. *Lancet* (1968), **i,** 108–110

13. Pearse, A.G.E., Polak, J.M. & Heath, C.M.: Polypeptide hormone production by "carcinoid" apudomas and their relative cytochemistry. *Virchows. Arch. path. Anat. B. Cell Path.* **16,** 95–109 (1974)

14. Rees, L.H. & Ratcliffe, J.G.: Ectopic hormone production by non-endocrine tumours. *Clin. Endocr.* **3,** 263–299 (1974)

15. Silva, O.L., Becker, K.L., Primack, A., Doppman, J. & Snider, R.H.: Ectopic secretion of calcitonin by oat cell carcinoma. *New. Engl. J. Med.* **290,** 1122–1123 (1974)

16. Whitelaw, A.G.L. & Cohen, S.L.: Ectopic production of calcitonin. *Lancet* (1973), **ii,** 443

Extra-thyroidal Origin of the Heaviest Fractions of Circulating Human "Calcitonin"

B. Argémi

Various authors have already emphasized the non-specific interference of plasma proteins in the radioimmunoassay of human calcitonin (hCT) (1, 2, 3, 5, 8). This interference reduces the sensitivity of most of the techniques already published, and in addition, probably causes the discrepancies found in results obtained by different laboratories.

In the radioimmunoassay technique developed in our laboratory (1), this phenomenon is particularly clear when using a plasma concentration of 20%, and prevents any physiological study of circulating CT, for in order to obtain an accurate standard curve with plasma from a thyroidectomized subject, the plasma concentration must not exceed 2%. In this case, the sensitivity range of our method is 300—600 pg/ml, whereas most normal plasmas contain less than 100 pg/ml, of CT.

The aim of this study has been to identify the plasma factors interfering in the immunological reaction.

MATERIAL AND METHODS

1) Human calcitonin M (hCT M) radioimmunoassay

The method used has already been described. For the assay of the column effluent after gel filtration, the sample volume used varied from 0.1 to 0.2 ml (the detection limit being 60 and 30 pg per tube respectively).

2) Plasma calcitonin extraction

25 mg QUSO G 32 (microfine precipitated silica, Bie & Berntsen, 7 Sandbækvej, DK 2610 Rødovre) were added to 1 ml plasma. This mixture

Laboratoire de recherches sur le métabolisme phospho-calcique. Clinique médicale. Groupe hospitalier de la Timone, Marseille.

was stirred and centrifuged, and then the supernatant was discarded and the pellet eluted twice with acetone-acetic acid-distilled water (40—1—59). The two supernatants were mixed, and the acetone was evaporated off at 65°C for 15 min under nitrogen, while the mixture was shaken. The solution remaining was freeze-dried and then dissolved for assay with 1 ml phosphate buffer 0.1 M, pH 7.25, containing 0.02% neomycine, and 0.1% human serum albumin.

3) Gel filtration

This was performed on Sephadex G 75 (column size 35 x 1.6 cm; flow rate: 15 ml/h), and on Sephadex G 200 (column size 65 x 1.6 cm; flow rate: 6 ml/h). The elution buffer was the same as for assays, but without albumin.

RESULTS

1) Nature of the inhibition

Several plasmas from thyroidectomized (TX) or thyroparathyroidectomized (TPTX) subjects were found to have an inhibiting effect (variable, but always considerable) on the binding of radioiodinated CT to the antibody used. The binding fall was between 40 and 60%, and was similar to, or slightly greater than the fall produced by most normal plasmas.

a) Habener *et al.* (9) have advanced the hypothesis of an enzymatic degradation of the hormone by plasma, but in assay conditions plasma CT appeared stable (4); moreover, heating the plasma to 65°C for 30 min or to 80°C for 10 min, did not change its inhibiting effect.

b) The hypothesis of hormone binding to plasma proteins must also be rejected: labelled or unlabelled CT submitted to gel filtration on Sephadex G 75 or G 200, eluted at the same position when diluted in the buffer and after incubation in the plasma. Moreover, plasma saturation by up to 10 ng/ml salmon CT, which has no cross-reaction with hCT for this antiserum, produced no change in its inhibiting effect.

c) Plasma interference is due to the presence of one or several thermostable proteins whose immunological characteristics are identical with certain characteristics of hCT M.

— Labelled hormone displacement by plasma varied from one antiserum to another; unfortunately, the antibodies with the least plasma interference were also the ones with the least affinity for CT M, and could not be used for measuring physiological circulating levels.

— TX or TPTX plasma extraction by QUSO made it possible to

eliminate the inhibiting effect entirely in the supernatant. The labelled hormone binding to the antibody was the same, whether the extract or the buffer was used, thus proving that there was not enough CT M to be detected. The standard curves after extraction by this method and those obtained in the buffer alone were identical, with a 70—80% recovery of the unlabelled hormone previously added to the plasma.

2) Characteristics of the immunoreactive material

Plasmas from a TPTX subject, and a patient with C-cell carcinoma secreting large amounts of CT, were each submitted to gel filtration on Sephadex G 75 and then on Sephadex G 200: the immunoreactive material was assayed in each tube of effluent.

a) Gel filtration on Sephadex G 75 (Fig. 1): For the TPTX plasma, an immunoreactive peak was seen in the void volume of the column. The total amount of eluted CT-like material was the same as that measured in the pure plasma assay. The labelled CT M added as a marker eluted significantly later. In plasma from a normal subject, a smaller immuno-

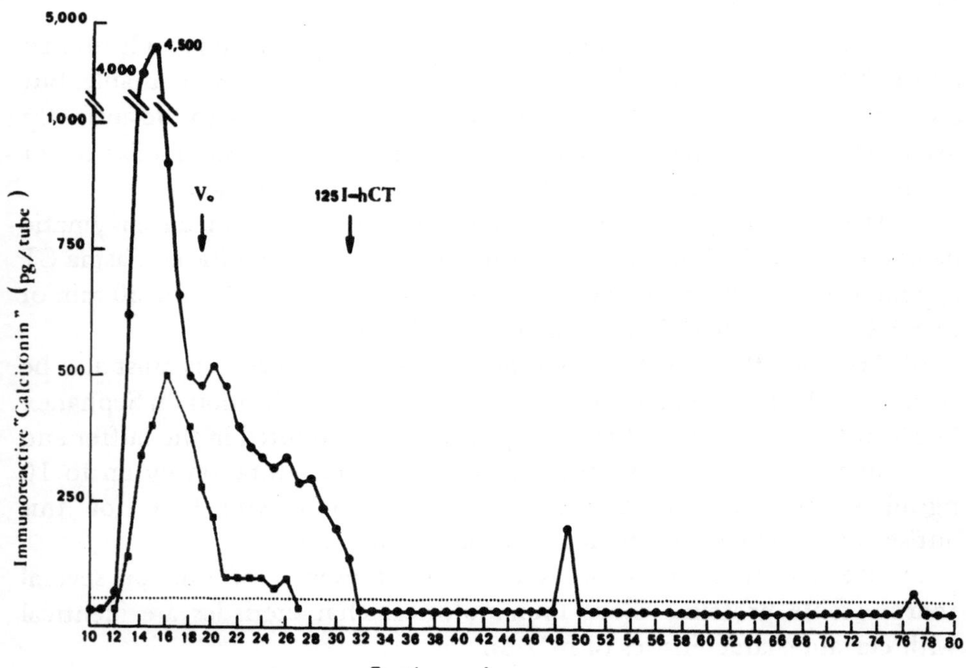

Figure 1. Gel filtration on Sephadex G 75, of 1 ml plasma from a thyroparathyroidectomized subject (●- - - -●) and from a patient with C-cell carcinoma (●————●). Volume of each fraction: 2 ml. Sample volume for assay: 0.1 ml. Detection limits of the essay: 60 pg/tube (horizontal dotted line).

reactive peak was found in the same position; no CT M could be detected because the samples were over-diluted. Gel filtration of the plasma from a subject with C-cell carcinoma resulted in a much greater immunoreactive peak in the same void volume position, and then in a series of compounds occurring between this "Big Calcitonin" and CT M, and decreasing in concentration. No marked peak was found in the CT M position, but there were two small peaks, which probably represent fragments of low molecular weight.

b) Gel filtration of these plasmas on Sephadex G 200 (Fig. 2) made it possible to analyse the immunoreactive material in greater detail. In the TPTX plasma, this "Big Calcitonin" appeared to be composed of three fractions:

The first one (I) appeared in the void volume of the column and had a very high molecular weight, higher than that of fibrinogen ($>$ 350,000).

The second one (II) eluted with the Immunoglobulins G (M = 150,000).

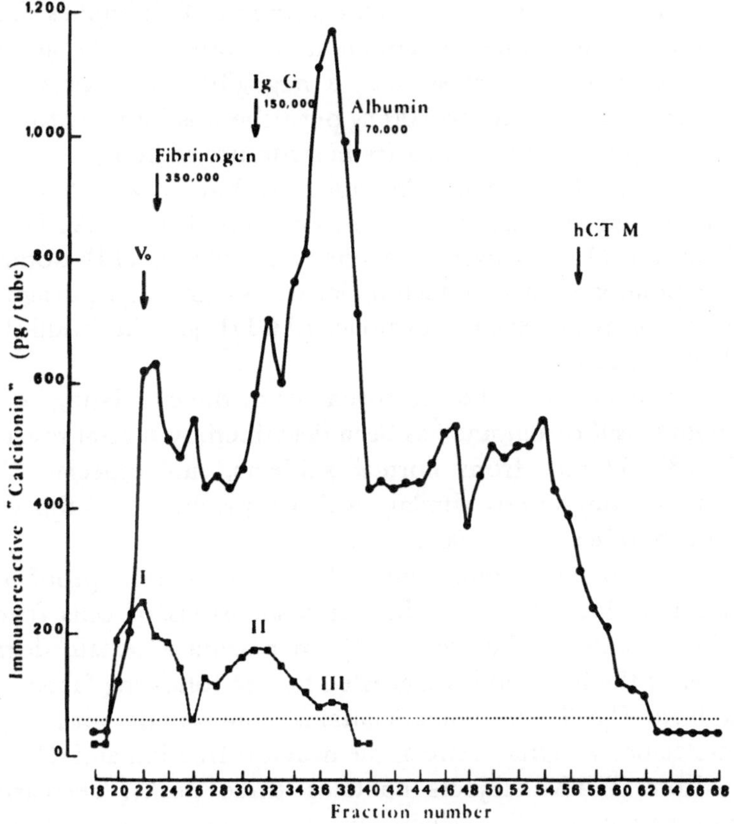

Figure 2. Gel filtration on Sephadex G 200, of the same plasmas as in Fig. 1, with identical characteristics.

The third one (III) much less concentrated, eluted slightly before albumin: its molecular weight was roughly 70,000—80,000.

Plasma from C-cell carcinoma was also composed of three fractions which eluted in slightly different positions:

The first fraction coincided exactly with TPTX plasma peak I.

The second one, much more concentrated, spread over TPTX plasma peaks II and III.

The third one was located between peak III and CT M, in a wide range of molecular weights between 3,600 and 70,000.

DISCUSSION

Plasma from thyroidectomized and thyroparathyroidectomized patients contains at least three CT-like substances with molecular weights ranging from 70,000 to more than 350,000. The same products are found much more abundantly in patients with C-cell carcinoma. This implies that part of their molecular structure is common to both CT M and these substances, since in this assay system, a negligible or inexistant cross-reaction for concentrations up to 100 ng per tube was found with porcine or salmon CT, in spite of their similarity in structure to hCT.

Several hormones have been described as having a heterogeneous circulating aspect in normal or pathological conditions: ACTH (18), gastrin (19), insulin (17), growth hormone (7), prolactin (14). Sometimes they have compounds with very high molecular weights e.g. glucagon (15, 16). The heavy forms of certain hormones (ACTH, gastrin, insulin), have been described as precursors.

As far as CT is concerned, the heterogeneity of the circulating hormone in patients with C-cell carcinoma has been described by several researchers (6, 10, 12, 13). Plasma from normal subjects, and especially TPTX subjects, contains compounds similar to those produced by this cancer, and this discovery raises several questions.

1) Are these substances prohormones? So far, there is no proof of this; but a prohormone has been described in suspensions of cells from the ultimobranchial glands of the trout (11). In plasma a certain degree of polymerization of the hormone is suggested by the following facts:

— Plasma from C-cell carcinoma contains compounds with regularly decreasing molecular weights between the heaviest fraction and CT M.

— Plasma from TPTX subjects contains three peaks, decreasing in concentration and molecular weight. In this connection, it should be noted that a small proportion of labelled CT (3%) elutes on Sephadex G 200 in the same position as plasma TPTX peak I.

2) These immunoreactive proteins have an extrathyroidal origin. They could be produced by authentic C-cells after migrating to other areas: but the high concentration of these substances, compared with the low level of CT M produced by thyroidal C-cells would seem to imply that these C-cells can be found elsewhere in far greater numbers — which has never been proved. It is much more tempting to think that these substances may be produced by other cells of the APUD system: it is known

— that these different cells have the same embryological origin and histochemical characteristics;

— that C-cells produce serotonin; and

— that certain digestive carcinoid tumours can produce a CT-like material.

3) We have only been able to extract assayable amounts of CT M from plasma of subjects whose thyroid gland was intact. Thus the presence of thyroidal C-cells seems necessary for the production of CT M.

CONCLUSIONS

Two important practical consequences of these findings, concerning the detection and biological control of C-cell carcinoma of the thyroid gland, are as follows:

1) The physiological CT M levels and their variations can be studied by extracting the plasma CT with QUSO G 32, while avoiding interference from CT-like material; however, this method is not effective for detecting C-cell carcinomas, for CT M occurs in very small quantities compared with the heavy compounds, and only 10% or less of the total immunoreactive material of these plasmas can be recovered by extraction. In order to diagnose these carcinomas, one must use an antiserum with a strong affinity for the heavy compounds, and a technique making it possible to measure them completely.

2) During biological control of patients treated for a C-cell carcinoma, the discovery of an assayable level of CT-like material does not necessarily mean that there is a relapse or a metastasis. Such a discovery is significant only if the level is considerably higher than that found in thyroidectomized subjects.

ACKNOWLEDGEMENTS

We thank Dr. Szpilfogel from Organon Laboratories, who kindly provided us with the synthetic hCT M; Pr. M. Delaage for his valuable advice; and Mrs M.C. Hours for her excellent technical assistance.

REFERENCES

1. Argémi, B.: Le dosage radioimmunologique de la Thyrocalcitonine humaine. Technique et résultats chez le sujet normal. *Ann. Endocr.* (Paris) **34**, 290–292 (1973)

2. Bieler, E.U., van Royen, R.J., De Bruin, E.J.P. & Hoog, J.M.C.: A radioimmunoassay technique for human calcitonin in plasma avoiding the non specific interference of plasma proteins on the tracer binding. *Horm. Metab. Res.* **5**, 231 (1973)

3. Clark, M.B., Byfield, P.G.H., Boyd, G.W. & Foster, G.V.: A radioimmunoassay for human calcitonin M. *Lancet* (1969) ii, 74–77

4. Deftos, L.J.: Immunoassay for human calcitonin. I: Method. *Metabolism* **20**, 1122–1128 (1971)

5. Deftos, L.J., Bury, A.E., Habener, J.F., Singer, F.R. & Potts, J.T. Jr.: Immunoassay for human calcitonin. II: Clinical studies. *Metabolism* **20**, 1129–1137 (1971)

6. Deftos, L.J., Roos, B.A., Bronzert, D. & Parthemore, J.G.: Immunochemical heterogeneity of calcitonin in plasma. *J. clin. Endocr.* **40**, 409–412 (1975)

7. Dimond, R.C., Wartofsky, L. & Rosen, S.W.: Heterogeneity of circulating growth hormone in acromegaly. *J. clin. Endocr.* **39**, 1133–1137 (1974)

8. Frölich, M., Kassenaar, A.A.H. & Smeenk, D.: Radioimmunoassay of human plasma calcitonin. *Horm. Metab. Res.* **3**, 297–298 (1971)

9. Habener, J.F., Singer, F.R., Deftos, L.J. & Potts, J.T. Jr.: Immunological stability of calcitonin in plasma. *Endocrinology* **90**, 952–960 (1972)

10. Moukhtar, M.S., Julienne, A., Taboulet, J., Calmettes, C., & Milhaud, G.: Hétérogénéité de la calcitonine immunoréactive dans le plasma du cancer médullaire. *Colloque I.N.S.E.R.M.* Lyon 1974 (Abstr)

11. Roos, B.A. Okano, K. & Deftos, L.J.: Evidence for a pro-calcitonin. *Biochem. Biophys. Res. Comm.* **60**, 1134–1140 (1974)

12. Singer, F.R. & Habener, J.F.: Multiple immunoreactive forms of calcitonin in human plasma. *Biochem. Biophys. Res. Comm.* **61**, 710–716 (1974)

13. Snider, R.H., Silva, O.L., Becker, K. & Moore, C.F.: Heterogeneity of calcitonin. *Lancet* (1975) i, 49–50

14. Suh, H.K. & Franz, A.G.: Size heterogeneity of human prolactin in plasma and pituitary extracts. *J. clin. Endocr.* **39**, 928–935 (1974)

15. Valverde, I., Villanueva, M.L., Lozano, I. & Marco, J.: Presence of glucagon immunoreactivity in the globulin fraction of human plasma ("Big plasma glucagon"). *J. clin. Endocr.* **39**, 1090–1098 (1974)

16. Weir, G.C., Knowlton, S.D. & Martin, D.B.: High molecular weight glucagon-like immunoreactivity in plasma. *J. clin. Endocr.* **40**, 296–302 (1975)

17. Yalow, R.S.: Radioimmunoassay methodology: application to problems of heterogeneity of peptide hormones. *Pharmacol. Rev.* **25**, 161–178 (1973)

18. Yalow, R.S. & Berson, S.A.: Characteristics of "big ACTH" in human plasma and pituitary extracts. *J. clin. Endocr.* **36**, 415–423 (1973)

19. Yalow, R.S. & Wu, N.: Additional studies on the nature of Big-Big Gastrin. *Gastroenterology* **65**, 19–27 (1973)

Analogues of Human Calcitonin

R. MAIER

Ample evidence has been accumulated to support the assumption that calcitonin preparations obtained from ultimobranchial bodies are generally 20 to 40 times as potent in their hypocalcaemic activity in the rat than those of thyroidal origin. Of the ultimobranchial peptides, only salmon calcitonin (SCT_1) has been synthesized so far. The amino-acid sequence of eel calcitonin has been elucidated, and many other avian, reptilian and piscine calcitonins have been extracted and assayed in more or less purified form. All have been found to be very active. Of the four known thyroidal peptides, three displayed low potency of comparable magnitude; as regards the fourth — ovine calcitonin — no activity has been reported.

This seemingly general disparity in hypocalcaemic potency led us to study the role of certain amino acids in strategic positions along the peptide chain (2, 3). A number of analogues were synthesized by the fragment condensation method (5) and tested for biological activity. Their hypocalcaemic potency was determined as described earlier (2). In our experiments, the potency of synthetic SCT_1 was 30 to 40 times greater on a weight-for-weight basis than that of human calcitonin (HCT).

In addition to their potency, the duration of action of these analogues was also compared, since SCT_1 elicits a considerably more persistent reduction of serum calcium in the rat than HCT. The hypocalcaemic effect of 30 μg/kg SCT_1 lasted up to 20 h, whereas that of the same dose of HCT had subsided after roughly 4 h.

SCT_1 differs in 16 amino-acid positions from HCT. In the course of our studies, two isohormones — SCT_2 and SCT_3 — were isolated from salmon ultimobranchials (1) and subsequently synthesized (4). SCT_2 differs in 4 and SCT_3 in 5 amino acids from SCT_1. It is interesting to note that these 4 and 5 amino acids are the same as those occurring in the human sequence, except for those in positions 29 and 31, which are in reversed

Research Department, Pharmaceuticals Division, Ciba-Geigy Ltd., Basle.

order in the isohormones. SCT_2 is equal in potency to SCT_1; SCT_3, differing from SCT_2 only in position 8, is only one third as active. The analogues of HCT presented here were substituted in amino-acid positions 8, 22, 29 and 31, the positions in which the isohormones differ from SCT_1, asparagine in position 15 exempted. Their formulae are shown in Fig. 1.

Introduction of valine for methionine in position 8 (analogue I, Fig. 1) increased the hypocalcaemic potency about 5-fold. This increased activity implies that in position 8 the less polar side chain of valine has a greater affinity for the receptor site than methionine. The influence of polarity is further documented by the fact that in both HCT and SCT_3 oxidation of the methionine to its sulphoxide drastically reduces biological activity. Thus valine[8] has clearly some bearing on the increased potency of SCT_1. However, the methionine-sulphoxide of porcine calcitonin, in which the methionine is located in position 25, exhibits practically the same activity as the native peptide. This indicates that the nature of the amino acid in position 25 is less critical with regard to its interaction with the receptor than that in position 8. The introduction of tyrosine[22] in place of phenylalanine (analogue II, Fig. 1) also improved the hypocalcaemic potency of HCT about 5 times. Here, the more polar tyrosine appears better suited for interaction with the target site. All eight known calcitonin sequences have an aromatic amino acid in position 22; in ultimobranchial peptides — salmon and eel — a tyrosine occurs and in thyroidal peptides phenylalanine is present. Exceptions in the former

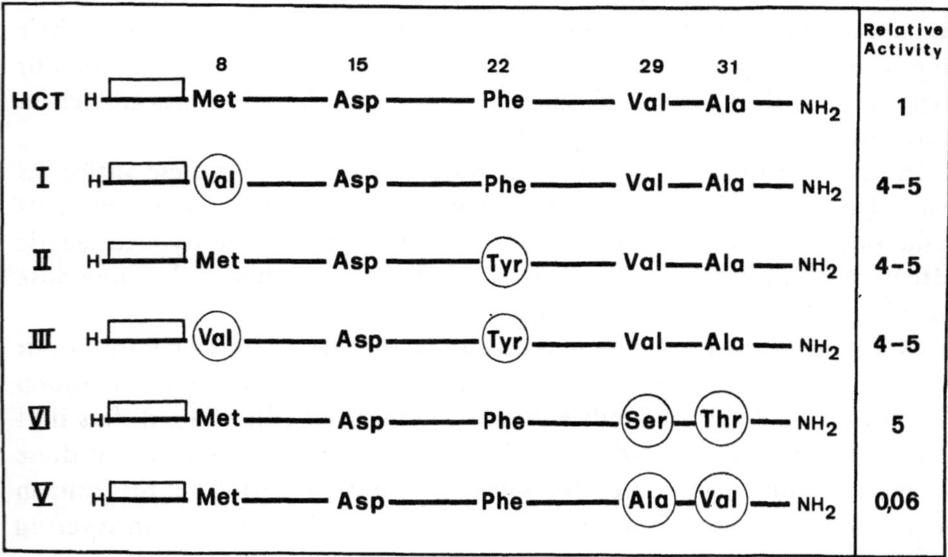

Figure 1. Five analogues of human calcitonin (see text).

group are the salmon isohormones and in the latter ovine calcitonin. The activity of the isohormones does not seem to depend on the presence of a tyrosine residue in posotion 22. Unfortunately, no data on the biological activity of ovine calcitonin are available.

The fact that the doubly substituted analogue with valine[8] and tyrosine[22] (analogue III, Fig. 1) showed no further increase in activity demonstrates that the effects of the single substitutions are not necessarily additive, and that the influences exerted by single amino acids can be altered by the introduction of further acids.

All three analogues I, II and III of Fig. 1 displayed identical activity in prolonging hypocalcaemia and were about twice as effective as HCT in terms of the areas subtended by the time-curves. Introduction of serine in position 29 and threonine in position 31 (analogue IV, Fig. 1) as they occur in SCT_1 improved the activity of HCT about 5-fold. But, reversal of the sequence of the two amino acids found in HCT to alanine[29] and valine[31] (analogue V, Fig. 1), as they occur in the two isohormones, practically nullified the biological activity of HCT. A similar effect was also evident in the duration of action. The area under the curve of analogue IV was about 1.6 times greater than that of HCT, and the area subtended by the curve of analogue V was roughly half that of HCT. In view of the results obtained with the analogues of HCT, it is of interest to compare them with those of SCT_2 (5), as shown in Fig. 2.
The potencies are given in relation to that of HCT which was taken as 1. Reversal of the two amino acids in positions 29 and 31 (analogue I, Fig. 2) did not affect the potency of SCT_2. Similarly, the introduction of

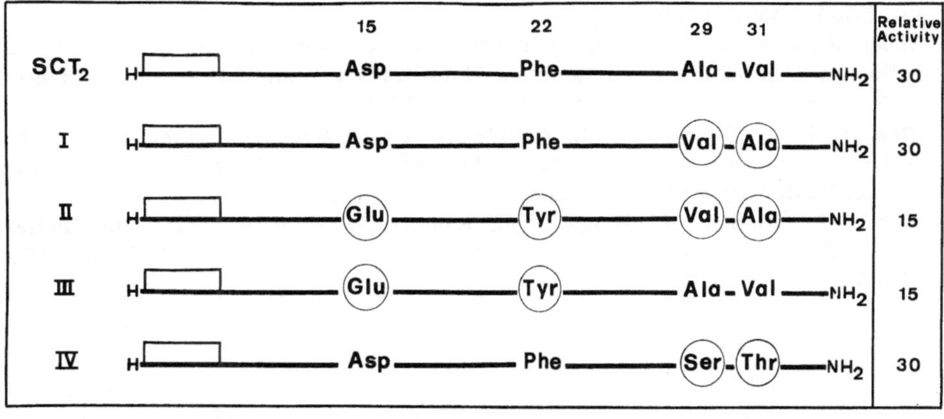

Figure 2. Analogues of salmon calcitonin (SCT_2), see text.

serine[29] and threonine[31] (analogue IV, Fig. 2), as they occur in SCT_1, caused no change in activity. Replacement of phenylalanine[22] together with asparagine[15] by tyrosine and glutamic acid, again the acids of SCT_1, reduced the activity to half irrespective of the sequence of the two amino acids in positions 29 and 31 (analogues II and III). This last result is somewhat contradictory to the findings made with HCT, in which case the substitution of tyrosine 22 for phenylalanine increased the potency. Unfortunately, no singly substituted analogues with tyrosine 22 and glutamic acid[15] have been synthesized.

CONCLUSION

From the data shown it is evident that the introduction of one or more amino acids into calcitonins with low biological activity, e.g. HCT, in the position in which they occur in peptides of high activity, e.g. SCT_1, and vice versa, must not necessarily elicit the expected changes in potency, i.e. either an improvement or a decrease. Moreover, effects of single substitutions are most often not additive when combined.

REFERENCES

1. Keutmann, H.T., Parsons, J.A., Potts, J.T. & Schlueter, R.J.: Isolation and chemical properties of two calcitonins from salmon ultimobranchial glands. *J. biol. Chem.* **245**, 1491—1496 (1970)
2. Maier, R., Riniker, B. & Rittel, W.: Analogues of human calcitonin. I. Influence of modifications in amino-acid positions 29 and 31 on hypocalcaemic activity in the rat. *FEBS Letters* **48**, 68—71 (1974)
3. Maier, R., Kamber, B., Riniker, B. & Rittel, W.: Analogues of human calcitonin. II. Influence of modifications in amino-acid positions 1, 8 and 22 on hypocalcaemic activity in the rat. *Horm. Met. Res.* 7, 511—514 (1975)
4. Pless, J., Bauer, W., Bossert, H., Zehnder, K. & Guttmann, S.: Synthesis of highly active analogues of salmon calcitonin. In: *Endocrinology 1971, Proc. 3rd Intern. Symp.*, W. Heinemann Ltd., London, pp 67—70 (1972)
5. Sieber, P., Riniker, B., Brugger, M., Kamber, B. & Rittel, W.: Menschliches Calcitonin. VI. Die Synthese von Calcitonin M. *Helv. chim. Acta* **53**, 2135—2150 (1970)

Combined Diphosphonate and Calcitonin Therapy for Paget's Disease of Bone

D.J. HOSKING & O.L.M. BIJVOET

Paget's disease of bone is a relatively common condition of unknown aetiology characterized by an increase in bone turnover. Many forms of treatment have been advocated but at the present time only the calcitonins (2, 4, 7) and the diphosphonates (1, 3, 5, 11) offer practical forms of specific therapy.

The long-term administration of porcine or salmon calcitonin relieves pain and reduces elevated levels of serum alkaline phosphatase (AP) or urinary hydroxyproline (Hypro).

However, in a significant number of patients these indices of disease activity do not normalize but remain elevated or begin to rise towards pre-treatment levels (2).

This event is suggested to be due to antibody formation (6, 9) and represents the main disadvantage of calcitonin therapy and one which may not be entirely avoided by the use of human synthetic calcitonin (HCT).

The diphosphonate disodium etidronate (EHDP) produces a good clinical and biochemical response in Paget's disease but in high doses is usually associated with the production of defective mineralisation most prominent in the diseased bones (11, 12, 3). Although the clinical significance of this change is not certain it may represent a limitation to long-term diphosphonate therapy.

In an attempt to minimize these disadvantages we have treated Paget's disease of bone with combinations of EHDP and HCT in low dosage and in this paper wish to report our preliminary findings.

PATIENTS AND METHODS

A total of 34 patients have been studied comprising three groups:
I   Nineteen patients treated for 4—30 months with EHDP in doses of

Department of Endocrinology and Metabolism, University hospital, Leiden.

7.5 or 15.0 mg/kg/day. The results from this part of the study have been described in detail elsewhere (3).

II a EHDP in a dose of 15 mg/kg/day orally for 6 months followed by HCT (0.5 mg × 2/day sc) for 6 months (3 patients).

   b Doses as above but starting with HCT and being followed by EHDP (3 patients).

III   EHDP (7.5 mg/kg/day) together with HCT (0.5 mg sc daily): 5 patients.

In all cases the diagnosis of Paget's disease of bone was based upon biochemical, radiographic and histological criteria. Patients from groups II and III were assessed every 6 months under balance conditions where in addition to dietary and faecal analyses, calcium kinetics (^{47}Ca) bone histology, radiology and scanning were performed.

The methods used in this study are described in detail elsewhere (3).

RESULTS AND DISCUSSION

The results are summarized in Table I and those of group II (a) and III are also shown graphically in Fig. 1.

Although our experience with HCT is limited the findings are similar to those reported from larger series (4). Pain relief and an early reduction in AP and Hypro have been obtained although the latter was not maintained following 6 months continuous treatment (Table I).

Prior treatment with HCT does not seem to impair the responsiveness of bone to subsequent EHDP therapy in a dose of 15 mg/kg/day (group II b). Normalization of the still elevated AP and Hypro occurred during the 6 months of EHDP in the two patients in whom adequate data is available. At the present time it is not possible to assess the effect of prior EHDP therapy upon the response to calcitonin.

Pain relief begins in the early months of EHDP therapy and the biochemical response is predictable, sustained and dose dependent (3). The dose dependency of the effect is shown by the similar percentage reduction in AP and Hypro at 3 months with 15 mg/kg/day and at 6 months with 7.5 mg/kg/day (Table I). The results at the higher dose at 6 months are distorted because patients were withdrawn from the study once normalization of biochemical indices were achieved and the groups at 3 and 6 months are, therefore, not comparable. Those that are included were often near normal for much of the last 3 months and, therefore, the percentage reductions in AP and Hypro are small.

While the combination of low dose HCT and EHDP is equally effective in producing relief of pain, it is clear (Table I) that the biochemical response

Table I. Pretreatment values and percentage reduction in alkaline phosphatase and hydroxyproline.

Treatment Group (Dose/24 h)	Alk. Phos. (U Bessey)	Hydroxyproline (μmol/24 h)	n.	3 months Alk.phos. %	Hypro. %	n.	6 months Alk.phos. %	Hypro. %
EHDP (15 mg/kg)	263.75 ± 54.0	932.4 ± 233.5	10	46.8 ± 5.6 (2)	55.0 ± 8.7 (5)	6	63.5 ± 5.2 (2)	57.4 ± 14.8 (3)
(7.5 mg/kg)	263.0 ± 63.2	906.9 ± 172.1	9	22.0 ± 4.6 (1)	34.2 ± 8.4 (2)	8	45.7 ± 10.7 (3)	49.9 ± 11.2 (4)
HCT (0.5 mg × 2)	899.0 ± 718.5	2024.3 ± 1334.7	3	51.0 ± 5.5 (0)	45.8 ± 7.8 (0)	3	51.2 ± 10.1 (0)	29.9 ± 21.4 (0)
EHDP (7.5 mg/kg)+	572.36 ± 314.74	999.1 ± 308.8	5	69.5 ± 8.6 (1)	70.9 ± 4.2 (4)	5	83.6 ± 5.5 (3)	81.5 ± 1.9 (5)
HCT (0.5 mg × 1)	291.0 ± 20.73*			64.8 ± 8.5*				

Mean ± SEM
Figures ( ) represent numbers of patients achieving normal values
Excludes patient with initial alk.phos. = 1696 U/l
Normal Alk. Phos. $<$ 60 U/l; Hydroxyproline $<$ 300 μmol/24 h.

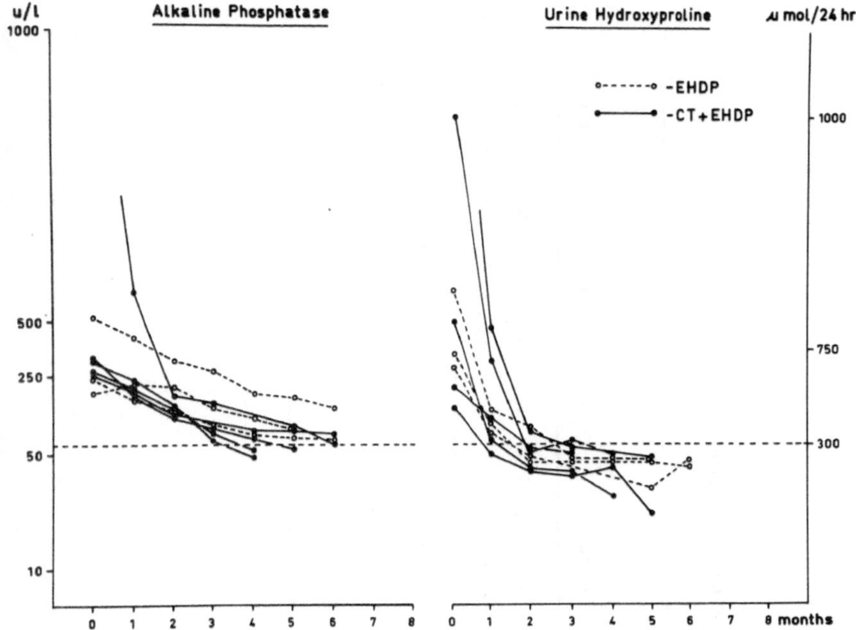

Figure 1. Changes in alkaline phosphatase and hydroxyproline with EHDP alone (15 mg/kg/day) or in combination (7.5 mg/kg/day) with calcitonin (0.5 mg/day sc).

obtained is superior to that achieved with either agent alone in higher dosage. Several patients had normalised by 3 months (Fig. 1) but all five were normal or near normal by 6 months.

Although the main pretreatment AP from this group is high due to a very elevated value from a patient with skull involvement, exclusion of this particular value results in a similar mean to the group given EHDP alone while leaving the per cent fall in AP with treatment almost unchanged.

Bone histology is only available from two patients treated with combined therapy for 6 months, one of whom had Paget's disease confined to the skull and spine. In the other case iliac crest bone biopsy showed marked reduction in the previously abnormal osteoblastic and osteoclastic activity, consistent with the normalization of AP and Hydro.

Unmineralized osteoid was not more extensive than that seen in the pretreatment biopsy and is in marked contrast to the severe mineralization defect found with high dose EHDP therapy (12, 3). It is not possible at present to assess whether the mineralization defect is more or less severe than that found with low dose EHDP (12) or HCT (8).

The mechanism of the enhanced action of EHDP and CT in combination is not clear. Since calcitonin is considered to act upon bone cells, particularly osteoclasts, while the action of the EHDP may be predominantly upon the mineral phase of bone, these combined effects may well be expected to be additive. This is clearly an oversimplification since the marked reduction of osteoblast and osteoclast number seen in EHDP treated Paget's disease indicate either a primary or secondary effect on bone cells.

However, differences do exist at a cellular level and are seen in tissue culture experiments where each agent has a different pattern of influence on intracellular enzymes (10).

Furthermore, there is an additive effect of HCT and EHDP in increasing the release of ^{45}Ca from prelabelled rat metaphysis (Lemkes *et al.*, unpublished).

In summary, therefore, the combination of low dose HCT and EHDP appears more effective than, and may avoid the disadvantages of, either agent given alone in high dosage.

REFERENCES

1. Altman, R.D., Johnston, C.C., Khairi, M.R.A., Wellman, H., Serafini, A.N. & Sankey, R.R.: Influence of disodium etidronate on clinical and laboratory manifestations of Paget's disease of bone (osteitis deformans). *New Engl. J. Med.* **289**, 1379–1384 (1973)

2. De Rose, J., Singer, F.R., Avramides, A., Flores, A., Dziadiw, R., Baker, R.K. & Wallach, S.: Response of Paget's disease to porcine and salmon calcitonins. Effects of long-term treatment. *Amer. J. Med.* **56**, 858–866 (1974)

3. De Vries, H.R. & Bijvoet, O.L.M.: Results of prolonged treatment of Paget's disease of bone with disodium ethane 1 hydroxy-1,1-diphosphonate (EHDP). *Neth. J. Med.* **17**, 281–298 (1974)

4. Greenberg, P.B., Doyle, F.H., Fisher, M.T., Hillyard, C.J., Joplin, G.F., Pennock, J. & MacIntyre, I.: Treatment of Paget's disease of bone with synthetic human calciton. *Amer. J. Med.* **56**, 867–870 (1974)

5. Gunçaga, J., Lauffenburger, T.H., Lentner, C.H., Dambacher, M.A., Haas, H.G., Fleisch, H. & Olah, A.J.: Diphosphonate treatment of Paget's disease of bone. A correlated metabolic, calcium kinetic and morphometric study. *Horm. Metab. Res.* **6**, 62–69 (1974)

6. Haddad, J.G. & Caldwell, J.G.: Calcitonin reistance: clinical and immunologic studies in subjects with Paget's disease of bone treated with porcine and salmon calcitonin. *J. Clin. Invest.* **51**, 3133–3141 (1972)

7. Kanis, J.A., Horn, D.B., Scott, R.D.M. & Strong, J.A.: Treatment of Paget's disease of bone with synthetic salmon calcitonin. *Brit. Med. J.* **III**, 727–731 (1973)

8. Rasmussen, H.R. & Bordier, P.: *The Physiological and Cellular Basis of Metabolic Bone Disease.* Williams and Wilkins, Baltimore, pp. 292–303, 1972

9. Rojanasathit, S., Rosenberg, E. & Haddad, J.G.: Paget's disease of bone: response to human calcitonin in patients resistant to salmon calcitonin. *Lancet* **II**, 1412–1415 (1974)

10. Russell, R.G.G., Mühlbauer, R., Williams, D.A., Reynolds, J., Morgan, B., Copp, H. & Fleisch, H.: Difference between the effects of calcitonin and of diphosphonates on bone resopriton induced by parathyroid hormone in tissue culture and on the osteoporosis of immobilization in rats. In: *Calcium, Parathyroid hormone and the calcitonins.* Talmage, R.V. and Munson, P.L. (eds.), Excerpta Medica International Congress, ser. No. 243, Amsterdam, pp. 440–441, 1972
11. Russell, R.G.G., Smith, R., Preston, C., Walton, R.J. & Woods, C.G.: Diphosphonates in Paget's disease. *Lancet* I, 894–898 (1974)
12. Smith, R., Russell, R.G.G., Bishop, M.C., Woods, C.G. & Bishop, M.: Paget's disease of bone. Experience with a diphosphonate (disodium etidronate) in treatment. *Quart. J. Med. NS* 42, 235–256 (1974)

CHAPTER VII
Phosphate and Pyrophosphate

Homeostasis of Inorganic Phosphate: An Introductory Review

H. FLEISCH, J.-P. BONJOUR & U. TROEHLER

Body phosphorus is to the greatest extent (80—90%) in the bone mineral as calcium phosphate. The rest is in the soft tissues and in the blood. In the latter, it is either in the red cells mostly as esters and phospholipids, or in the plasma as esters, lipids and inorganic phosphate. Only the homeostasis of inorganic phosphate will be considered in this presentation.

Plasma inorganic phosphate (P_i) depends on age, decreasing from about 11 mg per cent in the newborn to 3—4 mg per cent in the adult. The cause of this change is uncertain. Growth hormone might be involved. Man has the lowest values among the various animal species (8). Only about 60% of this phosphate is claimed to be free, the rest being bound either to proteins (17) or to other ions especially sodium, calcium and magnesium (18). Plasma P_i shows a definite circadian rythm, the maximum being attained in the evening and the minimum in the morning. Part of this rythm is due to food, some to PTH. However, neither of these factors explain the variation completely.

The plasma concentration is determined by the relative size of the various fluxes, namely net intestinal absorption, entry and exit into soft tissues and bone, renal handling and urinary excretion. It is believed that the renal handling, or more exactly the net tubular reabsorption is, in steady state and in healthy people, the most important parameter (2). In case of renal failure, however, the relative importance of the other fluxes increases. This explains why in renal failure thyroparathyroidectomy can actually lead to a decrease in plasma P_i and why in this disease dietary restriction of phosphate is so successful in lowering plasma P_i.

About 75% of the ingested phosphate is absorbed in the *intestine*, the net amount being a result of an absorption and a secretion process. The amount absorbed is closely dependent on the amount ingested, in

Department of Pathophysiology, University of Berne, Berne.

opposition to calcium where the amount absorbed is dependent only to a small extent on the dietary calcium. *In vitro* the absorption of P_i is highest in the duodenum, followed by the jejunum and ileum, and is lowest in the colon. *In vivo*, however, the absorption is largest in the ileum since the food stays longest in this part of the intestine. The intestinal transport is an active process and is increased by vitamin D (6). Various hormones such as gastrin, thyroxine, corticosteroids and growth hormone are thought to influence it. However, by far the most important factor is the amount of phosphate present in the diet.

In steady state the *urinary excretion* reflects the intestinal absorption and not the renal handling. Thus, attention has been devoted to other parameters to characterize renal handling. The clearance has been widely used. Since the clearance of P_i is dependent on glomerular filtration rate (GFR), the clearance ratio P_i/inulin is more accurate. Since this parameter is also dependent on plasma P_i, a correction factor has been introduced for the calculation of a value called the phosphate excretion index (11). Phosphate is reabsorbed, the reabsorption being a saturable process characterized by a maximum value (Tm) which can be evaluated using P_i infusion (12). However, P_i infusion is known to produce a rapid inhibition of PTH secretion (13), so that the values obtained will not necessarily reflect the actual PTH status of the patient. Since the Tm is also dependent on GFR, the value Tm/GFR has been used (2). It has been found to have an excellent correlation with plasma P_i and can be determined without P_i infusions (2).

In the *kidney* P_i is filtered at the glomerulus and is reabsorbed later in the tubule. Some P_i is also secreted (3). The difference between the filtered and the net reabsorbed mass is the amount excreted in the urine. The amount of phosphate filtered (the filtered load) is usually calculated as plasma P_i x GFR. This is only correct if the plasma P_i is completely ultrafiltrable, which is not true *in vitro* (17). Whether it is the case *in vivo* at the glomerular membrane is not known, although some micropunction studies at the glomerulus or very early proximal tubule suggest that P_i is 90—95% ultrafiltrable (5).

The largest part of the reabsorption occurs in the proximal tubule by means of an active process (15). Although it was previously believed that no reabsorption occurred in the distal nephron (14, 15), recent data in various laboratories including our own (1, 3) suggest strongly that this is not the case. As shown in Fig. 1 the fraction of P_i filtered which is not reabsorbed is lower in the final urine than in the distal tubule, suggesting that some reabsorption is occurring in the terminal nephron.
Whereas tubular secretion of P_i is a well known process in fish, amphibians, reptiles and birds, the possibility of a secretion of P_i in

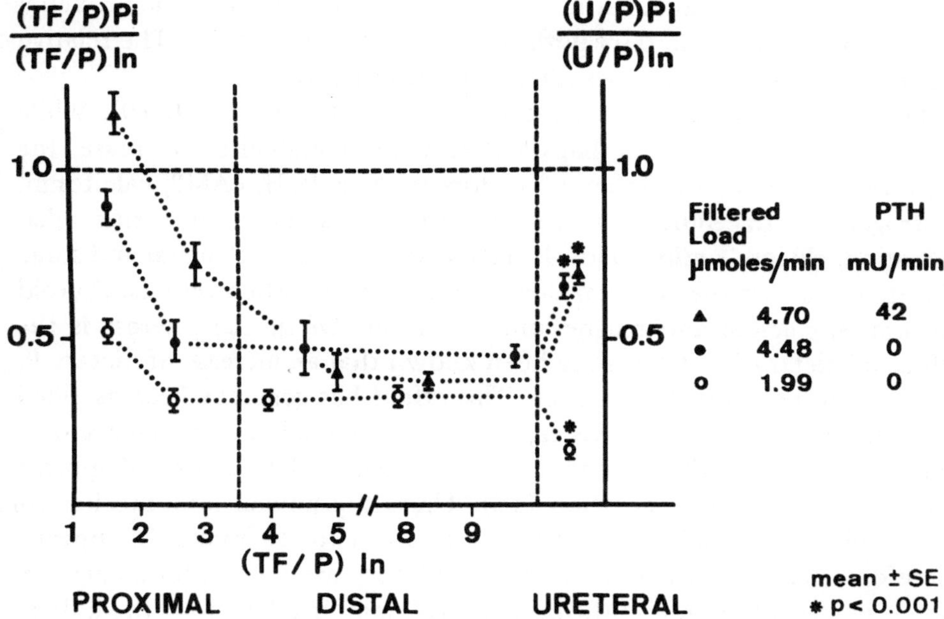

Figure 1. Micropuncture studies of P_i handling in the rat. ○ no P_i, no PTH; ● infusion of 6.66 μm/min P_i, no PTH; ▲ infusion of 6.66 μm/min P_i and 42 mU/min PTH. * Significance (p < 0.001) with late distal tubule. ○ shows a reabsorption in the terminal nephron; ● a secretion in the terminal nephron; ▲ a secretion in the terminal and also in the early proximal tubule. Adapted from Boudry *et al.* (3).

mammals has generally been negated. It is possible, however, that a blood — tubule P_i flux does exist, but that it is usually masked by the high reabsorption. Indeed, it has recently been reported that in X-linked hypophosphatemic rickets, where P_i reabsorption is drastically decreased, a clearance ratio P_i/inulin greater than 1 (that is, a net secretion) can be observed in over 50% of the cases (4). The refutation of secretion is often based on the presence of a Tm. However, Tm is not present in all animals. Thus, in the rat we found that when P_i is infused, reabsorption actually decreases at high plasma P_i values (16). Results in micropuncture studies depend upon the conditions. Thus, we found in rats (3) (Fig. 1) that while no secretion can be shown in normal rats, an increase in the non-reabsorbed fraction of P_i can be observed between the distal tubule and the terminal urine when P_i is infused. If PTH is administered i.v., entry is also seen in the proximal tubule since a recovered fraction above 1 is obtained. Furthermore, we found (3), in microperfusion of proximal tubules, an entry of P_i as shown by both an increase in cold phosphate and by a decrease in specific activity. Thus, it appears that there is a bidirectional tubular flux both in the proximal and in the terminal nephron. It, therefore, appears that renal handling of P_i is determined by the proximal and the distal tubule, and by both reabsorption and secretion. The relative role of these various processes remains to be evaluated.

The renal handling is influenced by many factors (7, 9, 10). While growth hormone and perhaps vitamin D metabolites increase the reabsorption, a large number of factors such as PTH, cAMP, calcitonin, oestrogens, extracellular volume expansion, vasodilatation, urine alkalinization, glucose, amino acids, PAH, bicarbonate, pentobartital and some diuretics decrease the reabsorption. The effect of corticosteroids, thyroid hormones, calcium and magnesium is unclear. Of special interest is the effect of dietary P_i. It has long been known that an increase of dietary P_i leads to an increased P_i clearance. This effect has generally been ascribed to a stimulation by PTH. Recent studies in our laboratory have shown that this effects is still present to a large extent in TPTX rats and appears within 3 days of the operation (16). Thus, the animal has a mechanism which allows it to adapt its urinary excretion to its intake. It appears, therefore, that while the homeostatic regulatory mechanism for changes in diet is in the intestine and the bone for calcium, it is largely in the kidney for P_i.

ACKNOWLEDGEMENTS

This work has been supported by the Swiss National Foundation for Scientific Research (3.121.73) and by the US Public Health Service (AM 07266).

REFERENCES

1. Amiel, C., Kuntziger, H. & Richet, G.: Micropuncture study of handling of phosphate by proximal and distal nephron in normal and parathyroidectomized rat. Evidence for distal reabsorption. *Pflügers Arch.* 317, 93—109 (1970)

2. Bijvoet, O.L.M. & Morgan, D.: The tubular reabsorption of phosphate in man. In: *Phosphate et Métabolisme Phosphocalcique, Int. Symp.*, Paris 1970, Hioco, D.J. (ed.), pp. 153—180, 1971

3. Boudry, J.-F., Troehler, U., Touabi, M., Fleisch, H. & Bonjour, J.-P.: Secretion of inorganic phosphate in the rat nephron. *Clin. Sci.*, (In Press)

4. Glorieux, F. & Scriver, C.R.: Loss of a parathyroid hormone-sensitive component of phosphate transport in X-linked hypophosphatemia. *Science* 175, 997—1000 (1972)

5. Harris, C.A., Baer, P.G., Chirito, E. & Dirks, J.H.: Composition of mammalian glomerular filtrate. *Am. J. Physiol.* 227, 972—976 (1974)

6. Harrison, H.E. & Harrison, H.C.: Intestinal transport of phosphate: Action of vitamin D, calcium, and potassium. *Am. J. Physiol.* 201, 1007—1012 (1961)

7. Knox, F.G., Schneider, E.G., Willis, L.R., Strandhoy, J.W. & Ott, C.E.: Site and control of phosphate reabsorption by the kidney. *Kidney Internat.* 3, 347–353 (1973)
8. Marshall, R.W. & Nordin, B.E.C.: The state of inorganic phosphate in plasma and its relation to other ions. In: *Phosphate et Métabolisme Phosphocalcique, Int. Symp.*, Paris 1970, Hioco, D.J. (ed.), pp. 127–138, 1971
9. Massry, S.G., Friedler, R.M. & Coburn, J.W.: Excretion of phosphate and calcium. Physiology of their renal handling and relation to clinical medicine. *Arch. Intern. Med.* 131, 828–859 (1973)
10. Mudge, G.H., Berndt, W.O. & Valtin, H.: Tubular transport of urea, glucose, phosphate, uric acid, sulfate, and thiosulfate. In: *Handbook of Physiology*, Orloff, J. and Berlinger, R.W. (eds.), American Physiological Society, Washington, pp. 587–652, 1973
11. Nordin, B.E.C. & Fraser, R.: Assessment of urinary phosphate excretion. *Lancet* (1960) I, 947–951
12. Pitts, R.F. & Alexander, R.S.: The renal reabsorptive mechanism for inorganic phosphate in normal and acidotic dogs. *Am. J. Physiol.* 142, 648–662 (1944)
13. Reiss, E., Canterbury, J.M., Bercovitz, M.A. & Kaplan, E.L.: The role of phosphate in the secretion of parathyroid hormone in man. *J. clin. Invest.* 49, 2146–2149 (1970)
14. Staum, B.B., Hamburger, R.J. & Goldberg, M.: Tracer microinjection study of renal tubular phosphate reabsorption in the rat. *J. clin. Invest.* 51, 2271–2276 (1972)
15. Strickler, J.C., Thompson, D.D., Klose, R.M. & Giebisch, G.: Micropuncture study of inorganic phosphate excretion in the rat. *J. clin. Invest.* 43, 1596–1607 (1964)
16. Troehler, U., Bonjour, J.-P. & Fleisch, H.: Inorganic phosphate metabolism: Renal adaption to the dietary intake in intact and thyroparathyroidectomized rats. Submitted.
17. Walser, M., Ford, M.J. & Butler, S.: Protein-binding of inorganic phosphate in plasma of normal subjects and patients with renal disease. *J. clin. Invest.* 39, 501–506 (1960)
18. Walser, M.: Ion Association. VI. Interactions between calcium, magnesium, inorganic phosphate, citrate and protein in normal human plasma. *J. clin. Invest.* 40, 723–730 (1961)

Calcium Uptake by Cultured Bone Cells: The Role of Phosphate, Calcitonin and 1,25-(OH)$_2$ D$_3$

A. HARELL, I. BINDERMAN & M. GUEZ

Hypocalcaemia produced by CT is assumed to result from its action on bone. Direct and indirect evidences suggest that, in addition to the inhibitory effect of CT on bone resorption, this hormone also increases bone formation. Previous studies in our laboratory have demonstrated that the extent of CT action on isolated bone cells is directly related to the concentration of calcium in the medium (3). We were able to show that, in response to CT, cultured bone cells accumulate calcium during the first two weeks of culture.

Talmage and co-workers proposed that CT is primarily a phosphate-regulating hormone (7). Raisz and Niemann found an additive inhibitory effect of CT and phosphate on resorption of cultured bone rudiments (4). Our present study was undertaken in order to determine the relationship between calcium and phosphorus accumulation in cultured bone cells and to study the possible dependence of CT and 1,25-(OH)$_2$ D$_3$ action on calcium transport, on intracellular phosphate concentration.

MATERIALS AND METHODS

Rat periosteum cells, derived from embryonic rat calvaria, cultured during two weeks, were used. The technique of culturing isolated periosteum cells was recently extensively described (2).

The culture medium was replaced by a less calcium Earle's B.S.S. supplemented with and without CT or 1,25-(OH)$_2$ D$_3$ or with both hormones and incubated in the same culture conditions for 45 min. Radioactive calcium (^{45}Ca in 0.3 mM CaCl$_2$) was then added to all treated and non-treated cells and incubated for an additional 45 min. At

Hard Tissues Unit, Department of Endocrinology Ichilov Medical Center and the "Sackler" School of Medicine, University of Tel Aviv.

the end of the experiment, the bathing solution was decanted and cells were washed three times with cold less calcium Earle's B.S.S. The cells were detached from the plates sonicated in bidistilled water and analyzed for calcium, phosphate and protein. Radioactive calcium was counted both in medium and cells. CT hormone was a gift from Armour and used in concentrations of 1 mU and 100 mU (SCT Batch K600—014 Cl, 5,000 MRC U/mg); 1,25-(OH)$_2$D$_3$ metabolite was prepared by Dr. S. Edelstein, Nutrition Laboratory, Ichilov Medical Center, Tel Aviv and kindly provided in concentration of 100 ng/ml. In our experiments, we have used this vitamin-D metabolite in a concentration of 1 ng/ml medium.

RESULTS

In our earlier studies, we have shown that cultured periosteum cells respond to CT hormone during the first two weeks in culture (1). From Fig. 1, it is obvious that calcium and phosphate accumulate in bone cells in an interdependent way.

Intracellular increase in phosphate enhances the calcium content in the cells as well. The addition of CT hormone activates accumulation of

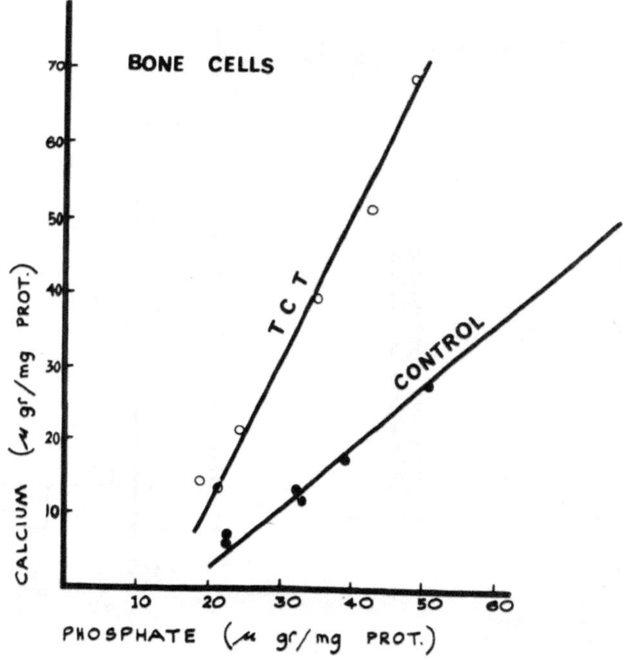

Fig. 1

334

calcium within the cells relatively to the intracellular phosphate concentration. In lower concentrations of phosphate, CT has a very limited and not significant effect on calcium accumulation in cells. However, an elevated intracellular concentration of phosphate enhanced signivicantly the intracellular content of calcium, leading to a change in Ca/P ratio, from 0.43 in the untreated, to 1.14 in the CT treated cells (Fig. 1). The data in Fig. 1 were accumulated from different experiments in which culture conditions determined the various phosphate and calcium concentrations in the cells.

When cultured bone cells were treated with CT in low calcium and phosphate in the medium, as described above, a rise in intracellular calcium was detected. However, a concentration of 100 mU CT was less effective on intracellular calcium content compared to cells treated with 1 mU CT. Also, the specific activity of calcium measured as $^{45}Ca/^{40}Ca$

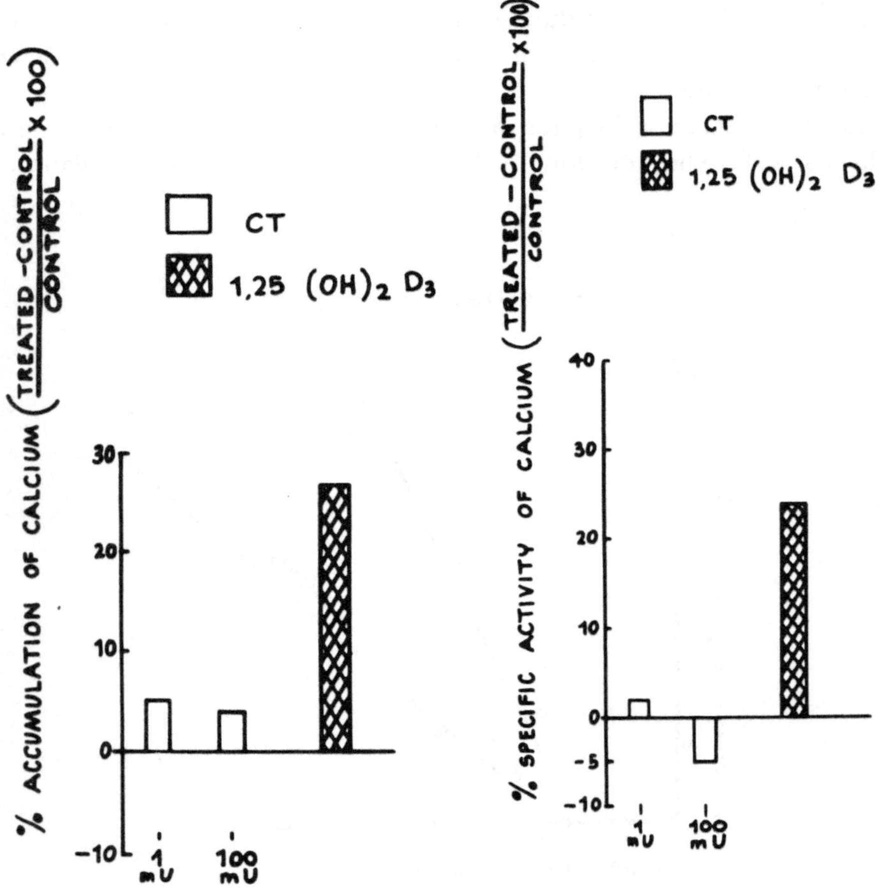

Figs. 2 and 3.

ratio within the cells was unaffected by the low CT concentration (1 mU) but a decrease of about 5 per cent in the specific activity was measured in response to 100 mU CT treatment.

These data suggest a possible inhibition in the per cent exchange of calcium and intracellular retention of calcium in cultured bone cells (Figs. 2 and 3). They support our previous contention, which also fits the model of Borle that calcitonin causes a sequestering of Ca in the intracellular pool, possibly in the mitochondria.

Intracellular inorganic phosphate content was uninfluenced by 1 mU CT but diminished by 16 per cent in response to 100 mU CT (Fig. 4). This may be due to the fact that only inorganic phosphorus was measured and it just may be that CT causes a shift of intracellular phosphate, making it unavailable for determination by the methods used. We are now investigating whether this shift is due to the formation of other

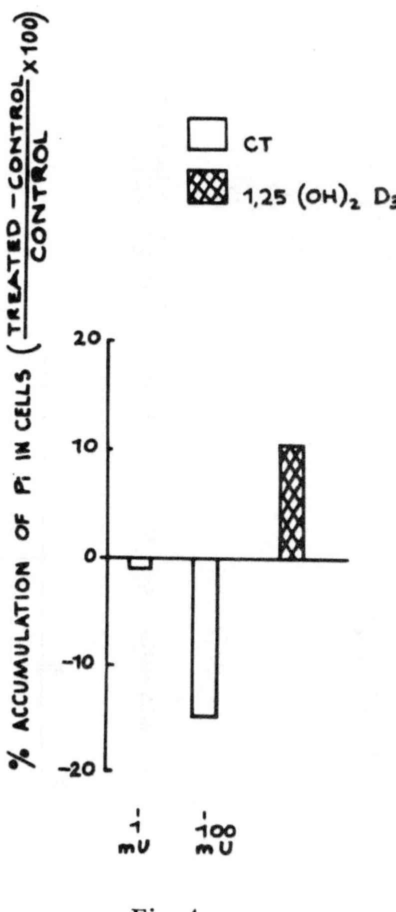

Fig. 4

phosphorus-complexes.

The subsequent experiments in which cultured bone cells were treated with $1,25\text{-}(OH)_2 D_3$ suggest strongly that this hormone is involved in intracellular calcium and phosphate regulations. Figs. 2 and 4 show that cultured bone cells treated with $1,25\text{-}(OH)_2 D_3$ alone significantly increase their intracellular content of calcium and phosphate. Also, specific activity of calcium ($^{45}Ca/^{40}Ca$) in the cells increase and uptake and per cent exchange of calcium in the cells rise (Fig. 3). The regulation of calcium and phosphate in cultured bone cells, when treated by CT and $1,25\text{-}(OH)_2 D_3$ simultaneously, is possibly predominated by the $1,25\text{-}(OH)_2 D_3$ hormone effect. Calcium and phosphate within the cells increase in response to CT and $1,25\text{-}(OH)_2 D_3$ and specific activity of calcium ($^{45}Ca/^{40}Ca$) is significantly elevated (Figs. 5 and 6).

The results presented indicate that, while CT hormone regulates mainly

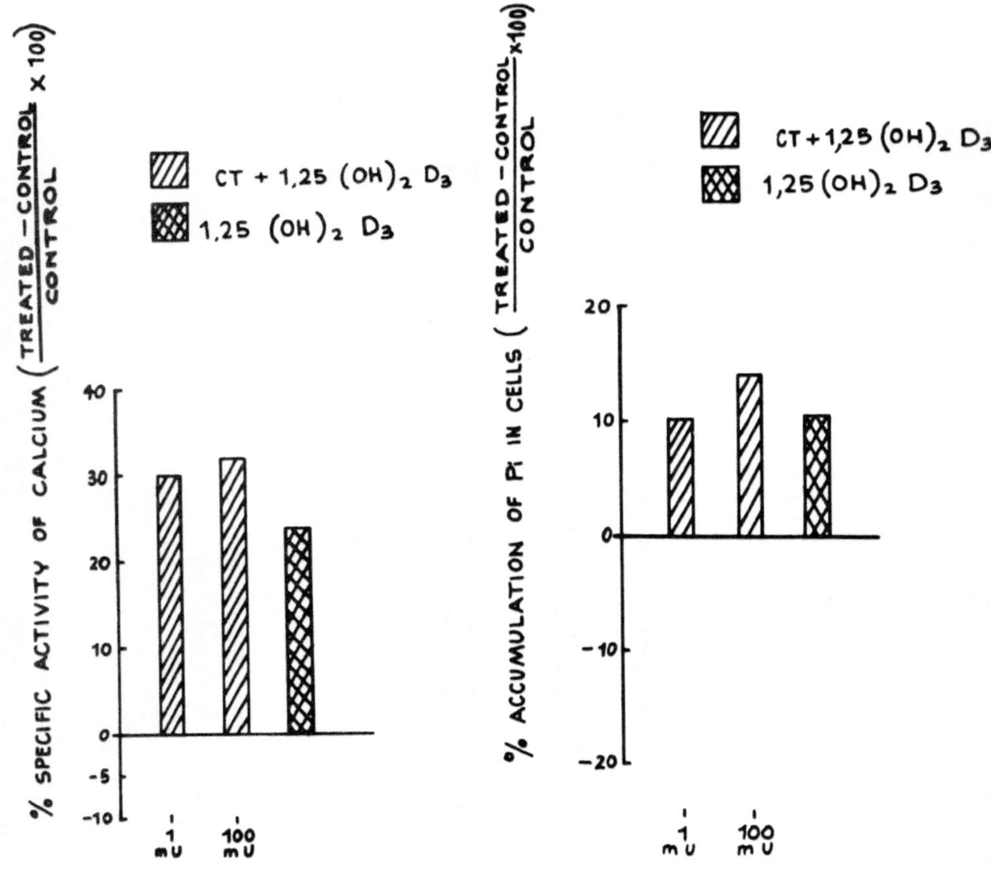

Fig. 5 and 6.

intracellular shifts of calcium and probably is activated by elevated concentrations of phosphate in the cells, $1,25\text{-}(OH)_2 D_3$ hormone increase calcium and phosphate turnover in bone cells and, therefore, may stimulate CT action in cultured bone cells.

DISCUSSION

The mode of action of CT at the cellular level is still under discussion. The expectation that CT is at any level of its actity antagonistic to PTH was not confirmed. If both PTH and CT enhance the intracellular calcium content and if the calcium content in bone cells is a major regulator of cell function as suggested by Rasmussen and Bordier (6), then one may expect that their mode of action differs on the subcellular level. Our data are consistent with a model in which CT affects calcium retention within the bone cells. However, when more intracellular calcium is available, phosphate efflux is inhibited as can be seen from the simulating actions of CT and $1,25\text{-}(OH)_2 D_3$ hormones.

Until recently, it was uncertain whether the vitamin-D (or its metabolites) affect the regulation of calcium or phosphate transportation in bone cells. Raisz and co-workers (5) showed that $1,25\text{-}(OH)_2 D_3$ hormone enhances bone resorption in organ culture. Our data show that cultured bone cells respond significantly to $1,25\text{-}(OH)_2 D_3$ in short-term experiments by increasing the uptake and per cent exchange of calcium and by increasing the intracellular content of calcium and phosphate. It seems, therefore, that $1,25\text{-}(OH)_2 D_3$ hormone has a direct effect on bone cells by the enhancement of calcium accumulation to much greater extent than CT. This short-term effect of $1,25\text{-}(OH)_2 D_3$, similar to the effect of aldosterone on the natrium/kalium pump, supports the hypothesis that calcium transport may act as messenger of this hormone's action.

Although CT itself decreased phosphate content in bone cells under the experimental condition, the addition of $1,25\text{-}(OH)_2 D_3$ hormone reverted this effect, probably because of the effect of $1,25\text{-}(OH)_2 D_3$ on the increase in intracellular phosphate in the cultured bone cells.

If the model system of cultured bone cells represents, to some extent, active osteoblasts and osteocytes, it would be possible to hypothesize that CT and $1,25\text{-}(OH)_2 D_3$ influences the regulation of calcium and phosphate transport in the cells. $1,25\text{-}(OH)_2 D_3$ hormone might stimulate calcium and phosphate uptake by bone cells and the CT effect on calcium uptake might be mainly dependent on the intracellular concentration of phosphate. The CT action on uptake of calcium and phosphate seems to depend on the rate of turnover of calcium in bone cells.

CONCLUSION

Our data present evidence that in cultured bone cells
1) A rise of phosphate concentration in the medium and the subsequent rise of the *intra*cellular phosphate content stimulate calcium and phosphate uptake by bone cells and enhance the calcitonin effect, which seems to be dependent on the intracellular concentration of phosphate.
2) Calcitonin predominantly influences the intracellular shift of calcium.
3) $1,25\text{-}(OH)_2 D_3$ has a direct effect on cultured bone cells: it increases the calcium and phosphate turnover and, therefore, stimulates the calcitonin action.

ACKNOWLEDGEMENT

This study was supported by Grant No. 322 of the Binational Israel-U.S.A. Foundation and by grant of the Israel Ministry of Health.
We thank Ms. Aliza Treves for her excellent technical assistance.

REFERENCES

1. Binderman, I., Duksin, D., Sachs, L. & Harell, A.: Thyrocalcitonin and accumulation of calcium by cultured bone cells. *Abstr. IVth Internat. Congr. Endocr.* **256**, 181 (1972)

2. Binderman, I., Duksin, D., Harell, A., Katzir, E. & Sachs, L.: Formation of bone tissue in culture from isolated bone cells. *J. cell. Biol.* **61**, 427–439 (1974)

3. Harell, A., Binderman, I. & Rodan, G.A.: The effect of calcium concentration on calcium uptake by bone cells treated with thyrocalcitonin (TCT) hormone. *Endocrinology* **92**, 550–555 (1973)

4. Raisz, L.G. & Niemann, I.: Effect of phosphate, calcium and magnesium on bone resorption and hormonal responses in tissue culture. *Endocrinology* **85**, 446–452 (1969)

5. Raisz, L.G., Trummel, C.L., Holick, M.F. & DeLuca, H.F.: 1,25-dihydroxycholecalciferol: A potent stimulator of bone resorption in tissue culture. *Science* **175**, 768–769 (1972)

6. Rasmussen, H. & Bordier, P.: Cell function: Ionic and hormonal control. In: *The Physiological and Cellular Basis of Metabolic Bone Diseases*, Rasmussen, H. and Bordier, P. (eds.), Williams and Wilkins, Baltimore, p. 169, 1974

7. Talmage, R.V., Anderson, J.J.B. & Cooper, C.W.: The influence of calcitonins on the disappearance of radiocalcium and radiophosphorus from plasma. *Endocrinology* **90**, 1185–1191 (1972)

Effects of a Diphosphonate (disodium etidronate; EHDP) on Phosphate Metabolism in Paget's Disease of Bone, Primary Hyperparathyroidism and Type I Hypophosphataemic Rickets

R.J. WALTON, R. SMITH & R.G.G. RUSSELL

Disodium etidronate (disodium ethane-1-hydroxy-1,1-diphosphonate; EHDP) is structurally related to inorganic pyrophosphate but contains a P-C-P bond and is therefore resistant to hydrolytic enzymes. Because of its ability to slow bone turnover in experimental animals (9), it is under investigation in man as a treatment for Paget's disease of bone (1).

Two side effects of EHDP have been reported. Firstly, there is an increase in unmineralised osteoid, probably related to inhibition of the growth and dissolution of apatite crystals as has been demonstrated *in vitro* (7, 8). Secondly, EHDP can increase plasma inorganic phosphate, an effect seen only in man and not in other species so far studied. Both of these effects occur quite consistently at a daily oral dose of EHDP of 80 μmol (20 mg) kg^{-1} but are less marked or absent at lower doses. The increase in plasma phosphate is real and not due to EHDP being measured as inorganic phosphate or interfering in some way with its assay (13). There is no increase in the urinary excretion rate of phosphate and no reduction in the ultrafiltrable fraction of plasma phosphate (10). Thus, the clearance of plasma phosphate by the kidney is reduced during EHDP administration, suggesting a renal mechanism for the hyperphosphataemia.
The renal handling of phosphate has been examined directly in six patients with Paget's disease off and on EHDP (13) by measuring urine phosphate excretion rates and simultaneous plasma phosphate concentrations during intravenous infusion of phosphate (2). The results of this study are summarised in Fig. 1. Glomerular filtration rate (GFR; measured as inulin clearance) was not significantly changed, but in each patient EHDP induced a large increase in the maximum rate of renal tubular reabsorption of phosphate (Tm_p). The ratio Tm_p/GFR, which is probably

Metabolic Unit, Nuffield Departments of Medicine and Orthopaedic Surgery, University of Oxford.

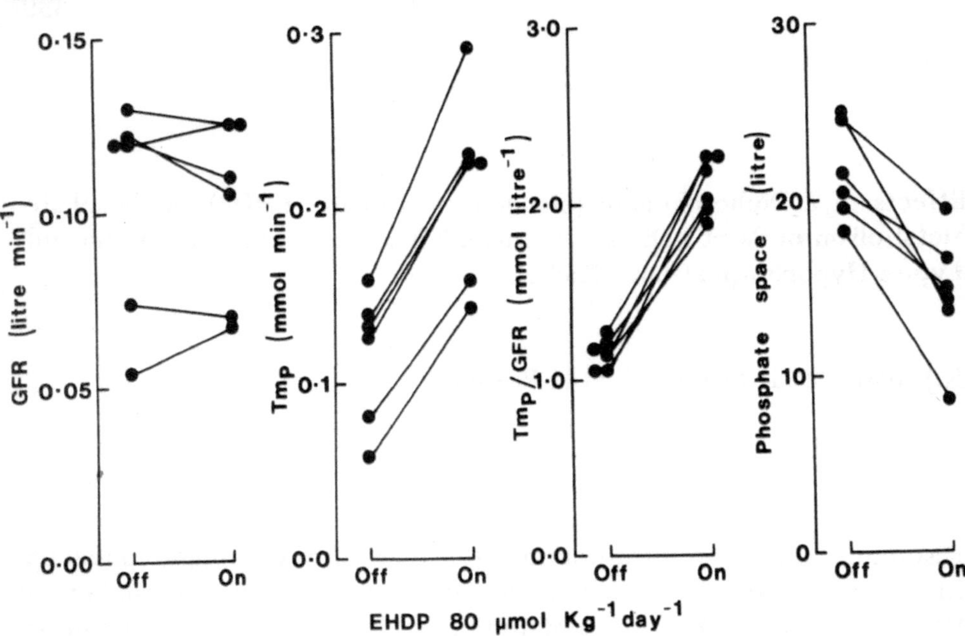

Figure 1. Effect of EHDP on the renal and extrarenal handling of phosphate. Results taken from Walton, Russell and Smith (13) and obtained in 6 patients with Paget's disease of bone, infused with phosphate and inulin over two h, both off and on EHDP. For explanation see text.

the best index of renal tubular reabsorption of phosphate since it is independent of filtered load (3), had a mean value of 1.15 mmol litre^{-1} (SEM 0.04) off EHDP compared with 2.10 mmol litre^{-1} (SEM 0.07) on EHDP.

Thus, EHDP appears to re-set net tubular reabsorption of phosphate at a higher level. This could represent either an increased flux from tubular lumen to capillary or a reduced flux in the opposite direction. Recent work suggests that both inorganic phosphate (5) and EHDP (12) may be secreted by the rat renal tubule. Since inorganic phosphate and EHDP are chemically related, they might be expected to compete for transport mechanisms. If EHDP induces hyperphosphataemia by competing with phosphate for a common renal tubular secretory pathway, the competition is unlikely to be a simple direct one since following intravenous injection of EHDP, Tm_p/GFR does not begin to rise for two to three hours (unpublished observations).

When the drug is given by mouth at a daily dose of 80 μmol kg^{-1}, there is a small but significant increase in fasting plasma phosphate and Tm_p/GFR (measured by the method of Bijvoet and Morgan (4)) the day after the first dose. The increase is then progressive over two to three weeks, after which Tm_p/GFR and plasma phosphate remain more or less constant whilst treatment is continued (13).

Fig. 2. compares the rise in Tm_p/GFR during the first two weeks of EHDP

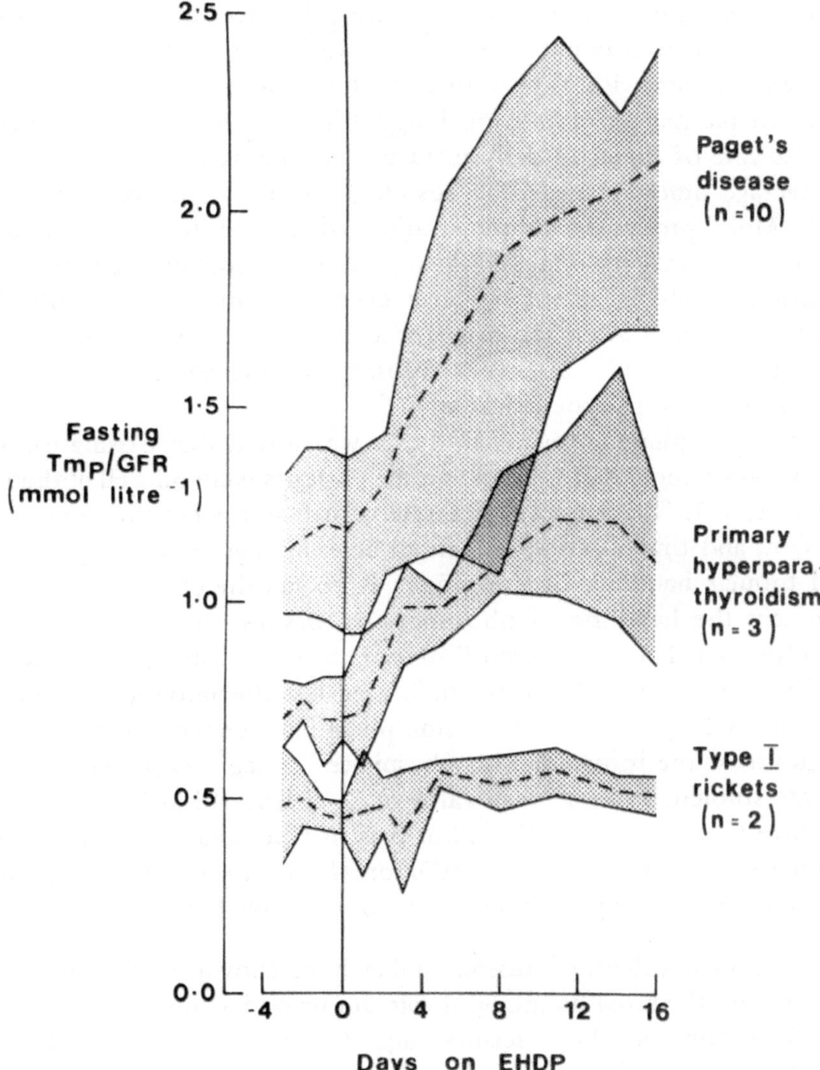

Figure 2. Changes in Tm_p/GFR during first 16 days of EHDP administration. Results in 10 patients (3 female; 7 male) with Paget's disease of bone, 3 (2 female, 1 male) with primary hyperparathyroidism and 2 (both female) with type I hypophosphataemic rickets. EHDP was given as a once daily oral dose of 80 μmol kg^{-1} to all patients, except one of the patients with type I rickets, who was given an equivalent daily intravenous dose of 5 μmol kg^{-1}. Fasting Tm_p/GFR was calculated by the method of Bijvoet and Morgan (4) and for each group of patients, the limits of the dotted areas denote the ranges, and the broken lines the mean values observed.

administration in ten patients with Paget's disease, three with surgically proven primary hyperparathyroidism, and two with type I (6), hypophosphataemic, vitamin D resistant rickets. The increase in Tm_p/GFR in primary hyperparathyroidism was marked (although less so than in Paget's disease), indicating that the drug can exert its effect on the renal tubule in the face of high circulating levels of parathyroid hormone. However, in the two patients with type I rickets, this short course of EHDP had no effect on plasma phosphate or Tm_p/GFR, despite the drug having been given to one of them by daily intravenous injection (the dose given being the average amount of EHDP absorbed from an oral dose of 80 μmol kg^{-1}). More prolonged administration of EHDP to detect a possible delayed increase in Tm_p/GFR was not considered justifiable since alterations in bone histology have been seen in patients with Paget's disease who have taken the drug for as little as two weeks. It is possible that EHDP normally affects a reabsorptive mechanism for phosphate which is deficient in type I rickets.

In the phosphate infusion study, it was noted that plasma phosphate concentration increased more when the patients were on EHDP than when they were off. The amount and rate of infusion were the same on each occasion, and this observation cannot be explained in terms of the altered renal tubular handling of phosphate. It seems therefore that EHDP also influences the handling of phosphate by tissues other than the kidney. This effect has been measured ("phosphate space" in Fig. 1), taken to be equal to the amount of phosphate infused less the amount excreted in the urine divided by the rise in plasma phosphate concentration during the two hours of the infusion. This phosphate "space" fell in each of the six patients studied, from a mean value of 21.7 litre (SEM 1.3) off EHDP to 14.8 litre (SEM 1.6) on EHDP, but it is not possible to say whether this represents a general effect of EHDP on the transport of phosphate, or a specific effect in one particular tissue, for example bone.

By contrast, calcium infusion studies have shown no significant effect of EHDP on the renal handling of calcium or on its volume of distribution (14). Not only do these results suggest a specific effect of EHDP on phosphate metabolism, but also provide evidence that EHDP does not produce hyperphosphataemia by interfering with the secretion or renal action of parathyroid hormone. Circulating levels of parathyroid hormone, calcitonin, growth hormone, cortisol and thyroid hormones are unchanged by EHDP (Walton, Russell, Woodhead, Heynen, Alberti and McVitie, unpublished observations).

ACKNOWLEDGEMENTS

This work has been supported by the Proctor & Gamble Company, the Wellcome Trust and the National Fund for Research into Crippling Diseases. R.J.W. holds the Mary Goodger Research Scholarship.

REFERENCES

1. Altman, R.D., Johnson, C.C., Khairi, M.R.A., Wellman, H., Serafini, A.W. & Sankey, R.R.: Influence of disodium etidronate (EHDP^{TM}) on clinical and laboratory manifestations of Paget's disease of bone (osteitis deformans). *New Engl. J. Med.* **289**, 1379–1384 (1973)

2. Anderson, J.: A method for estimating Tm for phosphorus in man. *J. Physiol. (Lond.)* **130**, 268–277 (1955)

3. Bijvoet, O.L.M.: Relation of plasma phosphate concentration to renal tubular reabsorption of phosphate. *Clin. Sci.* **37**, 23–36 (1969)

4. Bijvoet, O.L.M. & Morgan, D.B.: The tubular reabsorption of phosphate in man. In: *Phosphate et Métabolisme Phosphocalcique. Regulation Normale et Aspects Physiopathologiques*, Hioco, D.J., (ed.), Sandoz, Paris, pp 153–179 (1971)

5. Boudry, J.F., Troehler, U., Touabi, M., Fleisch, H. & Bonjour, J-P.: Evidence for renal tubular secretion of inorganic phosphate in the rat. *Clin. Sci. Molec. Med.* **48**, 475–489 (1975)

6. Dent, C.E.: Rickets and osteomalacia from renal tubule defects. *J. Bone Jt. Surg.* **34B**, 266–274 (1952)

7. Fleisch, H., Russell, R.G.G. & Francis, M.D.: Diphosphonates inhibit hydroxy-apatite dissolution *in vitro* and bone resorption in tissue culture and *in vivo*. *Science* **165**, 1262–1264 (1969)

8. Francis, M.D., Russell, R.G.G. & Fleisch, H.: Diphosphonates inhibit formation of calcium phosphate crystals *in vitro* and pathological calcification *in vivo*. *Science* **165**, 1264–1266 (1969)

9. Gasser, A.B., Morgan, D.B., Fleisch, H. & Richelle, L.J.: The influence of two diphosphonates on calcium metabolism in the rat. *Clin. Sci.* **43**, 31–45 (1972)

10. Recker, R.R., Hassing, G.S., Lau, J.R. & Saville, P.D.: The hyperphosphataemic effect of disodium ethane-1-hydroxy-1,1-diphosphonate (EHDP^{TM}): renal handling of phosphorus and the renal response to parathyroid hormone. *J. Lab. Clin. Med.* **81**, 258–266 (1973)

11. Russell, R.G.G., Smith, R., Preston, C., Walton, R.J. & Woods, C.G.: Diphosphonates in Paget's disease. *Lancet* (1974) **I**, 894–898

12. Troehler, U., Bonjour, J.P. & Fleisch, H.: Renal secretion of diphosphonates in rats. *Kidney Internat.* In press

13. Walton, R.J., Russell, R.G.G. & Smith, R.: Changes in the renal and extrarenal handling of phosphate induced by disodium etidronate (EHDP) in man. *Clin. Sci. Molec. Med.* **49**, 45–56 (1975)

14. Walton, R.J., Russell, R.G.G., Smith, R. & Warner, G.T.: Some effects of disodium ethane-1-hydroxy-1, 1-diphosphonate (EHDP) on calcium metabolism in man. *Clin. Sci. Molec. Med.* **47**, 13P (1974)

The Role of Matrix Vesicles in Calcification

R. FELIX & H. FLEISCH

The presence of matrix vesicles in calcifying tissue such as the epiphyseal growth plate and the osteoid was described by Bonucci (4) and Anderson (3). Since initial calcium phosphate crystal deposition seemed to be morphologically related to these organelles, it was proposed that they are involved in the calcification process. The mechanism, however, is still unknown. A large amount of alkaline phosphatase, pyrophosphatase and ATPase activity (1), which probably arises from the same enzyme molecule, is associated with these structures, and it was suggested (2) that the ATPase activity might be the important one in the active transport of calcium into the vesicles. Another possibility is that the pyrophosphatase activity of matrix vesicles is important since PP_i, an inhibitor of calcification (6), would be destroyed. It is also possible that the matrix vesicles stimulate calcification through a nucleation process and thus their structural composition, and not the enzyme activities, would be important (7, 8). In this work we have studied these various possible mechanisms using isolated vesicles.

METHODS

The matrix vesicles were isolated according to the method of Ali *et al.* (1) as modified by R.E. Wuthier (personal communication) from the epiphyseal growth plate of 7 weeks old Leghorn chickens, using collagenase and hyaluronidase for digestion of the tissue. The vesicle fraction was separated from cells and cell debris by centrifugation. For studying Ca transport the technique of Ali and Evans (2) was used. Vesicles were incubated *in vitro* with 5 mM ATP in the presence of ^{45}Ca. The solution was then filtered through a Millipore filter and the amount of ^{45}Ca retained on the filter was measured. Pyrophosphatase and ATPase

Department of Pathophysiology, University of Berne, Berne.

activity of the vesicle bound enzyme were studied as described earlier (5). To study the nucleation activity of matrix vesicles, they were incubated for 19 h at 37°C in solutions buffered at pH 7.6 with 10 mM HEPES and containing an increasing $Ca \times P_i$ concentration, according to a modified method of Fleisch and Neuman (6), in order to determine the minimum $Ca \times P_i$ product necessary for precipitation.

RESULTS AND DISCUSSION

Ca transport

As did Ali and Evans (2), we found an increasing amount of ^{45}Ca retained on the filters when an increasing amount of vesicles were incubated. This has been interpreted (2) as a Ca transport, since no uptake of ^{45}Ca was found when the vesicles had been heated previously. However, we also found an increasing amount of ^{45}Ca on the filters when an increasing amount of solubilized intestinal alkaline phosphatase was incubated with the heated vesicles. This strongly suggests that the increase of ^{45}Ca on the filter is not due to Ca transport through a membrane bound ATPase. An enzymatic hydrolysis of ATP (leading to a high concentration of P_i so that calcium phosphate precipitates) is more likely. Indeed, a $Ca \times P_i$ product up to 16 $(mM)^2$ was obtained. Furthermore, when the Ca^{2+} in the medium was decreased to 0.3 mM, a concentration at which no calcium phosphate should precipitate, no significant amount of ^{45}Ca on the filter could be observed. Thus, we did not find an active Ca transport into matrix vesicles, the earlier described results being probably due to a precipitation of ^{45}Ca.

Pyrophosphatase and ATPase activity of matrix vesicles

The properties of the pyrophosphatase and ATPase activity bound onto matrix vesicles were found to be very similar to the ones earlier found for the pyrophosphatase and ATPase activity of solubilized and purified calf bone alkaline phosphatase (5). This suggests that it is not important whether the enzyme is bound onto the membrane or free in solution. Of special interest is that when the substrate concentration was reduced from 2 mM PP_i to 3.5 μM, the plasma concentration (9), the pH optimum shifted from 8.5 to 7.0, into the physiological range. Thus, conditions would be favourable for the alkaline phosphatase of matrix vesicles to act as a pyrophosphatase and destroy PP_i, an inhibitor of calcium phosphate precipitation (6). The ATPase was activated by Mg^{2+}, the optimal concentration being 5 mM at pH 7.5; higher concentrations inhibited slightly. Addition of Ca^{2+} at the optimal Mg^{2+} concentration did not stimulate the enzyme, but Ca^{2+} activated the ATPase in the absence of Mg^{2+} or at suboptimal Mg^{2+} concentrations, although not as efficiently as Mg^{2+}. The ATPase of matrix vesicles is, therefore, not a Ca-ATPase in the sense that at the optimal Mg^{2+} concentration Ca^{2+} does not increase the activity further.

Table 1

MINIMAL $Ca \times P_i$ CONCENTRATION FOR CRYSTAL FORMATION

TREATMENT	$Ca \times P_i$ $(mg\%)^2$
No vesicles	70
Not treated vesicles	30
5xfrozen and thawed	30
5xfrozen and thawed at pH 6.0	40
5xfrozen and thawed at pH 6.0 + acetone treatment	70

Nucleation by matrix vesicles
As seen in Table I, addition of matrix vesicles reduced the minimal $Ca \times P_i$ product necessary for precipitation from 70 $(mg\%)^2$ (absence of matrix vesicles) to 30 $(mg\%)^2$. When the vesicles were frozen and thawed at pH

Table 2

Ca AND P CONTENT OF MATRIX VESICLES PER GRAM OF CARTILAGE $\pm$ SE (n)

Treatment	Ca	P	molar ratio
	μg	μg	
pH 7.4	3.11 ± 0.11 (3)	1.47 ± 0.25 (3)	1.66
pH 6.0	0.34 ± 0.2 (3)	0.28 ± 0.12 (3)	0.93

6.0 the nucleation activity was reduced. This treatment at pH 6.0 should dissolve calcium phosphate, which might be present in the preparation. As seen in Table II the Ca and P content was decreased by the pH 6.0 treatment. The significant nucleation activity left after pH 6.0 treatment was destroyed by acetone treatment, which should dissolve lipids and might also destroy the membrane structure. These results suggest that the matrix vesicles are able to stimulate the calcium phosphate precipitation in the absence of any hydrolysable substrate. A large part of this nucleation activity seems to originate from calcium phosphate present in the preparation, however, a significant part originates from the vesicles themselves.

SUMMARY

Vesicles isolated from epiphyseal chick cartilage have been studied *in vitro*. They did not actively transport calcium. The alkaline phosphatase, pyrophosphatase and ATPase activities are similar whether solubilized or membrane bound. The ATPase is not activated by calcium. The vesicles can induce precipitation of calcium phosphate, this nucleation activity being partly due to the vesicles themselves and partly to their calcium phosphate content.

ACKNOWLEDGEMENTS

We are very grateful to Dr. R. Schenk and Mr. W. Herrmann of the Institute of Anatomy, University of Berne, for making electromicrographs. This work has been supported by the Swiss National Foundation for Scientific Research (grant No. 3.121.73) and by the US Public Health Service (grant No. AM 07266).

REFERENCES

1. Ali, S.Y., Sajdera, S.W. & Anderson, H.C.: Isolation and characterisation of calcifying matrix vesicles from epiphyseal cartilage. *Proc. nat. Acad. Sci. (Wash.)* **67**, 1513–1520 (1970)
2. Ali, S.Y. & Evans, L.: The uptake of (^{45}Ca) calcium ions by matrix vesicles isolated from calcifying cartilage. *Biochem. J.* **134**, 647–650 (1973)
3. Anderson, H.C.: Calcium-accumulating vesicles in the intercellular matrix of bone. In: Hard Tissue Growth, Repair and Remineralization. *Ciba Foundation Symposium* II (New Series), Elsevier, Excerpta Medica, North-Holland, London, New York, pp. 213–246, 1973
4. Bonucci, E.: Fine structure and histochemistry of "calcifying globules" in epiphyseal cartilage. *Z. Zellforsch.* **103**, 192–217 (1970)
5. Felix, R. & Fleisch, H.: The pyrophosphatase and (Ca^{2+}-Mg^{2+})-ATPase of purified calf bone alkaline phosphatase. *Biochim. Biophys. Acta* **350**, 84–94 (1974)

6. Fleisch, H. & Neuman, W.F.: Mechanisms of calcification: role of collagen, polyphosphates and phosphatase. *Amer. J. Physiol.* **200**, 1296–1300 (1961)
7. Fleisch, H., Felix, R., Hansen, T. & Schenk, R.: Role of organic matrix in calcification. In: *Extracellular Matrix Influences on Gene Expression*, Slavkin, H.C. & Greulich, R.C. (eds.), Acad. Press, San Francisco, London, pp. 707–711, 1975
8. Howell, D.S., Madruga, J. & Pita, J.C.: Evidence for an organic nucleational agent of calcium phosphate mineral forms in endochondral plates. In: *Extracellular Matrix Influences on Gene Expression*, Slavkin, H.C. & Greulich, R.C. (eds.), Acad. Press, San Francisco, London, pp. 701–706, 1975
9. Russell, R.G.G., Bisaz, S., Donath, A., Morgan, D.B. & Fleisch, H.: Inorganic pyrophosphate in plasma in normal persons and in patients with hypophosphatasia, osteogenesis imperfecta, and other disorders in bone. *Clin. Invest.* **50**, 961–969 (1971)

Extrusion of Pyrophosphate into Extracellular Media by Osteoarthritic Cartilage Incubates

D.S. HOWELL, O. MUNIZ, J.C. PITA & J.E. ENIS

Information concerning fundamental biochemical defects which would explain the deposition of the mineral phase, calcium pyrophosphate dihydrate (CaPPi) (7) found in articular cartilage of patients with chondrocalcinosis is meager. McCarty has proposed that the articular cartilage is the most likely site of initial mineral deposition (8). This view with minor reservations implicates some error in local cartilage metabolism specifically which predispose it to deposit calcium pyrophosphate (CaPPi) — even when hydroxyapatite in the same person is deposited ectopically in the aorta or other non-articular sites (7).

In one recent study we found a high concentration ratio of PPi synovial fluid/plasma in patients with chondrocalcinosis as shown earlier by others (8, 9), but unexpectedly a similar finding was present in osteoarthritis patients (2); this result has been confirmed by Silcox (10).

Also, in a recent report, we observed that incubates of articular cartilage from young rabbits, but not mature rabbits as well as growth plates cartilage released PPi into incubation media during a 4 h period. Control rabbit ear cartilage and synovial membrane elaborated negligible amounts of PPi. The PPi was shown to be non-dialyzable (3).

An explanation of these phenomena might bring insight concerning peculiarities of cartilage metabolism relating to chondrocalcinosis and the characteristic deposition of CaPPi. In particular, elucidation of PPi handling by normal and osteoarthritic cartilage seemed a logical control starting point because of the much greater opportunity for sampling human osteoarthritic than chondrocalcinotic cartilage.

Rheumatology Section, Veterans Administration Hospital, Departments of Medicine, Orthopedic Surgery and Rehabilitation, University of Miami School of Medicine, Miami, Florida.

METHODS

Clinical Material

All patients received a complete medical history and physical examination during the years 1972—74 at Jackson Memorial Hospital, and Miami Veterans Administration Hospital; 16 patients were judged to have primary osteoarthritis based on exclusion of other diseases. Grading of knee and hip X-rays by the criteria of Kellgren and Lawrence (4) on each patient revealed moderate to severe disease (grades II to IV). Experimental cartilage samples were obtained at orthopedic surgery for total knee or hip replacement, taken from sites in joints on the margins of ulcerated cartilage *in a weight bearing site.* Control cartilage was obtained from hips of patients who have suffered femoral neck fractures and a patient at the time of above knee-amputation ("normals" Table I). Cartilages were promptly dissected fresh and incubated within 10 minutes of removal after processing as described below.

The material (50—90 mg wet weight) obtained from each fresh micro dissection was equally distributed among six 1-ml beakers, each containing 300 μl of the incubating Basal Eagle medium. Five per cent CO_2-filtered air with 85% humidity was circulated for 30 seconds, in the upper part of the beaker which was tightly covered with parafilm. Incubations were conducted at 38°C and maintained for 0, 1/2, 1, 2, 3, 4 and 6 h. The enzymic hydrolysis of PPi in the incubation was terminated by removal of the cartilage and addition of Cleland's reagent.

Solution pH was checked with microelectrodes as reported previously and pH values ranged 7.1—7.5 during the incubations.

Enzymatic determination of PPi

The enzymatic assay for PPi described in the current study was adapted from the three basic reactions.

1) UDP — Glucose + PPi ⇌ UTP + G—1—P
2) G—1—P ⇌ G—6—P
3) G—6—P + NADP$^+$ ⇌ Gluconolactone —6—P + H$^+$ + NADPH

Reaction 1) is catalyzed by UDP—Glucose Pyrophosphorylase, 2) by phosphoglucomutase and 3) by glucose-6-phosphate dehydrogenase.

Our methods for the determination of pyrophosphate, DNA, alkaline phosphatase, testing tissue viability, and histologic grading of samples have been described before (3).

Table I. PPi Elaboration by Incubates of Human Osteoarthritic Cartilage.

Disease	Pts.	Average Age	Males	PPi	Alkaline Phosphatase (pH 10.2)	DNA	Histological grade[c]
	(No.)	(Yrs.)	(No.)	(Conc.)	(Units)	(mg/g(dry wt))	(Range)
SYNOVIAL FLUID							
Normal[a]	3	67.5	3	0.36 ± 0.14[b]			
Osteoarthritis[a]	11	61.4	9	1.26 ± 0.70			
TISSUE							
"Normal"	2	65.5	1	<0.1[b]	0.2 ± 0.1[d]	4.1	0–1
Osteoarthritis	16	70.2	12	0.94 ± 0.26	1.78 ± 2.3	3.9 ± 1.3	5–12
Fracture	8	72.1	3	<0.1	0.1 ± 0.3	4.2 ± 1.1	3–8
Rheumatoid arthritis	3	67.0	1	<0.1	0.20 ± 0.1	4.5 ± 0.8	-
Avascular necrosis	3	70.0	0	<0.1	0.33 ± 0.0	4.2 ± 0.4	-

[a] Controls from a prior report (2).
[b] Concentration 10^{-5} M/L in synovial fluid and 10^{-11} M/hr/mg dry tissue in incubates. n = no. samples tested = 2 x no. patients.
[c] Mankin criteria.
[d] 10^{-2} units/hr/mg dry tissue.

RESULTS

Osteoarthritic cartilage samples from 16 patients all put out PPi at about 10 pmol/mg/h/dry weight or 2.5 pmol/mg DNA/h. Patients with the diagnosis of hip fracture, rheumatoid arthritis or avascular necrosis of the femoral heads have failed to produce in this small series any measurable PPi output (Table I). Alkaline phosphatase extrusion was significantly elevated above control (p < 0.01) although variations were extremely wide. Controls were of comparable ages (Table I), and cartilage cellularity indicated from DNA content was quite comparable between experimental and control groups. Histological examination of a portion of each incubated cartilage revealed the presence in radial and transitional zones of cell cloning and deep fissures, grade 3 to 12 by histological criteria (3). No histological evidence of the calcified cartilage layer or tidemark was found in these specimens. No inflammatory infiltrates were seen except in the rheumatoid samples.

DISCUSSION

To the authors' knowledge, this data (Table I) provides the first direct evidence in support of the hypothesis that metabolism of pyrophosphate in local articular cartilages of osteoarthritis is disturbed.

In respect to these human studies, a consistent finding in the laboratory data was simultaneous release of PPi and phosphatase into the incubating media; attempts to dissociate this enzyme release from that of PPi were unsuccessful in previous rabbit experiments, and so far human cartilages have released minimal phosphatase unless PPi was also released. In previous reports, bone formation has been associated with relatively high alkaline phosphatase content of growth cartilage (proliferating and hypertrophic cell zones) (5). Invasion of capillaries into the radial cartilage zone and new endochondral calcification of the base of the osteoarthritic ulceration at sites of bony remodelling is prevalent in elderly populations (6). Also, a high alkaline phosphatase content measured in cartilage in osteoarthritic sites (1), can be interpreted to indicate a preliminary stage to calcification.

A plausible interpretation of our previous rabbit cartilage data is that the release of PPi and phosphatase into the incubating medium relates also, to processes of calcification (3). In the human cartilage samples, the extension of this interpretation is that osteoarthritic cartilages, but not the other cartilages so far studied, are preparing for remodelling of underlying subchondral bone, and cells programmed in preparation to calcify release phosphatase and PPi.

REFERENCES

1. Ali, S.Y. & Bayliss, M.T.: Enzymic changes in human osteoarthritic cartilage. In: *Proceedings of Symposium on Normal and Osteoarthritic Articular Cartilage.* Ali, S.Y., Elves, M.W. and Leaback, D.H. (eds.), Institute of Orthopaedics Publication, London, pp. 189–205, 1974
2. Altman, R.D., Muniz, O., Pita, J.C. & Howell, D.S.: Articular chondrocalcinosis microanalysis of pyrophosphate (PPi) in synovial fluid and plasma. *Arthr. and Rheum.* 16, 171–178 (1973)
3. Howell, D.S., Muniz, O.E. & Pita, J.C.: Extrusion of pyrophosphate into extracellular media by cartilage incubates. In: *Proceedings of Symposium on Normal and Osteoarthritic Cartilage.* Ali, S.Y., Elves, M.W. & Leaback, D.H. (eds.), Institute of Orthopaedics Publication, London, pp. 177–187, 1974
4. Kellgren, J.H. & Lawrence, J.S.: Radiological assessment of osteoarthrosis. *Ann. rheum. Dis.* 16, 494–502 (1957)
5. Kuhlman, R.E.: Phosphatases in epiphyseal cartilage; their possible role in tissue synthesis. *J. Bone Jt. Surg.* 47A, 545–550 (1965)
6. Lane, L.B., Villacin, A. & Bullough, P.: A study of endochondral ossification in the adult: a mechanism for continuous joint remodelling. Orthopedic Research Society meeting, San Francisco, California, Feb. 27 – Mar. 1, 1975. *J. Bone Jt. Surg.* 57A, 576 (1975) (abstr)
7. McCarty, D.J. Jr., Hogan, J.M., Gatter, R.A. & Grossman, M.: Studies on pathological calcifications in human cartilage. I. Prevalence and types of crystal deposits in the menisci of two hundred fifteen cadavera. *J. Bone Jt. Surg.* 48A, 309–325 (1966)
8. McCarty, D.J., Solomon, S.D., Warnock, M.L. & Paloyan, E.: Inorganic pyrophosphate concentrations in the synovial fluid of arthritic patients. *J. Lab. clin. Med.* 78, 216–229 (1971)
9. Russell, R.G.G., Bisaz, S., Fleisch, H., Currey, H.L.F., Rubinstein, H.M., Dietz, A.A., Boussina, I., Micheli, A. & Fallet, G.: Inorganic pyrophosphate in plasma, urine, and synovial fluid of patients with pyrophosphate arthropathy (chondrocalcinosis or pseudogout). *Lancet* (1970) II, 899–902
10. Silcox, D.C. & McCarty, D.J. Jr.: Elevated inorganic pyrophosphate concentrations in synovial fluids in osteoarthritis and pseudogout. *J. Lab. clin. Med.* 83, 518–531 (1974)

CHAPTER VIII
Osteoporosis. Newer Methods for Quantitation

Bone Histomorphometry as Applied to Research on Osteoporosis and to the Diagnosis of "Hyperosteoidosis States"

P. MEUNIER, P. COURPRON, J.M. GIROUX, C. EDOUARD, J. BERNARD & G. VIGNON

Bone histomorphometry consists of measuring the bone tissue components and of counting the bone cells. The tetracycline double labelling process also permits quantitation of dynamics phenomenons such as the calcification rate of osteoid or the osteoblastic appositional rate. Thus, these methods are able to furnish vital information on the amount and the dynamics of bone tissue, but they have not yet demonstrated all their possibilities for the following reasons:

a) too few normal values have been collected for each parameter taking into account age and sex.

b) the representativness of a bone biopsy sample in comparison to the bone biopsied, the opposite bone and the whole skeleton is not well-established.

c) a diversity of parameters is being measured by different groups and a sometimes confusing vocabulary is being used producing a dynamic language to describe morphological changes.

d) the limited use of the tetracycline double labelling process proposed by Frost (5).

We shall illustrate some possibilities of bone histomorphometry keeping in mind the above mentioned remarks and emphasizing the problem of osteoporosis (OP) and the significance of the osteoid parameters.

Physiological senile osteopenia and osteoporosis as a disease

MATERIALS AND METHODS

All histological research on osteoporosis as a disease requires two

Laboratoire de Recherches sur l'Histodynamique Osseuse. 8 av. Rockefeller. 69008 Lyon, France.

conditions:

a) knowledge of the magnitude and of the pattern according to sex of *physiological senile osteopenia*, i.e. the bone rarefaction of normal ageing. For this, we analysed 285 autopsied control subjects, all having suffered violent deaths and without any macroscopic lesion at their autopsy.

b) a comparison between the data collected in control subjects and in *osteoporotic patients*, all having at least one crushed vertebra. For this, we analysed bone biopsies performed on 120 osteoporotic patients, out of which 31 had a secondary OP and 89 an apparently primitive OP.

As most other authors (1, 2, 8, 9, 17) we have used the iliac crest as our reference because (a) it belongs to the axial skeleton, (b) it can be easily biopsied, (c) it permits a clear separate analysis of cortical and spongy bone.

In control subjects the sampling area coincided with the site where the horizontal transfixing bone biopsy is performed in patients. The undecalcified sections were obtained according to methods published elsewhere (4, 10) and were stained with solochrome cyanin. All the parameters indicating the amount of bone were measured by using an image analysing computer (Quantimet 720), in a zone situated from 12 to 18 mm from the summit of the iliac crest (3). These parameters were: (a) the mean thickness of the iliac bone samples, (b) the total volume of the organ, necessary to know in order to express in terms of density the amounts of total cortical or trabecular bone, (c) the total density of bone tissue in the organ, (d) the density in cortical and subcortical bone, (e) the trabecular bone density expressed in percentage of the organ volume, (f) the trabecular bone volume (TBV), expressed in percentage of the spongy bone space. The TBV is also measurable manually by a point counting method (2). We have previously shown (4) that the latter method gave the same results as the automatic one.

RESULTS

Control subjects

(1) The mean thickness of the iliac samples, higher in males than in females, does not change significantly with age and the organ volume may be considered as a constant.

(2) The TBV decreases with age (x) in a different way for each sex and this fact prohibits the grouping of males and females in a global study (Fig. 1):

356

— in males (176 cases) the decrease is linear: TBV = (— 0.10 x + 24.17) ±
0.72. At 80 years of age the male has lost 27% of the amount of spongy
bone that he had at 20.

— in females (109 cases), after a slow decrease until 50 years of age, where
TBV = (— 0.06 x + 24.90) ± 1.24, there is a clearly defined break in the
curve between the ages of 50 and 69, and the rate of the bone loss is
multiplied by 5 during this phase: TBV = (— 0.30 x + 35.49) ± 1.80.
After 69, the decrease is again very slow: TBV = (— 0.04 x + 17.22) ±
1.20. Because of the post-menopausal rapid rarefaction, the female has

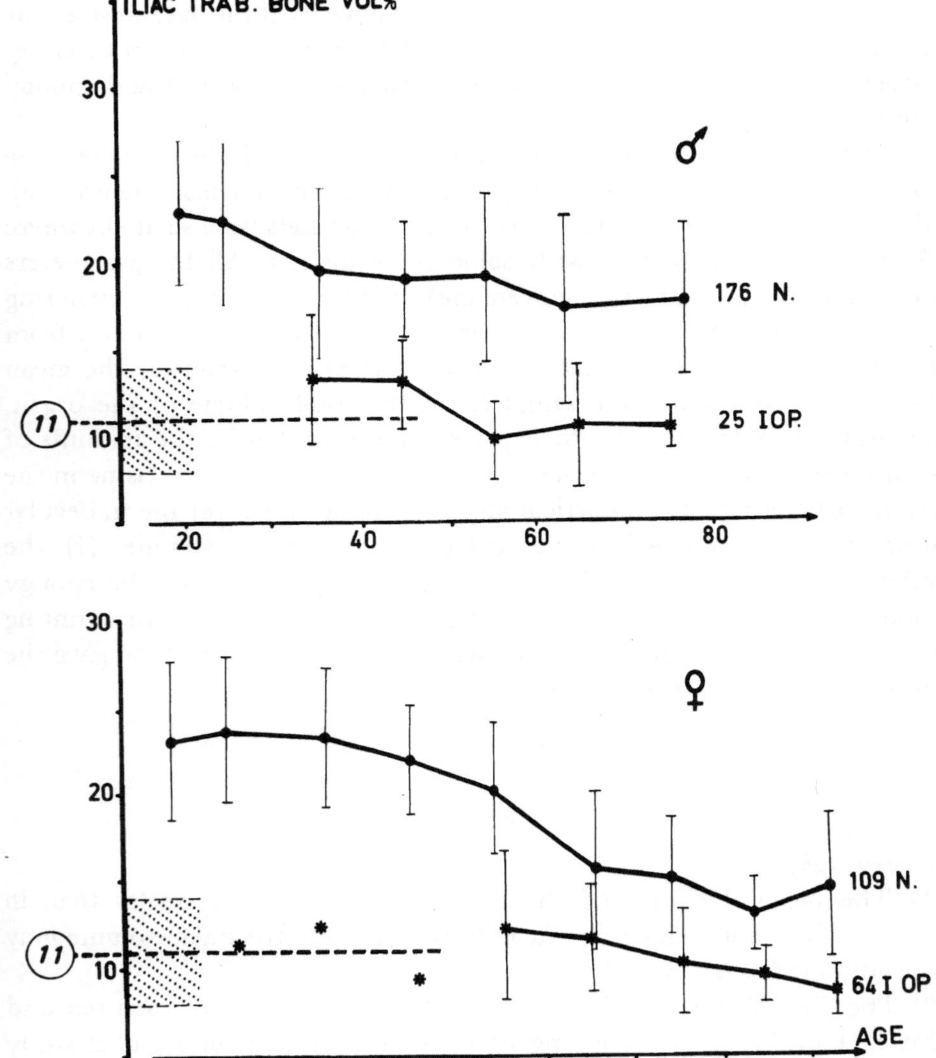

Figure 1.

lost at 80 years of age 41% of the amount of spongy bone that she had at 20.

Our results are in agreement with those noted on shorter series by Wakamatsu and Sissons (17) and Merz and Schenk (9) who have grouped males and females together. In both sexes, it is important to note that:

a) in each person, at the end of the growth period, there is a variable quantity of bone. These initial values of TBV are inversely related in males to the height of the individuals (7), and are determined in part by genetic factors, as shown by Smith (13).

b) the bone rarefaction begins immediately after the end of the bone growth.

c) the standard-deviation of mean TBV values, calculated for each 10 years of age division, does not increase with age in both sexes (Fig. 1). Our data correspond with the hypothesis of Newton-John and Morgan (12) in which all individuals lose with age the same absolute quantity of bone, no matter what their initial amount of bone was.

d) the density measurements showed that between 20 and 80 years of age the total density of bone tissue in the iliac crest decrease of cortical bone density is the same in males (− 34%) and females (− 35%). Conversely, the female loses much more trabecular bone with ageing (− 30%) than the male (− 12%).

e) the TBV (− 27% in males; − 41% in females) decreases much more than the trabecular bone density. This means that the TBV decrease is due partly to the proper atrophy of trabeculae, but also to the erosion of the cortices at their inner faces which enlarges the " spongy space" and dilutes bone tissue. In males about one half and in females about one third of the decreases of TBV depends on this thinning of cortices. Thus, TBV appears as a sensitive parameter because it reflects both cortical and trabecular atrophies. It is also a reliable parameter because its values, measured in 44 controls, do not differ significantly between symmetrical zones of right and left iliac crests, nor in the different 6 mm in height areas of a zone delimited by the summit of the iliac crest and a limit situated 24 mm below. These facts allow us to compare TBV values collected on successive iliac biopsies (7).

f) the iliac TBV is representative of the vertebral TBV, measured on horizontal sections of the third lumbar vertebra in 19 control subjects and of the sternal TBV measured in 16 controls (7).

Osteoporotic patients

The measurements of the iliac TBV allowed us:

1) to determine the value of the iliac TBV above which vertebral collapses

occur. This "vertebral fracture threshold" is 11% ± 3 (10.9% ± 2.8 in 89 primary OP; 11.1% ± 3.1 in 31 secondary OP) and does not change with sex (10).

2) to note that the iliac TBV is significantly lower in osteoporotics than in control subjects, in males throughout their lives, but in females only before 60 years of age. Beyond 60, there is an overlapping in the TBV values of female controls and those of women having a crushed vertebra.

3) to calculate from the range of the iliac TBV values observed before 29 years of age, the percentage of the control subjects crossing the vertebral collapse threshold at given ages. These predicted percentages in females are 1.8% at 50 years of age, 4.0% at 60, 18.0% at 70 and 80, 24.0% at 90. In males they are 0% at 50 and 60, 2.0% at 70 and 80. These calculated percentages are close to the observed ones regarding the frequency of crushed vertebra in the elder (6, 16, 14, 15).

All these observations suggest that post-menopausal OP could be only the *clinical and radiological expression of the lower fringe of physiological senile osteopenia.* In order to test this hypothesis we have quantified in control subjects and in patients suffering a primitive OP the extent of osteocalstic resorption surfaces, the size of periosteocytic lacunae and the relative trabecular osteoid volume. With these three parameters we have not observed any significant differences between the controls and the patients (10). This observation is in agreement with a unity conception of senile osteopenia and pathological OP.

Measurement and significance of the osteoid parameters

Osteoid tissue can be quantified in trabecular bone by measuring two parameters on undecalcified sections stained with solochrome cyanin or Von Kossa's method: (1) trabecular osteoid surfaces (O.S.), expressed as a percentage of the total trabecular surfaces. (2) relative osteoid volume (O.V.), expressed as a percentage of the total bone tissue volume. But the changes in these two parameters are difficult to interpret without using two other parameters:

1) the mean thickness of the osteoid layers, quantified by an index (i) calculated by the ratio O.V./O.S. x 100 (11).

2) the calcification rate of the osteoid, measured by using the tetracycline double labelling process and expressed in microns per day (5).

If (i) is normal according to age and sex, we can assume that the apposition rate (A.R.) is equal to the calcification rate (C.R.). If (i) is higher than normal, it proves that the A.R. is faster than the C.R. If (i) is low, the A.R. is slower than the C.R.

The analysis of 108 controls showed that there is a significant

	OSTEOID VOLUME	OSTEOID SURFACES	THICKNESS INDEX	CALCIFIC. RATE	APPOSITION RATE
_OM__alacia_	↗ +++	↗ +++	↗ +++	↘ ---	↘ -
Paget	↗ +	↗ ++	↘ -	↗ ++	↗ +
Paget + EHDP *	↗ +++	↗ +++	↗ ++	↘ ---	→
Prim. HPT	↗ +	↗ ++	↘ -	↘ -	↘ ---

Fig. 2   • 20 mg /kg/day. 6 months

difference between both sexes for the values of O.S. and O.V., higher in males than in females (11). There is a slight decrease with age of O.S., O.V. and (i) until 60 years and we noted no significant difference between data collected on right and left iliac crests. From the study of 45 osteomalacic females, it has been possible to define the osteoid parameter values above which Looser's zones occur: approximatively 20% for the O.V. and 40% for the O.S.

An increase of O.V., of O.S., or both, are not synonymous of osteomalacia because in Paget's disease or in primary hyperparathyroidism for example, both of these parameters are increased. The specific evidence of osteomalacia may only be proven by the association of an increase of (i) and a decrease of the C.R. Fig. 2 demonstrates the morphometric and histodynamic patterns of some bone diseases involving an "hyperosteo-idosis state".

This data can help us to avoid any dynamical misinterpretation of the morphometric results collected on osteoid tissue.

ACKNOWLEDGEMENT

Work supported by grants: CNRS ATP 5399.03; INSERM ATP 72.1422.7 and CL 72.1060.5.

REFERENCES

1. Beck, J.S. & Nordin, B.E.C.: Histological assessment of osteoporosis by iliac crest biopsy. *J. Path. Bact.* **80**, 391–397 (1960)
2. Bordier, P., & Tun Chot, S.: Quantitative histology of metabolic bone disease. *Clinics Endocrinol. Metab.* **1**, 197–215 (1972)

3. Courpron, P., Meunier, P., Edouard, C., Bernard, J., Bringuier, J.P. & Vignon, G.: Données histologiques quantitatives sur le vieillissement osseux humain. *Rev. Rhum.* **40**, 469–483 (1973)

4. Courpron, P., Giroux, J.M., Bringuier, J.P. & Meunier, P.: Histomorphométrie de l'os spongieux iliaque. Influence des techniques de préparation et de lecture des coupes histologiques sur la détermination du volume trabéculaire osseux iliaque. *Lyon med.* **232**, 515–522 (1974)

5. Frost, H.M.: Tetracycline-based histological analysis of bone remodeling. *Calc. Tiss. Res.* **3**, 211–237 (1969)

6. Gershon-Cohen, J., Rechtman, A.M., Schraer, N. & Blumberg, N.: Asymptomatic fractures in osteoporotic spines of the aged. *J. Amer. Med. Ass.* **153**, 625–628 (1953)

7. Giroux, J.M.: Histomorphométrie de l'ostéopénie physiologique sénile. Thesis University Claude Bernard. 69008 Lyon (France), 1975

8. Jowsey, J., Kelly, P.J., Riggs, B.L., Bianco, A.J., Scholz, D.A. & Gershon-Cohen, J.: Quantitative midroradiographic studies of normal and osteoporotic bone. *J. Bone Jt. Surg.* **47A**, 785–806 (1965)

9. Merz, W.A. & Schenk, R.K.: A quantitative histological study on bone formation in human cancellous bone. *Acta Anat.* **76**, 1–15 (1970)

10. Meunier, P., Courpron, P., Edouard, C., Bernard, J., Bringuier, J. & Vignon, G.: Physiological senile involution and pathological rarefaction of bone. Quantitative and comparative histological data. *Clinics Endocrinol. Metab.* **2**, 239–256 (1973)

11. Meunier, P., Edouard, C., Courpron, P. & Toussaint, F.: Morphometric analysis of osteoid in iliac trabecular bone. Methodology. Dynamical significance of the osteoid parameters in Vitamin D and problems related to Uremic Bone Disease. A.W. Norman and All. ed., de Gruyter, Berlin, pp. 149–155, 1975

12. Newton-John, H.F. & Morgan, D.B.: The loss of bone with age, osteoporosis and fractures. *Clin. Orthop. Related Res.* **71**, 229–252 (1970)

13. Smith, D.M., Nance, W.E., Kang, K.W., Christian, J.C. & Johnston, C.C.: Genetic factors in determining bone mass. *J. Clin. Invest.* **52**, 2800–2808 (1973)

14. Smith, R.W. & Rizek, J.: Epidemiologic studies of osteoporosis in women of Puerto Rico and South Eastern Michigan with special reference to age, race, national origin and to other related or associated findings. *Clin. Orthop. Related Res.* **45**, 31–48 (1966)

15. Urist, M.R.: Observations bearing on the problem of osteoporosis. In "Bone as a tissue" ed. by P. Fourman, Mac Graw-Hill Book Company, New York, pp. 18–45, 1960

16. Vignon, G., Marin, A. & Megard, M.: Etude radiologique de la colonne vertébrale du vieillard. *Rev. Lyon med.* **5**, 861–865 (1956)

17. Wakamatsu, E. & Sissons, H.A.: The cancellous bone of the iliac crest. *Calc. Tiss. Res.* **4**, 147–161 (1969)

Vertebral and Total Body Bone Mineral Content by Dual Photon Absorptiometry

M. MADSEN, W. PEPPLER & R.B. MAZESS

Single-photon absorptiometry has been widely accepted for biomedical research on bone (1, 4, 8). It has been apparent since the advent of single-photon absorptiometry that a dual-photon approach might be used to measure bone mineral content. The chief advantage of a dual-photon approach is the lack of the requirement for surrounding the measured bone in a constant thickness of soft-tissue. Concequently a dual-photon approach could be used on areas such as the hip or spine which are otherwise almost impossible to measure. Our initial work in this area consisted of definition of the various sources of error and their magnitude (2). Pioneering work on application of dual-photon absorptiometry to scanning of the spine was done by Reed (5) and greatly amplified by Roos and his co-workers (6, 7). We have done parallel work on spinal scanning, and have expanded to measure total body mineral as well.

METHODS

Measurements are made on a modified Ohio-Nuclear whole body rectilinear scanner. A high-intensity (1 Ci) source of ^{153} Gadolinium (4 mm bead) is mounted below the scanning table and a scintillation detector on the yoke above the subject. The detector has a slit collimator measuring 1 x 3 cm. The beam size at the body level is about 1 cm. The source and detector move simultaneously in a raster pattern, and the x and y-positions are indicated by encoders, and outputed together with the count rates at each data collection point.

The ^{153} Gd source emits at two energies (44 and 100 keV) and attenuation is essentially linear when correction is made for the spillover

Department of Radiology (Medical Physics), University of Wisconsin Hospital, Madison, Wisconsin.

from the higher energy channel into the lower. Correction is also made for system deadtime at the high count rates used.

Vertebral bone mineral is measured from about L4 to about T12. The scanner passes in a rectilinear raster at 0.25 cm/sec with 1.25 cm steps covering a distance of about 10 cm. The mineral content in this area is not uniform and the variation in a series of scans on a subject was about 15%; the variation in weights of these vertebrae on skeletons was about 13%. The mean value of a series is taken to represent the mineral content of the area so that the standard error of the spinal measure was about 5%.

For measurement of total body mineral the entire body area is scanned at a speed of 1 cm/sec with step intervals of 2.5 cm.

The radiation dose from these measurements has been measured using LiF thermoluminescent dosimetry and NBS calibrated ionization chambers. The exposure dose from a typical measurement of the lumbar vertebrae is about 2 mrad and that from a total body measurement is 0.1 mrad. The dose to the ovaries would be 0.09 mrad and 0.03 mrad respectively. The lower dose for the total body scan results from the faster scanning speed and larger step increment used in that procedure.

VERTEBRAL BONE MINERAL

The precision of vertebral scans *in vivo* was evaluated in 4 subjects measured 4 times each. The bone mineral content showed a coefficient of variation ranging from 0.7 to 3.8% (average 2.3%). Roos (7) found a variation of from 3 to 6% in his subjects. Measurements were made on a vertebral pahntom on 35 occasions over a 3-month period; the variation was 1.7%. The larger variation in humans probably reflects difficulties of relocation in this anatomically variable area.

The accuracy of vertebral measurement was assessed on phantoms imersed in a water bath; the standard error of estimate in predicting mineral content was 1.2%. Roos (7) also found his technique to have high accuracy.

Measurements were made on 41 female and 5 male subjects. Measurements were also made on the radius and ulna of the same subjects. There were only moderate (0.6 to 0.7) correlations between the bone mineral content in the radius and ulna and that in the lumbar vertebrae; the standard error of estimate was about 15 to 17%. Several of the elderly women in the sample had diagnosed osteoporosis, and were low in both peripheral and spinal mineral. The rate of bone loss with age was greater for the peripheral bones than for the spine. Women with low radius bone mineral (below the criterion established by our laboratory for risk of osteoporosis) had a commensurately low spinal bone mineral. There was, therefore, no evidence of preferential loss of spinal bone mineral.

TOTAL BODY BONE MINERAL

We have examined the precision of measurement on one skeleton, and in one subject. The variation of total mineral in the skeleton measured on 6 occasions was only 1.7%. Repeat measurements were made on one subject on 12 occasions over a two-weeks period; the last 4 measurements were made using a body immobilizer to limit subject movement. There was only a small variation (1.9%) in measured mineral; the variability seemed to be reduced slightly by the immobilizer. Thus, precision *in vivo* approaches that obtained *in vitro*; precision in both cases could be improved by taking scans at smaller step increments, but the scanning time would then be increased about 60 minutes.

By scanning the total body it is also possible to specify the distribution of bone mineral content in particular areas, for example the limbs or the trunk. Fig. 1 shows the distribution of bone mineral content along the length of the body.

In two subjects total body mineral was estimated indirectly using (a) a regression equation based on the mineral content of the radius shaft established by Mazess (3), and (b) the assumption that total mineral was 5% of the lean body mass. The measured bone mineral content was very close to the average of the estimated values; the measured mineral was 2470 g and 2027 g compared to the respective estimates of 2477 g and 2021 g. Accuracy of these measurements is being investigated in a large series of subjects.

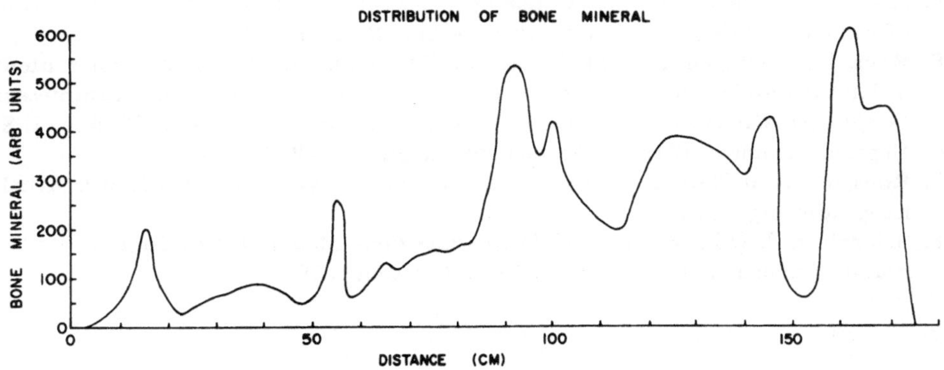

Figure 1. The distribution of bone mineral content along the axial length of a skeleton from a total body scan *in vivo.*

CONCLUSIONS

Techniques have been developed using dual-photon absorptiometry for precise and accurate measurement of bone mineral content in the lower spine and in the total body. A spinal measurement requires only 20 min and a total body scan about 60 min. The dose from these measurements is very low allowing them to be repeated at frequent intervals. We feel these improved techniques will prove of great value both for examining the skeleton in normal populations and for evaluation of osteoporosis and other skeletal abnormalities.

ACKNOWLEDGEMENTS

This research was supported by NASA-50-002-051 and NIH-AM-17892. We are grateful for the assistance of the University of Wisconsin Hospital Volunteers and the Sisters of Notre Dame who participated in this study.

REFERENCES

1. Cameron, J.R. (ed.): *Proceedings of the Bone Measurement Conference.* U.S. AEC Conf. 700515, CFSTI, Springfield, Va., 1970
2. Judy, P.F.: *A Dichromatic Attenuation Technique for the* In Vivo *Determination of Bone Mineral Content.* Ph.D. Thesis, University of Wisconsin, Madison, Wisconsin, 1971
3. Mazess, R.B.: Estimation of bone and skeletal weight by direct photon absorptiometry. *Invest. Radiol.* 6, 52–60 (1971)
4. Mazess, R.B. (ed.): *International Conference on Bone Mineral Measurement.* DHEW Publ. 75–863, U.S. Dept. of Health, Education and Welfare, Washington, D.C., 1974
5. Reed, G.W.: The assessment of bone mineralization from the relative transmission of Am-241 and Cs-137 radiations. *Phys. in Med. Biol.* 11, 174 (1966)
6. Roos, B., Rosengren, B. & Sköldborn, H.: Determination of bone mineral content in lumbar vertebrae by a double gamma-ray technique. In: *Proc. Bone Measurement Conference.* Cameron, J.R. (ed.), U.S. Atomic Energy Comm. Conf. 700515. U.S. Dept. of Commerce (CFSTI), Springfield, Va, pp. 243–254, 1970
7. Roos, B.: Dual Photon Absorptiometry in Lumbar Vertebrae. Akademisk Avhalning, Göteborg, 1974
8. Schmeling, P. (ed.): *Proceedings Symposium Bone Mineral Determinations.* Aktiebolaget Atomenergi Publ. AE-489. Studsvik, Sweden, 1974

Prospective and Cross-Sectional Study of Radial Bone Loss in Post-Menopausal Women

C. C. JOHNSTON JR., D. M. SMITH & M. R. A. KHAIRI

The development of the photon absorption technique has provided a precise and accurate method for measuring bone mineral mass in the appendicular skeleton (1). This technique has been utilized to study changes in bone mass in the radius. Scans were performed at midshaft and distal sites as previously reported (2, 3, 5). The measurement is reported as bone mineral content (BMC) in units of g/cm.

A group of 941 essentially healthy Caucasian females was scanned (6). There was an increase in BMC to age 20, a plateau, and a decrease beginning at approximately age 40. These data have been analyzed to determine if a sub-population of rapid losers — the potential osteoporotics — could be found. In some instances, normalization of data for size of bone (width measured from the scan was used) and age was performed. Regression equations were used to adjust to a standard width of 1.25 cm and age of 70 years (6).

A frequency distribution of midshaft BMC values was plotted and data both unadjusted and adjusted for width alone and for width and age was studied (6). The Chi-square test for goodness of fit to a normal distribution showed no difference between the observed values and the predicted normal frequency distribution. Similar results were obtained for distal mass measurements. In addition, probit analysis was performed on the data, grouped by decade, and support for normal distribution was found. Thus, no evidence for bimodality was found as might be expected if a large group of rapid losers were present in the population.

Should such a group of rapid losers be present, variance within the population should increase with age. This postulate was tested by examining the significance of the regression coefficients of the variance of BMC versus age, and no significant increase in variance was found (6).

From the Department of Medicine, Indiana University School of Medicine, Indianapolis, Indiana.

Such retrospective data as these and numerous similar reports in the literature (4) do not rule out the possibility of osteoporosis developing in a group of women who have low mass because of rapid loss. However, other models for bone loss better fit these data.

We next turned to a consideration of the relationship of fracture, the morbid event in osteoporosis, to bone mass.

Among 269 women over age 50 who had had X-rays of the spine, 108 had collapse fractures. Those with fracture had significantly less mass (midshaft mass — non-fracture 0.74 ± 0.16 g/cm, fracture 0.65 ± 0.14 g/cm, p < 0.001) but the rate of loss calculated here as the linear coefficient of BMC versus age was not significantly different (regression coefficient vs age — non-fracture 0.00656 ± 0.00076, fracture 0.00754 ± 0.00113, p N.S.) in the two groups (6). Again, no separate population could be found. The population was then divided into arbitrary levels of BMC and fracture. The slope of the plot of fracture versus BMC was significantly different from zero, and the hypothesis for linear trend in proportions could not be rejected (6). Thus, in this retrospective study, fracture rate was inversely related to bone mass values.

A group of 278 women has been followed for an average period of 1.7 years, a total follow-up of 470 subject years. The relationship of bone mass to subsequent development of non-traumatic fractures was determined. Fractures occurred in 31 subjects, and were of all varieties, although stress fractures of the metatarsal were the most frequent under age 65. Again, there was an inverse relationship between mass and fracture and again it is significant (6). These data, although preliminary, indicate that measurement of bone mass may be of value in determining who is at risk of subsequent fractures.

The majority of studies presented here and published to date have been cross-sectional, but it is only through prospective studies that the pathogenesis of low bone mass and fracture can actually be determined. The photon absorption technique because of its precision can be utilized in such studies.

A study was undertaken to determine rates of radial bone loss in a population of post-menopausal women. Since the true rate of loss is unknown and cannot be measured by other techniques at present, accuracy cannot be determined. Thus, the precision of measurement will determine its usefulness. Providing the rates of loss are linear, the standard error of the regression coefficient is an index of precision. Precision will increase with more frequent measurements over longer periods of time, and a practical compromise must be struck in relationship to the availability of the population to be studied.

Twenty-four women age 51 to 64 with a mean age of 57 who were determined to be free of significant disease were followed with measurements every three months for at least a three-year period (7). The midshaft rate stabilizes after two years, but the distal rate did not stabilize in the period of follow-up. The variability found in early rates could be due to non-linear rates of loss, but no quadratic component was found and it is assumed that the variability is due to imprecision of the method.

Table I.

Mean Age	Number of Subjects	Mean Years	Rate (g/cm/yr) ± S.E.	
			Midshaft	Distal
57.3	24	3.7	-0.01083 ± 0.00148	-0.01935 ± 0.00229
82.3	29	2.4	-0.00503 ± 0.00370	-0.01188 ± 0.00323

Table I shows the actual rates and standard errors. The rates for a group of 29 older women, mean age 82, followed for an average of 2.4 years are also shown. These rates have not as yet stabilized. However, it can be seen that the average rate of loss is less, but the variance is greater.

Using the intercept at time 0, the rate of loss can be expressed as per cent of initial mass. In Table II the per cent loss found in the prospective study of the younger women is compared to that calculated from the cross-sectional study mentioned above. The agreement is relatively good.

Table II.

	Rates of Change	
	Prospective Study	Population Survey
Midshaft mass	-1.37%/year	-1.07%/year
Distal mass	-2.35%/year	-1.34%/year

Thus, it is possible to measure rates of loss in the midshaft radius in a period of two years with four measurements each year. Either a longer period of time or more frequent measurements are necessary to precisely determine the distal rates.

With such data as these, it is possible to determine the number of subjects necessary for a therapeutic trial of an agent which is designed to slow bone loss, such as estrogen (7). If it is desired to demonstrate a significant reduction in the rates of loss at the 5% level with a certainty of 95% when the true reduction is 50% of the control group, the number of subjects in each group would be 40, provided the midshaft site is used and the subjects are followed for 3.7 years with four measurements a year.

The ability to determine rates of loss should aid in the study of the pathogenesis of osteoporosis. To determine if the rates among individuals in the groups were significantly different, an F test for homogeneity of regression coefficients was used. At 3.7 years of follow-up the test was highly significant for both distal and midshaft rates, indicating that differences in rates among individuals are, indeed, real and cannot be accounted for by variance about the regression lines alone. Thus, different individuals are losing radial bone mass at different rates.

It has been shown in the cross-sectional study that the average rate of loss for the population remains constant and there is no increased variance in the older population. Yet, from the prospective study, individuals lose at different rates. Several models may be proposed to reconcile these apparent differences. The rate of loss may be directly related to the initial mass so that rates vary among individuals but variance does not increase. Also, individuals might lose at different rates from time to time and yet variance not increase with age. Of course, other models might also be constructed.

The photon absorption technique utilized for such prospective studies as those currently underway, can provide solutions to this problem. It will also be possible to study individuals with rapid and slow loss to detect underlying biological differences which might explain the difference in rates found, such as in levels of parathyroid hormone or estrogen. It will also be possible to determine the relationship of rate of loss and absolute mineral mass to the subsequent development of fractures so that the diagnosis of osteoporosis could be made before fracture occurs.

ACKNOWLEDGEMENT

Supported in part of USPHS Grant AM 0726.

REFERENCES

1. Cameron, J.R. & Sorenson, J.: Measurement of bone mineral *in vivo*: An improved method. *Science* 142, 230–232 (1963)
2. Johnston, C.C. Jr., Smith, D.M., Yu, P. & Deiss, W.P. Jr.: *In vivo* measurement of bone mass in the radius. *Metabolism* 17, 1140–1153 (1968)
3. Johnston, C.C. Jr., Smith, D.M., Nance, W.E. & Bevan, J.: Evaluation of radial bone mass by the photon absorption technique. In: *Clinical Aspects of Metabolic Bone Disease*, Frame, B., Parfitt, A.M. and Duncan, H. (eds.), Int. Congr. Ser. No. 270, Excerpta Medica (Amst.), pp. 28–36, 1973

4. Newton-John, H.F. & Morgan, D.B.: The loss of bone with age, osteoporosis and fractures. *Clin. Orthop.* 71, 229–252 (1970)
5. Smith, D.M., Johnston, C.C. Jr. & Yu, P.: *In vivo* measurement of bone mass. *JAMA* 219, 325–329 (1972)
6. Smith, D.M., Khairi, M.R.A. & Johnston, C.C. Jr.: The loss of bone mineral with aging and its relationship to risk of fracture. *J. Clin. Invest.* 56, 311–318 (1975).
7. Smith, D.M., Norton, J.A. Jr., Khairi, R. & Johnston, C.C. Jr.: The measurement of rates of mineral loss with aging. *J. Lab. clin. Med.* (In press)

Quantitation of Bone Mass in Osteoporosis: Recent Advances

C.H. Chesnut III, W.B. Nelp & T.K. Lewellen

Knowledge of the pathophysiology and treatment of osteoporosis and other metabolic bone disease is enhanced by accurate measurement of bone mineral mass and precise estimation of bone mass change. Ideally, a non-invasive measurement of bone mass in the osteoporotic patient would precisely quantitate both regional bone mass at selected sites and total skeletal bone mass, since it may be that some skeletal sites change more rapidly than others when bone loss or gain occurs; consequently, regional measurements of one site may not reflect the state of other sites or of the entire skeleton.

To date, no one proven methodology provides all the desired information regarding bone mass with the necessary accuracy and precision; radiographic techniques including X-ray densitometry and morphometry are prone to imprecision, and provide information from only one site. The comparatively new procedure of total body calcium (TBC) determination by neutron activation analysis (NAA) (^{48}Ca (n, α) ^{49}Ca), however, provides precise ($\pm$ 2%) (8) and accurate ($\pm$ 5%) (9) assessment of *total* bone mass; the photon absorptiometry technique also provides precise ($\pm$ 3%) quantitation of *regional* bone mass (RBM), primarily appendicular. Together these latter methods have assumed primary importance in bone mass assessment in the osteoporotic patient.

364 NAA procedures have been performed in our laboratory in 134 patients with metabolic bone disease including 60 patients with osteoporosis. In addition, TBC has been quantitated in 50 normal individuals (age and sex distributed), forming the initial portion of a projected cross sectional population study of age-related bone mass loss. These preliminary data, although limited, provide considerable comparative information between normal and compression fracture osteopototic individuals.

Until now, we have not developed an ideal method to normalize the

Division of Nuclear Medicine, University of Washington School of Medicine, Seattle, Washington.

TBC measurement with the variables of age, sex and height, especially in the osteoporotic where height is abnormally reduced. The inclusion of the variable lean body mass (as assessed by total body potassium-TBK) may, however, aid in the normalization of the TBC data (4) although TBK and TBC may be lost in equal proportions in osteoporosis. Currently, it is uncertain whether a measurement of TBC alone can be used to clearly identify the individual with clinically significant osteoporosis. Previously we noted (2) a correlation between TBC and height (TBC = 0.203 Height^3, r = 0.97, derived from 8 normal male subjects age 21—41). In the larger sample of the normal population, this expression does not appear ideal with an r value of 0.80 and a greater variance.

TBC/NAA method is ideally suited to assess response of bone mass to therapy in osteoporosis; a recent study (5) in our laboratory assessed the value of intravenous calcium infusion in 5 post-menopausal compression fracture osteoporotic females age 45—66. Measurement of per cent change in TBC with time revealed a mean loss in the treated patients of 4% per year as compared to a 1.5% per year loss in untreated osteoporotic control subjects. Bone biopsy data also supported lack of benefit of this form of osteoporosis therapy. A second study (3), double blinded and controlled, assessed the effect of the synthetic anabolic steroid methan-drostenolone in 26 post-menopausal osteoporotic females. There was a

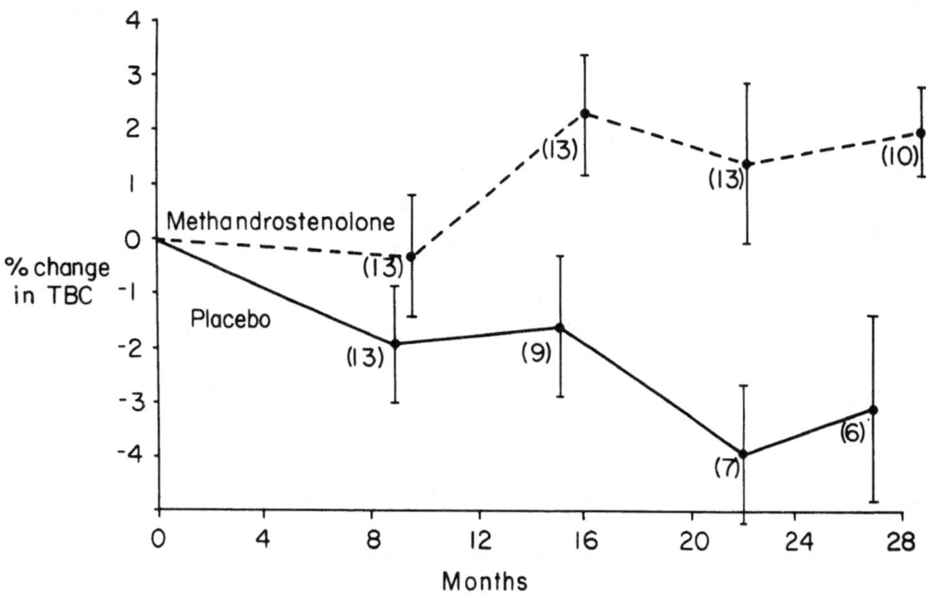

Figure 1. **Per cent change in TBC in treated (methandrostenolone) and placebo patients with time; the data points represent the mean (± S.E.M.) of the results and the parenthesis ( ) indicate the number of patients remaining in the study at each NAA procedure.**

significant 2% TBC gain and 3% TBC loss over 26 months in treated and placebo groups respectively, suggesting efficacy of this form of therapy (Fig. 1).

Although current photon absorptiometry techniques may precisely quantitate regional bone mass at predominantly appendicular sites, the relationship of this measurement to total bone mass, and more importantly to vertebral column bone mass, remains unclear. In our laboratory correlation of TBC to RBM (distal radius) in 14 compression fracture osteoporotic patients was high (r = 0.94, Fig. 2) but with an 8% error (S.E.E. ± 60 g) in predicting TBC from RBM (7). In addition, high correlations (r = 0.84 to 0.93, p < 0.001) between five other sites of radius, ulna, humerus, and TBC, were noted; also high correlations with r values from 0.89 to 0.95, p < 0.001 were found between RBM of the distal radius and RBM of the five other sites. In contrast, correlations between RBM determined by radiographic morphometry and TBC were weak or non-significant. If TBC and RBM were expressed as per cent of their mean value, the slopes of the estimating equations (describing the relationship between TBC and RBM) were essentially the same. The slopes, however, were essentially less than one which is the value of the slope expected if TBC and RBM had changed at the same relative rate. From these relationships, it was concluded that

1. the rate of change in RBM was similar in the six sites examined, and

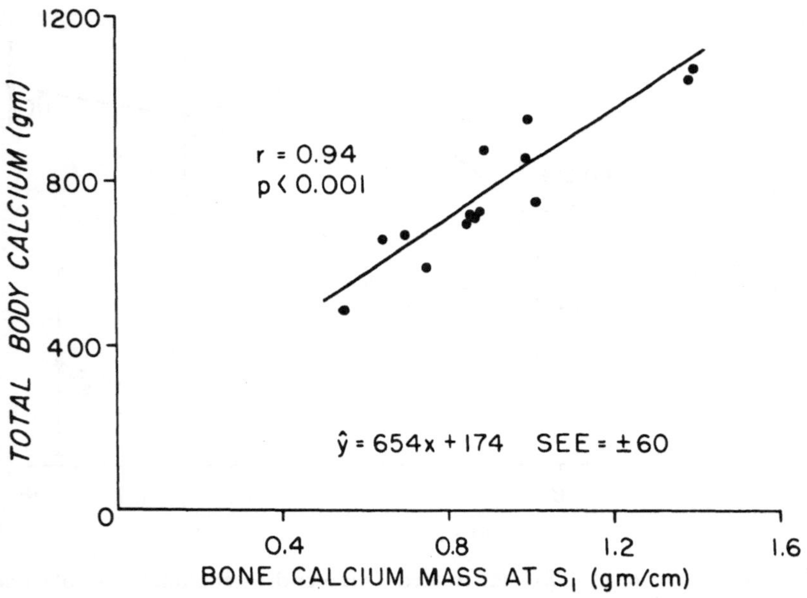

Figure 2. Correlation between total body calcium and regional bone mass (bone calcium mass) at the Sl site (distal radius).

2. that the rate of change in the six sites was relatively more rapid than the change in TBC. Whether the change in rate of RBM was related to rate of increase during attainment of maximum RBM or subsequent rate of loss of RBM or both remain to be determined; obviously, longitudinal (rather than cross-sectional) comparisons of rates of change of TBC and RBM are needed. Some initial work has been performed in this latter area (1).

In our normal population, the relationship between TBC and RBM at the Sl site (distal radius) was again linear with, however, an r value or 0.81 (S.E.E. ± 107 g) giving an approximate 10% error in the prediction of TBC from a RBM measurement.

Newer methods under investigation for quantitating bone mass include dual isotope photon densitometry techniques, partial body NAA, and NAA with 14 MeV neutrons to produce argon-37 from calcium-40 (^{40}Ca (n, α) ^{37}Ar). In this technique TBC can be quantitated by counting the ^{37}Ar which is exhaled in the breath (6). The ^{37}Ar technique, developed in our laboratories, has been applied to 16 individuals undergoing simultaneous determination of TBC by the ^{48}Ca-^{49}Ca technique. The maximum rate of ^{37}Ar excretion occurs immediately (5–20 min) post irradiation; by six hours the excretion rate has decreased to approximately 1% of the maximum value. The correlation between TBC and rate of ^{37}Ar excretion at 30 min post irradiation is linear with an r value of 0.97. The ^{37}Ar technique has the advantage of lower radiation dosage (10 millirads) than previous NAA procedures (allowing more frequent measurements) and may provide both total bone mass and vertebral column regional bone mass. The latter measurement would be of obvious value in the assessment of osteoporosis. The technique will be applied to the problem of space flight related bone loss; data from recent Skylab missions predicts significant calcium (i.e. bone mass) loss in long term zero gravity space travel (10). The degree and site of loss may be defined by ^{37}Ar quantitation.

Great advances in bone mass quantitation have been made; equally great challenges remain.

ACKNOWLEDGEMENTS

Supported in part by Atomic Energy Commission grant #AT (45–1)–2225 and NASA grant (NAS 9–13029). Dr. Chesnut is the recipient of a Picker Scholar Award.

REFERENCES

1. Aloia, J.F., Ellis, K., Zanzi, I. & Cohn, S.H.: Photon absorptiometry and skeletal mass in the treatment of osteoporosis. *J. Nucl. Med.* 17, 196—199 (1975)
2. Chesnut, C.H. III, Nelp, W.B., Denney, J.D. & Sherrard, D.J.: Measurement of total body calcium (bone mass) by neutron activation analysis: Applicability to bone-wasting disease. *Clin. Aspects of Metabolic Bone Disease*, Excerpta Medica (Amst.), pp. 50—54, 1972
3. Chesnut, C.H. III, Nelp, W.B. & Baylink, D.: Efficacy of Dianabol in osteoporosis as determined by changes in total bone mineral mass (Abstract): *Clin. Res.* 22, 464a (1974)
4. Cohn, S.H., Ellis, K.J., Wallach, S., Zanzi, I. *et al.*: Absolute and relative deficit in total skeletal calcium and radial bone mass in osteoporosis. *J. Nucl. Med.* 15, 428—435 (1974)
5. Dudl, R.J., Ensinck, J.W., Baylink, D., Chesnut, C.H. III *et al.*: Evaluation of intravenous calcium as therapy for osteoporosis. *Amer. J. Med.* 55, 631—637 (1973)
6. Lewellen, T.K., Chesnut, C.H. III, Nelp, W.B., Palmer, H.E. *et al.*: Preliminary observations on the excretion of argon-37 from man following whole body neutron activation — an indicator of total body calcium. *J. Nucl. Med.* 16, 672—675 (1975)
7. Manzke, E., Chesnut, C.H. III, Wergedal, J.E., Baylink, D.J. *et al.*: Relationship between local and total bone mass in osteoporosis. *Metabolism* 25, 605—615 (1975)
8. Nelp, W.B., Palmer, H.E., Murano, R., Pailthorp, K. *et al.*: Measurement of total body caldium (bone mass) *in vivo* with the use of total body neutrom activation analysis. *J. Lab. clin. Med.* 76, 151—162 (1970)
9. Nelp, W.B., Denney, J.D., Murano, R., Hinn, G.M. *et al.*: Absolute measurement of total body calcium (bone mass) *in vivo. J. Lab. clin. Med.* 79, 430—437 (1972)
10. Wheden, G.D., Lutwak, L., Rambaut, P.C., Whittle, M.W. *et al.*: Mineral and nitrogen metabolic studies. *Proceedings of the Skylab Life Sciences Symposium Vol. 1*, NASA Technical Memorandum (eds.), pp. 353—371, 1974

Quantitation of the Degree of Osteoporosis by Measure of Total-body Calcium Employing Neutron Activation

S. H. COHN, I. ZANZI, A. VASWANI, S. WALLACH, J. ALOIA & K. J. ELLIS

A basic assumption underlying the measurement of bone mass for the evaluation of osteoporosis is that there exists a relationship between the level of bone mineral and the occurrence of spontaneous fractures characteristic of the condition. If the hypothesis that the structural integrity of the skeleton, (particularly of the vertebrae), is associated with a critical level of mineral is accepted, it is then essential to determine the threshold value and to establish degrees of osteoporosis in terms of the deficiency of calcium from a "normal" value. Since osteoporosis occurs chiefly in post-menopausal women, age 50 or greater, who "normally" exhibit an increase in the rate of Ca loss associated with the regression of the gonadal hormones, the problem of distinguishing osteoporotic individuals from normal women in the same age range is a particularly difficult one.

In the present study, two techniques for measuring the amount of Ca in the total skeleton were employed: total-body neutron activation analysis (TBNAA) and the determination of the mineral content of a bone of the appendicular skeleton (absorptiometric measurement of the radius, BMC).

METHODS

Thirty-six women were selected on the basis of having either one or more compression fractures of the vertebrae or radiological evidence of severe osteoporosis. Twenty-three "normal" women, manifesting no clinical evidence of osteoporosis, and ranging in age from 50 to 78 years, served as controls. The patients were uniformly exposed to a beam of partially moderated fast neutrons which induces the reaction ^{48}Ca (n, γ)^{49}Ca. The induced ^{49}Ca was then measured with the Brookhaven whole-body

Brookhaven National Laboratory, Medical Research Center, Upton, New York.

counter which gives an absolute measurement of the ^{49}Ca. From these data, absolute levels of total-body calcium were calculated with an accuracy of ±5% and a precision of ±1% (1 SD) (5). The radiation dose to the patient in this technique is 0.028 rad (or 0.28 rem). The bone mineral content (BMC) of the radius and width (W) were measured with a Norland-Cameron absorptiometer (2,3). Measurements were made at the 8 cm site on the left radius.

RESULTS AND DISCUSSION

The range in absolute levels of TBCa in the osteoporotic patients was very large (344–865 g), reflecting the effects of the parameters of body habitus, age and sex (see Table I). In order to assess the relative deficit of TBCa in an individual, it is necessary that the calcium level be compared with a "normal" value for that individual. This variability in TBCa diminishes the usefulness of the average as the basis for comparison. For this reason, a previously developed algorithm was used to predict the normal TBCa in each subject based on lean body mass (potassium), height, sex and age (6, 7). The formula used was:

$$Ca_p = \alpha \, HK^{1/2} \tag{I}$$

$$
\begin{aligned}
\text{where } Ca_p \;&=\; \text{predicted total-body Ca (g)}\\
H \;&=\; \text{height (m)}\\
K \;&=\; \text{total-body potassium (g)}\\
\alpha \;&=\; 56.62{-}0.38\,(A{-}55),\ \text{where } A = \text{age in years,}\\
&\quad\ \text{for subjects} > 55 \text{ years}\\
\alpha \;&=\; 56.62 \text{ for subjects} \leqslant 55 \text{ years}
\end{aligned}
$$

The measured total-body Ca expressed in terms of the predicted normal value (Ca_p) is referred to as the calcium ratio ($TBCa/Ca_p$). The difference between the value of this ratio and I is referred to as the Ca deficit.

In the group of female osteoporotics, the mean calcium ratio was 0.858 ± 0.110 (SD); the mean was 0.996 ± 0.072 (SD) for the normal contrast group (Table I). In patients with kyphosis, correction was made for loss of stature due to the kyphotic changes by estimation of the height from a radiograph of the tibia length (7). Correction was also made for the abnormally low K values observed in a few patients. The predicted normal K was calculated from a previously derived relationship involving height and weight (9).

The BMC values varied from 0.362 to 1.028 g/cm in the osteoporotic population. To facilitate intercomparison of bone mineral content (BMC),

Table I. Skeletal Calcium and Bone Mineral Content in Adult Females.

No.	Age	Wt	Ht	TBCa	$\dfrac{TBCa}{Ca_p}$	BMC	W	BMC/W	$\dfrac{BMC}{BMC_p}$
		kg	cm	g		g/cm	cm	g/cm²	
Osteoporotic Females (44–81 y)									
36	65.7	56.9	153.3	589.8	0.858	0.646	1.234	0.528	0.917
		±24.3*	± 5.4	± 18.4	±12.8	±23.8	±12.5	±22.4	±16.9
Normal Females (50–79 y)									
23	61.7	64.4	156.3	728.3	0.996	0.752	1.173	0.638	0.993
		±14.7	± 4.6	± 13.6	± 7.2	±16.3	±10.9	±13.2	±10.7

* Coefficient of variation (%SD)

TBCa — total body calcium, measured
Ca_p — total body calcium, predicted
BMC — bone mineral content of radius (g/cm), measured
BMC_p — bone mineral content of radius (g/cm), predicted
W — radius width (cm)

measured by absorptiometry, the data must be normalized for size and age as was necessary for the TBCa data. While division of the BMC value by the width of the radius (W) tends to reduce the variability in the group, the procedure is not satisfactory as a normalization. Inasmuch as TBCa and BMC measurements correlate well, BMC data can also be normalized by a modification of the same algorithm employed for the normalization of the TBCa data (8). The new algorithm with an age correction factor was derived as follows:

$$BMC = 0.00112\ TBCa - 0.0664 \qquad (II)$$
$$TBCa = \alpha\ HK^{1/2} \qquad (III)$$
$$BMC_p = 0.0635\ HK^{1/2} - 0.0664 \text{ for age} \leqslant 55 \qquad (IV)$$
$$BMC_p = [0.0635 - 0.00043\ (age - 55)]HK^{1/2} - 0.0664 \qquad (V)$$
$$\text{for age} > 55$$

The normalized TBCa and BMC values are shown in Fig. 1. The values of 75% of all the calcium ratios for the osteoporotic subjects were more than 1 SD below the normal mean, while 70% of the BMC ratios of the osteoporotic subjects fell within 1 SD of the mean of the normal subjects.

A highly significant correlation was found between the directly

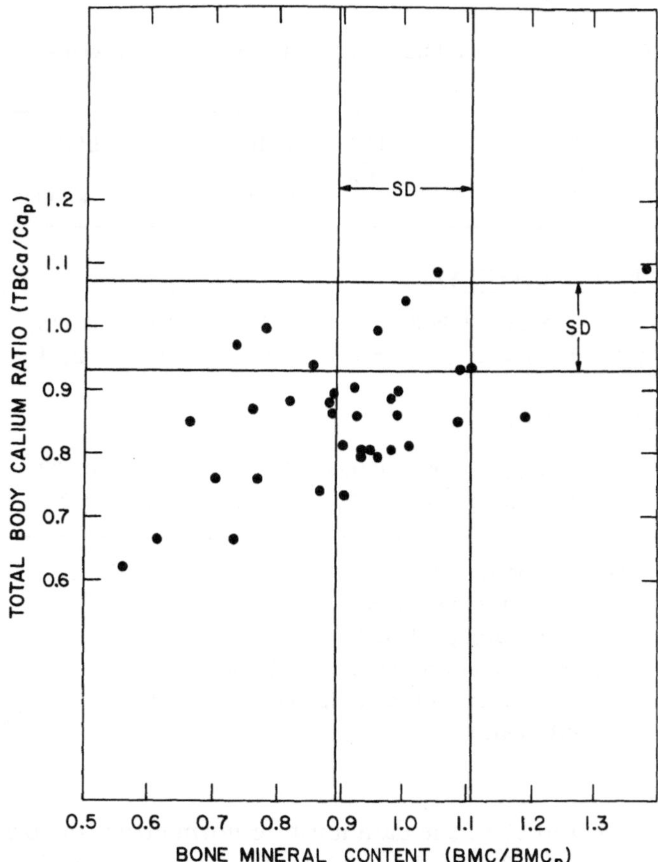

Fig. 1

measured BMC and TBCa values. The correlation coefficient was 0.826 (p < 0.001); it was almost the same as that found for the age matched normal population (0.813). This finding is surprising in light of the probable differential rates of loss in different bones in osteoporotic patients. Ca loss is known to occur primarily from the trabecular bone of the vertebrae in the osteoporotic patients, and only secondarily from the radius. The correlation coefficient relating TBCa and BMC reported here is in good agreement with the results reported by Chesnut (4).

Thirty-four of these osteoporotic patients underwent various therapeutic regimes: calcitonin, calcium, estrogen and growth hormone administration, and were studied both before and following therapy by both TBNAA and photon absorptiometry.

The correlation coefficient between the change in TBCa (Δ%) and the change in BMC (Δ%) was very poor (r = 0.17) (9). These findings suggest that changes in radial BMC values, at the 8 cm site, do not always reflect changes in skeletal mass in response to treatment of osteoporosis.

As previously reported, there is a large statistical variability in the BMC of the peripheral skeleton in normal populations up to 18% (10). This variability, even in data normalized for sex, age and size of the individual, makes it unsatisfactory to diagnose osteoporosis solely on the basis of the measurement of radial BMC. The TBCa value, normalized for sex, age and skeletal size, appears to be statistically more reliable than the BMC values of the radius as an index for quantitating the degrees of osteoporosis in an individual. The normal loss of calcium with age in post-menopausal women often makes it difficult to distinguish between osteoporotic and non-osteoporotic individuals. The degree of osteoporosis in an individual can be assessed in terms of their calcium deficit, as measured by TBNAA, but only on a statistical basis.

ACKNOWLEDGEMENT

Supported by the U.S. Energy Research and Development Administration.

REFERENCES

1. Aloia, J.F., Ellis, K.J., Zanzi, I. & Cohn, S.H.: Photon absorptiometry and skeletal mass in the treatment of osteoporosis. *J. Nucl. Med.* 16, 196–199 (1975)
2. Cameron, J.R. & Sorenson, J.: Measurement of bone mineral *in vivo*: an improved method. *Science* 142, 230–232 (1963)
3. Cameron, J.R., Mazess, R.B. & Sorenson, J.: Precision and accuracy of bone mineral determination by direct photon absorptiometry. *Invest. Radiol.* 3, 141–150 (1968)
4. Chesnut, C.H., Manske, E. & Baylink, D. *et al.*: Correlation of total-body calcium (bone mass) and regional bone mass in osteoporosis. *J. Nucl. Med.* 14, 386 (1973)
5. Cohn, S.H., Shukla, K.K. & Fairchild, R.G.: Design and calibration of a "broad-beam" ^{238}Pu, Be source for total-body neutron activation analysis. *J. Nucl. Med.* 13, 487–492 (1972)
6. Cohn, Shukla, K.K. & Ellis, K.J.: A multivariate predictor of total-body calcium. *Int. J. Nucl. Med. Biol.* 1, 131–134 (1974)
7. Cohn, S.H., Ellis, K.J., Wallach, S., Zanzi, I., Atkins, H.L. & Aloia, J.F.: Absolute and relative deficit in total-body skeletal calcium and radial bone mineral in osteoporosis. *J. Nucl. Med.* 15, 428–435 (1974)
8. Cohn, S.H. & Ellis, K.J.: An algorithm for predicting radial bone mineral content in normal subjects. *Int. J. Nucl. Med. Biol.* 2, 53–57 (1975)
9. Ellis, K.J., Shukla, K.K. & Cohn, S.H.: A predictor for total-body potassium in man based on height, weight, sex and age: Application in metabolic disorders. *J. Lab. clin. Med.* 83, 716–727 (1974)
10. Johnston, C.C., Smith, D.M. & Yu, P.L. *et al.*: *In vivo* measurement of bone mass in the radius. *Metabolism* 17, 1140–1153 (1968)

Measurement of Skeletal Blood Flow in Normal Man and in Patients with Paget's Disease of Bone

R. WOOTTON, J. REEVE* & N. VEALL

INTRODUCTION

The metabolism of any tissue is dependent on its blood supply, and the possibility exists that control of its blood supply forms one method of regulating the metabolism of bone. However, it has not previously been possible to measure skeletal blood flow (SBF) in man, for lack of a suitable technique. Use of the plasma clearance of a bone-seeking tracer (8) for instance, lacks full validation.

In acute experiments in the rabbit, in which the uptake in bone of an intra-arterial injection of ^{18}F was compared with the uptake of a simultaneously-administered injection of radioactive microparticles, the single-passage extraction ratio of ^{18}F in bone was shown to approximate to 1.00 (9). Skeletal clearance of ^{18}F therefore, may be equated with SBF. However, the calculation of skeletal clearance as the difference between total plasma clearance and renal clearance (8) ignores the possibility of reflux of the tracer from bone. In the case of ^{18}F this reflux is known to be significant (2) and it will be shown that this leads to a serious underestimate of SBF by Van Dyke's method.

METHOD

The technique described here for deriving the skeletal clearance of ^{18}F depends on obtaining the response function of the extravascular ^{18}F pool to a theoretical pulse injection into this pool from the plasma.

^{18}F is injected intravenously. From measurements on serial blood samples, the fraction of the tracer dose remaining in the blood pool as a function of time is obtained, $I(t)$, and the cumulative urinary excretion

MRC Clinical Research Centre, Harrow, Middlesex.
*) MRC Clinical Research Fellow.

of tracer is measured, U(t). From these data, the fraction of the tracer dose in the extravascular pool at any given time, Q(t), is obtained by difference.

Mathematical deconvolution of the blood pool input function from the bone/ECF retention function yields the single-passage impulse response function, H(t), describing the response of the bone/ECF pool to a pulse input of tracer. The initial value of this function, before there is time for backflow of tracer, gives the tracer clearance into the bone/ECF pool (7).

The impulse response function for ^{18}F can be described by the sum of two exponentials over the first 2–3 h from injection. Extracellular tracers, such as ^{82}Br or ^{51}Cr-EDTA, on the other hand, have only the first component when deconvoluted in this way, so evidently the second slow component represents the slow release of ^{18}F from bone. Subtraction of the extravascular, extracellular response function for ^{51}Cr-EDTA or ^{82}Br from that for ^{18}F, yields an estimate of the skeletal ^{18}F content. By extrapolating the residual second exponential back to zero time, the intercept is obtained, which is equal to the vascular clearance or SBF.

1. *Normal SBF.* Resting SBF was measured in eight normal volunteers from the scientific staff of the Clinical Research Centre. In order to assess the reproducibility of the technique, SBF was remeasured in these subjects after an interval of one week to six months.

2. *SBF in Paget's disease.* Resting SBF was measured in six informed, consenting patients with severe (plasma alkaline phosphatase > 30 King-Armstrong units/100 ml), untreated Paget's disease of bone, and in three further patients with less severe disease (alkaline phosphatase between 18 and 30 K-A u/100 ml). SBF was remeasured in one patient after three months' treatment with porcine calcitonin and was also measured in a single adult with progressive diaphyseal dysplasia (PDD) or Engelmann's disease.

Technique

100 μCi of ^{18}F — supplied by the MRC Cyclotron Unit, Hammersmith Hospital, London — and 50 μCi ^{51}Cr-EDTA were injected intravenously, and blood was sampled at intervals over 2 h through an indwelling venous cannula. Urine was collected by voluntary voiding. Plasma volume was measured by dilution analysis with 1 μCi ^{125}I-human serum albumin. Use of ^{51}Cr-EDTA enabled the glomerular filtration rate (GFR) to be measured (1), with which a correction was made for any urinary retention. The total body radiation dose from these procedures is about 2 mrad.

RESULTS

1. Normal SBF. In eight normal, male volunteers, aged between 23 and 57 years, total skeletal blood flow was 4.05—5.39% blood volume/min, with a mean value of 4.89% (S.D. 0.55%). This mean flow was equivalent to 278 ml/min (S.D. 44), 3.45 ml/min/kg body weight (S.D. 0.75) or 236 ml/min/1.73 m^2 body surface area (S.D. 39). The precision of the method, estimated by the replicate observations, was 16.4%.

2. *SBF in Paget's disease.* In comparison with the normal volunteers, SBF was significantly increased in the six patients with untreated Paget's disease, ranging from 7.8 to 13.9% blood volume/min. The three patients with less severe disease had a normal SBF. There was a linear correlation of SBF with plasma alkaline phosphatase (Fig. 1) ($r = 0.73$, $0.02 > p > 0.05$), but no significant correlation with urinary hydroxyproline excretion (Fig. 2).

After three months' treatment with calcitonin, SBF fell significantly in the one patient remeasured (Fig. 3). Resting SBF in the case of Engelmann's disease (PDD) was 20.9% blood volume/min.

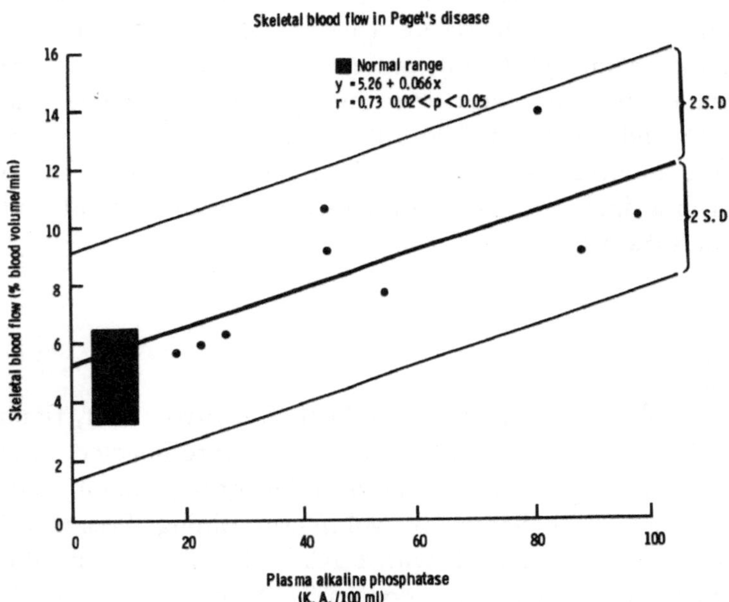

Figure 1. Relation between skeletal blood flow and plasma alkaline phosphatase levels in 9 patients with untreated Paget's disease of bone.

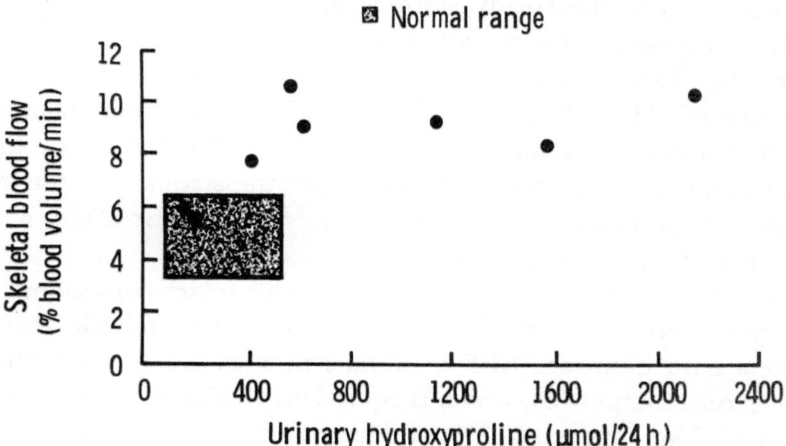

Figure 2. Relation between skeletal blood flow and rate of urinary excretion of total hydroxyproline in 8 patients with untreated Paget's disease of bone.

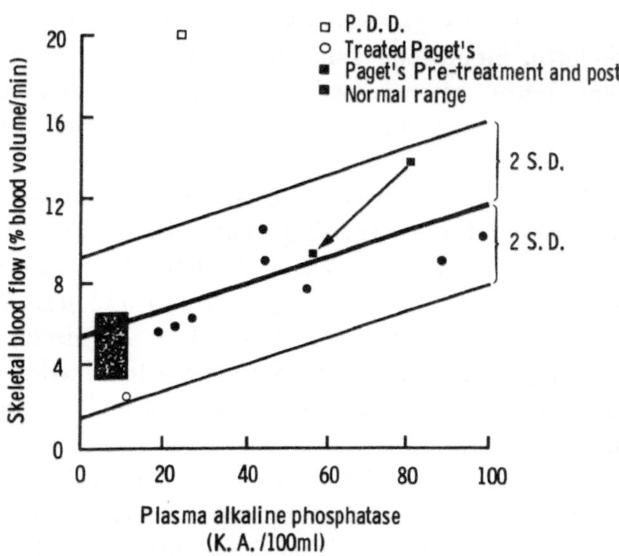

Figure 3. Relation between skeletal blood flow and plasma alkaline phosphatase levels in 9 patients with untreated Paget's disease of bone, in two patients after treatment with porcine calcitonin and in one patient with progressive diaphyseal dysplasia (PDD).

DISCUSSION

The technique for the measurement of SBF described in this paper depends on several assumptions. It is assumed that the system under observation is in the steady state, since use is made of the principle of linear superposition. Correction of the observed excretion of ^{18}F by the use of the plasma clearance of ^{51}Cr-EDTA also assumes a constant renal clearance of ^{18}F, and this necessitates a urine flow of more than 5 ml/min (3). Derivation of the initial skeletal uptake of ^{18}F depends on the use of an adequate analogue for extravascular, extraosseous ^{18}F; the similarity of the results obtained by use of either ^{82}Br or ^{51}Cr-EDTA is evidence that EDTA is an adequate analogue.

The normal range of SBF reported in this paper corresponds with an average rate of perfusion of cortical bone of some 5 ml/min/100 g. Other workers have estimated SBF in normal man as 2.4 ml/min/100 g (6) or 3.3% cardiac output (8), which is equivalent to about 2.4 ml/min/100 g in a man of 7 kg bone mass. The techniques upon which these estimates are based, however, the plasma clearance of bone-seeking isotopes, involve unproven assumptions about the *in vivo* behaviour of the tracers. Thus, the fall with time of the ^{18}F response function indicates the reflux of this tracer from bone and explains the underestimate of SBF by the latter method.

The measurements of SBF in Paget's disease represent an average value for the whole skeleton including its unaffected parts. Quantitation of the well known phenomenon of increased blood flow in individual affected bones must depend on a combination of this technique with quantitative scanning.

In contrast with the more severe cases of Paget's disease, the extremely high value of bone blood flow in the single case of PDD was accompanied by a relatively modest increase in plasma alkaline phosphatase (to 24.4. K-A u/100 ml), hydroxyproline excretion rate (to 515 μmol/24 h) and accretion rate (A_5 by Marshall's (4) method = 47 mmol/day). This indicates that the suggested relationship between osteoblastic activity and SBF in the patients with Paget's disease can be assumed valid only in this latter condition.

The development of a method for the clinical measurement of SBF should promote an improved understanding of the factors regulating skeletal metabolism both in health and in diseases affecting bone. Furthermore, Ray *et al.* (5) have shown that tracer calcium is at least 50% extracted in a single passage through bone, which implies that the activity of osteocytes in moderating influx and efflux of bone surface calcium may be critically dependent on SBF. It also raises the possibility that bone blood flow, besides playing a facilitative role, may be involved in active processes regulating skeletal metabolism.

REFERENCES

1. Chantler, C., Garnett, E.S., Parsons, V. & Veall, N.: Glomerular filtration rate measurement in man by the single-injection method using ^{51}Cr-EDTA. *Clin. Sci.* **37**, 169—180 (1969)
2. Costeas, A., Woodard, H.Q. & Laughlin, J.S.: Depletion of ^{18}F from blood flowing through bone. *J. nucl. Med.* **11**, 43—45 (1970)
3. Hosking, D.J. & Chamberlain, M.J.: Studies in man with ^{18}F. *Clin. Sci.* **42**, 153—161 (1972)
4. Marshall, J.H.: Theory of alkaline earth metabolism. In: *Techn. Reports* Ser. No. 32, I.A.E.A., Vienna, pp. 21—33, 1964
5. Ray, R.D., Kawabata, M. & Galante, J.: Experimental study of peripheral circulation and bone growth. *Clin. Orthop.* **54**, 175—185 (1967)
6. Shim, S.S., Mokkhavesa, S., McPherson, G.D. & Schweigel, J.F.: Bone and skeletal blood flow in man measured by a radioisotopic method. *Canad. J. Surg.* **14**, 38—41 (1971)
7. Stephenson, J.L.: Integral equation description of transport phenomena in biological systems. In: *Proc. 4th Berkeley Symposium on Mathematical Statistics and Probability*, Univ. Calif. Press, pp. 335—345, 1960
8. Van Dyke, D., Anger, H.O., Parker, H., McRae, J., Dobson, E.L., Yano, Y., Naets, J.P. & Linfoot, J.: Markedly increased bone blood flow in myelofibrosis. *J. nucl. Med.* **12**, 506—512 (1971)
9. Wootton, R.: The single-passage extraction of ^{18}F in rabbit bone. *Clin. Sci. molec. Med.* **47**, 73—77 (1974)

Skeletal and Lean Body Mass in Alcoholics with and without Cirrhosis

M.S. ROGINSKY, I. ZANZI & S.H. COHN

INTRODUCTION

The state of the skeleton has not been carefully studied in chronic alcoholism. A decrease in bone mass has been reported in both chronic alcoholics and in patients with cirrhosis. Although this decrease has been attributed to osteoporosis, the recent discovery of the importance of the liver in the metabolic transformation of vitamin D, makes increasingly possible a role for vitamin D defiency in this effect of alcohol and liver disease on the bone.

The present study was undertaken to evaluate most precisely the skeletal mass in two groups of alcoholics, one with and the other without Laennec's cirrhosis. The skeletal mass was determined by the measurement of total body calcium (TBCa) using total body neutron activation analysis (TBNAA) and by bone mineral content (BMC) of the radius, using the photon absorption technique of Cameron and Sorenson. Lean body mass, which is proportional to total body potassium (TBK), was estimated from the measurement of ^{40}K by whole body counting. This determination, was of particular interest, in view of our prior observation of a close relationship between TBK and TBCa in normal man.

In order to compare individual values of TBK or TBCa it was necessary that the absolute values of these measurements be normalized for the variables of age, sex and body size. Previously developed algorithms were used to predict the expected TBCa (Ca_p,) and expected TBK (K_p). The relationship of the observed to the predicted values gave a figure we call the calcium ratio ($TBCa/Ca_p$) and potassium ratio (TBK/K_p), respectively.

Nassau County Medical Center, East Meadow, New York and Brookhaven National Laboratories, Upton, New York.

RESULTS

The TBCa/Ca$_p$, tables I and II, for the cirrhotic group ranges from 0.841 to 1.154 with a mean of 1.083 ± 0.129; in the males the mean was 1.006 ± 0.109 and in females 1.092 ± 0.061. A ratio of 1.00 indicates that any measured value equals the predicted value for a person of that age, sex, and body size. The TBCa/Ca$_p$ for the non-cirrhotic alcoholics, all males, ranges from 0.848 to 1.161 with a mean of 1.023 ± 0.104. The difference between the TBCa/Ca$_p$ of either alcoholic groups was not significantly different from the normal controls (12 males, 0.997 ± 0.056, and 12 females 0.987 ± 0.045). By contrast the TBCa/Ca$_p$ of an osteoporotic group (14 males 0.867 ± 0.067 and 36 females, 0.790 ± 0.100) was unquestionably lower than both groups of alcoholics (p < 0.005).

The TBK/K$_p$ for the cirrhotic group ranges from 0.677 to 1.049 with a mean of 0.879 ± 0.094. In males the mean was 0.899 ± 0.102 and in females 0.859 ± 0.053. On the other hand the TBK/K$_p$ for the non-cirrhotics ranges from 0.864 to 1.181 with a mean of 0.988 ± 0.084.

Table I. Total Body Calcium and Potassium Measurements in Group I Chronic Alcoholics with Cirrhosis.

Subject	Age	Sex/Race	Weight (kg)	Height (cm)	TBK (g)	TBCa (g)	TBK/K$_p$	TBCa/Ca$_p$
P.G.	36	F/B	59	166	77.4	993	0.874	1.122
L.H.	32	F/B	65	172	81.7	967	0.807	0.981
C.S.	52	F/B	73	160	75.3	931	0.850	1.080
A.E.	55	F/W	56	152	66.0	672	0.954	1.154
P.L.	34	F/W	52	158	61.8	884	0.811	1.127
Mean	41.8 ± 9.7		60.7 ± 7.2	161.4 ± 6.9	72.4 ± 7.4	889 ±115	0.859 ±0.053	1.092 ±0.061
W.B.	38	M/B	53	175	90.8	1142	0.812	1.130
E.R.	39	M/B	65	184	130.0	1343	0.957	1.172
R.W.	51	M/B	56	174	106.6	1127	0.981	1.154
W.E.	49	M/W	116	169	148.9	948	0.995	0.841
H.C.	59	M/W	69	162	90.8	846	0.932	0.997
S.D.	59	M/W	71	166	106.0	1048	0.969	1.122
J.J.	49	M/W	77	169	89.3	1006	0.736	0.990
P.M.	55	M/W	68	170	96.8	910	0.859	0.928
D.M.	53	M/W	86	170	135.1	1128	1.049	1.046
J.M.	55	M/W	58	176	76.2	940	0.677	0.922
C.O.	65	M/W	61	161	84.0	707	0.894	0.848
H.P.	37	M/W	75	177	117.7	1144	0.869	1.020
A.S.	44	M/W	68	167	108.0	861	0.955	0.908
Mean	50.2 ± 8.4		71.1 ±15.6	170.9 ± 6.1	106.2 ±20.7	1012 ±160	0.899 ±0.102	1.006 ±0.109

Table II. Total Body Calcium and Potassium Measurements in
Group II Chronic Alcoholics without Cirrhosis.

Subject	Age	Sex/Race	Weight (kg)	Height (cm)	TBK (g)	TBCa (g)	TBK/K_p	TBCa/Ca_p
W.F.	39	M/B	75	173	130.4	1265	1.175	1.026
J.B.	35	M/B	77	181	151.9	1431	1.181	1.053
J.P.	53	M/B	65	175	102.2	1119	0.864	1.161
M.O.	35	M/W	76	181	129.9	1025	0.911	0.900
P.N.	28	M/W	86	180	166.8	1109	0.876	1.083
S.N.	60	M/W	102	170	144.8	996	0.892	1.056
F.W.	54	M/W	80	171	127.6	1045	0.993	1.028
G.F.	34	M/W	81	178	126.0	1128	1.039	0.885
R.D.	40	M/W	85	177	148.5	1210	1.028	1.037
S.D.	46	M/W	96	183	161.4	1322	1.045	1.013
G.M.	31	M/W	76	191	136.1	1289	1.063	0.848
C.F.	42	M/W	73	169	128.3	952	1.074	0.922
Mean	41.4		81.0	177.4	137.8	1157	0.988	1.023
	± 9.5		±10.1	± 6.3	±17.0	±140	±0.084	±0.104

Only the alcoholics with cirrhosis, both male and female, differed significantly ($p < 0.001$) from the control group, males 1.019 ± 0.069, and females 0.963 ± 0.074. The TBK/K_p for the osteoporotics, males 0.954 ± 0.087, and females 0.959 ± 0.109 interestingly enough did not differ significantly from the control group.

The Bone Mineral Content (BMC) of the radius has been reported by us to correlate well with TBCa in normal subjects ($r = 0.973$). In both groups of alcoholics a good correlation between BMC and TBCa was also observed ($r = 0.931$), whereas in osteoporosis this close relationship was not observed. However, the osteoporotics did differ significantly from alcoholics, cirrhotics and normals in the BMC determination. (Table III).

When the ratio TBK/TBCa was examined (Table IV) it was apparent that the cirrhotics differ from all others, normals, osteoporotics and non-cirrhotic alcoholics in this ratio that represents the relationship of lean body mass to skeletal mass ($p < 0.05$).

In summary, these findings do not support a loss of skeletal mass in alcoholics with or without Laennic's cirrhosis. In those studies in which this observation had been made the condition of the entire skeleton was inferred by extrapolation from samples taken at highly selected skeletal sites. The accuracy of this extrapolation is open to question.

This finding is of special interest to us in view of our observation of a significant decrease in the serum 25-Hydroxy vitamin D (25-OHD) concentration of chronic alcoholics with cirrhosis when compared to alcoholics without cirrhosis, 12.3 ± 6.9 ng/ml versus 26.6 ± 11.5 ng/ml (p

Table III. Bone Mineral Content of the Radius of Chronic Alcoholics.

	BMC (g/cm)	B.W. (cm)	BMC/B.W. (g/cm^2)
Group I (Cirrhotics)			
Males (7)	1.159	1.488	0.788
Females (3)	±0.194	±0.159	±0.097
Females (3)	0.936	1.312	0.718
	±0.094	±0.187	±0.033
Group II (Non-Cirrhotics)			
Males (12)	1.305	1.619	0.808
	±0.102	±0.086	±0.071
Control Subjects			
Males (12)	1.192	1.469	0.813
	±0.188	±0.143	±0.107
Females (12)	0.893	1.204	0.749
	±0.093	±0.153	±0.069
Osteoporotics			
Males (4)	0.879	1.406	0.625
	±0.136	±0.078	±0.092
Females (36)	0.646	1.234	0.528
	±0.151	±0.152	±0.116

(   ) number of subjects studied.
BMC bone mineral content of radius.
B.W. width of radius.

< 0.0005). The normal control value was 29.9 ± 9.9 ng/ml. This failure of
a low serum level of 25-OHD in the cirrhotics to lead to a decrease in
TBCa, the latter an objective manifestation of vitamin D-deficiency, can
be explained by a variety of reasons. One unexpected explanation may be
the low values for immunoreactive parathyroid hormone in the serum of
the cirrhotic subjects, 200 ± 97.3 pg/ml versus 425 ± 137.6 pg/ml in the
non-cirrhotic alcoholics (p < 0.0005), the latter value not different from
our range in normal controls, 365 ± 200 pg/ml. Our present explanation
for this finding was the low state of body magnesium in the cirrhotics as
reflected by the low serum magnesium concentration of 1.4 ± 0.3 mg/dl in
this group versus 2.0 ± 0.2 mg/dl in the non-cirrhotics (p < 0.0005).
Ongoing studies may further clarify the significance of these findings.

Table IV. The Relative Deficit in Calcium and Potassium in Chronic Alcoholics.

	$TBCa/Ca_p$	TBK/K_p	TBK/TBCa
Group I (Cirrhotics)			
White Males (10)	0.962 ±0.089*	0.894 ±0.115	0.111 ±0.021
White Females (2)	1.140 ±0.014	0.882 ±0.071	0.084 ±0.014
Black Males (3)	1.152 ±0.021	0.917 ±0.091	0.093 ±0.009
Black Females (3)	1.061 ±0.059	0.844 ±0.027	0.081 ±0.043
Group II (Non-Cirrhotics)			
White Males (9)	0.974 ±0.069	0.991 ±0.084	0.127 ±0.013
Black Males (3)	1.172 ±0.008	0.981 ±0.083	0.100 ±0.024
Control Subjects			
White Males (12)	0.997 ±0.056	1.019 ±0.069	0.122 ±0.008
White Females (12)	0.987 ±0.045	0.963 ±0.074	0.100 ±0.007
Osteoporotic Patients			
White Males (4)	0.867 ±0.067	0.954 ±0.087	0.120 ±0.017
White Females (36)	0.790 ±0.100	0.959 ±0.109	0.109 ±0.021

( ) number of subjects.
* S.D.
TBCa total body calcium (g).
Ca_p predicted total body calcium (g).
TBK total body potassium (g).
K_p predicted total body potassium (g).

The alteration in the cirrhotics of the constant relationship of TBK to TBCa we have observed in both normals and osteoporotics is not inconsistant with many other reports of a decrease in body potassium in cirrhosis, with a variety of factors possibly responsible. The absence of this loss of potassium in the alcoholics without cirrhosis, on the other hand, was somewhat surprising and remains in conflict with other reports of such an observation.

ACKNOWLEDGEMENTS

Research supported in part by The Energy and Research Development Agency.

The Use of Total Body *in vivo* Neutron Activation Analysis (TBIVNAA) in Balance Studies in Rodents

D.A. SMITH[1], R.L. LINDSAY[1], K. BODDY[2], A. ELLIOTT[2], I. HOLLOWAY[2] & J. ANDERSON[1]

In the investigation of animals subject to alteration in diet or other metabolic experiments, the measurements of change in body calcium, phosphorus, sodium and nitrogen are of considerable interest. However, conventional balance studies are tedious and subject to both random and cumulative error, necessitating as they do accurate estimates of dietary intake and faecal and urinary output. The object of the present study was to determine the usefulness of total body *in vivo* neutron activation analysis (TBIVNAA), used at the beginning and end of the experimental period, as an alternative to conventional balance techniques.

METHODS AND MATERIALS

Twelve Wistar female rats were divided randomly into two groups of six. One group was fed on a standard diet containing 794 mg calcium, 537 mg phosphorus, 95 mg magnesium per 100 g of diet. The second group was fed on a diet containing 11.5 mg calcium, 197 mg phosphorus and 1012 mg of magnesium per 100 g of diet. During the period of the study the animals were kept in individual metabolic cages. The experiment was run for 44 days.

At the start of the experiment, the natural body radioactivity of each animal was measured in a fixed position using a Shadow Shield Whole Body Counter (3). The animals were then irradiated with 14 MeV electrons for 60 sec at 12 cm using a Phillips 18602 Tube Neutron Generator (3, 4). Sixty seconds later, the radioactivity induced (^{28}Al

1) University Departments of Medicine & Orthopaedic Surgery, The Western Infirmary, Glasgow.
2) The Scottish Universities Research & Reactor Centre, East Kilbride, Glasgow.

from phosphorus and ^{13}N) was measured in a Whole Body Counter. On the following day, the whole body activity was again measured and then the animals were irradiated with neutrons from a Nuclear Reactor (1, 2), and counted for Ca and Na in a Whole Body Counter. At the end of the 44-days period the procedure was repeated and the animals then sacrificed. The right femurs were then removed from five of the animals and the left femurs from five others. The mass per unit length of bone (K value) was measured using a Gamma-ray Densitometer (5). The femora were then ashed and the total ash, calcium and phosphorus determined.

RESULTS

The control animals (Table I) showed a significant increase in the total calcium (35.5%), phosphorus (17.3%), sodium (108.9%), and nitrogen (26.4%), counts over the period of the study. The experimental group showed no significant change in total accounts for calcium (6.4%), and phosphorus (1.9%). There were, however, significant increases in the sodium (112%) and nitrogen (18.6%) counts. In Table II, this data is shown recalculated as the mineral to nitrogen ratios in order to allow for increase in mean body mass over the period of the study. Expressed in this way, the control animals showed a small, but not significant, increase in the Ca:N ratio (7.1%) and a small, significant fall in the P:N ratio (7.0%). There were significant increases in the Na:N (64.8%) and Ca:P (15.2%) ratios. The experimental group showed a small but significant falls in the

Table I.

The Change in Mean Whole Body Counts in the Control and Experimental Groups at the Beginning and End of the 44-days Experimental period.

Group	Initial mean whole body count	Final mean whole body count	Mean change (counts)	Significance (Wilcoxon's test)
Control Calcium	2449.33	3318.17	+ 868.84	< 0.01
Control Phosphorus	42385.67	49722.33	+ 7336.67	< 0.01
Control Sodium	170.92	357.63	+ 186.71	< 0.01
Control Nitrogen	101933.33	128884.17	+26950.84	< 0.01
Experimental Calcium	2481.00	2639.67	+ 158.67	n.s.
Experimental Phosphorus	40410.67	39636.67	+ 774.00	n.s.
Experimental Sodium	182.10	397.23	+ 215.13	< 0.01
Experimental Nitrogen	99963.50	118508.83	+18545.33	< 0.01

n.s. = not significant

394

Table II.
The Change in the Mean Mineral:Nitrogen Ratio (x 10^{-3}) and Ca:P Ratio
in the Control and Experimental Groups at the Beginning and End of the 44-days
Experimental Period.

Group	Initial mean ratio	Final mean ratio	change	Mean Mean change	Significance (Wilcoxon's test)
Control Ca:N	2.413	2.584	+	0.171	n.s.
Control P:N	4.156	3.866	−	0.290	< 0.02
Control Na:N	1.686	2.779	+	1.093	< 0.01
Control Ca:P	5.813	6.697	+	0.884	< 0.05
Experimental Ca:N	2.480	2.239	−	0.241	< 0.05
Experimental P:N	4.042	3.345	−	0.397	< 0.01
Experimental Na:N	1.824	3.335	+	1.511	< 0.01
Experimental Ca:P	6.161	6.689	+	0.528	n.s.

n.s. = not significant

Ca:N (9.7%) and P:N (9.8%) ratios. There was a significant increase in the Na:N ratios (82.8%) and Ca:P (8.6%). The calcium and phosphorus ratios increased by a small amount in both groups of animals (Table II). However, this only reached statistical significance in the control group.

There was a highly significant relation between the total body counts for calcium and phosphorus and the mass per unit length of femur measured by gamma-ray densitometry ($y = 0.12x - 124.1$, $r = 0.89$, $t = 5.5$; and $y = 0.008x - 134.1$, $r = 0.93$, $t = 7.0$). There was also a highly significant relation between the total body counts for calcium and phosphorus and the calcium and phosphorus content of the femur determined chemically ($y = 0.045x - 39.01$, $r = 0.88$, $t = 5.3$; and $y = 0.002x - 33.7$, $r = 0.94$, $t = 7.97$).

DISCUSSION

In the present study, it must be remembered that the whole body estimations of calcium, phosphorus, sodium and nitrogen, refer to isotope counts and not to molar quantities. The results show that the control groups of animals on a normal diet had a significant increase in these constituents over the period of the study. There was no significant increase in calcium and phosphorus in the experimental group, though significant increases in sodium and nitrogen occurred. When the changes were estimated as mineral:nitrogen ratios in the control group, no significant change in the Ca:N ratio was seen, though a small fall in the P:N ratio occurred. The experimental group showed a significant fall in both the Ca:N and P:N ratios. The results indicate a significant decrease in the amounts of calcium and phosphorus in the experimental group in relation to mean body mass, and, therefore, represent a true failure of the experimental group to gain mineral on the experimental diet. Both groups showed a significant increase in Na:N ratio and a very slight increase in Ca:P ratio, which only reached significance in the control group.

However, measured changes in whole body constituents represent possible changes in both soft tissue and in the skeleton. The object of the present study was to determine mineral changes in bone. The whole body changes were, therefore, correlated with the bone mass per unit length of femora measured by photon absorption, and to the calcium and phosphorus content of the femora determined from the ashed samples. These measurements were found to be highly significantly related.

In this study, the measurement of relative changes in total calcium, phosphorus, sodium and nitrogen using TBIVNAA in rodents confirmed the many advantages of the technique. It is both a consistent and rapid method of determining when changes in bone have reached statistical significance so that studies can be concluded at the appropriate point in time. Moreover, the technique is not subject to cumulative errors, which inevitably occur due to the difficulties arising in conventional balance techniques. There is the added advantage of being able to relate the changes in whole body mineral to nitrogen, and thus to mean body mass. This is an especially important and useful measurement in growing animals, where dietary or therapeutic regimes may cause non-specific changes in the rate of growth without proportional changes in skeletal mass.

ACKNOWLEDGEMENT

We gratefully acknowledge the financial support of The National Fund For Research Into Crippling Diseases.

REFERENCES

1. Boddy, K.: *In vivo* activation analysis of iodine in the thyroid gland — a preliminary study. *Proc. 7th Symposium on Radioactive Isotopes in Clinical Medicine and Research*, Urban und Schwarzenberg, München, p. 377, 1966
2. Boddy, K. & Alexander, W.D.: Clinical experience of *in vivo* activation analysis of iodine in the thyroid gland: An assessment of the problems. *Proc. Symposium on Nuclear Activation Techniques in the Life Sciences*, I.A.E.A. Vienna, p. 583, 1967

3. Boddy, K., Holloway, I., Elliott, A., Glaros, D., Robertson, I. & East, B.W.: Low cost facilities for partial-body and total-body *in vivo* activation analysis in the clinical environment. *Proc. Symposium on Nuclear Activation Techniques in the Life Sciences*, I.A.E.A. Vienna, p. 589, 1972
4. Boddy, K., Holloway, I. & Elliott, A.: Preliminary results of measuring total-body calcium with a new facility for total-body *in vivo* activation analysis. I.A.E.A. Vienna, p. 163, 1973
5. Shimmins, J., Smith, D.A., Aitken, M., Anderson, J.B. & Gillespie, F.C.: The accuracy and reproducibility of bone mineral measurements *in vivo*. (b) Methods using sealed isotope sources. *Clin. Radiol.* 23, 47—51 (1972)

Variations in Bone Mass and Bone Activity within the Mandible

NINA VON WOWERN

Bone mass and bone activity are depending on the function of the bone. When bone function is reduced, bone mass is decreased (7). The function of the mandible varies in the three regions: the incisive (I-) — premolar (P-) — and molar (M-) region. The function depends on the dentition, since it is changed after extraction of teeth. Also, the attachments of the muscles differ from region to region, causing great variations in the shape within the mandible, in the course of the trajectories, and the thickness of the cortices from region to region.

The purpose of the present study is to analyse the variations in bone mass between cortices in I-, P-, and M-region, and between buccal and lingual cortices of the same region of the mandibular body, and to estimate possible differences in variations in bone mass between groups of mandibles with different dentition (dentate, partially dentate, edentulous).

MATERIAL AND METHODS

The material consisted of 24 autopsy specimens of half mandibles and blocks of iliac crests two cm behind the anterior spine in the right side from the 24 normal subjects who in the period May-August 1972: 1) were suddenly dead without previous illness, and on whom 2) removal of autopsy specimens was possible. The subjects consisted of 8 woman and 16 men in the age of 26 to 87 years ($\bar{x}$ = 53 years). None were osteoporotic neither clinically nor according to Nordin's 9-point scale (2).

Eight mandibles were dentate, 8 mandibles were partially dentate, and 8 mandibles were edentulous. The dentition in the maxilla corresponded to the state of dentition in the mandible. The latter groups were all

Department of Oral Surgery, Royal Dental College and Department of Oral Surgery and Oral Medicine, Rigshospitalet, Copenhagen.

wearing dentures in both jaws. Each mandible was cut in three blocks, consisting of the I-, P-, and M-region. The mesial side of each block was perpendicular to a plane through the top of the alveolar process and the lowest border of the mandibular base in the region. The mesial side was placed just in front of the first tooth in the different regions. In edentulous mandibles the mesial side of the blocks in the I-region was placed in the midline; in the P-region it was placed where the two cortices began to be parallel and in the M-region: 5 mm distal for the most anterior point of the mental foramen. All bone blocks were radiographed, no unhealed tooth sockets were found. The bone-blocks were embedded in methylmetacrylate (6). Three ground sections of 250 μ of the I-, P-, and M-regions were cut with a rotating saw parallel to and 2 mm from the mesial side of the blocks and with 1 mm between each section. Two sections from the corresponding iliac crest autopsies were cut *ad modum* Nordin (2). The sections were ground under water to 100 μ in thickness. For microradiography a Machlett 50 O:E:G X-ray tube with wolfram anode generated at 12 e.E. with 12 m.A. was used, exposures being made on Kodak spectroscopic plates 649—0 for 10 min at a focus-film distance of 15 cm and a developing time of 4 min. Each microradiogram of the sections from the mandible was divided of 4 lines (*a, b, c,* and *d* on Fig. 1). The line *a* is going through the middle of the top of the alveolar process and the lowest point of the mandibular base; *b, c,* and *d* are perpendicular to *a*. The line *b* is the tangent *to* the endosteal surface in

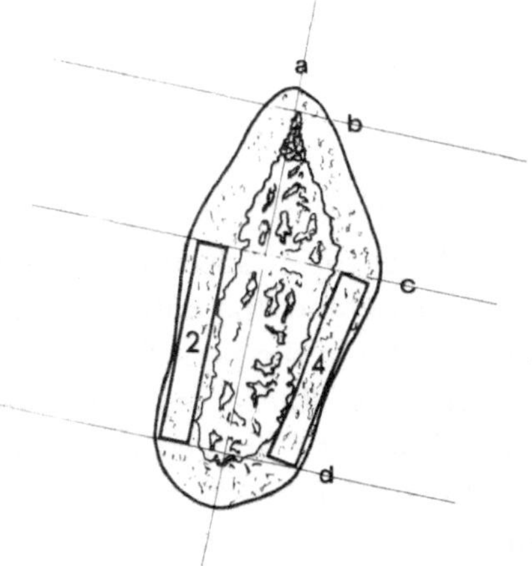

Figure 1. A buccal-lingual section of the molar-region of the mandible. The line *a* is going through the middle of the top of the alveolar process and the lowest point of the mandibular-base. Above the line *c* is the alveolar process, between *c* and *d* is the mandibular body, in which the rectangular areas 2 and 4, respectively, cover the buccal and lingual cortex.

endentulous alveolar processes; c is a line through the apices of teeth, or in edentulous regions: a line, where the distance to the lowest point of the base corresponds to the distance between the mesial point of the mental foramen and the base; d is a tangent to the endosteal surface of the base.

Leitz Classimat[R], an automatic image analysis system was used for estimation of bone mass in the cortices of the mandibular body. The microradiograms were placed under the microscope and the microscopic picture was projected in 50 x linear magnification on the television camera screen. The rectangular mask on the television camera screen was adapted to cover the maximum of the cortices of the mandibular body (Fig. 1). The mask was placed between the periostal circumferentel lamellae and the outer points of the endosteal surface. The bone mass in the cortices was estimated in per cent of the area of the mask (Fig. 1, areas 2 and 4).

RESULTS

I. Variation in bone mass within the buccal and lingual cortex in the incisive, premolar, and molar region
The means of the three values for bone mass in per cent in buccal and lingual cortices, measured on the three sections from same region were calculated, as well as the average standard deviation $s = 2.57$.

The means (named I, P, and M) were used as an expression for bone mass in the I-, P-, and M-region of each cortex.

II. Variation in bone mass between the incisive-, premolar-, and molar-region in the buccal and in the lingual cortex
The means of bone mass in the I-, P-, and M-region were graphically compared two and two, I-P, P-M, and I-M, respectively in the buccal and the lingual cortex.

In diagrams of the relations I-P and I-M buccally (Fig. 2a), six points were displaced to the left for the rest of the points. These points corresponded to small values of I in the buccal cortex. Five of the cases were the eldest subjects in the material. The sixth case was a 26-year-old man, showing great differences between the buccal measurements in the I-region due to a defect here.

The rest of the points in the I-P and I-M diagrams (Figs. 2a, b), as well as all 24 points in the P-M diagram were placed in a continuous group. This structure was also observed in the three diagrams for the lingual cortex.

In the diagram of I-P buccally, and the lingual I-P, P-M, and I-M, the points were grouped around the line y = x, while the points on the

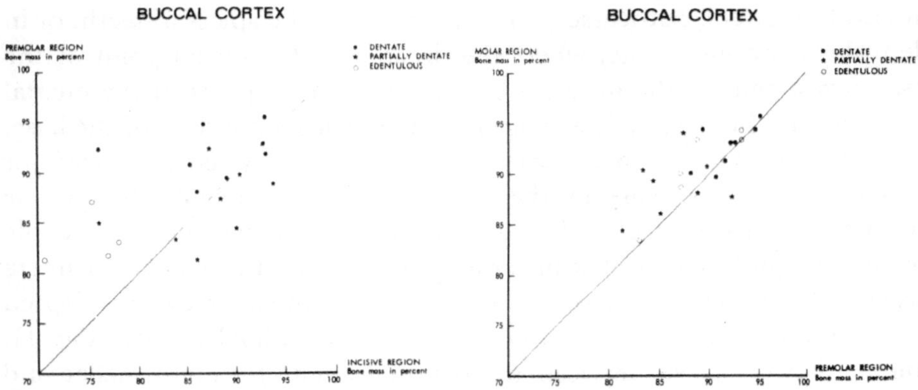

Figure 2a. Diagram on the relations between bone mass in per cent in buccal cortices of incisive- and premolar-regions. Notice the six points, displaced to the left, and the rest of the points placed in a continuous region around the line y = x.

Figure 2b. Diagram on the relations between bone mass in per cent in buccal cortices of premolar- and molar-regions. Notice that the points is displaced from the line y = x.

diagrams for P-M and I-M buccally (Fig. 2b) were clearly displaced from the identity line. For both cortices a slight correlation was found between P and M, but no correlation between I and P, and I and M.

There were no systematic difference between the placing of the points for the different groups of dentition, dentate, partially dentate and edentulous.

Fig. 3 shows the total means of bone mass in the I-, P-, and M-regions of

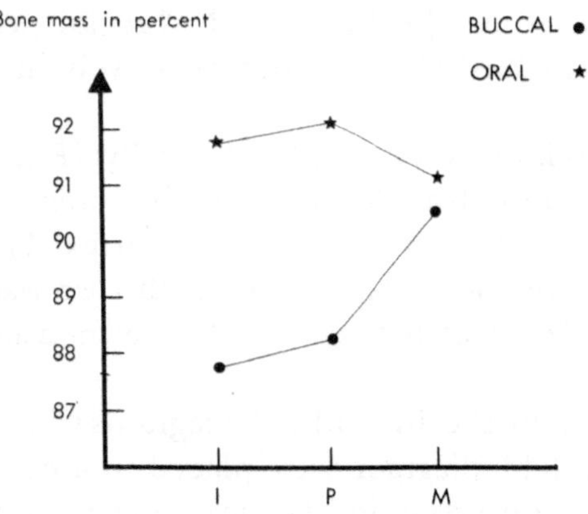

Figure 3. Diagram on the total means of bone mass in the incisive (I), premolar (P), and molar (M)-region of the buccal and lingual (ORAL) cortex of the mandibular body.

the cortices in the entire material. The figures are corrected for exclusion of single observations by means of the estimation technique of 2-way-analysis of variance with unequal number of cells. The diagram shows that the average percentage of bone mass, lingually lies on the same level in I-, P-, and M-regions. Further, the average percentage of bone mass for the buccal I and P-regions lies on the same level, but lower than the level for the M-value, which is on the same level as the average values for the lingual cortex. Tests for homogeneity of the three lingual regions gave: $F = 1.28$, $(2/42)$. $P = 29\%$, i.e. no significance. Corresponding test for the three buccal regions gave: $F = 5.57$, $(2/34)$, $P = 0.81\%$, i.e. a high significance.

The statistical analysis did not reveal any differences between the three groups of dentition with respect to the mean structure in this diagram.

III. Variation in bone mass between buccal and lingual cortices in the three regions
Graphical comparison of buccal and lingual measurements of I-, and P-regions produced diagrams of the same appearance as Fig. *2b*, while the points in the diagram for the M-region were grouped around the y = x line. These observations are reflected in the differences between buccal and lingual bone mass in the three regions in Fig. 3. Tests for differences between lingual and buccal bone mass showed a high significance in region I- and P-, while there was no significance between buccal and lingual M-values.

DISCUSSION

Several quantitative analyses of variations in bone mass in different bones of the skeleton have been made (5, 6, 7, 10). A few have been dealing with the jaws (1, 3, 4, 8), but have only been concerning the age-bound variations in bone mass in single parts of the jaw. The present analysis, therefore, is the first systematic comparative analysis of the variations in bone mass on a series of ground sections of dentate, partially dentate, and edentulous mandibles.

Quantitation of bone mass with Leitz Classimat[R] on 100 μ thick ground sections has been chosen, because histoquantitation with an electronic point-counting-system is the quickest and most exact method for quantitation (9, 11, 12). Ground sections of 100 μ were chosen, since they give the best contrast on microradiograms (6). Also, they are found most suitable for comparative analysis, resulting in the highest repro-ducibility of the bone mass, analysed by such systems (12). Microradio-grams of ground sections were prefered to died undecalcified 5—8 μ thin sections or died decalcified sections, as shrinkage, defects, distortions, and unadequate contrast on the black-white television camera screen may occur.

As to the results of the analyses the only low values for bone mass was found in the buccal cortex of the I-region of the five eldest subjects. This may be caused by a change in function of the mental muscle. By exclusions of these values, no significant difference was found between the means of I and P buccally. Contrary to this there was a high significance in differences between the means of P and M (Fig. 3). This seems to show, that bone mass in the buccal cortex of a mandibular body lies on the same level in the I- and P-region, while the buccal cortex is more dense in the M-region, than in the I- and P-region (Fig. 3). In old subjects, though, a low density in the buccal cortex of the I-region may be expected.

In the present material no significant differences were found between the lingual I, P, and M means (Fig. 3). Therefore, the lingual cortex of the entire mandibular body may be expected to show the same degree of density.

Significant differences, respectively, between the means of the buccal and lingual I and P means were found, but none between the two M means (Fig. 3). This indicates that the lingual cortex in the I- and P-region is more dense than the corresponding buccal cortex, while the cortices of the M-region have the same degree of density. In the present material no differences were found between the three groups of dentition with respect to the mean structure. This indicates that the variations in bone mass within a single mandible may be independent of the state of dentition. On the other hand, this does not exclude the possibility that minor local changes in the bone density, or general changes in the density, may occur after extraction of teeth.

Conclusively, for comparative analysis of pathological conditions, or sex and age-bound variations in bone mass in mandibular bodies, biopsies must be taken either in the premolar or the molar region in all cases. Further analyses are necessary to show if the state of dentition has a marked influence upon the density of the mandibular cortices.

ABSTRACT

A study of the intermandibular variations in bone mass between the buccal and lingual cortices in the incisive-, premolar-, and molar-region, and between buccal and lingual cortices of the same region of mandibular bodies has been carried out, as well as an estimation of possible differences in the intermandibular variations in bone mass between groups of mandibles with different dentition (dentate, partially dentate, edentulous). The material consisted of 24 autopsy specimens of half mandibles and iliac crests from normal subjects, suddenly dead without previous illness. Microradiograms of ground sections of 100 μm were used. Quantitation of bone mass in per cent within the cortices was done by an electronic point-counting system (Leitz Classimat(R)). The analysis showed that: 1) the density of lingual cortices seems to be independent of the region, 2) except for old subjects, the density of buccal cortices generally lies on the same level in the incisive- and the premolar-region, while the density in the molar-region is higher and on the same level as lingually, 3) the density of the buccal cortex in the incisive- and premolar-region is lower than the density of the corresponding lingual cortex, and 4) the variations in bone mass within the cortices of mandibular bodies seem to be independent of the state of dention. This does not exclude that extractions of teeth cause changes in bone density.

ACKNOWLEDGEMENTS

The autopsy specimens were obtained from the University Institute of Forensic Medicine, The University of Copenhagen. The technical part of the work was carried out at the Bone Laboratory, Orthopedic Department, Rigshospitalet, Copenhagen, Denmark. The help from these departments is gratefully acknowledged.

Supported by grants from the Danish Medical Research Counsil, No. 512—2686, 512—3049.

REFERENCES

1. Atkinson, P.J. & Woodhead, C.: Changes in human mandibular structure with age. *Arch. oral. Biol.* 13, 1453—1463 (1968)
2. Beck, J.S. & Nordin, B.E.E.: Histological assessment of osteoporosis by iliac crest biopsy. *J. Path. Bact.* 80, 391—397 (1960)
3. Henrikson, P.-Å. & Wallenius, K.: The mandible and osteoporosis (I). *J. oral Rehab.* 1, 67—74 (1974)
4. Henrikson, P.-Å. & Wallenius, K.: The mandible and osteoporosis (II). *J. oral Rehab.* 1, 75—84 (1974)
5. Jowsey, J.: Variations in bone mineralisation with age and disease. In: *Bone Biodynamics.* Frost, H.E. (ed.), Churchill, London, pp. 461—479, 1964
6. Jowsey, Jenifer, Kelly, P.J., Riggs, B.L., Bianco, A.J., Scholz, D.A. & Gershon-Cohen, J.: Quantitative microradiographic studies of normal and osteoporotic bone. *J. Bone Jt. Surg.* 47A, 785—806 (1965)
7. Little, K.: Degenerative condition. In: *Bone Behavior.* Little, K. (ed.), Academic Press, p. 303, 1973
8. Manson, J.D. & Lucas, R.B.: A microradiographic study of age changes in the human mandible. *Arch. oral Biol.* 7, 761—769 (1962)
9. McQueen, C.M., Smith, D.A., Monk, I.B. & Horton, P.W.: A television scanning system for the measurement of the spatial variation of microdensity in bone sections. *Calc. Tiss. Res.* 11, 124—132 (1973)

10. Meunier, P.: *La Dynamique du remaniement osseux*. Thesis. Lyon, 1967
11. Williams, E.D.: Automated histoquantitation studies of bone. *Proc. roy. Soc. Med.* **65**, 539–541 (1972)
12. Wowern, Nina v.: Histoquantitation of ground sections of human mandibles. *Scand. J. dent. Res.* **81**, 567–571 (1973)

CHAPTER IX
Osteoporosis. Etiology, Patogenesis, Diagnosis, and Treatment

Osteoporosis – A Clinical Review

F. KUHLENCORDT

In my clinical review I shall treat questions concerning definition, classification, course of disease, and therapy. I shall try to address not only the clinicians but also the theorists, hoping that they will more easily understand the clinical problems than clinicians sometimes can understand when listening to the theorists.

DEFINITION

The theoretical problems
Osteoporosis was defined mainly according to pathological and anatomical principles by Pommer in 1885 (7), and again by Albright and Reifenstein in 1948 (1). This has held good until now, and we, therefore, speak of a deficit in bone-mass in comparison with a healthy control group of corresponding age and sex, while the quality of bone substance is normal. According to this definition, the term of osteoporosis clearly marks a pathological condition, which has to be sharply distinguished from physiological bone loss with increasing age. This age-dependent atrophy is sometimes called physiological osteoporosis, opposed to a pathological osteoporosis (4). Besides, another definition of osteoporosis, based on clinical and symptomatic aspects, is used, which led to the term of crush-fracture-syndrome (5). The disadvantage of such a name lies in the fact that here obviously the later stages of a surely long-standing disease with negative bone balance are called osteoporosis.

An early diagnosis of primary osteoporosis will remain Utopian because of the wide range of variation of bone mass in healthy people of all age-groups. It is unimportant from which kind of examination conclusions about bone mass are drawn. This is true for both histomorphometry of bone biopsies and the numerous radiologic methods for determination of

Department Clinical Osteology – I. Medical University Clinic, Hamburg.

mineral content, or whole body calcium by neutron activation analysis. This situation cannot be improved by increasing the precision of measurement of the different methods.

PATHOGENESIS AND CLASSIFICATION

When looking at the clinical classification of osteoporosis by Albright and Reifenstein (1) one can find etiological as well as pathogenetic principles. By now, it is generally agreed that pathogenesis of osteoporosis can uniformly be regarded under the aspect of a pathologically negative bone balance. A loss of bone mass without a negative balance of bone remodelling is inconceivable. It has to be taken into account that the different parts of the skeleton are not affected to the same degree. I am thinking of the fact that there are cases of osteoporosis where almost exclusively the spine is affected, whereas the periphery remains practically unconcerned. Microscopic examinations have shown that the pathological bone remodelling of primary osteoporosis affects above all the endosteal surfaces, whereas the Haversian systems of the corticalis and periost are hardly ever involved (2).

Table I. Etiology of Osteoporosis.

OSTEOPOROSIS

I.   GENERALIZED

      A.   PRIMARY   (UNKNOWN ETIOLOGY)

      B.   SECONDARY (KNOWN ETIOLOGY)

II.   LOCALIZED

      A.   PRIMARY   (UNKNOWN ETIOLOGY)

      B.   SECONDARY (KNOWN ETIOLOGY)

In spite of the uniform pathogenesis of osteoporosis, etiology can be very different (Table I). In case it is found, often after extensive clinical examinations, we speak of a secondary osteoporosis. The large group of those with unknown etiology remains, called primary or idiopathic osteoporosis (6) (cases of generalized as well as of localized forms, although we cannot be sure about a primary form of localized osteoporosis).

Table II. Types of Generalized Osteoporosis.

GENERALIZED   OSTEOPOROSIS

A. PRIMARY	B. SECONDARY
SYN. : IDIOPATHIC	
IUVENILE	ENDOCRINE
PRESENILE	RENAL
POSTMENOPAUSAL	GASTROINTESTINAL
SENILE	etc.

In the following I shall confine myself to generalized osteoporosis (Table II). Numerous diseases may account for the secondary form. They are listed as endocrine, gastro-intestinal, renal, etc. Here I shall not go into details about the various diseases that might be considered. By far the largest group is that of the primary or idiopathic osteoporosis. Four out of five of our patients suffer from primary osteoporosis and only one from secondary osteoporosis. The so-called juvenile, presenile, post-menopausal, and senile osteoporosis are also primary osteoporoses. These adjectives do not say anything about etiology, but only inform us about the moment of diagnosis or clinical manifestation. This fact is often not considered when one is dealing with the so-called post-menopausal osteoporosis. Per definition, osteoporosis is a pathological condition with a bone mass below the lower level of the physiological range of variation, which is — thank God — not reached by all menopausal women. Menopause with decreasing oestrogen production, however, will only be important for the increased physiological bone loss at this age.

COURSE OF THE DISEASE

It is to be assumed that the disease can take various courses (3), which I will here show schematically (Fig. 1). In this graph you can see the physiological reduction of bone mass with increasing age, represented as a linear correlation. Considering the normal range of variation, the condition above the mark must be referred to as hyperostosis, below the mark as osteoporosis. The dashed line in the lower part delineates the

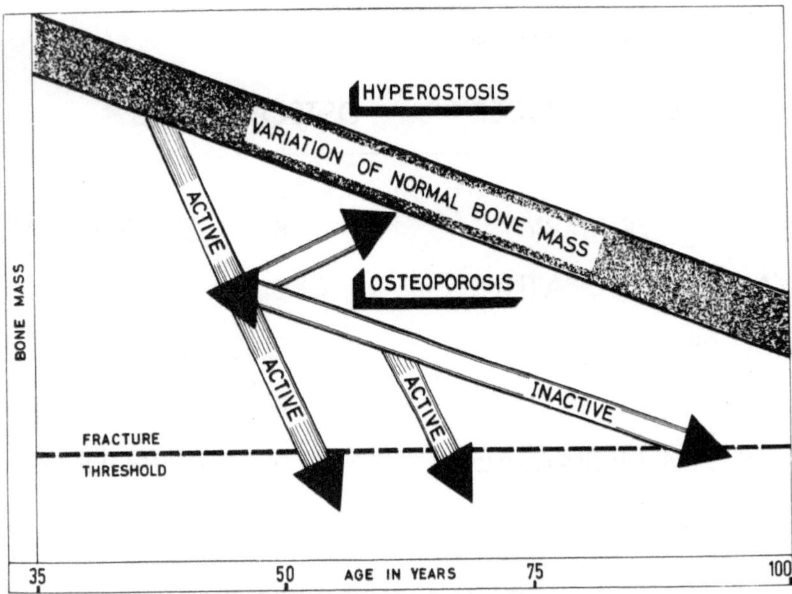

Figure 1. Time Course of Osteoporosis.

limit below which a reduction of bone mass causes spontaneous fractures. This applies to the spine as well as to the peripheral skeleton.

As a first possibility I show an osteoporosis which develops in two phases and is cured at the end. In the first phase pathologically negative bone balance, a loss of bone mass, occurs. We are thinking here of a secondary osteoporosis caused by e.g. hyperthyroidism. After treating the basic disease and reaching euthyroidism, the following development is not yet quite clear. We might imagine a phase of reparation – as shown here – where normal bone mass is attained again. Whether this is possible, probably depends on the degree of osteoporosis reached and on the age of the respective patients. In most cases of secondary osteoporosis, bone remodelling normalized after successful treatment. Thus, in the second phase, a physiological bone loss develops on a lower level. A similar process must be supposed to occur also in many cases of primary osteoporosis. This conclusion can be drawn from the finding that in some of the patients with manifest osteoporosis a physiological bone remodelling is detected by bone histology. The pathological reduction of bone mass must, therefore, have taken place in an earlier phase. However, until to-day it could not be ascertained whether more than one such gradually deteriorating period of pathological bone resorption can occur. As far as our experience goes, this last course of an osteoporosis is probably the one that is most rarely met with. It surely occurs in a secondary form, if the basic disease has not been diagnosed and thus remained untreated for years. The graph (Fig. 1) presents a synopsis of all the possibilities discussed, from which a number of consequences appears. Among other things, the figure explains why the occurrence of a spontaneous fracture need not necessarily be a sign of a quick progression of osteoporosis. The fracture limit can be reached in the phase of a pathologically negative bone balance as well as in physiological bone resorption. Various questions concerning the evaluation of cases of secondary osteoporosis appear in a different light when considered on the basis of these concepts.

THERAPY

In our opinion, the ideas developed here concerning the possible courses which osteoporosis can take, are of fundamental importance for the choice of an effective therapy. During the active phases with quickly decreasing bone mass, several therapeutic principles will perhaps be necessary. On the other hand, one will have to search intensively for a

Table III. Principles for Treatment of Osteoporosis.

OSTEOPOROSIS THERAPY

I. BASIC THERAPY

    a) TREATMENT OF PAIN
    b) PHYSICAL THERAPY
    c) DIET

II. ETIOLOGIC THERAPY

    a) TREATMENT OF BASIC DISEASE IN SECONDARY OSTEOPOROSIS
    b) UNKNOWN OF PRIMARY OSTEOPOROSIS

III. PATHOGENETIC THERAPY

    a) POSITIVE INFLUENCE ON BONE BALANCE
      1. STIMULATION OF BONE FORMATION
      2. DIMINUTION OF BONE RESORPTION
      3. COMBINATION OF 1. + 2.

    b) POSITIVE INFLUENCE ON CALCIUM BALANCE
      1. STIMULATION OF INTESTINAL CALCIUM – ABSORPTION
      2. DIMINUTION OF RENAL CALCIUM – EXCRETION

IV. ADDITIONAL MEASURES

    a) ORTHOPEDIC TREATMENT
    b) SURGICAL TREATMENT

410

basic disease causing secondary osteoporosis. However, when diagnosing an inactive osteoporosis with physiological bone remodelling, it is inconceivable that at the moment of examination a pathophysiological process is taking place (secondary osteoporosis). Prognosis in this case is comparatively favourable, as progression does not exceed the physiological limits. We are subdividing the therapeutic principles into four groups: basic therapy, etiological treatment, pathogenetic treatment, and additional measures (Table III) (3).

I am not going to discuss basic therapy and additional measures at this moment. Since in secondary osteoporosis there exists a disease outside the skeletal system, etiological therapy in these cases consists in a treatment of the basic disease. As the etiology of primary osteoporosis is unknown per definition, no etiological treatment of this largest group of osteoporosis is possible. Only a therapy based on pathogenetic principles can be considered, aimed at a positive influence on bone balance on the one hand, and on calcium balance on the other. This method of treatment is, of course, also possible in cases of secondary osteoporosis, if the basic disease can only be treated insufficiently or if no restitution of osteoporosis is to be expected after the end of treatment. Which of the above mentioned possibilities of treatment based on pathogenetic principles should be chosen, of course depends on the respective results of examination. It is hardly imaginable that an essential therapeutical influence on the course of the disease can be gained in case of an equilibrated calcium balance and a physiological bone remodelling. The only exception is perhaps the therapy aimed at supporting bone formation, a therapy that has developed promisingly within the last years.

For many years scientists of various disciplines have worked on the problems presented here. In this condensed survey, I have on purpose abstained from mentioning any names in particular. The concept of osteoporosis I developed for you, is based on a synthesis of the many experiences of others and of my own team.

ACKNOWLEDGEMENT

Supported by Deutsche Forschungsgemeinschaft, SFB 34 "Endocrinology".

REFERENCES

1. Albright, F. & Reifenstein, E.C.: *The Parathyroid Glands and Metabolic Bone Disease.* Williams and Wilkins, 1948

2. Frost, H.M.: The spinal osteoporoses. *Clinics in Endocrinology and Metabolism* 2, 155—158 (1973)

3. Kuhlencordt, F. & Kruse, H.-P.: Was ist gesichert in der Therapie der Osteoporose und Osteomalacie? *Internist (Berl.)* 15, 588—593 (1974)

4. McLean, F.C. & Urist, M.R.: *Bone. Fundamentals of the Physiology of Skeletal Tissue.* 3rd Ed., University of Chicago Press, 1968

5. Nordin, B.E.C.: Introduction, *Clinics in Endocrinology and Metabolism* 2, 257—275 (1973)

6. Nordin, B.E.C.: *Metabolic Bone and Stone Disease*, Churchill Livingstone, 1973

7. Pommer, G.: *Untersuchungen über Osteomalacie und Rachitis.* Vogel, Leipzig, 1885

Impaired Binding of Estradiol to Vaginal Mucosal Cells in Post-menopausal Osteoporosis

F.J. BARTIZAL, CAROLYN B. COULAM, T.A. GAFFEY, R.J. RYAN & B.L. RIGGS

Strong circumstantial evidence implicates the menopause as an important factor in pathogenesis of both age-related bone loss in the general population of post-menopausal women and in women with the clinical syndrome of post-menopausal osteoporosis. First, osteoporosis is up to ten times more common after age 50 in women than in men (1). While it is true that total skeletal mass before bone loss begins is greater in young men than in young women, the volume of trabecular bone (the type of bone primarily lost in osteoporosis) is actually initially greater in young women than in young men (2). Second, epidemiologic studies evaluating changes in bone density with age have shown that beginning after age 50 women lose more bone than men (2, 6). A relationship to the menopause is particularly suggested by the study of Meema and co-workers (5). They separated the variables of chronologic age and years elapsing since menopause by comparing 112 pre-menopausal women, 56 post-menopausal women who had had a physiologic menopause and 35 young women who had been surgically castrated for benign gynecologic conditions. The young surgical castrates had significantly less bone (p < 0.001) than their pre-menopausal peers; however, when assessed as a function of years since menopause the amount of bone loss was the same as in post-menopausal women. They concluded from this data that the important variable was the menopause and not chronological age. Third, institution of chronic estrogen therapy slows the rate of bone loss as assessed by peripheral bone densitometry (7, 8). Finally, short-term therapy with physiologic doses of estrogen in women with post-menopausal osteoporosis reduces bone turnover (6, 9, 10).

However, it seems unlikely that the menopause is the only etiologic factor causing post-menopausal osteoporosis. All post-menopausal women are estrogen dificient, yet only some of them develop osteoporosis. In 27 post-menopausal women with and 27 post-menopausal women without

Mayo Clinic and Mayo Foundation, Rochester, Minnesota.

	Non-osteoporotic*	Osteoporotic*	P
Age (yr)	65.6 ± 1.15	66.2 ± 1.6	NS
Bone density (g/cm)			
Midradius	0.89 ± 0.02	0.74 ± 0.03	<0.001
Distal radius	0.81 ± 0.02	0.68 ± 0.03	<0.001
Serum estrogen (ng/100 ml)	3.95 ± 0.37	4.63 ± 0.35	NS
Plasma testosterone (ng/100 ml)	50.3 ± 4.2	51.6 ± 4.5	NS
Serum FSH (μg LER/100 ml)	234.7 ± 17.3	252.6 ± 13.5	NS
Serum LH (μg LER/100 ml)	34.5 ± 2.3	35.8 ± 2.4	NS

*) Values are mean ± SE

clinical evident osteoporosis we found no significant difference in the degree of post-menopausal production of sex hormones. (Table)

Also, if estrogen deficiency were the sole etiologic factor then it would be expected that a physiologic replacement dose of estrogen would normalize the abnormality in bone turnover in post-menopausal osteoporosis but, in fact, this does not occur. Women with untreated osteoporosis have increased values for bone resorption surfaces and normal values for bone formation surfaces (microradiography of bone biopsies) (11). After chronic estrogen therapy bone resorbing surfaces decrease to within or slightly above normal levels but bone forming surfaces decrease to within or slightly above normal levels but bone forming surfaces decrease to subnormal levels (9). Radiocalcium kinetic studies have shown similar results (10). This suggests that there is an additional abnormality of bone cell function or modulation that is not completely corrected by estrogen therapy.

These observations suggested to us that patients who either have, or who are prone to develop, excessive post-menopausal bone loss have a resistance to the biologic effect of estrogens so that the low post-menopausal serum concentrations of estrogen have less of a protective effect on bone than do similar concentrations in non-osteoporotic post-menopausal women. Tissue resistance to sex steroids (estrogens and perhaps androgens) would account for the failure of full replacement doses of estrogens to normalize the abnormality of bone turnover and would also explain the clinical observations that, compared with normal peers, women with post-menopausal osteoporosis have a greater reduction in the number of cornified cells in the vaginal epithelium, have thinner and more transparent skin, and have less axillary hair (6, 12, 13).

These considerations led us to consider strongly the possibility that a defect in estrogen binding to receptors in target tissues including bone

414

could cause resistance to the action of sex steroids and may be one of the fundamental abnormalities responsible for bone loss and osteoporosis in post-menopausal women. Also, an analagous defect in estrogen binding to bone cells could contribute to the pathogenesis of bone loss and osteoporosis in older men with apparently normal gonadal function.

METHODS

Estrogen binding to vaginal mucosa (an easily accessible estrogen sensitive tissue) was studied in 15 osteoporotic and in 15 normal post-menopausal women. Mean age and number of years since the menopause did not differ significantly. Vaginal epithelial cells, removed with a metal spatula, were immediately placed in a nutrient broth and divided into experimental and control samples. Both samples were treated identically except that the control sample was preincubated with an excess of unlabelled estrogen to saturate binding sites and to correct for non-specific (non-saturable) binding. 0.05 μCi of high specific activity ^{3}H-E$_2$ (40 to 60 Ci/mM), was added and both mixtures equilibrated for 1 h. The suspended cells were centrifuged and washed three times. The cellular sediment was then drop-plated on slides which were coated with an autoradiographic emulsion (Kodak Nuclear Tract NTB–2). The procedures were all done at $20°$C. The slides were then exposed at $4°$C in a vacuum desiccator for 30 days. They were developed and then stained with a Papanicolaou stain. The estrogen binding was quantitated by counting the number of dots of darkened emulsion ("grain counting") due to β-disintegrations from the ^{3}H-E$_2$ in the underlying cells. Slides were coded so that the processes of cell selection and quantitation were done without identification in order to eliminate observer bias. Under microscopic examination, 10 nucleated precornified cells on Each of the control and experimental autoradiographs were selected randomly and photographed. Grain counts made from the control sample were subtracted from those made from the experimental sample to derive counts from specific ^{3}H-E$_2$ binding.

RESULTS

The grain counts per cell showed significantly less ^{3}H-E$_2$ binding (P < 0.001) in the osteoporotic (47 + 7 ± 7.1) than in the non-osteoporotic (123.6 ± 14.0) post-menopausal women.

DISCUSSION

Sex steroid action is related to tissue concentrations of specific receptor proteins (14). The initial step in the interaction of these hormones with cells of target tissue is the binding to a specific receptor in the cytosol. This binding which is not found in non-target tissue, produces a change that allows the receptor-steroid complex to migrate to specific receptor sites on the nuclear chromatin, which in turn alters or modifies the rate of nuclear synthesis of RNA. Our data suggest but do not prove that post-menopausal osteoporotic women have decreased binding of estrogen to specific receptors in target tissue. Confirmation will require isolation and quantitation of tritiated-estrogen bound to cytosol receptors using sucrose-gradient centrifugation or related techniques. Studies investigating this possibility are underway.

ACKNOWLEDGEMENT

Supported in part by grants AM8658, RR585 and HD3726 from the National Institute of Health, Public Health Services.

REFERENCES

1. Albright, F., Smith, P.J. & Richardson, A.M.: Post-menopausal osteoporosis: Its clinical features. *J. Am. Med. Assoc.* **116**, 2465 (1941)
2. Meunier, P., Courpron, P., Edouard, C., Bernard, J., Bringuier, J. & Vignon, G.: Physiological senile involution and pathological rarefaction of bone: Quantitative and comparative histological data. *Clin. Endocr. and Metab.* **2**, 239–256 (1973)
3. Vose, G.P., Hoerster, S.A. & Mack, P.B.: New techniques for the radiographic assessment of bone density. *Am. J. Electron.* **3**, 181 (1964)
4. Nordin, B.E.C., MacGregor, J. & Smith, D.A.: The incidence of osteoporosis in normal women: Its relation to age and the menopause. *Quart. J. Med.* **35**, 25–28 (1966)
5. Meema, H.E. & Meema, S.: Prevention of post-menopausal osteoporosis by hormone treatment of the menopause. *Can. Med. Assn. J.* **99**, 248–251 (1968)
6. Gallagher, J.C. & Nordin, B.E.C.: Calcium metabolism and the menopause. In: *Biochemistry of Women.* Curry, A.S. and Hewitt, J.C. (eds.), CRC Press, Cleveland, Ohio, 1974
7. Davis, M.E., Strandfjord, N.M. & Lanzl, L.H.: Estrogens and the aging process. *J. Am. Med. Assn.* **196**, 129–134 (1966)
8. Aitken, J.M., Hart, D.M. & Lindsay, R.: Oestrogen replacement therapy for prevention of osteoporosis after oophorectomy. *Brit. Med. J.* **3**, 515–518 (1973)
9. Riggs, B.L., Jowsey, J., Goldsmith, R.S., Kelly, P.J., Hoffman, D.L. & Arnaud, C.D.: Short- and long-term effects of estrogen and synthetic anabolic hormone in post-menopausal osteoporosis. *J. Clin. Invest.* **51**, 1659–1663 (1972)

10. Harris, W.H. & Heaney, R.P.: Skeletal renewal and metabolic bone disease. *N. Engl. J. Med.* **280**, 193–311 (1969)
11. Jowsey, J., Kelly, P.J., Riggs, B.L., Bianco, A.J. Jr., Scholz, D.A. & Gershon-Cohen, J.: Quantitative microradiographic studies of normal and osteoporotic bone. *J. Bone Jt. Surg. Am.* **47**, 785 (1965)
12. McConkey, B., Bligh, A.S., Fraser, G.M. & Whiteley, J.: Transparent skin and osteoporosis. *Lancet* (1973) **I**, 693–695
13. Smith, R.W. Jr.: Dietary and hormonal factors in bone loss. *Fed. Proc.* **26**, 1737–1746 (1967)
14. King, R.J.B. & Mainwaring, W.I.P.: Steroid-cell interactions, University Park Press, Baltimore, 1974

Acid-induced Osteoporosis: An Experimental Model of Human Osteoporosis

U.S. BARZEL

SUMMARY

Adult animals, male or female, react to ammonium chloride ingestion with a non-hormonal, generalized, slow but progressive and unrelenting mobilization of bone and develop osteoporosis which seems indistinguishable in all parameters measured from human osteoporosis. (Table I).

Table I.

Comparison of Human Osteoporosis and Experimental (Acid-induced) Osteoporosis.

Parameter	Human osteoporosis	Animal model
Blood calcium & phosphorus	normal	normal
Bone histology	osteoporosis	osteoporosis
Bone microradiography	increased resorption	increased resorption
	normal formation	normal formation
Bone per cent ash	normal	normal
Bone calcium content	diminished	diminished
Bone crystallography	normal	normal

Osteoporosis can be defined as a condition in which generalized, progressive diminution in bone density renders bones increasingly vulnerable to fracture. Clinically, osteoporosis has a slow natural history and an unpredictable course: it is virtually impossible to predict if and when any given patient is about to have a fracture (21), and in some cases the progression of the disease seems to have been arrested altogether (19). This slow and unpredictable nature of osteoporosis makes it desirable to use experimental animal models in its study. The present paper will

Metabolic Endocrine Laboratory, Department of Medicine, Montefiore Hospital and Medical Center, Albert Einstein College of Medicine, Bronx, New York.

summarize available information on the model of acid-induced osteo-porosis (12, 3, 4, 5, 6).

THEORETICAL CONSIDERATIONS AND MECHANISMS

Since the end of the last century, bone has been recognized to contain large amounts of alkaline salts and has been thought likely to participate in the maintenance of acid base balance of the organism. The fact that during acid loading urinary calcium excretion always increases gave support to this idea (2). Recent studies in normal man (14, 15), in renal acidosis (10, 16), and in starvation acidosis (18) gave additional support to the concept that bone was mobilized in acid loading and its salts made available to augment the buffering capacity of the extracellular fluid.

In 1969 we suggested, on the basis of experimental observations, that cellular mechanisms involved with bone formation and bone resorption were responsive to changes in acid base balance: microradiographic studies demonstrated that the chronic ingestion of ammonium chloride caused an increase in bone resorption surfaces without change in bone formation surfaces, and, conversely, the chronic ingestion of an alkali solution caused an increase in bone formation surfaces without change in bone resorption surfaces (4). Delling and Donath (9) demonstrated by electronmicroscopy that the bone resorption which occurs as a direct consequence of the dietary ingestion of excess acid is indeed mediated by bone cells. They demonstrated that osteoclasts and osteocytes were activated in acid loaded animals even in the absence of the parathyroid glands. Acid-induced bone resorption was shown by Reed and Beck to be independent of gastrointestinal and renal factors and was observed in animals after nephrectomy and after removal of the entire GI tract (17). Cuisinier-Gleizes *et al.* also found that ammonium chloride ingestion caused bone resorption in the presence and in the absence of the parathyroid glands (8). This group, however, did not find histologic evidence of cellular mediation of this response. They believe that the resorption of bone is not mediated by cells and also that bone formation is depressed in ammonium chloride fed animals.

The observations cited above further strengthen the hypotesis that resorption of bone is a *bona fide* physiologic compensatory response to the ingestion of acid.

THE EXPERIMENTAL MODEL

Adult animals given ammonium chloride solution as their drinking fluid, or fed ammonium chloride in the diet, develop histologic osteoporosis (12, 4, 5, 9, 8), with a ratio of osteoblasts to osteoclasts comparable to that of control animals (8). Ammonium chloride ingestion does not affect the length or volume of the femur (4, 5, 6). There is a decrease in femoral density as measured gravimetrically, by Archimedes' principle (4, 5, 6), or by radiograph densitometry. These density measurements correlate closely with each other and with chemical analyses (7). The latter show that bone fat free weight, bone ash, bone matrix, and the calcium content of the femur all decrease as a result of ammonium chloride ingestion* but the ratio of ash to matrix and the relationship of fat free weight to calcium content of the bone remain unchanged (4, 5, 6). (Figs. 1 and 2). Chrystallographically, the size of the appatite crystals and the per cent of crystalline and amorphous fractions are not affected by ammonium chloride ingestion (5).

Other observations made in animals chronically fed ammonium chloride reveal that blood calcium and phosphorus are normal (4). Blood sodium and potassium are also normal (4), and osteoporosis develops in the presence of normal arterial blood pH (4, 8). The blood chloride tends to be high and the blood carbon dioxide content slightly depressed (4). Clearly, the organism makes available stored base buffer, the bone carbonate *and* the hydroxyappatite, to help maintain acid base neutrality in the face of excessive ingestion of acid.

Observations are essentially the same when male or female rats are studied (4, 5, 6). The degree of osteoporosis that is observed seems to depend on the dose of ammonium chloride used and the duration of its administration.

Low calcium diet has been shown previously to produce osteoporosis by a parathyroid dependent mechanism (11, 13). The effect of ammonium chloride is evident in animals given a low calcium diet to the same degree as in animals given a regular laboratory diet. The combination of ammonium chloride and a low calcium diet is additive in its effect, causing a more profound degree of osteoporosis than either ammonium or a low calcium diet alone (4, 5). (Figs. 1 and 2).

Another potential mechanism for the development of osteoporosis is the withdrawal of estrogenic hormones (1). The effect of ammonium

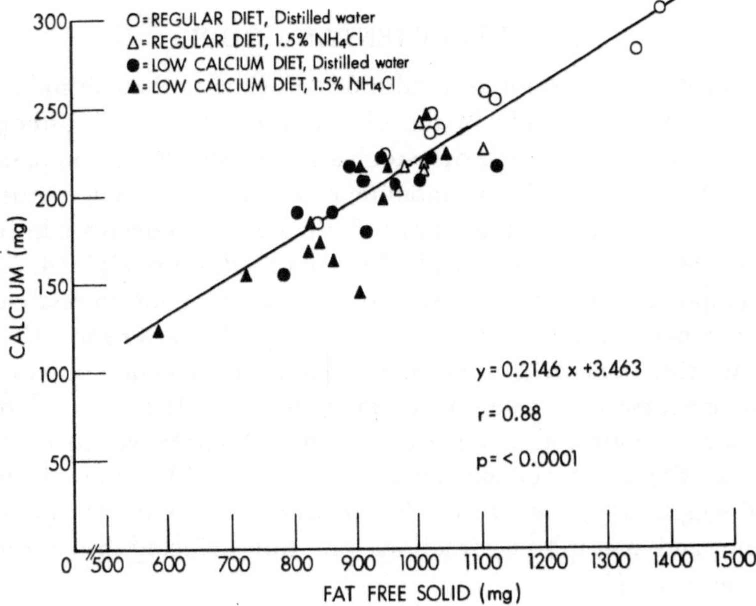

Figure 1. Fat free solid weight and calcium content of the right femur of 250 g male rats fed a regular or a low calcium diet and distilled water or 1.5% ammonium chloride for 330 days. (Modified from Barzel & Jowsey (4)).

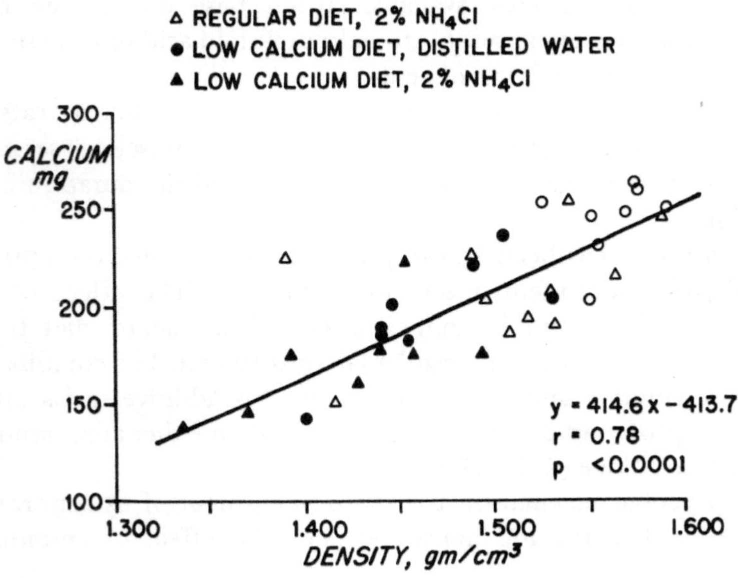

Figure 2. Density and calcium content of the right femur of 550 g male rats fed a regular or a low calcium diet and distilled water or 2% ammonium chloride for 6 months. (Modified from Barzel (6)).

chloride was compared with that of oophorectomy in adult female rats. Oophorectomized animals, allowed a regular diet *ad lib.*, failed to develop any change in bone density, fat free weight, ash and calcium content of the femur, except when given ammonium chloride. The oophorectomized animals were not more sensitive to ammonium chloride than control animals (6).

REFERENCES

1. Aitken, J.M., Armstrong, E. & Anderson, J.B.: Osteoporosis in the mature female rat and the effect of oestroge and/or progesterone replacement therapy in its prevention. *J. Endocr.* **55**, 79–87 (1972)
2. Albright, F.A. & Reifenstein, E.C.: *The Parathyroid Gland and Metabolic Bone Disease: Selected Studies*, Williams and Wilkins, Baltimore, p. 241, 1948
3. Barzel, U.S.: Acid base balance and bone metabolism: A study in the etiology of osteoporosis. *Clin. Res.* **14**, 485 (1966)
4. Barzel, U.S. & Jowsey, J.: The effect of chronic acid and alkali administration on bone turnover in adult rats. *Clin. Sci.* **36**, 517–524 (1969)
5. Barzel, U.S.: The effect of excessive acid feeding on bone. *Calc. Tiss. Res.* **4**, 94–100 (1969)
6. Barzel, U.S.: Studies in osteoporosis: The long term effect of oophorectomy and of ammonium chloride ingestion on the bone of mature rats. *Endocrinology* **96**, 1304–1306 (1975)
7. Colbert, C. & Barzel, U.S.: Radiographic determination of rat bone mineral loss in dietary-induced osteoporosis. *Invest. Radiol.* **7**, 339–340 (1972)
8. Cuisinier-Gleizes, P., Benest, D., George, A. & Thomasset, M.: Thyroparathyroid glands, bone and acid-base homeostasis. *XIth Europ. Symp. Calc. Tiss.*, May 25–29, Elsinore, Denmark, Abstract book, p. 70, 1975
9. Delling, G. & Donath, K.: Morphometrische, electronen-mikroskopische und physikalisch-chemische Untersuchungen *uber die experimentelle Osteoporose bei chronischer Acidose.* Virchows Arch. Path. Anat. Abt. A **358**, 321–330 (1973)
10. Goodman, A.D., Lemann, J. Jr., Lennon, E.J. & Relman, A.S.: Production, excretion, and net balance of fixed acid in patients with renal disease. *J. clin. Invest.* **44**, 495–506 (1965)
11. Harrison, M. & Fraser, R.: Bone structure and metabolism in calcium-deficient rats. *J. Endocr.* **21**, 197–204 (1960)
12. Jaffe, H.L., Bodansky, A. & Chandler, J.P.: Ammonium chloride decalcification as modified by calcium intake: relation between generalized osteoporosis and ostitis fibrosa. *J. exp. Med.* **56**, 823–834 (1932)
13. Jowsey, J. & Raisz, L.G.: Experimental osteoporosis and parathyroid activity. *Endocrinology* **82**, 384–396 (1968)
14. Lemann, J. Jr., Lennon, E.J., Goodman, A.D., Litzow, J.R. & Relman, A.S.: The net balance of acid in subjects given large loads of acid or alkali. *J. clin. Invest.* **44**, 507–517 (1965)
15. Lemann, J. Jr., Litzow, J.R. & Lennon, E.J.: The effects of chronic acid loads in normal man: Further evidence for the participation of bone mineral in the defense against chronic metabolic acidosis. *J. clin. Invest.* **45**, 1608–1614 (1966)

422

16. Litzow, J.R., Lemann, J. Jr., & Lennon, E.J.: The effect of treatment of acidosis on calcium balance in patients with chronic azotemic renal disease. *J. clin. Invest.* **46**, 280–286 (1967)
17. Reed, S.W. & Beck, N.: Acute metabolic acidosis on calcium mobilization from bone and calcium excretion, and on parathyroid hormone actions. *Clin. Res.* 23, 328 A (1975)
18. Reidenberg, M.M., Haag, B.L., Channick, B.J., Shuman, C.R. & Wilson, T.G.G.: The response of bone to metabolic acidosis in man. *Metabolism* 15, 236–241 (1966)
19. Schenk, M., Rodstein, M. & Barzel, U.S.: Frequency and rate of progression of osteoporosis. Israel *J. Med. Sci.* 10, 1471 (1974)
20. Smith, D.A., Lindsay, R., Boddy, K., Elliott, A. & Anderson, J.: The use of total body *in vivo* neutron activation analysis in balance studies in rodents.
21. Urist, M.R., Gurvey, M.S. & Fareed, D.O.: Long term observations on aged women with pathologic osteoporosis. In: *Osteoporosis*, Barzel, U.S. (ed.), Grune and Stratton, New York, pp. 3–37, 1970

Effect of Weightlessness on Mineral Metabolism; Metabolic Studies on Skylab Orbital Space Flights

G.D. WHEDON[1], L. LUTWAK[2], P. RAMBAUT[3], M. WHITTLE[3], CAROLYN LEACH[3], JEANNE REID[1] & M. SMITH[3]

After eight years of preparation, which took place before and during the Apollo-Moon space-flight series, the National Aeronautics and Space Administration in May 1973 launched into orbit a cylinder 22 feet in diameter and 48 feet in length called the Skylab Orbital Workshop. The living quarters and storage area of this cylinder contained nearly all of the life support and scientific experimentation equipment and supplies for three separately launched three-man astronaut crews. The first crew remained in orbit for 28 days, the second for 60 days, and the last crew for 84 days.

Of the more than 100 scientific experiments or observations in the Skylab Program, 16 were devoted to the medical or physiological performance of man in the weightless state in space. This report provides the principal results of one of these studies, that on mineral and nitrogen balance. This "experiment" required of the cooperating astronauts fairly constant dietary intake, continuous 24-h urine collections and total fecal collections for 21–31 days before each flight, throughout each flight and for 17–18 days post-flight for a total of 909 man days of metabolic study.

Prediction that the various stresses of space flight, particularly weightlessness, would bring about significant derangements in the metabolism of the musculoskeletal system has been based on various mineral and nitrogen balance study observations of long immobilized or inactive bedrest. The earliest was that of Deitrick, Whedon and Shorr (1) in 1948.

[1]) National Institute of Arthritis, Metabolism, and Digestive Diseases, National Institutes of Health, Bethesda, Maryland.

[2]) University of California at Los Angeles School of Medicine, Veterans Administration Hospital, Sepulveda, California.

[3]) Biomedical Research Division, Johnson Space Center, NASA, Houston.

Immobilization of four healthy young men in body casts for 6 to 7 weeks led to marked increases in urinary calcium and significantly negative calcium balances, and there were related losses of nitrogen and phosphorus. Several subsequent bedrest studies of normal subjects confirmed these substantial metabolic derangements (2). The longest observation (3) showed that although the elevated urinary calcium subsided partially during the third and fourth months of bedrest, it nevertheless remained significantly higher than control levels for as long as bedrest was continued (for 7 months) and did not fall to normal until the subjects were put back on their feet.

The only attempt — not fully successful — at controlled metabolic observations in space flight prior to Skylab was performed by us (4) in conjunction with the 14-day Gemini-VII flight in 1965. That relatively short study revealed quite modest losses of calcium and phosphorus and varied changes in the metabolism of other elements.

PROCEDURE

A cardinal principle of metabolic study is that changes in the excretion of key nutrient elements, such as calcium or nitrogen, can only be interpreted as due to the influence or agent under test if environmental factors are kept as constant as possible from phase-to-phase and from day-to-day. One of the most important of these factors in metabolic study is the dietary intake. We had relatively little control over the selection by NASA, for various reasons of stability and acceptability, of the 70-odd space food items, the main types of which were solid foods in plastic packages, sticky semi-liquid foods in metallic tins, magnetically held in flight to a warming tray, and drinks in closed plastic cylinders. Selection was constrained for most items by the requirement of stability at room temperature in space for more than a year; only seven frozen food items could be used. In addition, in an effort at best acceptability, many items were mixtures of foods and thus not conducive to exactness of composition in their production. Although these foods were far from ideal for balance studies, nevertheless by skillful, lengthy consultations with the astronaut crew members, our dietitians developed for each crewman, sequences of six menus of similar elemental composition which were rotated steadily throughout the pre-flight, in-flight and post-flight study phases. Whenever a particular food could not be consumed, a system of rapid calculation and provision of supplement tablets for pertinent elements helped to maintain dietary element constancy. During the flight phase, the crew's evening report to Ground Control included the relatively infrequent dietary omissions; rapid calculations were then made for deficits in key elements; the correct number of supplement tablets or capsules were prescribed up to the flight; then from previously stowed supplies, the prescribed supplement tablets (or capsules) were taken next morning.

Twenty-four-hour urine collections were made throughout the studies, but in-flight, because of limitations in return weight and volume, approximately 120 ml aliquots of each days' urine collection were taken, frozen and returned to Earth, using a very complex system because of the absence of gravity. In weightlessness there are difficult technical problems of collecting urine, separating liquid from air and taking a well-mixed measured aliquot, all without the aid of gravity which we so take for granted. In addition, because volume cannot be measured in the weightless state in the same way as on Earth, in-flight 24-h urine volumes were determined by a tracer dilution technique, using lithium chloride, preinjected into the 24-h collection bags. In-flight stool samples were dried in the Workshop and returned to Earth *in toto*. No metabolic balance study was ever perfect, and this extraordinary venture under bizarre circumstances certainly lived up to that tradition; but the studies seem to have been done well enough to have provided some quite clear answers.

RESULTS

Urinary creatinine excretion revealed considerably more fluctuation than is found under ideal urine collection conditions, but the values were consistent enough to indicate that average 24-h urinary creatinine excretion was not changed by space flight.

Urinary calcium during the 28-day flight (SL-2) increased steadily to a plateau in virtually the same pattern as in bedrest studies. As seen in bedrest, inter-individual variation occurred in degree of loss, but the peak reached during the latter part of flight was from 80% greater to more than double the control, pre-flight levels. During recovery, post-flight, urinary calcium excretion subsided promptly toward control levels. In the 60-day flight (SL-3), the same pattern of gradual rise occurred in two crewmen and a rather abrupt rise in the third; there was also inter-individual variation in degree of loss, which in one was to much more than double control levels. Urinary calcium data in the 84-day flight (SL-4) showed the same characteristics, plus the added point of interest of no suggestion of decline toward the end of the flight in the high level of excretion (Fig. 1). Urinary hydroxyproline increased in flight with considerable inter-

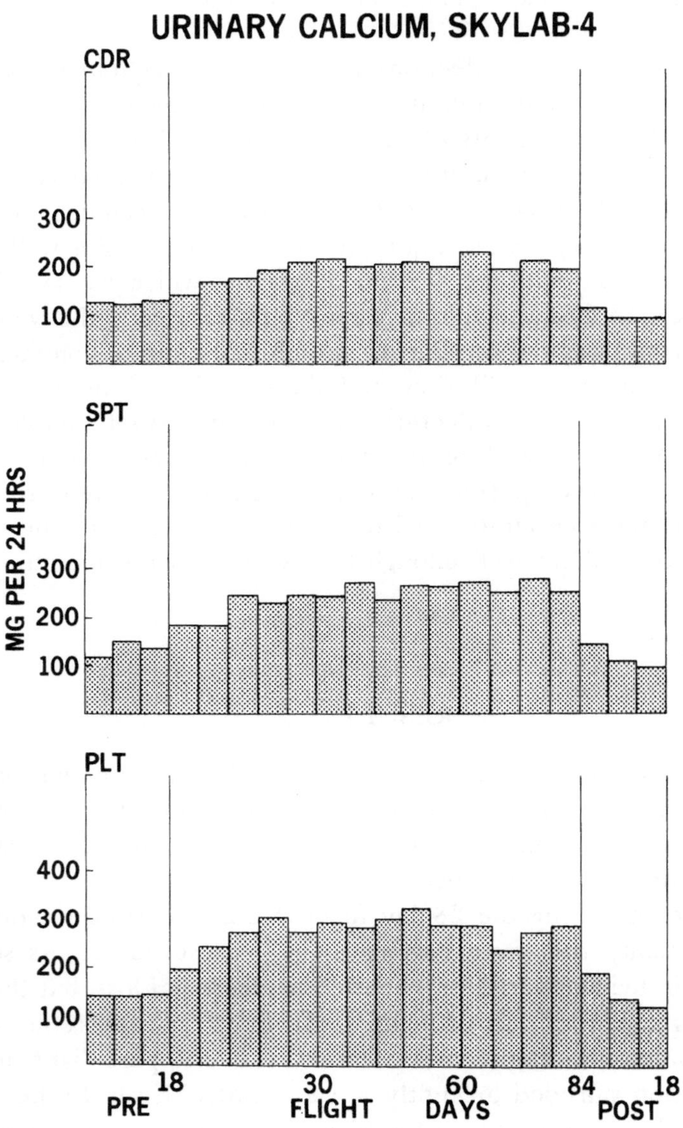

Figure 1. Effect of space flight on urinary calcium excretion in the three astronauts of the 84-day flight (SL-4).

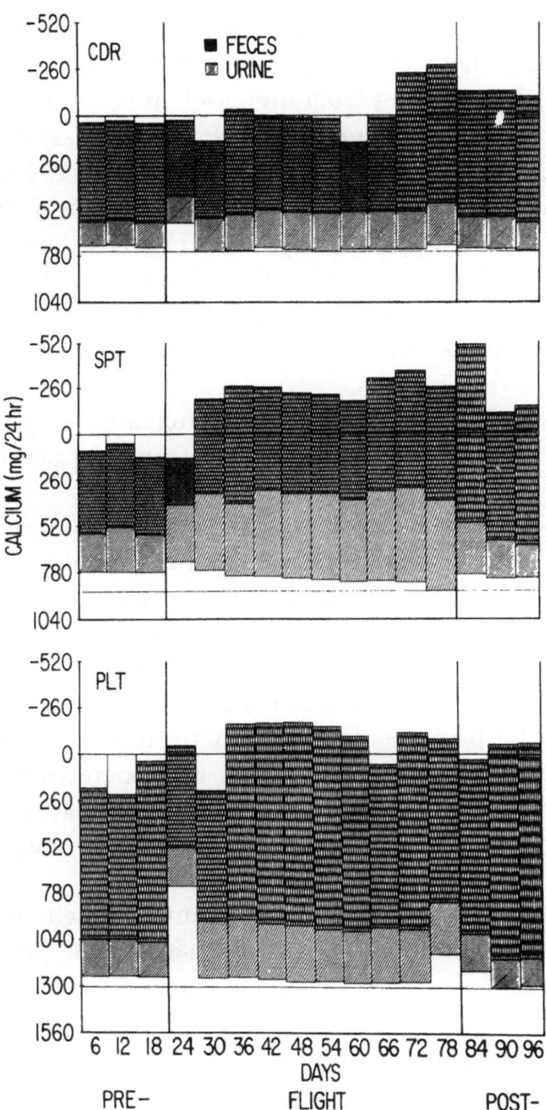

Figure 2. Calcium balances before, during and after space flight in the astronauts of the 60-day flight (SL-3). In this balance graph the data are plotted in conventional Albright-Reifenstein style, the intake downward from the zero base-line, then urinary (light shading) and fecal (heavy shading) excretion upward from the intake lines; shaded areas above the zero base-line indicate negative balance or loss.

individual differences; the mean increase for the six crewmen of the first two flights was 33%.

For the 28-day flight fecal calcium increased during flight in one crewman (CDR) and decreased slightly in the other two; calcium balance became negative in two crewmen and changed imperceptively in the third (SPT). For the 60-day flight, the negative shift in calcium balance was more apparent (Fig. 2), resulting from increases in both urinary and fecal calcium. The mean *shift* in calcium balance for all six crewmen from control phase to the last 16—18 days in-flight was minus 184 mg/day. The mean negative calcium balance during the second month in space for the three astronauts on the 60-day flight was 140 mg/day. Analyses of fecal calcium for the 84-day flight are undergoing checking and re-analysis. In this longest flight significant decreases in the bone density of the os calcis occurred in the two astronauts with the greatest increases in urinary calcium excretion.

Phosphorus balance data showed a distinct increase in-flight in urinary phosphorus, a small increase in fecal phosphorus, and negative balance in all.

Nitrogen balance data revealed during flight a pronounced increase in urinary nitrogen excretion, while fecal nitrogen remained characteristically unchanged. In the 60-day flight the highly negative balance of the first 6-day period was due to the lowered intake resulting from marked anorexia during the first two to three days in the new weightless environment; nitrogen balance continued negative for a few weeks and then was only slightly positive despite high protein and calorie intake levels. The mean shift in nitrogen balance (for the six crewmen of the first two flights) from pre-flight phase to flight was 4.5 gm/day. In the 84-day flight (SL-4) increase in nitrogen excretion of similar magnitude were observed.

With respect to other elements, significant, though variable and generally small losses occurred in-flight in magnesium, potassium and sodium.

COMMENT

The increases in urinary calcium were strikingly similar in both pattern to and nearly as great in degree as the rises in urinary calcium seen in immobile bedrest. In addition, as compared with immobile bedrest (1), the negative shift in calcium balances during flight in the six SL-2 and SL-3 crewmen was of the same magnitude as in immobile bedrest and the mean actual calcium loss of the three 60-day flight crewmen was virtually identical. Although the total calcium loss rate generated by the second month in space (approximately 4 gm per month or 0.3–0.4% of total body calcium per month), appears small in relation to the whole skeleton, the similarity to bedrest in pattern and degree and failure to show any tendency to abatement in 3 months' time makes it necessary to deal with an assumption that mineral loss might continue for a very long time, presumably many months. By 6 months of weightless flight, the calcium loss would amount to about 2.0% of total body calcium. This is the amount of loss at which in studies of paralytic poliomyelitis (5) X-ray visible rarefaction appeared in the distal tibiae. Assuming that such excess excretion would continue, then, the calcium loss rate of 0.3–0.4% per month observed in Skylab takes on clearer and more ominous significance when it is realized that flights to Mars and return, when ultimately conducted, will take from one and one-half to three years.

The increased excretion of nitrogen and phosphorus, also similar to that in bedrest, reflected substantial loss of muscle tissue, which was clearly observed in the astronauts' legs. Both muscle and mineral loss occurred despite an exercise regimen on all flights, which was extremely vigorous on the second and third flights.

We concluded, then, that it seems reasonable to predict musculoskeletal "safety" in space flight for at least 6 months and probably for 9 months, but that musculoskeletal function might be seriously impaired in crews on space flights of extreme duration *unless protective measures can be developed.* The likelihood of need for protective measures in space flight is accentuated by the following consideration: although the bone losses thus far observed have been reversible upon return to normal gravity (or to ambulation after bedrest), no observations are available to permit estimation of a magnitude of loss that would represent a "point of no return". Thin trabeculae in bone can be returned to normal thickness but, from our present understanding of the adult skeleton, completely lost trabeculae cannot be restored. In anticipation of these results, NASA about 5 years ago began sponsoring studies, using bedrest as the analogue for weightlessness, designed to test a variety of hormonal, dietary and physical measures which might protect the skeleton in long space flight. So far, there have been no convincingly positive results in these time-consuming studies, but several leads for further investigation have been developed.

In a final comment, these observations in space may have significance for Earth medicine. In reminding us of the deleterious effects of disuse on bone mass, they re-emphasize the importance of direct physical longitudinal stress (weight bearing) to the integrity of bone. In continuing search for effective protective or even therapeutic measures in osteoporosis, more attention might well be given to determining just how weight bearing affects the balance between bone formation and loss. Once we know the specific characteristics of that mechanism, perhaps we can take practical advantage of that knowledge in dealing with osteoporosis more effectively.

REFERENCES

1. Deitrick, J.E., Whedon, G.D. & Shorr, E.: Effects of immobilization upon various metabolic and physiologic functions of normal man. *Amer. J. Med.* 4, 3–26 (1948)
2. Birge, S.J. Jr. & Whedon, G.D.: Bone. In: *Hypodynamics and Hypogravics*. McCally, M. (ed.), Academic Press, New York, p 213 (1968)
3. Donaldson, C.L., Hulley, S.B., Vogel, J.M. Hattner, R.S., Bayers, J.H. & McMillan, D.: Effect of Prolonged Bed Rest on Bone Mineral. *Metabolism* 19, No. 12, 1071–1084 (1970)
4. Lutwak, L., Whedon, G.D., Lachance, P.A., Reid, J.M. & Lipscomb, H.S.: Mineral, Electrolyte and Nitrogen Balance Studies of the Gemini-VII Fourteen-Day Orbital Space Flight. *J. clin. Endocr.* 29, 1140–56 (1969)
5. Whedon, G.D. & Shorr, E.: Metabolic studies in paralytic acute anterior poliomyelitis. II. Alterations in calcium and phosphorus metabolism. *J. clin. Invest.* 36, 966–981 (Part II) (1957)

Inhibition of cAMP Accumulation in Epiphyseal Cartilage Cells Exposed to Physiological Pressure

L. A. BOURRET & G. A. RODAN

It has long been known that mechanical forces "mold" the shape and change the internal architecture of living bone (16). These changes occur through bone remodeling and involve cell proliferation and cytodifferentiation. Since both processes have been suggested to be under cyclic nucleotide control (2, 5, 3, 8) we have measured the effect of compressive forces on the 3'5' cyclic AMP (cAMP) accumulation in chick epiphyseal cartilage and isolated cartilage cells using a method we have recently developed (11). We found that the application of a compressive force of physiological magnitude (60 g/cm^2) to a 16-days old chick embryo tibia in culture inhibits the accumulation of cAMP in the epiphyses (12). We also found that the response of isolated cells to pressure varies according to their site of origin ("state of differentiation"). The cells separated from the proliferative zone of the epiphyseal cartilage showed a decrease in cAMP accumulation when exposed to mild hydrostatic pressure. In this study we proceeded to investigate the mechanism of this effect.

METHODS AND MATERIALS

Cells were obtained from the proliferative (P) and the hypertrophic (H) zones of the epiphyses of 16-days old chick embryo tibiae by a method previously described (10). 1 x 10^6 cells were incubated in 250 μl Krebs-Ringer-glucose for 15 min at 37°C in sealed polypropylene tubes under a nitrogen pressure of 60 g/cm^2. The cells were processed as previously described (12) and cAMP was determined by radioimmunoassay (15).

Calcium incorporation was determined in similarly incubated cells by the method of Martonosi and Feretos (7). Sodium uptake was measured

University of Connecticut, Health Center, Farmington, Conn.

432

by an identical method.

Membrane-bound adenyl cyclase preparations were obtained by a two-phase separation procedure according to Brunette and Till with omission of $ZnCl_2$ stabilization (1). The final pellet was suspended in 0.01 Tris (pH 7.6). Adenylate cyclase activity was assayed according to the method of Salomon *et al.* (13). $40-100\,\mu g$ protein (6) were incubated for 15 min at 37°C. 10^{-4} M ethylenebis (oxyethylenenitrilo) tetraacetic acid (EGTA) was used to obtain free calcium concentrations of 4×10^{-7} M to 6.7×10^{-5} M based on apparent association constant of 1.0×10^{6} M^{-1}.

Collagenase (CLS II) and hyaluronidase were purchased from Worthington Biochemicals, the cAMP radioimmunoassay from Collaborative Research, Waltham, Mass., Ca^{45}, α-P^{32}-ATP, Na^{22} and H^{3}-cAMP from New England Nuclear, Dextran (M.W. 500,000), ATP, cAMP, alumina, creatine phosphokinase, and phosphocreatine from Sigma Chemical Co., and polyethylene glycol (M.W. 6000) from Polysciences Inc., Warrington, Pa. Incubated eggs were obtained from Spafas Inc., Norwich, Conn. AG-50 from Bio-Rad Laboratories; calcium ionophore A23187 from Mr. David Wong at Eli Lilly.

RESULTS

The cells obtained by collagenase digestion from the three segments of epiphyseal cartilage were similar in appearance under phase microscope examination except for size. The P cells measured $4-6\,\mu$ in diameter whereas the H cells varied between $8-10\,\mu$ in diameter. As shown in Table I exposure of the P cells to a hydrostatic pressure of 60 g/cm^2 caused a significant reduction in cAMP content. A similar change was produced by the addition of the calcium ionophore A23187 to the medium. The effects of the two were not additive, suggesting that they may act through the same mechanism. EGTA significantly increased the cAMP level of the P cells and abolished the pressure effect. Unlike the P cells, the cAMP content of the H cells was unaffected by pressure, ionophore or EGTA.

Since the above findings were consistent with the hypothesis that the effect of pressure on cAMP levels is mediated by calcium inhibition of the adenyl cyclase we measured the effect of pressure on cellular calcium uptake. The 60 g/cm^2 hydrostatic pressure increased the calcium uptake in the P cells from 0.82 ± 0.19 (SEM) nmoles per 10^6 cells to 2.07 ± 0.28 nmoles per 10^6 cells (9 determinations on 3 different cell preparations) and in the H cells from 0.75 ± 0.12 nmoles per 10^6 cells to 1.52 ± 0.15. The sodium uptake measured with ^{22}Na under the same conditions was less than 1 nmole/10^6 cells in both the P cells and the H cells and was

Table I EFFECT OF A23187 AND EGTA ON PRESSURE INDUCED DECREASE
IN cAMP ACCUMULATION IN EPIPHYSEAL CARTILAGE CELLS

(picomoles cAMP/10^6 cells)

	CONTROL		A23187		EGTA	
	PRESSURE	CONTROL	PRESSURE	CONTROL	PRESSURE	CONTROL
P CELLS	3.21 ±.3 6* (12)	4.08 ± .37 (12)	3.48 ± .42 (12)	3.47 ± .47 (12)	6.79 ± .11 (6)	6.93 ± .31 (6)
H CELLS	3.87 ± .90 (8)	3.61 ± .71 (8)	4.80 ±1.13 (8)	5.27 ±1.43 (8)	3.01 ± .17 (8)	3.39 ± .19 (8)

* $p < 0.01$

unaffected by the application of pressure. The difference between the P
cells and the H cells suggested that the adenyl cyclase of the latter was not
susceptible to calcium inhibition. To check this hypothesis we investigated
the effect of calcium on the adenyl cyclase activity of plasma membranes
separated from P cells and H cells respectively. Electron microscopical
examination of the membrane preparations showed vesicular structures
with typical trilaminar appearance and some rough endoplasmic reticulum. No mitochondria or nuclei were present. The mean adenyl cyclase

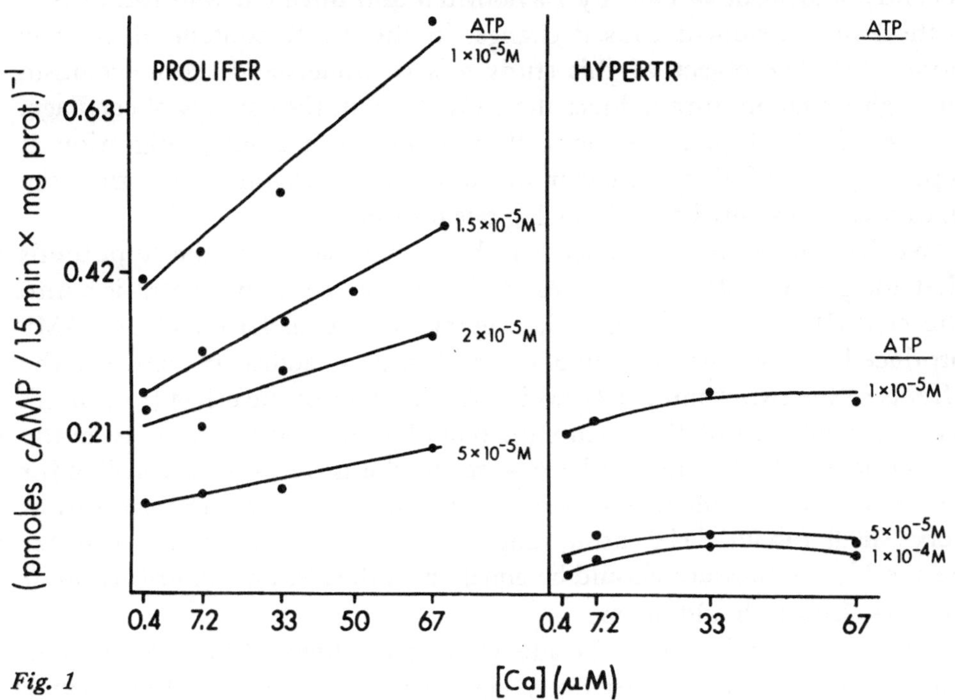

Fig. 1

434

activity of both the P and the H preparations were stimulated about two fold by 10^{-8} M parathyroid hormone (1—34 human) and were partially inhibited by 1 u/ml somatomedin. They differed, however, substantially in their response to calcium. As seen in Fig. 1 a change in calcium concentration from 4×10^{-7} M to 6.7×10^{-5} M had hardly any effect on the adenyl cyclase activity of the H preparation but substantially inhibited that of the proliferative epiphyseal cartilage. Kinetic analysis showed that the inhibitory effect of calcium is non-competitive with respect to ATP. No change in the Km of the enzymatic reaction occurred (7×10^{-5} M) whereas the Vmax was reduced about two fold. A Hill plot gave a constant of $n_H = 0.5$, suggesting negative cooperativity for the calcium effect.

DISCUSSION

Mechanical effects on bone remodeling imply the existence of a cellular transduction mechanism, since the stimulus is physical and the transfer of information at the genetic level is most probably chemical in nature. It was, therefore, of interest to find that mechanical forces affect the cAMP content of chick tibia epiphyses (11) possibly by acting on the ubiquitous cell membrane trasducer, the adenyl cyclase. A similar observation was recently also made *in vivo* by Davidovitch and Shanfeld who found that orthodontic treatment caused changes in the cAMP content of cat jaw bones. (4). The object of this study was to understand the mechanism through which pressure reduces the cAMP level in the epiphyseal cartilage. If the effect of pressure on cells isolated by enzymatic digestion is representative of their behavior in the tissue, it would appear that the matrix is not essential for producing this response.

We have presented evidence which strongly supports the hypothesis that the pressure effect is the result of increased calcium penetration into the cell: (I) enhanced radiocalcium uptake (II) a similar effect on cAMP produced by the calcium-ionophore and lack of additivity between the effects of the two stimuli (III) EGTA abolishment of the effect (IV) direct calcium inhibition of the membrane bound adenyl cyclase activity in the susceptible cells. The molecular changes in the membrane responsible for the increased calcium uptake are totally unknown to us. The pressure is very small (about 1/10 atm.) and since the cell membrane is water permeable the pressure should be equal on both sides and no distortion or volume changes should occur.

The effect of calcium on the adenyl cyclase activity of the P cell plasma membranes was non-competitively inhibitory. In this respect the enzyme seems to be similar to that of other cells, especially the turkey erythrocyte (14) and the human platelet (in preparation). The Hill coefficient of 0.5 indicates that the inhibitory effect (or binding) of calcium is proportional to the square root of its concentration. The physiological implication is that the enzyme responds in an on/off fashion to changes in calcium concentrations, hardly dependent on the extent of the change. This is an additional example of a relationship between calcium and cAMP in the regulation of a biological system as suggested by several investigators (8, 9).

In many systems including chick fibroblasts (2, 5, 3), a reduction in cAMP or adenyl cyclase activity was shown to correlate with the proliferative stage of the cell cycle. The pressure-induced decrease in cAMP could provide a stimulus to cell proliferation as part of the bone remodeling process. We have, indeed, observed an increase in the incorporation of ^{14}C-thymidine into DNA under these conditions (11).

The difference between the P and H cells is one of the most interesting findings of this study. It shows that changes which occur during the maturation of the cells include membrane alterations which affect the responsiveness of the adenylate cyclase to modulating factors. It emphasizes the role of the state of differentiation of the cell in determining the response of the tissue to an extracellular stimulus. It also illustrates a possible mechanism to account for the difference and offers support to the view that (cAMP) is involved in the control of differentiation.

REFERENCES

1. Brunette, D. & Till, J.E.: A rapid method for the isolation of L-cell surface membranes using an aqueous two-phase polymer system. *J. Membrane Biology* 5, 215—224 (1971)

2. Burger, M.M., Bombik, B.M., Breckenridge, B.M. & Sheppard, J.R.: Growth control and cyclic alterations of cyclic AMP in the cell cycle. *Nature, New Biology* 239, 161—163 (1972)

3. Bürk, R.: Reduced adenyl cyclase activity in a polyoma virus-induced cell line. *Nature* 219, 1272—1275 (1968)

4. Davidovitch, Z. & Shanfeld, J.: Cyclic AMP levels in alveolar bone of orthodontically-treated cats. *Arch. oral Biol.* 20, 567—574 (1975)

5. Johnson, S.J., Friedman, R. & Pastan, J.: Restoration of several morphological characteristics of normal fibroblast in sarcoma cells treated with cAMP and its derivative. *Proc. nat. Acad. Sci. (Wash.)* 68, 425—429 (1971)

6. Lowry, O.H., Rosebrough, N.J., Farr, A.L. & Randall, R.J.: Protein measurement with the folin phenol reagent. *J. biol. Chem.* 193, 265—275 (1951)

7. Martonosi, A. & Feretos, R.: Sarcoplasmic reticulum. I. The uptake of Ca^{++} by sarcoplasmic reticulum fragments. *J. biol. Chem.* 239, 648—658 (1964)

8. McMahon, D.: Chemical messengers in development: A hypothesis. *Science* **185**, 1012–1021 (1974)

9. Rasmussen, H.: Cell communication, calcium ion, and cyclic adenosine monophosphate. *Science* **170**, 404–412 (1970)

10. Rodan, S.B. & Rodan, G.A.: The effect of parathyroid hormone and thyrocalcitonin on the accumulation of cyclic adenosine 3':5'-monophosphate in freshly isolated bone cells. *J. biol. Chem.* **249**, 3068–3074 (1974)

11. Rodan, G.A., Mensi, T. & Harvey, A.: A quantitative method for the application of compressive forces to bone in tissue culture. *Calc. Tiss. Res.* **18**, 125–131 (1975)

12. Rodan, G.A., Bourret, L.A., Harvey, A. & Mensi, T.: 3',5'-cyclic AMP and 3',5'-cyclic GMP: Mediators of the mechanical effects on bone remodeling. *Science* **189**, 467–469 (1975)

13. Salomon, Y., Londos, C. & Rodbell, M.: A highly sensitive adenylate cyclase assay. *Analyt. Biochem.* **58**, 541–549 (1974)

14. Steer, M.L. & Levitzki, A.: The control of adenylate cyclase by calcium in turkey erythrocyte ghosts. *J. biol. Chem.* **250**, 2080–2084 (1975)

15. Steiner, A.L., Kipnis, D.M., Utiger, R. & Parker, C.W.: Radioimmunoassay for the measurement of adenosine 3',5'-cyclic monophosphate. *Proc. nat. Acad. Sci. (Wash.)* **64**, 367–373 (1969)

16. Wolff, J.: Das Gesetz der Transformation der Knochen. A. Hirschwald, Berlin, p. 152, 1882.

Synthesis by the Liver of a Glycoprotein which is Concentrated in Bone Matrix

J.T. TRIFFITT, U. GEBAUER & M.E. OWEN

INTRODUCTION

Our previous studies have shown that a glycoprotein of α-electrophoretic mobility which can be isolated from bovine and rabbit bone matrix is present also in the blood plasma (2, 8). Relative to plasma albumin the α-glycoprotein is concentrated in bone and dentine but it is not present above the expected plasma ratio in a number of other tissues. This material appears to originate from the plasma because a large proportion of the radioactivity in bone is in this glycoprotein following injection of total plasma glycoproteins labelled by using ^{14}C-glucosamine (8). However, there is also the possibility that its presence in plasma could be the result of its synthesis and release by the bone tissue.

The α-glycoprotein is present in a relatively high concentration in plasma and the majority of the plasma proteins of quantitative importance are manufactured by the liver (7). Therefore, the liver was tested for its ability to synthesize the α-glycoprotein.

METHODS

Young rabbits weighing 300 g were used and their livers perfused *in situ* by a modification of the method of Hems *et al.* (5) using a total of 50 ml perfusion medium and a flow rate of 27 ml/min. The perfusion medium contained 2.6 g per cent (w/v) bovine plasma albumin in bicarbonate buffer (6) with bicarbonate added to 25 mM concentration, and human erythrocytes to give 2.5% (w/v) haemoglobin. 30 ml perfusion medium was used to wash out the blood contained in the liver before addition of

Bone Research Laboratory, Nuffield Department of Orthopaedic Surgery, Nuffield Orthopaedic Centre, Oxford.

20 μCi D-[1-^{14}C] glucosamine (The Radiochemical Centre, Amersham, Bucks, U.K.) to the perfusate. Previous work has shown that D-[1-^{14}C] glucosamine is incorporated into proteins mainly as glucosamine and sialic acid and, therefore, it is a useful substrate for the study of glycoprotein synthesis (9). Samples (0.5—1.0 ml) of perfusate were taken at various time intervals up to 4 h after addition of labelled glucosamine. These samples were centrifuged at 2000 G for 30 min at 4°C to remove erythrocytes, and the supernatants collected. Portions (100 μl) of the supernatants were dissolved in 0.5 ml soluene (Packard Instrument Ltd., Wembley, Middx., U.K.) and assayed for radioactivity in an automatic liquid scintillation spectrometer after addition of 5 ml scintillation fluid (3). Samples (0.1 ml) of the supernatant media were added to 1.0 ml 12% (w/v) aqueous trichloroacetic acid (TCA) and allowed to precipitate at 4°C overnight. The protein precipitates were collected by centrifugation at 1800 G for 15 min at 4°C and washed three times with 1 ml 12% (w/v) aqueous TCA containing 1 mg/ml glucosamine. The washed precipitates and individual supernatants were dissolved in soluene (1 ml) and the contents of radioactivity determined. The macromolecular contributions to the radioactivity present in the media was determined also by using gel chromatography on Biogel P30 (Bio-Rad Labs. Ltd., Bromley, Kent, U.K.).

Samples of perfusion media were taken for immunoelectrophoresis followed by autoradiography (8) and for polyacrylamide disc gel electrophoresis followed by autoradiography (4). The isolation of, and the preparation of antibodies to, the rabbit α-glycoprotein present in bone was performed as described previously (8).

RESULTS AND DISCUSSION

The amount of radioactive label present in the perfusion medium with time after administration of [^{14}C] glucosamine is shown in Fig. 1, together with the proportion of the total radioactivity which was TCA precipitable. The total amount of radioactive label in the medium showed a rapid fall to about 50 min as the liver removed radioactive precursor. Thereafter, this value increased steadily to the end of the perfusion period (4 h). The proportion of the radioactivity which was precipitable by TCA increased from a value of 5% at the 20 min time points to 77% at 4 h.

Column chromatography of medium samples on Biogel P30 confirmed that TCA precipitability represented macromolecular radioactivity as similar results were obtained by both methods. Radioactivity associated with macromolecules first appears in the medium between 10—20 min

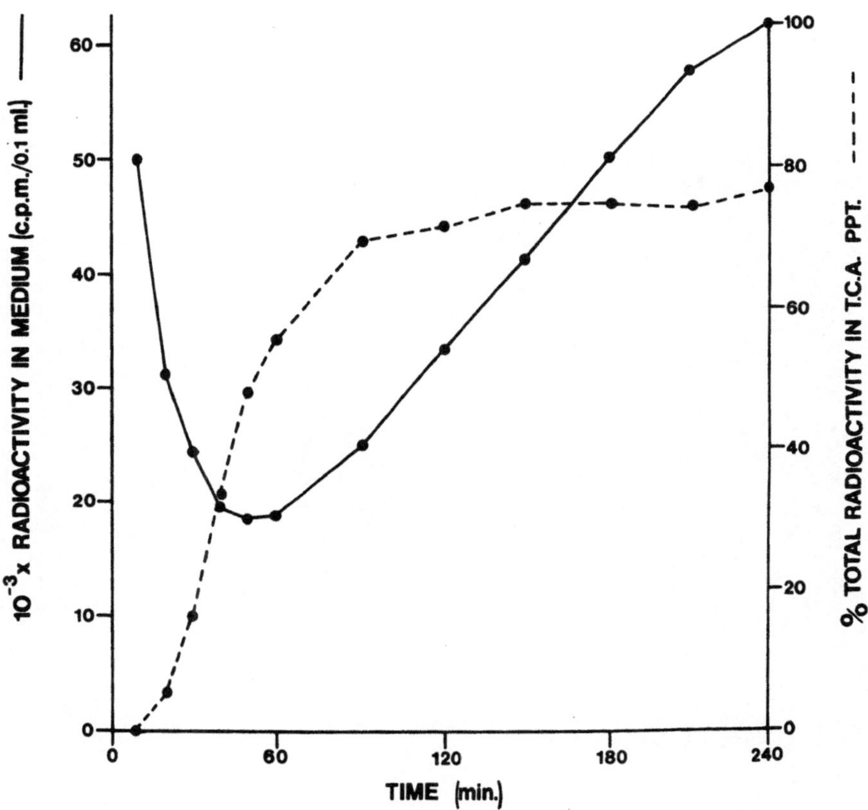

Figure 1. Graph of total radioactivity in liver perfusion medium, and the proportion TCA-precipitable, with time after introduction of ^{14}C-glucosamine into the perfusate.

Total radioactivity (cpm x 10^{-3}/0.1 ml sample) in perfusion medium (————) per cent of total radioactivity in perfusion medium which was precipitable by TCA (- - - - -).

after addition of labelled precursor and this time-lag indicates the time taken for uptake, synthesis and secretion of the protein by the liver (1). Therefore, the perfused liver is demonstrated to be synthesizing macromolecules and the nature of this material was investigated by disc gel- and immuno-electrophoresis.

Electrophoresis of medium samples on 7% polyacrylamide gels was followed by analysis of the radioactivity present in slices by liquid scintillation counting. It was shown that a large proportion of the radioactivity present had α-electrophoretic mobility in this system and was present between transferrin and albumin. Direct autoradiography of the gel confirmed the presence of radioactively labelled material in the position of authentic α-glycoprotein isolated from bone.

Therefore, immunoelectrophoresis of the perfusion medium was per-

440

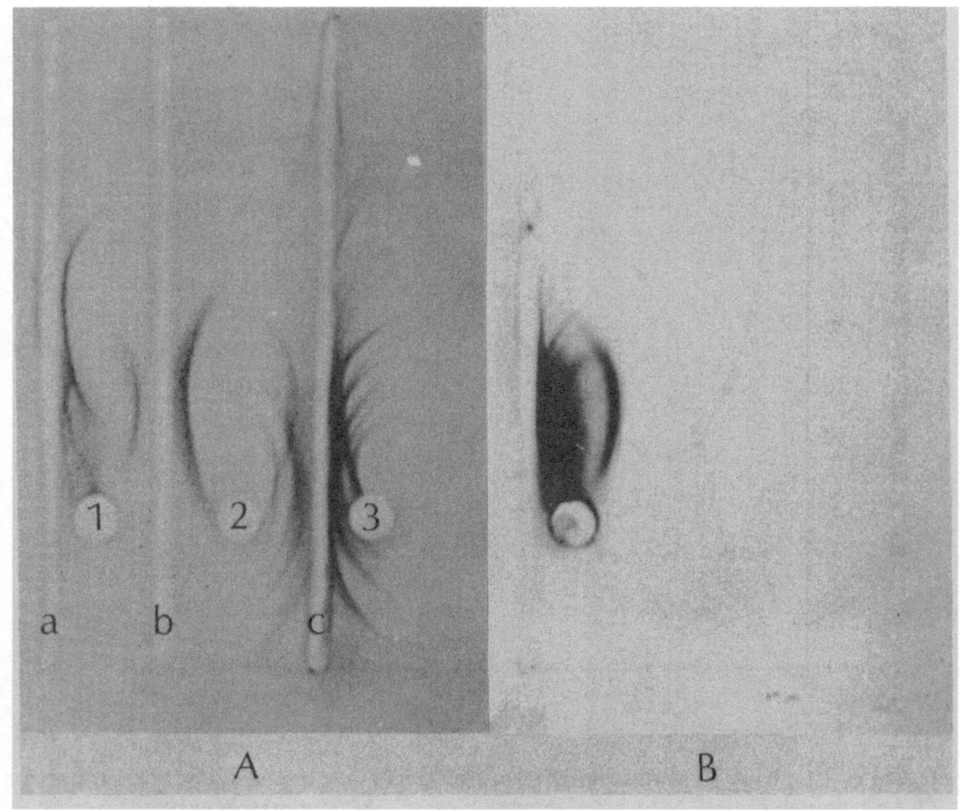

Figure 2. Immunoelectrophoretogram of rabbit liver perfusate and rabbit plasma against antiserum to rabbit plasma and antiserum to rabbit bone α-glycoprotein (A), and its autoradiograph (B).
Well 1) Perfusion medium, 240 min after ^{14}C-glucosamine addition
Well 2) Rabbit plasma to which an amount of ^{14}C-glucosamine equivalent to the radioactivity in well 1) was added.
Well 3) Rabbit plasma, non-radioactive.
Troughs a) & c) antiserum to rabbit plasma.
Trough b) antiserum to rabbit bone α-glycoprotein.

formed to test whether any of the proteins synthesized by the liver reacted immunologically with the antiserum prepared against the α-glycoprotein from rabbit compact bone. The results are given in Fig. 2 together with an autoradiograph of the immunoelectrophoretogram. The liver perfusate contained many proteins which reacted against the polyvalent anti-rabbit plasma and their synthesis during perfusion is demonstrated by the autoradiograph. The liver synthesized also a component showing immunological identity with the α-glycoprotein isolated from bone tissue. Free-glucosamine did not interact with the immunoprecipitated protein arcs; at least not after the staining and washing procedure. This was demonstrated by the addition of D-[1-^{14}C] glycosamine to normal rabbit plasma in an amount equivalent to the radioactivity present in the final perfusion medium, followed by immunoelectrophoresis and autoradiography (Fig. 2).

The results show that an α-glycoprotein component, which makes up 3.9% (w/v) of the rabbit cortical bone non-collagenous organic matrix, can be synthesized by the liver. Conclusive proof that bone tissue cannot synthesize this glycoprotein is lacking and experiments are in progress to investigate the synthesis of glycoproteins by bone in tissue culture. However, the present results strongly support the suggestion that the α-glycoprotein is taken up by bone from the plasma during bone formation (8).

ACKNOWLEDGEMENT

The authors are greatly indebted to Professor Sir Hans Krebs for helpful advice and encouragement and to Mr. Philip Gregory of Professor Krebs' Laboratory, Oxford for technical expertise in performance of the perfusions.

REFERENCES

1. Anker, H.S.: The biosynthesis of plasma proteins. In: *The Plasma Proteins*, Putman, F.W. (ed.), Academic Press, N.Y. & London, vol. II, pp. 267—307, 1960

2. Ashton, B.A., Triffitt, J.T. & Herring, G.M.: Isolation and partial characterization of a glycoprotein from bovine cortical bone. *Europ. J. Biochem.* 45, 525—533 (1974)

3. Bray, G.A.: A simple efficient liquid scintillator for counting aqueous solutions in a liquid scintillation counter. *Analyt. Biochem.* 1, 279—285 (1960)

4. Fairbanks, G., Levinthal, C. & Reeder, R.H.: Analysis of ^{14}C-labelled proteins by disc electrophoresis. *Biochem. Biophys. Res. Commun.* 20, 393—399 (1965)

5. Hems, R., Ross, B.D., Berry, M.N. & Krebs, H.A.: Gluconeogenesis in the perfused rat liver. *Biochem. J.* 101, 284—292 (1966)

6. Krebs, H.A. & Henseleit, K.: Untersuchungen über die Harnstoffbildung im Turkörper. *Hoppe-Seyler's Z. physiol. Chem.* 210, 33 (1932)

7. Miller, L.L. & John, D.W.: Nutritional, hormonal and temporal factors regulating net plasma protein biosynthesis in the isolated perfused rat liver. In: *Plasma Protein Metabolism*, M.A. Rotschild and T. Waldmann (eds.), Academic Press, New York & London, pp. 207—222, 1970

8. Triffitt, J.T. & Owen, M.: Studies of bone matrix glycoproteins. Incorporation of [1-^{14}C] glucosamine and plasma [^{14}C] glycoprotein into rabbit cortical bone. *Biochem. J.* 136, 125—134 (1973)

9. Winzler, R.J.: Metabolism of Glycoproteins. *Clin. Chem.* 11, 339—347 (1965)

Calcium Absorption in the Elderly

B.E.C. NORDIN, R. WILKINSON, D.H. MARSHALL, J.C. GALLAGHER, A. WILLIAMS & M. PEACOCK

Calcium absorption can be measured in at least three ways — by the balance procedure, by intubation of the small intestine with a double or triple lumen tube and by the oral administration of radioactive calcium.

The balance procedure as developed by us takes two weeks. During an equilibration week on a constant calcium intake, polyethyleneglycol (PEG) is administered in a dose of 0.5 g t.d.s. with meals, and during the second week urine and faecal collections are made daily. Faecal calcium is calculated from the calcium/PEG ratio and urinary calcium from the calcium/creatinine ratio (1). The difference between calcium intake (i) and faecal calcium (f) is determined on a daily basis and the mean values calculated together with the standard error:

$$b = i - f$$

where b represents net calcium absorption.

The incubation procedure involves the passage of a triple lumen tube into the small intestine, perfusion with fluids of known calcium concentration containing PEG of known concentration, and the calculation of calcium absorption from the difference between the calcium/ PEG ratio in the perfused and recovered fluid (12). By using progressively increasing perfusate calcium concentrations, a progressive increase in the rate of calcium absorption is established and the relation between absorbed and perfused calcium is compatible with the existence of a two-component system in calcium absorption comprising an active saturable mechanism and a diffusion component. This relationship can be expressed in the following general terms:

$$a = \frac{Vmax \times (Ca)}{Km + (Ca)} + d\,(Ca) + e$$

where a is absorbed calcium, Vmax is the maximum capacity of the active

MRC, Mineral Metabolism Unit, The General Infirmary, Leeds.

transport component, (Ca) is calcium concentration, d represents the diffusion slope and e the efflux of calcium into the gastro-intestinal tract.

The radiocalcium absorption procedure involves the administration of 20 mg of calcium as calcium chloride in 250 ml of water to the fasting subject with 5 μCi of radioactive calcium, and the collection of blood samples at 15, 30, 45, 60, 90, 120 min. The rate of calcium absorption can be calculated from the plasma activity values by means of the following general term:

$$f = \frac{\alpha}{\beta - \alpha} \quad (e^{-\alpha t} - e^{-\beta t})$$

where f is the fraction of the dose circulating in the extracellular fluid (plasma activity/l x 15% of body weight), α is the rate of calcium absorption as a fraction of the radio-activity in the small intestine, and β is the exponential function representing the rate at which radioactivity is being removed from plasma. This procedure, derived from Marshall and Nordin (7), yields essentially the same information as the double isotope procedure suggested by other workers (11) and agrees well with the rate of calcium absorption calculated from 169 calcium balances (r = 0.69; p < 0.001).

RESULTS

The relation between ingested (i) and net absorbed calcium (b) in 212 calcium balances on normal subjects selected from the literature, has been calculated as follows:

$$b = 2.08 \log(i) - 2.10 \text{ mg/kg/day}.$$

In view of the dependence of absorbed calcium on ingested calcium, it is only possible to assess the normality of any given absorbed calcium value by relating it to the mean normal line and any given balance result is, therefore, most conveniently expressed in terms of standard deviations above and below the normal line. This form of calculation has been applied to 90 balance studies in 27 normal pre- and post-menopausel women and 63 women with osteoporosis and has shown a just significant fall in calcium absorption with age (r = 0.28 : p < 0.01) (Fig. 1).
The radiocalcium absorption procedure yields a more convincing fall in calcium absorption with age, starting at about the age of 60 (2). Alpha values in the normal population aged 20—60 are logarithmically distributed and the mean value is 0.64 (fraction of the dose absorbed per h) with a range of 0.30 — 1.50. In subjects over 60, there is a progressive fall in radiocalcium absorption (Fig. 2). Over the age of 80 the data are normally

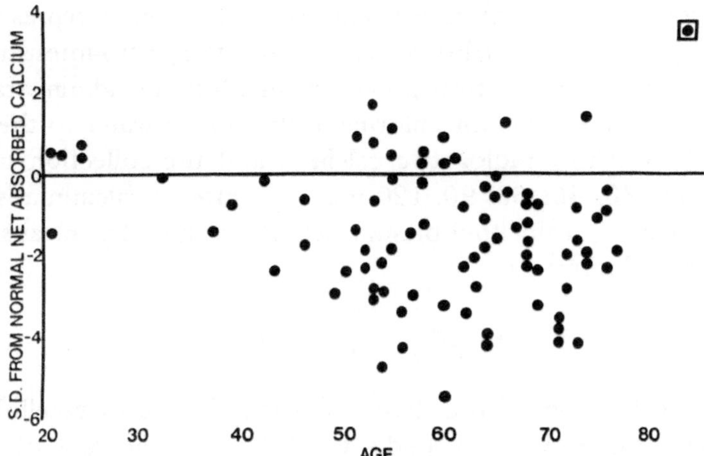

Figure 1. Absorbed calcium (in standard deviation units above and below the predicted normal mean) as a function of age in pre- and post-menopausal women. ($r = 0.28$; $p <$ 0.01). (The enclosed point was not included in the regression calculation.)

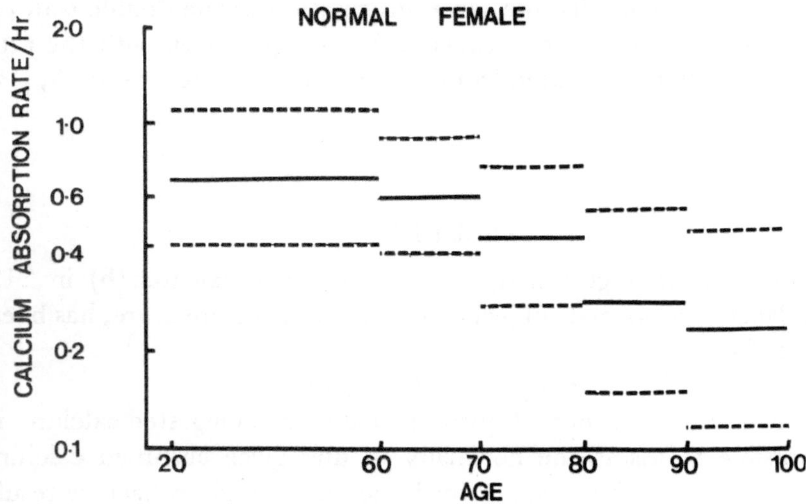

Figure 2. Mean radiocalcium absorption ± 1 S.D. as a function of age in normal men and women.

distributed but indicate severe malabsorption (mean α 0.29; range 0.08 to 0.90). Cases of fractured neck of femur resemble the over 80 group, although many of them are in fact less than 80 years old.

These observations confirm the development of calcium malabsorption in elderly subjects but do not tell us what component of the absorptive mechanism is affected. In rats, calcium absorption studies from an isolated jejunal loop (12) indicate not only that there is a fall in calcium

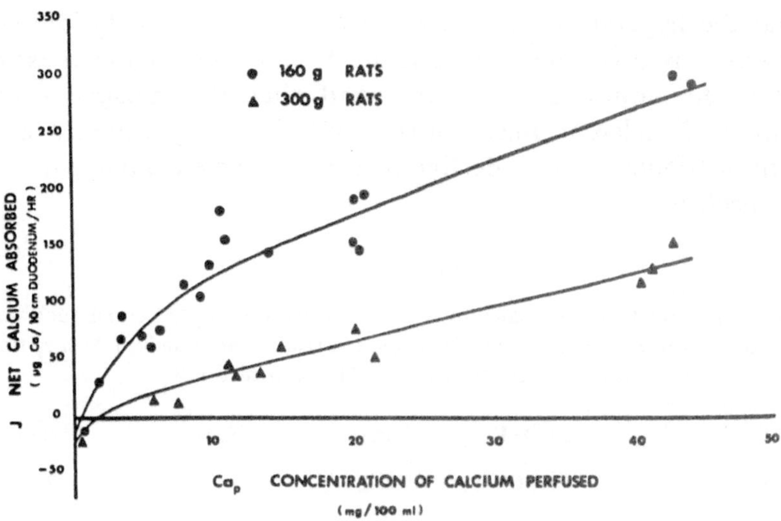

Figure 3. Calcium transport as a function of perfusate calcium concentration in young and old rats.

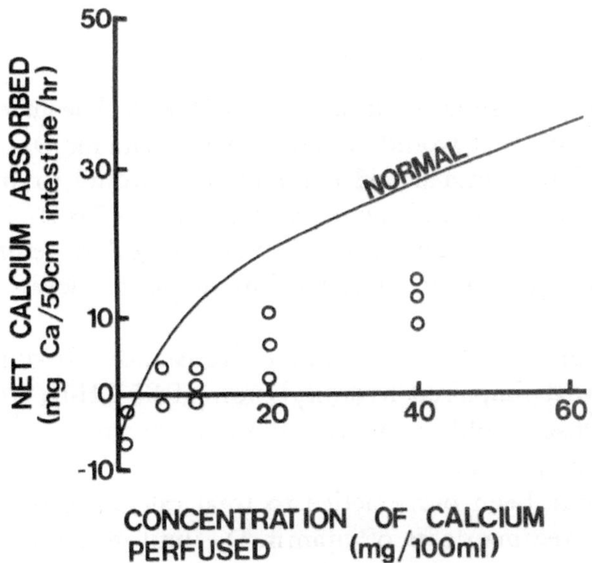

Figure 4. Calcium transport as a function of perfusate calcium concentration in 3 elderly subjects.

absorption with age but that this decline predominantly affects the active transport component of the absorption mechanism (Fig. 3).

A few studies in elderly human subjects suggest the same thing (Fig. 4). The various components of calcium absorption determined by intubation studies in different groups of subjects are indicated in Table I. This table shows that the hyperabsorption of hyperparathyroidism and "idiopathic hypercalciuria" is predominantly due to an increase in the maximum capacity of the saturable component and that the impaired calcium absorption of 5 calcium malabsorbers and 3 elderly subjects is predominantly attributable to a decline in the saturable component of the transport mechanism.

Table I.

Parameters of the Model for Net Calcium Absorption by the Triple Lumen Technique (see text for definitions) in Different Diagnostic Groups. Net Calcium Absorption is Measured in Units of mg Ca/50 cm intestine/h.

	SLOPE	Vmax	Km	Efflux
Controls (9)	0.30	28.2	7.2	− 7.8
Hyperpara (9)	0.60	42.2	4.2	− 8.0
Hypercalciuria (11)	0.47	33.6	7.0	− 6.8
Malabsorption (5)	0.48	-	-	− 3.4
Elderly (3)	0.49	-	-	− 4.1

Response to vitamin D

There are at least two possible explanations for the decline in calcium absorption with age. The first would be deficiency of vitamin D, possibly due to inadequate dietary intake and inadequate exposure to sunlight. This is supported by our own measurements of plasma 25-OH-D_3 levels in normal young and old subjects and cases of fractured neck of femur (Fig. 5) which show a highly significant reduction in plasma levels in old people.

The second explanation would be an acquired "resistance" to vitamin D, possibly associated with impaired hydroxylation of 25-OH-D_3 to 1,25-$(OH)_2$-D_3. The response of old people to vitamin D therapy suggests that this factor may also be operating.

It has for some time been our practice to treat calcium malabsorbers with progressively increasing doses of vitamin D_2 until malabsorption of calcium has been corrected. In many of these subjects, we have progressed through doses of 1,000, 10,000 and 20,000 units of vitamin D_2, repeating the radiocalcium absorption procedure after 6 weeks at each dose level. The result of such studies on patients with backache and/or fractures but without evidence of steatorrhea, osteomalacia, renal failure, or hyper- or hypoparathyroidism is summarised in Table II. Preliminary inspection of the data suggested that patients above and below the age of 70 responded differently to vitamin D therapy and they have been divided accordingly. In subjects below the age of 70, vitamin D_2 1,000 units daily significantly

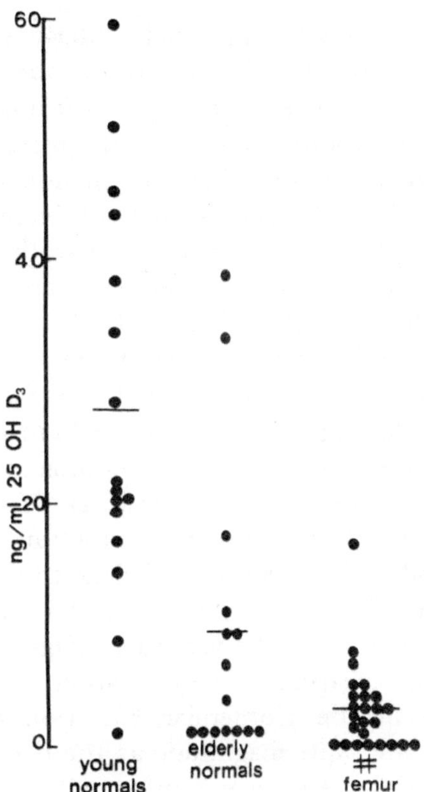

Figure 5. Plasma 25-OH-D₃ levels in normal young and old subjects and in cases of fractured neck of femur.

increases calcium absorption, but only slight further improvement is achieved by raising the dose to 10,000 units and no further improvement is achieved at dose of 20,000 units. In subjects over the age of 70, there is no response to 1,000 units daily, a significant response to 10,000 units daily and a further improvement in the response at 20,000 units daily. A few patients who failed to respond to 20,000 units were subsequently given doses of 40–50,000 units generally with a satisfactory response, but these have not been included in the table. Taken at their face value, these data suggest that patients below the age of 70 with malabsorption of calcium may well be suffering from vitamin D deficiency inasmuch as they frequently respond to relatively small doses of vitamin D_2. Over the age of 70, on the other hand, they seldom respond to small doses of vitamin D and generally require treatment with 10–20,000 units or even more. This suggests that the malabsorption of calcium in elderly subjects is not simply a matter of vitamin D deficiency but may include an element of vitamin D "resistance".

There are at least two ways in which elderly subjects, particularly women, might become "resistant" to vitamin D therapy in the sense that they require large doses to correct their malabsorption.

The first mechanism would result from the increased "sensitivity" of bone to parathyroid hormone after the menopause which we have postulated elsewhere (9, 5). We suggest that this increased "sensitivity" of bone tends to elevate plasma calcium, reduce parathyroid activity, reduce the activity of renal 1α-hydroxylase and so lead to malabsorption of calcium. This hypothesis implies decreased PTH activity in post-menopausal women in general. This mechanism may be important in post-menopausal osteoporosis and may also be a factor in the calcium malabsorption which develops in women over 70 inasmuch as evidence of severe oestrogen deficiency is increasingly common after this age.

Another possible factor in the vitamin D "resistance" of elderly subjects would be the decline in renal function which is known to accompany ageing and which may be sufficient to reduce renal hydroxylase activity. We have shown elsewhere (3) that the malabsorption of calcium in renal failure appears at a very early stage of renal insufficiency, i.e. when the plasma creatinine exceeds about 1.5 mg/100 ml. This represents a decline in the glomerular filtration rate of only about 40—50% which is of the same magnitude as the decline in G.F.R. which accompanies ageing (4). We have confirmed the marked decline in endogenous creatinine clearance with age (r = −0.51; p < 0.001), and also looked at the relation between radiocalcium absorption and endogenous creatinine clearance in our normal pre- and post-menopausal women and post-menopausal crush fracture cases. This analysis shows a highly significant correlation between the creatinine clearance and radiocalcium absorption. Thus, impaired renal function may be a factor in calcium malabsorption in the elderly.

It is entirely compatible with this concept that elderly subjects should require relatively large doses of vitamin D_2 to cure their malabsorption of calcium but will respond readily to what appear to be relatively small doses of 1α-OH-D_3. This is clearly indicated in Fig. 6 which shows the impressive effect of 1α-OH-D_3 (2—4 μg daily) on calcium absorption in a few elderly subjects previously treated with large doses of vitamin D_2 (10—40,000 I.U. daily). This is what would be expected if the malabsorption of calcium in elderly subjects were attributable to reduced renal hydroxylase activity. There is also some indication that 25-OH-D_3 (50—150 μg daily) is more effective than D_2 but this may be simply a matter of dosage.

This concept is also compatible with the failure of adaptation to low calcium diets which has been reported in elderly subjects by Ireland and

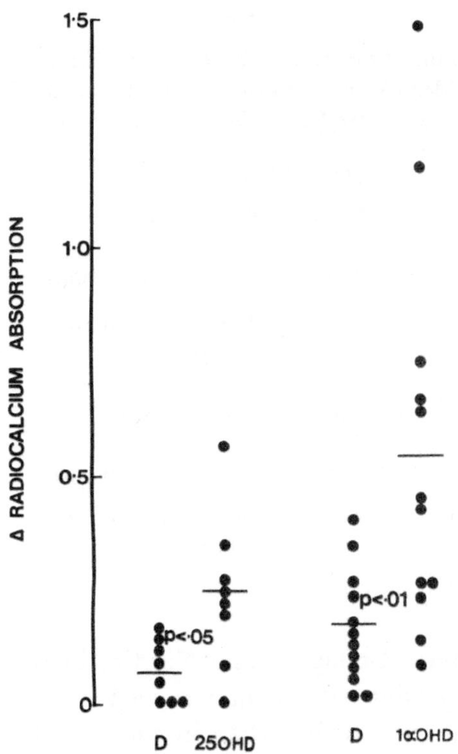

Figure 6. (Left) A comparison of the effect on radiocalcium absorption in 8 subjects of large doses of vitamin D_2 (o.25 to 1.0 mg daily) and small doses of 25-OH-D_3 (50 to 150 μg). (Right) A similar comparison in 12 subjects of the effect of large doses of vitamin D_2 and small doses of 1α-OH-D_3 (2 to 4 μg daily).

Fordtran (6). It is now reasonably well established that calcium deprivation stimulates calcium absorption, and it is likely that the mechanism involves a marginal reduction of plasma ionized calcium concentration, stimulation of parathyroid hormone secretion, activation of 1α-hydroxylase in the kidney and stimulation of the active transport component of the calcium absorption mechanism. If malabsorption of calcium in elderly subjects is attributable, at least in part, to impaired renal 1α-hydroxylase activity, it would be expected that elderly subjects would be less able than young subjects to increase their calcium absorption in response to calcium deprivation.

The significance of these observations to the pathogenesis of senile osteoporosis is fairly clear. It must not be thought that vitamin D deficiency can only produce osteomalacia. We have previously noted (8) that malabsorption of calcium *per se* is likely to produce osteoporosis rather than osteomalacia, and in this connection we have noted and will

Table II.

Changes in Radiocalcium Absorption in Response to Various Doses of Vitamin D_2
Showing Mean Values and the Significance of the Increase
between Dose Levels for Paired Observations.

VITAMIN D (units/day)

70 years of age and over

	C		1,000		10,000		20,000
α	0.28	(NS)	0.30	(NS)	0.36	($p<0.05$)	0.66
n		12 prs.		10 prs.		9 prs.	

Below 70 years of age

	C		1,000		10,000		20,000
α	0.25	($p<0.05$)	0.39	($p<0.02$)	0.50	(NS)	0.52
n		10 prs.		9 prs.		7 prs.	

be reporting elsewhere (Gallagher and Nordin, in preparation) that the most severe malabsorption of calcium is seen in femoral neck fracture patients with *osteoporosis* rather than in those with *osteomalacia*. This does not mean, of course, that osteomalacia is not generally speaking a disorder of vitamin D metabolism, but simply that osteomalacia is not due to calcium malabsorption *per se* though frequently associated with it. Osteomalacia represents a degree or type of vitamin D deficiency or resistance generally associated with hypocalcaemia, hypophosphataemia or both. These biochemical abnormalities are associated with impaired mineralization of new bone and, though generally associated with malabsorption of calcium as well, can lead to impaired mineralization of new bone whether or not calcium malabsorption is present. Insofar as malabsorption of calcium gives rise to bone disease it is liable to produce osteoporosis rather than osteomalacia and it, therefore, seems likely that senile malabsorption of calcium is a significant factor in the pathogenesis of senile osteoporosis.

REFERENCES

1. Bullamore, J.R., Marshall, D.H., Nordin, B.E.C., Oldfield, W.A. & Wilkinson, R.: Measurement of calcium balance and bone turnover by new techniques. *Calc. Tiss. Res.* **4**, (Supplement) 93–94 (1970)

2. Bullamore, J.R., Gallagher, J.C., Wilkinson, R., Nordin, B.E.C. & Marshall, D.H.: Effect of age on calcium absorption. *Lancet* (1970) ii, 535—537

3. Cochran, M., Bulusu, L., Horsman, A., Stasiak, L. & Nordin, B.E.C.: Hypocalcaemia and Bone didease in renal failure. *Nephron* 10, 113—140 (1973)

4. Davies, D.F. & Shock, N.W.: Age change in glomerular filtration rate, effective renal plasma flow and tubular excretory capacity in adult males. *J. clin. Invest.* 29, 496—507 (1950)

5. Gallagher, J.C., Aaron, J., Horsman, A., Marshall, D.H., Wilkinson, R. & Nordin, B.E.C.: The crush fracture syndrome in post-menopausal women. *Clin. Endocr. & Metab.* 2, 293—315 (1973)

6. Ireland, P & Fordtran, J.S.: Effect of dietary calcium and age on jejunal calcium absorption in humans studied by intestinal perfusion. *J. clin. Invest.* 52, 2672—2681 (1973)

7. Marshall, D.H. & Nordin, B.E.C.: Kinetic analysis of plasma radioactivity after oral ingestion of radiocalcium. *Nature (Lond.)* 222, 797 (1969)

8. Nordin, B.E.C.: Osteoporosis, osteomalacia and calcium deficiency. *Clin. Orthop.* 17, 235—258 (1960)

9. Nordin, B.E.C.: The clinical significance and pathogenesis of osteoporosis. *Brit. Med. J.* 1, 571—576 (1971)

10. Riggs, B.L., Jowsey, J., Kelly, P.J., Hoffman, D.L. & Arnaud, C.D.: Studies on pathogenesis and treatment in post-menopausal and senile osteoporosis. *Clin. Endocr. & Metab.* 2, 317—332 (1973)

11. Tothill, P., Dellipiani, A.W. & Calvert, J.: Plasma concentrations of radiocalcium after oral administration, and their relationship to absorption. *Clin. Sci.* 38, 27—39 (1970)

12. Wilkinson, R.: *Studies of Calcium Absorption by the Small Intestine of Rat and Man.* Ph.D. Thesis, University of Leeds, U.K., 1971

Quantitative Analysis of Amorphous and Crystalline Bone Tissue Mineral in Women with Osteoporosis

C.A. BAUD, J.A. POUËZAT & H.J. TOCHON-DANGUY

Bone mineral has been shown to consist of two calcium phosphate phases, an amorphous and a crystalline apatitic one (4). Various techniques have been used to evaluate them quantitatively; X-ray diffraction (5), infrared absorption spectrophotometry (8), electron spin resonance spectroscopy (7), and it has been found that the non-apatitic component represents about 40% of the total mineral in mature compact bone and an even higher percentage in younger bone. Experimentally induced nutritional or hormonal deficiency, however, can modify the amorphous/crystalline ratio of the bone mineral (9).

The only published study of bone mineral phases under pathological conditions in humans concerns the ectopic bone tissue of para-osteo-arthropathies in paraplegic patients (2). In these cases it was shown that the amorphous component varied little dirung the first 12 months of the development of the lesions, remaining at about 58% of the total mineral, but decreased gradually to only 40% after 30 months. The present study also concerns the proportions of amorphous and crystalline mineral in human bone tissue, in this case, however, of patients with osteoporosis.

MATERIAL AND METHODS

Iliac crest samples were obtained by biopsies before the beginning of treatment in 26 women with osteoporosis. New biopsies were taken from 15 of those after one year of treatment with 30 mg fluoride per day.

Control samples from the iliac crest were obtained surgically during graft removal from 8 non-osteoporotic women, and during autopsy from 17 women who died a violent death, and in whom no sign of disease that could affect the skeleton was found.

Department of Morphology, University of Geneva School of Medicine, Geneva.

For purposes of comparison, the femurs were removed from 36 mice that had been divided into two groups; one had been given drinking water supplemented with 100 ppm of fluoride from the time of weaning, the other (control group) was normally fed.

Percentages of amorphous and crystalline mineral in all samples except for those obtained by autopsy, were determined by infrared absorption spectrophotometry according to the method described by Blumenthal, Posner and Holmes (3). In order to insure satisfactory reproducibility of results, the technique was fully automatized; infrared spectrum analogue information was digitalized and stored on punched tape, the data was processed by computer, and filtering programs optimized the precision of results (± 1% coefficient of variation of repeated measurements of the same bone sample).

The degree of mineralization of bone tissue was measured by quantitative microradiography by means of a specially designed automatic scanning microdensitometer with a pulse height analyzer (1).

RESULTS

In control women over 60 years of age, it was found that the degree of mineralization of the compact bone tissue was 1.17 ± 0.08 g min/cm^3, and the percentage of mineral being in the crystalline state was 62.7 ± 1.7. In women with osteoporosis of the same age group (Fig. 1), the percentage of crystalline mineral of compact bone tissue was found to be higher than the average for control women in 77% of the cases; the degree of mineralization was higher in 73% of the cases; and both were higher in 62%.

A very low degree of mineralization was found in only 2 of the 26 patients. This low degree of mineralization was coupled with a high percentage of crystalline mineral in a patient suffering from rheumatoid arthritis who had had long-term treatment with cortisone. It was associated with a low percentage of crystalline mineral, however, in a patient whose diet had been very deficient in calcium.

In 15 osteoporotic patients treated with fluoride for 1 year (Fig. 2), the percentage of crystalline mineral increased from $63.4 \pm 1.5\%$ before treatment to $65.8 \pm 1.3\%$ after treatment and the difference between these two values was found to be highly significant ($p < 0.0001$).

This same effect of fluoride was also observed in mice, where the percentage of crystalline mineral was $62.2 \pm 1.2\%$ in the controls, 13 months of age, as compared with $69.0 \pm 1.3\%$ in mice treated for twelve

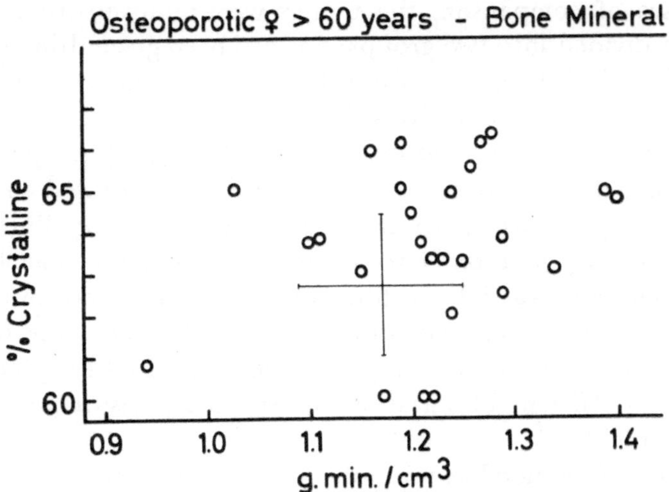

Figure 1. Relationships between degree of mineralization and percentage of crystalline mineral in bone biopsy specimens obtained from 26 osteoporotic women over 60 years of age (o). Comparison with values for controls of this same age group (1.17 ± 0.08 g min/cm³, $62.7 \pm 1.7\%$ crystalline), reveals that most of the cases are in the upper right quadrant of the graph indicating both high degree of mineralization and high percentage of crystalline mineral. A very low degree of mineralization was found in only 2 cases, one coupled with a high percentage and the other with a low percentage of crystalline mineral.

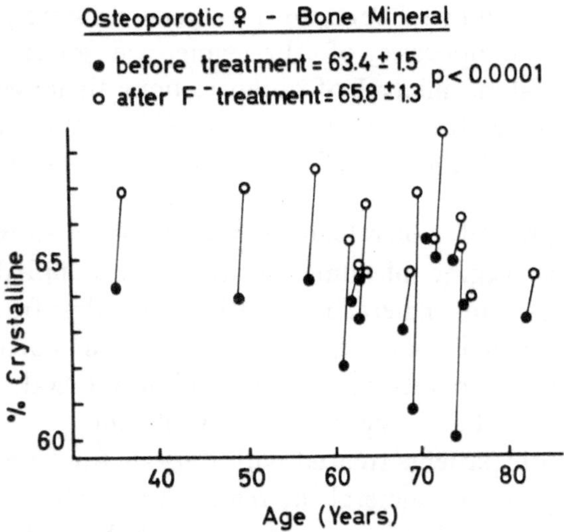

Figure 2. The effect of 12 months of fluoride treatment on the percentage of crystalline mineral as observed in biopsy specimens obtained from 15 woman with osteoporosis. The difference between the mean values obtained before (●) and after (o) treatment was found to be highly significant.

months after weaning. The difference between these two values is highly significant (p < 0.00005).

DISCUSSION

The results of this study would seem to indicate three different types of interrelationships between the crystalline/amorphous mineral ratio and the degree of mineralization of the bone tissue.

The first one is an increase in the percentage of crystalline mineral coupled with an increase in the degree of mineralization, which can be considered a "hypermaturation", i.e., an accentuation of the ageing process in bone tissue as has been observed in rats (8) as well as in humans (2).

The second one is an increase in the percentage of crystalline mineral associated with a decrease in the degree of mineralization mainly at the expense of the amorphous component, which is characteristic of halastatic diffuse demineralization. This phenomenon was first observed under experimental conditions where there was a hormone imbalance induced in eels by long-term administration of total pituitary extract (6).

The third type consists of low crystalline mineral percentage values associated with low degree of mineralization, which can be considered a hypomaturation, similar to that induced experimentally as a result of nutritional deficiencies in calcium and phosphorus (9).

Fluoride has been found to enhance the transformation of amorphous mineral into a crystalline apatitic phase, both *in vitro* (10) and *in vivo*, as observed in experimentally indeced fluorosis in mice. In osteoporotic patients, fluoride not only stimulates the formation of new bone tissue but also increases the percentage of stable crystalline mineral.

SUMMARY

The data on the percentage of crystalline mineral and on the degree of mineralization of bone tissue, considered together, indicate three different types of osteoporosis, i.e., with hypermaturation, hypomaturation or with halastatic demineralization. It is also shown that fluoride therapy enhances the transformation of bone mineral from the amorphous into the crystalline phase.

REFERENCES

1. Baud, C.A. & Baud, J.P.: Photométrie automatique de microradiographies. *Acta anat. (Basel)* **86**, 304 (1973)
2. Baud, C.A., Pouëzat, J.A. & Very, J.M.: Ultrastructure and distribution of mineral salts in POA tissue. *Paraplegia* **11**, 70–71 (1973)

3. Blumenthal, N.C., Posner, A.S. & Holmes, J.M.: Effect of preparation conditions on the properties and transformation of amorphous calcium phosphate. *Mat. Res. Bull.* 7, 1181—1190 (1972)

4. Eanes, E.D., Harper, R.A., Gillessen, I. & Posner, A.S.: An amorphous component in bone mineral. *Proc. 4th Europ. Symp. Calc. Tiss.*, Excerpta Medica Internat. Congress Series 120, 24—26 (1966)

5. Harper, R.A. & Posner, A.S.: Measurement of non-crystalline calcium phosphate in bone mineral. *Proc. Soc. exper. Biol. and Med. (N.Y.)* 122, 137—142 (1966)

6. Lopez, E., Lee, H.S. & Baud, C.A.: Etude histophysique de l'os d'un Téléostéen, *Anguilla anguilla L.*, au cours d'une hypercalcémie provoquée par la maturation expérimentale. *C.R. Acad. Sci. (Paris)* D 270, 2015—2017 (1970)

7. Termine, J.D.: *Amorphous calcium phosphate: the second mineral of bone.* PhD. Thesis. New York. Cornell Univ. (1966)

8. Termine, J.D. & Posner, A.S.: Infrared analysis of rat bone: age dependency of amorphous and crystalline mineral fractions. *Science* 153, 1523—1525 (1966)

9. Termine, J.D. & Posner, A.S.: Amorphous/crystalline interrelationships in bone mineral. *Calc. Tiss. Res.* 1, 8—23 (1967)

10. West, V.C. & Storey, E.: The *in vitro* effect of fluoride and tetracycline on phase transformation of a calcium phosphate. *Calc. Tiss. Res.* 9, 207—215 (1972)

Correlation of Clinical, Densitometric, and Histomorphometric Data in Osteoporosis

H.-P. KRUSE, F. KUHLENCORDT & J.-D. RINGE

There are different methods of examination for diagnosis and estimation of the severity of osteoporosis. The most usual methods are the conventional X-ray examination of the skeleton, direct radiological measurement of bone mineral content as well as histomorphometry of bone biopsies. Firstly, the value and importance of the above mentioned methods will be examined and secondly, findings which seem to be discrepant will be discussed.

We investigated 63 patients with osteoporosis, 35 women and 28 men. In 13 cases, i.e. 21%, secondary osteoporosis was diagnosed. These were mainly osteopathies caused by gastrointestinal or endocrinological disturbance. In the other 50 patients no basic disease could be found. So 79% of the patients had primary osteoporosis (Table Ia). The diagnosis of osteoporosis was in all cases made from clinical, biochemical, radiological and bone-histological data.

The bone mineral content was determined by photonabsorptiometry of the forearm according to Cameron and Sorenson (1, 6). The measured values were calculated in per cent mineral loss in relation to normalized data, with respect to age and sex (5, 6).

The bone biopsy was performed vertically through the iliac crest at about 2 cm behind the anterior superior iliac spine. In an undecalcified ground section embedded in methylmethacrylate volumetric density of cancellous bone was determined by means of an integrating eyepiece (3). The values were again calculated in per cent of bone loss regarding to age- and sex-depending normative data (2). The individual results of the measurements of mineral content and the determination of volumetric density of the iliac crest spongy bone are shown in correlation in Fig. 1. On the horizontal axis we marked the loss of spongy bone in per cent. The appropriate normal range of ± 9.5% is given as a hatched area on the left.

Department of Clinical Osteology, I. Medical University Clinic, Hamburg.

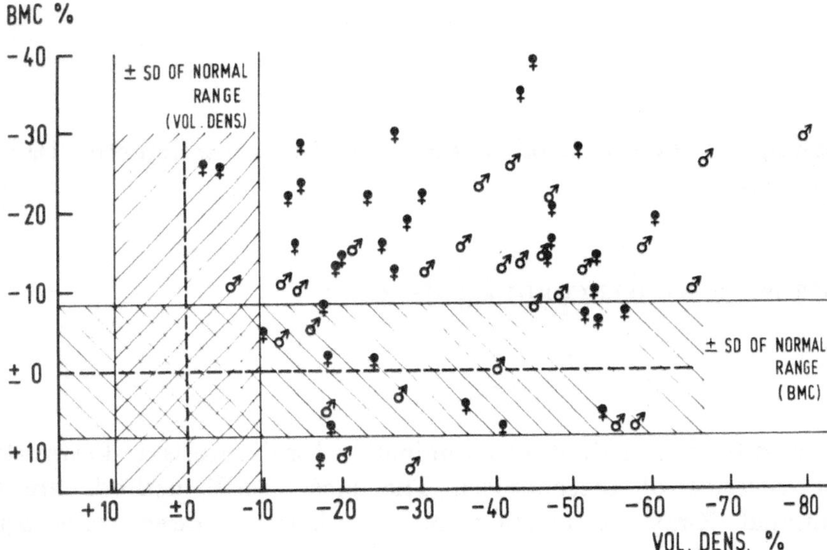

Figure 1. Bone mineral content and volumetric density of iliac crest in 63 cases of osteoporosis.

The vertical axis represents the loss of mineral content of the radius in per cent. The corresponding normal range is the horizontal hatched area from + 8.4 till − 8.4%.

All cases in the upper right area of the diagram outside the two hatched normal ranges have in both parameters pathological values, i.e. 60% of all osteoporoses. A linear correlation between mineral content and volumetric density could not be ascertained. The results are shown in Table Ia. As a whole, for both sexes the mineral content and the volumetric density was lowered in 60%, whereas in 35% only the volumetric density and in 5% only the bone mineral content was lowered. If one consideres men and women separately, it becomes obvious that both sexes behave almost

Table Ia. Sex-distribution, Diagnoses and Results of Examined Patients.

	♂ n	♂ %	♀ n	♀ %	♂ + ♀ n	♂ + ♀ %
Number of patients	35	100	28	100	63	100
Primary osteoporosis	26	74	24	86	50	79
Second. osteoporosis	9	26	4	14	13	21
B.M.C. ↓ Vol.dens. ↓	21	60	17	61	38	60
Vol.dens. ↓ B.M.C. ↔	12	34	10	36	22	35
B.M.C. ↓ Vol.dens. ↔	2	6	1	3	3	5

↓ = decreased, ↔ = normal.

similarly. In no case the mineral content of the radius and the volumetric density of the iliac crest were normal at the same time, so that all cases of osteoporosis could be diagnosed by the combination of the two methods.

The findings made us suppose that among patients with normal mineral content in the peripheral skeleton and with lowered volumetric density in the iliac crest cancellous bone there are cases, whose osteoporosis is active in the axial skeleton only. Therefore, for further analysis we took into consideration the X-ray findings as a third parameter. In order to differentiate different degrees of severity we defined X-ray indices from 0 to 3.

Index 0: No typical changes (3 patients).
Index 1: Increased radiolucency of the skeleton only (20 patients).
Index 2: Increased radiolucency, deformities and/or fractures of vertebrae (34 patients).
Index 3: Increased radiolucency, deformities and/or fractures of vertebrae and spontaneous fractures of peripheral bones (6 patients).

Comparing these X-ray-defined degrees of severity with the parameters examined above we obtained the following results: For the groups of patients (index 0 to 3) we calculated the mean loss of bone mineral content and of volumetric density of cancellous bone. The loss of cancellous bone in per cent $\pm$ s.d. was for index 0 -13.7 ± 0.4, for index 1 -29.7 ± 16.7, for index 2 -37.1 ± 16.2, and for index 3 -46.3 ± 23.5. For the bone mineral content the percentage loss $\pm$ s.d. was for index 0 -5.2 ± 4.6, for index 1 -10.8 ± 14.3, for index 2 -12.2 ± 10.3 and for index 3 -18.1 ± 13.8.

The differences did not prove to be significant by Student's t-test. Similarly in an earlier examination (4) we were not able to establish a correlation between the volumetric density of cancellous bone and the radiological index according to Barnett and Nordin. But considering all patients one can say that in general the lowest values of bone mineral content and volumetric density are found in patients with the most pronounced X-ray-changes of the skeleton.

On the other hand there are 12 cases among the 40 patients with X-ray-index 2 or 3 (i.e. with at least obvious impression fractures of vertebrae), that have a normal mineral content of the peripheral skeleton in spite of lowered volumetric density (Table Ib). They are 7 men and 5 women, representing 19% of the examined 63 patients. The volumetric density of cancellous bone was in relation to healthy persons on the average lowered by 37.9%, whereas the change of bone mineral content averaged + 0.98%. Obviously, those patients suffer from a type of osteoporosis that affects predominantly the axial skeleton. In these cases the changes of the iliac crest go together with those of the spine. It is remarkable that all these cases were primary osteoporoses.

Table Ib. Severe Osteoporosis of the Spine
without Participation of the Forearm

	♂	♀	♂ + ♀
n ≅ %	7 ≅ 25%	5 ≅ 14%	12 ≅ 19%
Age (mean)	48.0	55.0	50.9
Vol.dens. %	−40.2	−34.7	−37.9
B.M.C. %	+ 2.7	− 1.3	+ 0.98

CONCLUSION

1. Combining measurements of bone mineral content and of volumetric density of iliac crest spongy bone all cases of osteoporosis could be proven.
2. In 60% of the cases bone mineral content and volumetric density were lowered simultanously.
3. Though in general the most severe X-ray-changes go together with the lowest bone mineral content and the lowest density of cancellous bone, an estimation of the degree of severity of an osteoporosis in the individual case is only possible on the basis of all three methods.
4. In 19% of our patients the osteoporosis involved the axial skeleton only, without affecting the peripheral skeleton. All these patients had primary osteoporosis.

ACKNOWLEDGEMENT

Supported by Deutsche Forschungsgemeinschaft, SFB 34 "Endocrinology".

REFERENCES

1. Cameron, J.R. & Sorenson, J.: Measurement of bone mineral *in vivo*: an improved method. *Science* 142, 230−232 (1963)
2. Eger, W., Gerner, H.J. & Kämmerer, H.: Bau und Dichte der menschlichen Spongiosa in Rippe, Wirbel und Becken als Ausdruck der statischen Funktion. *Arch. orthop. Unfall-Chir.* 62, 97−112 (1967)

3. Kruse, H.-P.: Histologie der Osteoporose. *Verh. Dtsch. Orthop. und Traumatolog. Ges.*, *57. Kongress*, Enke, pp. 226–230, 1971

4. Kuhlencordt, F., Kruse, H.-P., Lozano-Tonkin, C., Wieners, H. & Bartelheimer, H.: Vergleichende röntgenologische und morphometrische Untersuchungen bei der Osteoporose. *Klin. Wschr.* 45, 1020–1023 (1967)

5. Kuhlencordt, F., Ringe, J.-D., Kruse, H.-P. & v. Roth, A.: Bone mineral determination of radius, ulna, and fingerbones by iodine 125-photon absorptiometry on healthy persons. In: *International Conference on Bone Mineral Measurement 1973*, U.S. Department of Health, Education, and Welfare, Publication No. 75–683, pp. 277–281, 1974

6. v. Roth, A., Ringe, J.-D., Kruse, H.-P. & Kuhlencordt, F.: Bestimmung des Knochenmineralgehaltes durch 125 J-Photonenabsorptionsmessung bei Gesunden. *Fortschr. Röntgenstr.* 121, 597–603 (1974)

Interrelationship Between Osteoporosis and Fractures of Neck of Femur

J. MENCZEL, M. MAKIN, G. ROBIN, R. STEINBERG & M. LENDER

Incidence of fracture of the proximal end of the femur in Jerusalem was subject to previous surveys (3, 2). Epidemiological studies of osteoporosis in Israel investigating nutritional factors involved in this disease were also carried out (1). The following is a report of a new study of prevalence of osteoporosis, as well as the incidence of fractures of the neck of femur studied simultaneously in a sample of the general population of Jerusalem. The correlation between these two conditions was investigated and the role of osteoporosis as a causative factor in fractures of neck of femur is discussed.

MATERIAL AND METHODS

The two surveys were carried out simultaneously between January 1967 and December 1971 among the Jewish population in Jerusalem and investigations were done by the same team of workers.

Osteoporosis

For the purpose of investigating the prevalence of this condition, a total of 3600 subjects aged between 45 and 84 years, chosen at random from the Jewish population in Jerusalem, were studied. Half of the random samples were women. All subjects were interviewed. Roentgenograms of the spine, pelvis, non-dominant hand and femur, were obtained, using a standardized technique. Osteoporosis was determined by qualitative examinations of these films in accordance with predetermined standard criteria, on a five grade scale according to Smith *et al.* (4):

    0. No abnormalities.

    1. Borderline changes, no vertebral body deformation.

Hebrew University, Hadassah Medical Center, Jerusalem.

2. Overall density loss and trabecular thinning, end plate accentuation and deformity.
3. Further loss of density and trabecular markings, end plate deformity and biconcavity, definite wedging of one or more vertebral bodies.
4. Severe demineralization, extensive biconcavities, marked wedging or collapse of several vertebral bodies.

Definite osteoporosis is regarded as stage 2 or more.

In all subjects osteoporosis was related to age, sex and ethnic origin.

Femoral neck fractures

The Jewish population in Jerusalem gets service by two hospitals which drain all trauma to the locomotor system including that of the upper femur. The present survey includes all patients admitted to the two hospitals during the period 1967–1971. The fractures under discussion include intracapsular and extracapsular fractures, the latter including subtrochanteric fractures that are within 1 cm of the lesser trochanter. All fractures that were the result of neoplastic conditions or other pathology in bone were excluded. A detailed questionnaire covering information pertinent to the study was completed for each patient and roentgenograms of the pelvis, lumbar spine, non-dominant hand and non-dominant shaft of femur were taken. A total of 570 patients were included in the survey. Unfortunately, in 181 cases the lumbar spine films were unavailable for the study due to inadequate processing technique standards, or, were lost. 389 cases were studied for the degree of osteoporosis.

RESULTS

The degree of osteoporosis by age and sex in general population sample in Jerusalem is given in Table I. It is demonstrated that the prevalence of definite osteoporosis in females is increasing progressively with age, from 3.2 per cent in the 45–54 years age group to 26.5 per cent in the 75–84 years age group. In males, the prevalence was smaller but increased with age similarly from 0.6 per cent in the first age group to 5.0 per cent in the latter group, respectively. Definite osteoporosis was found 5 times more frequently in females than in males (254 females and 51 males). Three hundred and eighty nine cases of femoral neck fractures were studied for the degree of osteoporosis and the results are compiled in Table II.

The prevalence of osteoporosis in this population is increasing with age in a similar pattern as in the general population. Females with definite osteoporosis were 5.6 times more frequent than males (106 females versus 19 males).

464

Table I.

Degree of Osteoporosis by Age and Sex in a Population Sample
in Jerusalem, 1967—1971.

Age groups	Female Degree of Osteoporosis			
	0	1	2—4	Total
45—54	310 (83.6)	49 (13.2)	12 (3.2)	371 (100)
55—64	335 (57.2)	185 (31.6)	66 (11.2)	586 (100)
65—74	217 (51.1)	131 (30.8)	77 (18.1)	425 (100)
75—84	114 (30.6)	160 (42.9)	99 (26.5)	373 (100)
Total	976 (55.6)	525 (29.9)	254 (14.5)	1755 (100)

Age groups	Male Degree of Osteoporosis			
	0	1	2—4	Total
45—54	311 (91.2)	28 (8.2)	2 (0.6)	341 (100)
55—64	453 (81.6)	90 (16.2)	12 (2.2)	555 (100)
65—74	370 (79.8)	78 (16.8)	16 (3.4)	464 (100)
75—84	278 (66.2)	121 (28.8)	21 (5.0)	420 (100)
Total	1412 (79.3)	317 (17.8)	51 (2.9)	1780 (100)

The percentages are given in parentheses.

Table II.

Degree of Osteoporosis by Age and Sex Among 389 Patients with Fracture
of Neck of Femur.

Age groups	Female Degree of Osteoporosis			
	0	1	2—4	Total
—54	11 (68.8)	3 (18.7)	2 (12.5)	16 (100)
55—64	29 (56.9)	14 (27.4)	8 (15.7)	51 (100)
65—74	25 (25.3)	42 (42.4)	32 (32.3)	99 (100)
75—84	12 (14.1)	25 (29.4)	48 (56.5)	85 (100)
85—	3 (10.3)	10 (34.5)	16 (55.2)	29 (100)
Total	80 (28.6)	94 (33.6)	106 (37.8)	280 (100)

Age groups	Male Degree of Osteoporosis			
	0	1	2—4	Total
—54	8	—	—	8
55—64	10	2	2	14
65—74	14	13	9	36
75—84	10	22	3	35
85—	7	4	5	16
Total	49	41	19	109

The percentages are given in parentheses. In small groups no percentages are given.

In Fig. 1 the prevalence among the general population and the femoral neck fracture population is compared. In both populations the prevalence of osteoporosis increases progressively with age. In the age groups under 54 years, osteoporosis was found in 3.2 per cent, whereas, in the femoral neck fracture population osteoporosis was 4 times more prevalent (12.5 per cent). In the oldest age groups, (75—84 years), the difference in prevalence of osteoporosis was smaller.

The degree of osteoporosis by place of birth is summarized in Table III. It should be noted, that in females born in Asian-African countries, osteoporosis is found almost twice as often as in females of European-American birth place.

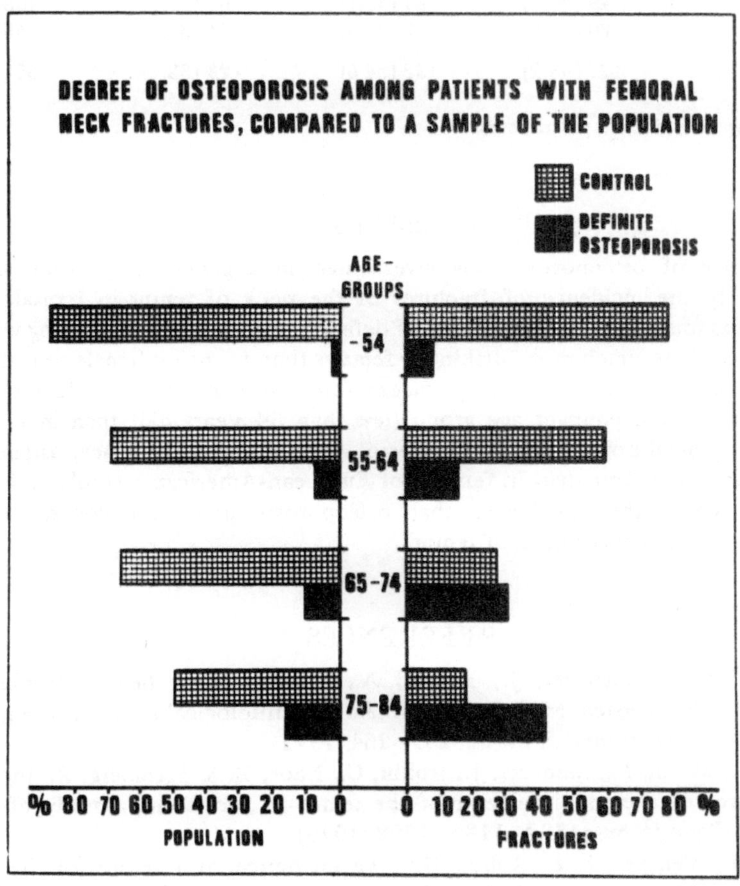

Figure 1. Degree of osteoporosis by age among patients with femoral neck fractures compared to a general population sample in Jerusalem, 1967—1971. As controls served the groups where no osteoporosis was found.

Table III.
Degree of Osteoporosis by Birth Place Among Patients with Fracture
of Neck of Femur Compared with a General Population Sample
in Jerusalem, 1967–1971.

General Population Degree of Osteoporosis

Country of birth	0	1	2–4	Total
Europe-America	1270 (72.9)	368 (21.1)	105 (6.0)	1743 (100)
Asia-Africa	785 (60.8)	341 (26.4)	166 (12.8)	1292 (100)
Israel	332 (66.5)	133 (26.7)	34 (6.8)	499 (100)
Total	2387 (67.6)	842 (23.8)	305 (8.6)	3534 (100)

Femoral-neck Fracture Population Degree of Osteoporosis

Country of birth	0	1	2–4	Total
Europe-America	58 (37.2)	56 (35.9)	42 (26.9)	156 (100)
Asia-Africa	49 (29.2)	59 (35.1)	60 (35.7)	168 (100)
Israel	20 (34.5)	17 (29.3)	21 (36.2)	58 (100)
Total	127 (33.2)	132 (34.6)	123 (32.2)	381 (100)

The percentages are given in parentheses.

SUMMARY

The prevalence of osteoporosis was investigated in a general population sample. Simultaneously the incidence of fractures of the neck of femur in Jerusalem was studied. It was found that the prevalence of definite osteoporosis is increasing with age in both populations, much more striking in females than in males; female to male ratio being 5.0–5.6:1.0. Osteoporosis is prevalent four times more in the femoral neck fracture group at the younger age group (less than 54 years old) than in the same group in the general population. In females of Asian-African birthplace, osteoporosis was twice more prevalent than in females of European-American birthplace. The two surveys strenghten the conclusion that osteoporosis plays a major role in the pathogenesis of fractures of neck of femur.

REFERENCES

1. Guggenheim, K., Menczel, J., Reshef, A., Schwartz, A., Ben Menachem, Y., Bernstein, D.S., Hegsted, M. & Stare, F.J.: An epidemiological study of osteoporosis in Israel. *Arch. environm. Hlth.* 22, 259–264 (1971)
2. Levine, S., Makin, M., Menczel, J., Robin, G., Naor, E. & Steinberg, R.: Incidence of fractures of the proximal end of the femur in Jerusalem, a study of ethnic factors. *J. Bone Jt. Surg.* 52A, 1193–1202 (1970)
3. Makin, M., Menczel, J. & Robin, G.: The incidence of fractures of the hip in Jerusalem (1957–66) as an index of osteoporosis. In: *Osteoporosis*, Barzel, U.S. (ed.), Grune & Stratton, New York, pp. 164–173, 1970
4. Smith, R.W., Eyler, W.R. & Mellinger, R.C.: On the incidence of senile osteoporosis. *Ann. intern. Med.* 52, 773–781 (1960)

Growth Hormone in Osteoporosis

H.G. HAAS, M.A. DAMBACHER, H. GÖSCHKE, J. GUNCAGA, TH. LAUFFEN-BURGER, CH. LENTNER, A.J. OLAH & H.R. WACKER

In 1972, Harris *et al.* (1) reported a marked increase of cortical bone formation and calcium accretion in adult dogs treated with high doses of growth hormone. Since in osteoporosis, bone formation is insufficient to compensate for the bone loss, human growth hormone (HGH) appears to be a promising substance for the treatment of this bone condition. Thus, a long-term treatment program in osteoporotics was started using a highly purified preparation of HGH*.

Nine post-menopausal women with overt osteoporosis were treated for up to one year with HGH, 16 I.U. i.m. every second day. The patients were studied before and at three months intervals after HGH. Investigations consisted of biochemical bone studies, e.g. determination of serum alkaline phosphatase (aPh) and urinary hydroxyproline (OHPr), as well as of calcium (Ca) balance investigations combined with Ca kinetics, and of histomorphometry of paired iliac crest biopsies.

On HGH, a marked increase of hydroxyproline of + 150% and an also significant increase of the serum phosphorus was observed. In contrast to this, aPh showed only a slight and transitory increase. Both the Ca accretion Vo+ and the Ca mobilization Vo– were increased significantly above the normal range, while the Ca balance remained unchanged. On treatment, the only statistically significant change found in the biopsies was an increase of osteoblasts after three months of treatment. However, osteoblasts remained within the normal range of this age group. The osteoid seams, osteoclasts and the volume density showed a slight increase during the first six months of treatment, the figures, however, were not statistically significant. Tetracycline labelling of the bones revealed a marked periosteocytic mineralization besides the extensive labelling of the

*) Supplied by courtesy of KABI Pharmaceuticals, Stockholm, Sweden.

1. Med. Universitätsklinik, 4004 Basel, Switzerland.

mineralization front.

It is concluded that exogenous HGH stimulates the metabolic activity of all bone cells, osteoblasts, osteoclasts and osteocytes, which in turn leads to a dramatic increase of the calcium- and matrix turnover without major changes of the calcium balance and the trabecular structure of bone.

REFERENCES

1. Harris, W.H., Heaney, R.P., Jowsey, J., Cockin, J., Akins, C., Graham, J. & Weinberg, E.H.: Growth Hormone: The effect on skeletal renewal in the adult dog. I. Morphometric studies. *Calc. Tiss. Res.* 10, 1–13 (1972)
2. Heaney, R.P., Harris, W.H., Cockin, J. & Weinberg, E.H.: Growth Hormone: The effect on skeletal renewal in the adult dog. II. Mineral kinetic studies. *Calc. Tiss. Res.* 10, 14–22 (1972)

Preliminary Trial of Low Doses of Human Parathyroid Hormone 1—34 Peptide in Treatment of Osteoporosis

J. REEVE, G.W. TREGEAR[1] & J.A. PARSONS[2]

It has recently been suggested that parathyroid hormone (PTH), administered chronically in low doses so as to induce a positive calcium balance, may be effective in the treatment of osteoporosis (11). This hypothesis was based in part on the relative sensitivity of the multiple effects of the parathyroids on calcium metabolism (12) and in part on the known anabolic effects of PTH in low doses on young rat bone (17, 5, 21). In the adult dog, which provides a better model than the rat of adult human calcium metabolism, Parsons and Reit demonstrated that chronic low-dose infusions of bovine PTH (0.025 to 0.1 µg/kg/h) resulted in increased calcium absorption without a concomitant rise in urinary calcium, in spite of moderate hypercalcaemia at the higher dose levels. These results provided the justification for a preliminary trial of low doses of a synthetic amino-terminal fragment of human PTH (hPTH 1—34 (8, 20)) in the treatment of primary osteoporosis.

PATIENTS AND METHODS

The patients studied were four post-menopausal women, with primary osteoporosis and crush fractures of one of more vertebrae. Other causes of bone rarefaction were excluded by histological examination of an iliac crest bone biopsy and investigations of blood chemistry. All patients gave informed consent to their participation in the trial in the manner accepted by the Hospital Ethical Committee.

Each patient was admitted to hospital for baseline studies so that she

Medical Research Council, Clinical Research Centre, Watford Road, Harrow Middlesex.
1) Endocrine Unit, Massachusetts General Hospital, Boston, Mass.
2) National Institute for Biological Standards and Control, Holly Hill Hampstead, London.

could act as her own control. For the three months prior to the first admission and continuously throughout the trial, each patient was given 500 I.U. of supplemental ergocalciferol daily by mouth to exclude marginal vitamin D deficiency (19). During this first admission, initial calcium and magnesium balance studies were carried out. Each patient's diet was designed to be as similar as possible to her home diet, and after a five day run-in period, three consecutive six-day balances were performed. From the day of admission, the patients were given 1.5 g/day of chromium sesquioxide as a faecal marker, and in most studies they were also given three 500 mg lactose capsules daily, containing 0.03 μCi $^{51}CrCl_3$ in about 1 mg of chromic chloride carrier. The radioactive marker was counted in each bulk faecal collection against a counting standard, so that the result could be compared for quality control purposes with the recovery of stable chromium, estimated from its concentration in an aliquot of the faecal homogenate. A minimum of six duplicate diets per admission were analysed, both diets and faecal collections being prepared for chemical analysis by standard methods (24). Calcium and magnesium were estimated by atomic absorption spectrophotometry in diets, faeces and urine (13) and calcium in plasma samples by an automated cresolphthaleine-complexone colorimetric method as modified by Gitelman (4). Inorganic phosphate in urine was estimated by an automated colorimetric method depending on reduction of phosphomolybdate (6) and creatinine by the method of Raabo and Wallöe-Hansen (14).

Tracer studies of calcium kinetics were conducted during each balance. A calcium absorption test was performed using a development of the double isotope method as described by Reeve, Hesp and Veall (15), except that the dose was given in 200 ml of milk containing added calcium chloride to give a total of 7.5 mmol calcium. 20 μCi of ^{45}Ca was added to the milk, and 10 μCi ^{47}Ca was given intravenously as the milk was drunk. The excreta were counted over the next 14–20 days, with a counting error of no greater than 2% in the case of the 24 h urine samples. Endogenous faecal loss of calcium was then calculated by the method of Bronner and Harris (3), and the addition (accretion) rate of calcium to the skeleton (A_t) calculated by the method of Marshall (7). An alternative non-compartmental method of calculation was also used. This method requires fewer *a priori* assumptions than previous methods and provides an addition rate (A) that is not time-dependent. (Reeve, Hesp and Wootton, submitted for publication). Results obtained with both methods had coefficients of variation of less than 10%.

The hPTH used in the trial was synthesized as described by Tregear *et al.* (20) and purified by ion exchange chromatography and gel filtration.

Automated sequence analysis by Edman degradation, end-group analysis and thin layer chromatography indicated that at least 95% of its peptide content consisted of a homogeneous tetratriacontapeptide of the sequence reported by Niall *et al.* (8) for the amino-terminal 34 residues of human PTH. Under conditions approved by the Department of Health and Social Security for therapeutic substances, it was ampouled in 100 μg aliquots from a solution containing mannitol (2.5 mg/ampoule), sterilised by membrane filtration, lyophilised, sealed and tested for sterility and freedom from pyrogens. Immediately before intramuscular injection, the contents of each ampoule were dissolved in 1 ml of a separately ampouled vehicle containing 0.1% heat-inactivated human serum albumen (Lister Institute) and 5 mg of epsilon-amino caproic acid to enhance absorption by inactivating tissue peptidases at the site of injection (Rafferty, Zanelli and Parsons, 1976, submitted for publication).

Analysis of balance data

In two of the patients, a preliminary study was carried out to establish an optimum dose schedule for hPTH. Three schedules were tested in sequence, each being held constant for eight days, subdivided into two four-day balances. Carmine markers were given orally at the beginning of each balance in order to accurately separate the stool collections. On one of the days at each dose level, a special time-course study was performed during which clearances of Ca, Mg, PO_4 and creatinine were monitored in two-hour urine samples and blood levels of calcium were estimated at two-hourly intervals following the hormone injection.

In the main study, the recoveries of ^{51}Cr and stable chromium (^{52}Cr) were compared in twenty-seven faecal collections from individual balances on four patients. The mean recovery ratio ^{51}Cr: ^{52}Cr was 1.03 ± 0.12 (1 S.D.). If an individual ratio fell outside the range $0.85 - 1.15$, the homogenisation and subsequent chemical analysis were repeated.

The confidence limits for a 3 x 6 day calcium balance study with 6 duplicate diet analyses were also estimated by calculating the differences in mmol Ca between individual analyses and the appropriate mean value for each 18-day study. Data were pooled from seven such 18-day studies conducted in the steady state in 5 osteoporotic patients, on diets ranging from 20—44 mmol calcium per day. The standard errors calculated for individual stools and diets were 1.12 and 2.08 mmol calcium respectively. These results give 95% confidence limits of $\pm$ 2.1 mmol/day for each 18 day balance study.

In the dose-level studies, where the diet was held constant and the dose of hPTH 1—34 raised after every second balance, a difference between two regimes was statistically significant at the 5% level when the difference was greater than 2.2 mmol/day. (In these calculations it was assumed that the error contribution from the estimation of urine calcium was trivial).

RESULTS

1. Dose-level studies

In both patients (cases 3 and 4 of Table II), absorption of calcium from the diet increased on all three dose schedules and calcium balance became more positive. However, at a dose level of 400 µg/day (given as 200 µg

Table I.
The Effects of Increasing Doses of h-PTH 1—34 on Calcium Metabolism.

		8 day periods		
	Baseline studies	Dose = 100 µg/day	Dose = 200 µg/day	Dose = 200 µg twice daily
Patient 3				
Calcium balance (mmol/D)	+ 1.2 (4 months gap)	+ 7.1	+ 4.6	+ 2.1
Urine calcium excretion (mmol/D)	6.8	6.8	7.7	11.1
Net absorption (mmol/D)	8.0	13.9	12.3	13.2
Urine hydroxyproline (µmol/D)	(275)	117	158 (n.s.)*	176 (0.01<P<0.02)*
Patient 4				
Calcium balance (mmol/D)	+ 2.0	+ 7.4**	+ 9.6	+ 3.9
Urine calcium (mmol/D)	2.1	2.1	2.0	3.6
Net absorption (mmol/D)	4.2	9.5	11.6	7.5
Urine hydroxyproline (µmol/D)	146	134 (n.s.)	160 (n.s.)	233 (P < 0.001)

*   Statistical comparisons made with the results at the low dose level.
** Second 4 days balance only; first 4 days balance affected by diarrhoea, result —4.0 mmol/D.

twice daily), urinary excretion of calcium and hydroxyproline increased significantly, and the calcium balance became significantly less positive than at the lower dose levels (Table I).

The time courses of urinary excretion following I.M. injection of 100, 200 and 400 μg PTH were analysed by expressing the calcium and magnesium data as ratios of their clearances to that of creatinine, and the phosphate data was expressed as Tm_{PO_4}/GFR (1). As expected, all three calculated variables tended to fall following I.M. PTH, each showing a maximal effect between 2 and 4 h and recovery which was nearly complete at 6 h (Fig. 1). Plasma calcium tended to rise only on the highest dose, and never exceeded the normal range.

2. Long-term studies on 100 µg hPTH 1–34 per day

In three of the four patients in this study (cases 1, 2 and 3 of Table II), the original balance and tracer protocol has been repeated after 5 to 6 months of uninterrupted treatment with hPTH 1–34.

It was found that patient 1 (six months therapy) had moved from a negative balance (– 1.3) to a positive balance of + 1.4 mmol calcium* per day. Patient 2, after 5 months, had moved from – 3.0 to + 1.6 mmol/day, and patient 3 after 6 months from + 1.2 to + 4.2 mmol/day. Both net and true absorption of calcium from the diet (9) were increased in all cases, there being no significant change in the excretion of endogenous

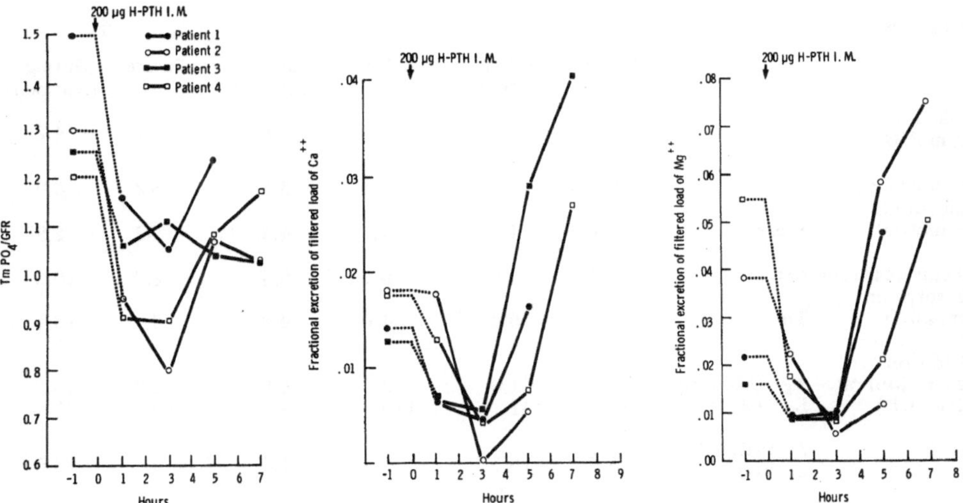

Figure 1. The renal handling of calcium, magnesium and phosphate following an intramuscular injection of 200 micrograms of hPTH (1–34) in all four patients. Qualitatively similar results were obtained in patients 3 and 4 in response to injected doses of 100 and 400 micrograms of hormone.

---

*) 1 mmol = 40 mg Ca.

faecal calcium. Magnesium balances were not significantly altered.

There were striking increases in the addition rate of calcium to the skeleton, which ever method of calculation was used. They exceeded the improvements in the calcium balance (thus indicating associated but smaller increases in the rate of bone resorption) and were accompanied by moderate increases in urinary calcium excretion.

The density of the lower third of the femur, measured by γ absorptiometry using an ^{241}Am source (23) showed no significant change over any of the 3 six-month periods of study. A 5—10% change would have been needed to achieve significance at the 95% confidence level.

It is noteworthy that in spite of the highly significant increases in absorption of dietary calcium indicated by the balance measurements, no consistent changes were seen in the results of the formal double-tracer calcium absorption test using a 300 mg carrier (Table II).

Table II.
The Long-term Effects of 100 μg/day h-PTH (1—34) on Calcium Metabolism.

Patient No.		1		2		3	
		Before	During treatment	Before	During treatment	Before	During treatment
Calcium balance (mmol/D)		- 1.3	+ 1.4	- 3.0	+ 1.6	+ 1.2	+ 4.2
Endogenous Calcium losses (mmol/D)	Urine	3.0	4.0	2.3	3.7	6.8	6.0
	Faeces	4.1	4.6	4.7	4.5	2.7	2.9
Dietary calcium Absorption (mmol/D)	Net	1.7	5.4	- 0.7	5.3	8.0	10.2
	True	5.8	10.0	4.0	9.8	10.7	13.1
Addition (accretion) rate (mmol/D)	A_5 (Marshall, E_5 1964)	14.0 139	21.0 141	9.9 115	16.0 114	11.8 109	18.2 120
	A, Authors' method	5.8	15.5	3.7	11.8	5.9	13.0
Hydroxyproline excretion rate (μmol/D)		254	280	175	156	275	265
Isotopic Calcium	% absorbed	18	22	25	25	29	33
Absorption test (7.5 mmol carrier)	Max. transfer rate (mmol/h)	0.58	0.48	0.65	0.62	0.96	1.42

DISCUSSION

These results demonstrate that chronic administration in low dosage of the amino-terminal fragment of parathyroid hormone improves the calcium balance and increases the accretion of calcium to the skeleton in primary osteoporosis. As can be seen from Table II, at the end of this 6 month period the calculated increase in skeletal calcium mass was 0.5 mole (20 g) per annum in the first two patients and 1.2 moles per annum in the third. The observations of West (22) on the relationship between femur density and skeletal calcium mass suggest that each patient's skeleton contained about 17 moles of calcium at the start of treatment. It is, therefore, clear that significant strengthening of the skeleton will require a prolonged period of treatment using the present regime.

A pointer to a possible improvement in the treatment protocol is provided by the renal clearance studies (Fig. 1), which indicate that blood levels must have begun to decline 3 h after injection in the vehicle used in the present study. Even this duration of action must reflect some delay in absorption because the circulating half lives of intact parathyroid hormone and its amino terminal fragment do not exceed 6 min (18, 16, 2).

The major physiological effects of parathyroid hormone have widely different time courses as well as differences in dose dependence (10). While the increased intestinal calcium absorption and the anabolic effect on bone appear to require one or more days to induce, the enhanced renal tubular reabsorption and osteolytic activity begin within minutes of an intravenous dose, and decline after a few hours. Therefore, administration of PTH so as to produce short periods of high hormone concentration has the twin disadvantages of inducing an osteolytic response which might be entirely avoided if the same dose were given by continuous infusion, and of allowing urine calcium excretion to escape as the blood hormone level declines. An improvement in therapeutic response may, therefore, be expected when delayed-release preparations of PTH become available. The possibility also exists of using PTH in combination with other agents which favour the increase of skeletal mass.

The conclusion that low doses of hPTH 1—34 provide an effective treatment for osteoporosis must remain provisional. Long-term demonstration of increased bone mass is required and it must be shown that the newly formed bone contributes to the strength of the skeleton. With currently available techniques, a statistically significant increase in bone mass cannot be expected within less than a year's continuous treatment. Nevertheless, these preliminary findings provide encouraging grounds for continuing investigation of the therapeutic role of the amino-terminal fragment of parathyroid hormone in osteoporosis.

ACKNOWLEDGEMENTS

The authors are grateful to Mrs C. Tait, Mr J.R. Green and Mr G.P. Gibbs for expert technical assistance. They would like to thank the technical staff of the Clinical Research Centre divisions of Clinical Chemistry and Computing and Statistics for their contributions to sample and data analysis respectively.

Thanks are also due to Mr L. Klenerman (who referred the patients), to Mr R. Hesp for bone densitometry measurements, and to Dr N. Veall for many helpful discussions in the course of this work. J. Reeve is a MRC Clinical Research Fellow.

REFERENCES

1. Bijvoet, O.L.M. & Van der Sluys Veer, J.: The interpretation of laboratory tests in bone disease. *Clin. Endocr. & Metab.* 1/1, 217–237 (1972)
2. Blum, J.W., Mayer, G.P. & Potts, J.T. Jr.: Parathyroid hormone responses during spontaneous hypocalcaemia and induced hypercalcaemia in cows. *Endocrinology* 95, 84–92 (1974)
3. Bronner, F. & Harris, R.S.: Absorption and metabolism of calcium in human beings studied with calcium-45. *Ann. N.Y. Acad. Sci.* 34, 314–325 (1956)
4. Gitelman, H.J.: An improved automated method for the determination of calcium in biological specimens. *Analyt. Biochem.* 18, 521–531 (1967)
5. Kalu, D.N., Pennock, J., Doyle, F.H. & Foster, G.V.: Parathyroid hormone and experimental osteosclerosis. *Lancet* (1970) i, 1363–1366
6. Lawrence, R.: Assay of serum inorganic phosphate without deproteinisation: Automated and manual micromethods. *Ann. Clin. Biochem.* 11, 234–237 (1974)
7. Marshall, J.H.: Theory of alkaline earth metabolism. In: *Technical Report No. 32*, I.A.E.A. Vienna, pp. 21–33, 1964
8. Niall, H.D., Sauer, R.T., Jacobs, J.W., Keutmann, H.T., Segre, G.V., O'Riordan, J.L.H., Aurbach, G.D. & Potts, J.T. Jr.: The amino acid sequence of the amino terminal 37 residues of human parathyroid hormone. *Proc. nat. Acad. Sci. (Wash.)* 71, 384–388 (1974)
9. Nordin, B.E.C. & Smith, D.A.: *Diagnostic Procedures in Disorders of Calcium Metabolism.* Churchill, London, 1965
10. Parsons, J.A. & Potts, J.T. Jr.: Physiology and chemistry of parathyroid hormone. *Clin. Endocr. & Metab.* 1/1, 33–78 (1972)
11. Parsons, J.A. & Reit, B.: Significance of entry rate in studies of hormone action. Chronic response of dogs to parathyroid hormone infusion. *Nature (Lond.)* 250, 254–257 (1974)
12. Parsons, J.A., Rafferty, B., Gray, D., Reit, B., Zanelli, J.M., Keutmann, H.T., Tregear, G.W., Callahan, E.N. & Potts, J.T. Jr.: Pharmacology of parathyroid hormone and some of its fragments and analogues. In: *Proceedings of the Vth Parathyroid Conference*, Talmage, R.V., Owen, M. and Parsons, J.A. (eds.), Excerpta Medica (Amst.), pp. 33–39, 1975

13. Pybus, J., Feldman, F.J. & Bowers, G.N.: Measurement of total calcium in serum by atomic absorption spectrophotometry with use of a strontium internal reference. *Clin. Chem.* **16**, 998–1007 (1970)

14. Raabo, E. & Wallöe-Hansen, P.: A routine method for determining creatinine avoiding deproteinisation. *Scand. J. clin. Lab. Invest.* **29**, 297–301 (1972)

15. Reeve, J., Hesp, R. & Veall, N.: Effects of therapy on rate of absorption of calcium from gut in disorders of calcium homeostasis. *Brit. med. J.* iii, 310–313 (1974)

16. Segre, G.V., Niall, H.D., Habener, J.F. & Potts, J.T. Jr.: Metabolism of parathyroid hormone, physiologic and clinical significance. *Amer. J. Med.* **56**, 774–784 (1974)

17. Selye, H.: On the stimulation of new bone-formation with parathyroid extract and irradiated ergosterol. *Endocrinology* **16**, 547–558 (1932)

18. Silverman, R. & Yalow, R.S.: Heterogeneity of parathyroid hormone; clinical and physiologic implications. *J. clin. Invest.* **52**, 1958–1971 (1973)

19. Stanbury, S.W., Mawer, E.B., Lumb, G.A., Hill, L.F., Holman, C.A., Taylor, C.M. & Torkington, P.: Vitamin D metabolism and renal bone disease. In: *Clinical Aspects of Metabolic Bone Disease*, Duncan, J. and Frame, B. (eds.), Excerpta Medica (Amst.), pp. 562–573, 1973

20. Tregear, G.W., Van Rietschoten, J., Greene, E., Keutmann, H.T., Niall, H.D., Reit, B., Parsons, J.A. & Potts, J.T. Jr.: Bovine parathyroid hormone: minimum chain length of synthetic peptide required for biological activity. *Endocrinology* **93**, 1349–1353 (1974)

21. Walker, D.G.: The induction of osteopetrotic changes in hypophysectomised thyroparathyroidectomised and intact rats of various ages. *Endocrinology* **89**, 1389–1406 (1971)

22. West, R.R.: The estimation of total skeletal mass from bone densitometry measurements using 60 KeV photons. *Brit. J. Radiol.* **46**, 599–603 (1973)

23. West, R.R. & Reed, G.W.: The measurement of bone mineral *in vivo* by photon beam scanning. *Brit. J. Radiol.* **43**, 886–893 (1970)

24. Wootton, I.D.P.: *Microanalysis in Medical Biochemistry*. Livingstone, Edinburgh, 1974

Oestrogens and Post-menopausal Osteoporosis

J. LINDE, TH. FRIIS & E. ØSTERGAARD

The prophylactic effect of oestrogens on post-menopausal osteoporosis has not been clearly demonstrated. Donaldson and Nassim (4) could not demonstrate any relation between the time of oophorectomi and development of osteoporosis and they had the opinion that osteoporosis was a phenomenon of age and had nothing to do with lack of oestrogens. In contrast to these authors, Aitken *et al.* (1) could demonstrate that osteoporosis was more frequent in oophorectomized patients than in other women. Salomon *et al.* (11) could not demonstrate any prophylactic effect of oestrogens, but Davis *et al.* (3) found a protecting effect of oestrogens on post-menopausal osteoporosis estimated by photon absorptiometri.

Bone histological studies have rendered interesting results. Villanueva *et al.* (12), Jowsey *et al.* (7) and Riggs *et al.* (10) have shown, by tetracycline labelling and microradiographic investigations, that patients with post-menopausal osteoporosis had greater bone resorption, and that oestrogen treatment for 2–4 months counteracted this, but that treatment for longer time, for instance a year, also inhibited bone formation so that a new steady state was attained.

In order to evaluate the prophylactic effect of oestrogens on post-menopausal osteoporosis, we have done the following:

MATERIAL

Retrospectively we have studied the effect of oestrogens in three groups of healthy women (Table I).

Twenty healthy women were pre-menopausal. None of these got oestrogens. Thirty-eight healthy post-menopausal women were not treated

From medical department E and gynaecological department F, Frederiksberg hospital, Copenhagen.

Table I.

MATERIAL		Number	Body weight (kg)	Age (years)
I	Premenopausal women.	20	61.9 ± 9.6	30.8 ± 5.6
II	Post-menopausal women. Untreated.	38	63.1 ± 6.8	61.3 ± 2.7
III	Post-menopausal women. Treated with oestrogens.	30	63.4 ± 11.4	60.5 ± 2.4

and 30 healthy post-menopausal women received oestrogens. The duration of treatment was 3 to 13 years (mean 8,5 years). The mean body weight was the same in all three groups and the age did not differ significantly in the groups 2 and 3.

The doses of oestrogens used were 1–2 mg oestradiol per day which is equivalent to only about a third of the pre-menopausal endogen oestrogen production. Six of the women received 4–6 mg oestradiol per day in sequentiel combination with gestagens corresponding to the conditions in the pre-menopausal age.

METHODS

The following studies were done (in the fasting state) in the morning:

Determination of serum calcium by atomic absorption spectrophotometry, serum ionized calcium, serum phosphorus, alkaline phosphatase in serum, calcium/creatinine-index in urine (8), phosphate excretion-index (PEI) *ad modum* Nordin (9), third metacarpal index *ad modum* Exton-Smith (5), and photon absorptiometry of the forearm and of the calcaneus (6) with Studsvik osteodensitometer using 1^{125} (27,5 KeV) as γ-source. The result of bone mineral content (BMC) is given in arbitrary units linearly correlated to the calcium content per cm^2 in the scan areas (mean of sex scans with an interval of 2 mm). Coefficient of variation (day to day measurements on the same person) was 3–4%.

RESULTS

The results are given in Table II.

Serum Calcium. Serum calcium was higher in the two post-menopausal groups (II and III) compared to the pre-menopausal group (I). In group III

Table II. Parameters for premenopausal (I), untreated post-menopausal (II) and oestrogen treated post-menopausal (III) women.

	n	se-Ca meq/l	se-ion. Ca meq/l	se-P mmol/l	bas. phosph. U/l	PEI	Ca/cr. index	Photon absorptiometry		
								forearm	calcaneus	3. metacarp index
I	20	4.71 ±0.21	2.73 ±0.13	1.12 ±0.13	1.20 ±0.38	0.03 ±0.09	0.08 ±0.08	19.5 ± 3.4	32.0 ± 6.4	52.3 ± 6.7
II	38	4.84 ±0.26	2.80 ±0.17	1.14 ±0.19	1.82 ±0.50	0.01 ±0.11	0.26 ±0.19	16.3 ± 3.6	23.8 ± 4.8	44.2 ± 7.3
III	30	4.89 ±0.19	2.85 ±0.09	1.07 ±0.15	1.74 ±0.41	0.03 ±0.09	0.17 ±0.09	18.4 ± 3.2	23.6 ± 4.2	46.0 ± 7.4
I vs. II					$p<0.001$		$p<0.001$	$p<0.001$	$p<0.001$	$p<0.001$
I vs. III		$0.01>p>0.001$	$0.01>p>0.001$		$p<0.001$		$0.01>p>0.001$		$p<0.001$	$0.01>p>0.001$
II vs. III							$0.05>p>0.002$	$0.02>p>0.01$		

Significance with Students' t-test between the groups I, II and III:
Only significant differences are shown.

the difference was highly significant compared to group I ($0.01 > p > 0.001$).

Ionized calcium: The alterations are parallel to the total serum calcium. A tendency to higher values in group two and significantly higher values in group three ($0.01 > p > 0.001$) was fould.

The *alkaline phosphatase* in serum in the three groups: the post-menopausal women had significantly higher values than the pre-menopausal group ($p < 0.001$). There was no difference between the two post-menopausal groups.

The calcium excretion (estimated by the calcium/creatinine index in urine): In the untreated post-menopausal group it was significantly higher compared to the pre-menopausal group ($p < 0.001$). The oestrogen-treated group had values a little lower than the untreated group, significant by Student's t-test ($0.05 > p > 0.02$), but not significant by the Wilcoxon-test.

In *the phosphate excretion-index* (PEI): no differences were found between the three groups.

The third metacarpal-index: there was lesser mineral content in the post-menopausal women compared to pre-menopausal women, but there were no differences between the treated and untreated group.

Photon absorptiometry of the right forearm showed that the untreated post-menopausal group (II) had significantly lower values compared to pre-menopausal women ($p < 0.001$). The treated group (III) showed significantly higher bone mineral content (BMC) than the untreated group ($0.02 > p > 0.01$). No difference was found between the pre-menopausal groups and the treated post-menopausal group.

Photon absorptiometry of the right calcaneus showed lower values for the post-menopausal women than for the pre-menopausal (p < 0.001). On the other hand there was no difference between the two post-menopausal groups.

The significance of the following factors were analyzed: a) duration of treatment, b) difference between synthetic and genuine oestrogens, c) dose and d) time of the beginning of treatment in relation to menopause.

ad a) Among the 30 treated women no information about the duration of treatment was obtained in two cases. 17 had treatment for 7 years or less, 11 had treatment for more than 7 years. No differences in photon absorptiometry was found.

ad b) 17 patients had genuine and 10 synthetic oestrogens. In 3 patients a combination with androgens was given. The patients who got genuine oestrogens had a little higher BMC value in the calcaneus than the persons who got synthetic oestrogens (24.6 ± 4.0 and 20.8 ± 3.0, 0.02 > p > 0.01). However, the age of the first group was 58.2 ± 2.6 and of the second 61.7 ± 1.6 years.

ad c) Among the 30 oestrogen-treated patients, six had large doses of oestrogens (4—6 mg oestradiol daily). BMC of the forearms showed values of 20.7 ± 3.0 in the persons who had large doses compared to 17.0 ± 3.2 in 24 persons who got smaller doses (0.02 > p > 0.01). The age of the first group was 57.8 ± 3.3, and of the second 61.2 ± 1.6 years.

ad d) In 16 cases the treatment was started within 3 years after the menopause, in 12 cases 3 years or more after the menopause. No difference in BMC was found between these groups.

DISCUSSION

Our findings that serum calcium is increased in post-menopausal women compared to pre-menopausal women are in accordance with the findings of Young and Nordin (13), who stated, that it is caused by an increased bone resorption in post-menopausal women. They found in contrast to us that oestrogen treatment reduced serum calcium. The tendency to a lowered calcium excretion is in accordance with the observations of Young and Nordin (13).

Our phosphate studies have not shown any differences in serum or urine excretion in contrast to the observations of Aitken *et al.* (1, 2), who found increased serum phosphorus and decreased phosphate excretion-index in oophorectomized patients. The authors also found stronger effect of oestrogen treatment, if this was started within 3 years after the menopause (2). We could not reproduce these findings.

As a whole the registrated differences between treated and untreated postmenopausal women are only slight. The reason for this may be that the doses of oestrogens used in the majority of cases have been comparatively low.

CONCLUSION

Pre-menopausal women compared to untreated post-menopausal women: Alkaline phosphatase in serum and calcium/creatinine-index were higher, BMC of the forearm and calcaneus and 3rd metacarpal-index lower in the untreated post-menopausal group compared to the pre-menopausal group.

Serum calcium and serum ionized calcium tended to be higher in the untreated post-menopausal group.

Post-menopausal women treated with oestrogens compared to untreated post-menopausal women: Calcium/creatinine-index tended to be lower and BMC of the forearm significantly higher in the treated group compared to the untreated post-menopausal group.

Serum calcium and serum ionized calcium were significantly higher in the treated groups compared to the pre-menopausal group.

In these studies the oestrogens seem to have a slightly protective effect against post-menopausal osteoporosis.

ACKNOWLEDGEMENTS

Supported by grants from Statens lægevidenskabelige forskningsråd, and Den lægevidenskabelige forskningsfond for Storkøbenhavn, Færøerne og Grønland.

REFERENCES

1. Aitken, J.M., McKay Hart, D. & Smith, D.A.: The effect of long-term mestranol administration on calcium and phosphorus homeostasis in oophorectomized women. *Clin. Sci.* 41, 233–236 (1971)
2. Aitken, J.M., Hart, D.M. & Lindsay, R.: Oestrogen replacement therapy for prevention of osteoporosis after oophorectomy. *Brit. Med. J.* 3, 515–518 (1973)
3. Davis, M.E., Lanzl, L.H. & Coxi, A.B.: Detection, prevention and retardation of menopausal osteoporosis. *Obstet. Gynecol.* 36, 187–198 (1970)
4. Donaldson, I.A. & Nassim, J.R.: The artificial menopause with particular reference to the occurrence of spinal osteoporosis. *Brit. Med. J.* 1, 1228–1230 (1954)
5. Exton-Smith, A.N., Millard, P.H., Payne, P.R. & Wheeler, E.F.: Method for measuring quantity of bone. *Lancet* II, 1153–1157 (1969)

6. Jensen, H., Christiansen, C., Lindbjerg, J.F. & Munck, O.: The mineral content in bone. Measured by means of 27.5 KeV radiation from 125-I. *Acta Radiolog.* **suppl. 313**, 214—220 (1972)

7. Jowsey, J., Kelly, P.J., Riggs, B.L., Bianco, A.J., Scholz, D.A. & Gershon-Cohen, J.: Quantitative microradiographic studies of normal and osteoporotic bone. *J. Bone Joint Surg.* **47A**, 785—806 (1965)

8. Nordin, B.E.C.: The pathogenesis of osteoporosis. *Lancet* **I**, 1011—1014 (1961)

9. Nordin, B.E.C. & Bulusu, L.A.: A modified index of phosphate excretion. *Postgrad. Med. J.* **44**, 93—97 (1968)

10. Riggs, B.L., Jowsey, J., Kelly, P.J., Jones, J.D. & Maher, F.T.: Effect of sex hormones on bone in primary osteoporosis. *J. clin. Invest.* **48**, 1065—1972 (1969)

11. Solomon, G.E., Dickerson, W.J. & Eisenberg, E.: *Geriatr.* **15**, 46—60 (1960)

12. Villanueva, A., Frost, H., Ilnicki, L., Frame, L., Smith, R. & Arnstein, R.: Cortical bone dynamics measured by means of tetracycline labelling in 21 cases of osteoporosis. *J. Lab. Clin. Med.* **68**, 599—616 (1966)

13. Young, M.M. & Nordin, B.E.C.: Effects of natural and artificial menopause on plasma and urinary calcium and phosphorus. *Lancet* (1967) **II**, 118—120

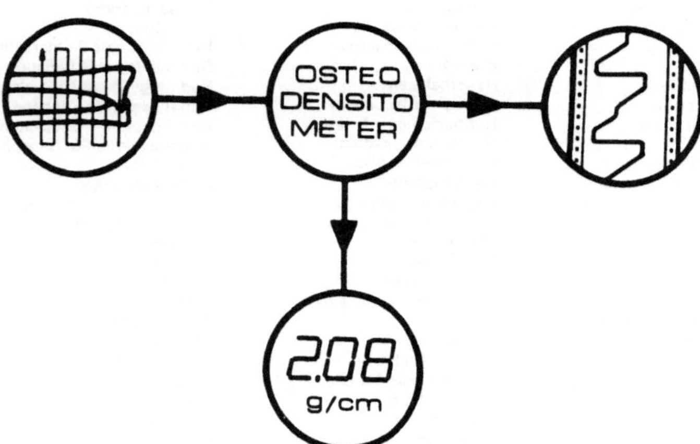

– in vivo determination of bone mineral content with 1% long term precision.
– now available in computerized version comprising automatical selection of measuring site and digital presentation of final result.

GAMMATEC·STORMLY 16·HARESKOV·DK 3500–TEL.02·98 2075

LEITZ-T.A.S.
Texture Analyzing System

The new farsighted concept of quantitative image analysis with one- and two-dimensional structuring elements for the determination of stereometric characteristics

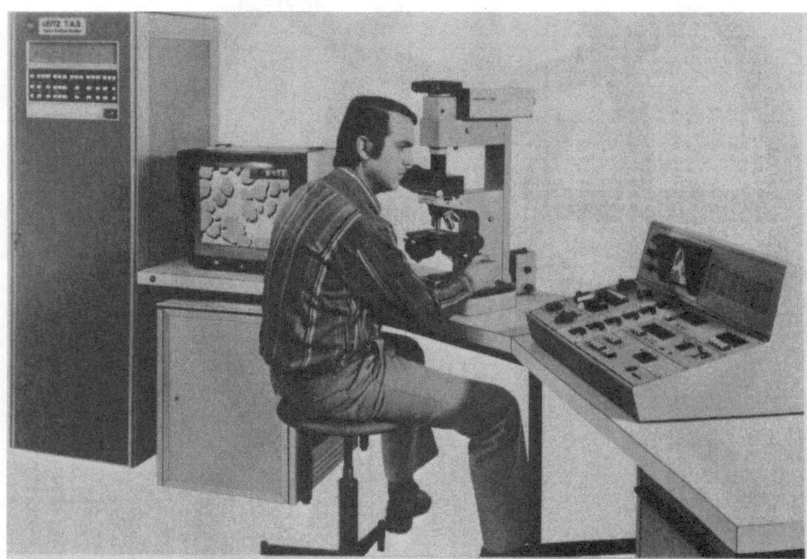

LEITZ production range

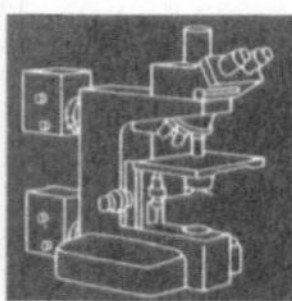

Microscopes
Microscopes of modern design for all investigations in transmitted, incident, and polarized light, fluorescence microscopes with incident-light or transmitted-light excitation.
Microscope accessories, such as phase contrast and interference contrast equipment, heating and cooling stages, universal rotating stages.
Instruments for special aspects of microscopy such as micromanipulator, transmitted-light interference microscope, stereo-microscopes, comparison macroscopes.
Equipment for photomicrography.
Television microscopes.
ORTHOMAT® W fully automatic microscope camera.
4x5" large-format camera with fully automatic exposure control

Microtomes
Microtomes for research and routine laboratories; in freezing chamber as HISTOKRYOTOM

Photographic equipment
LEICA® 35mm camera
LEICA lenses and accessories
LEICAFLEX® 35mm single-lens reflex camera
LEICAFLEX lenses and accessories
Accessories for scientific and technical photography
Enlargers
LEICINA® SUPER 8mm cine-camera

Binoculars
TRINOVID® for sport, travel, hunting.

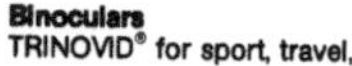

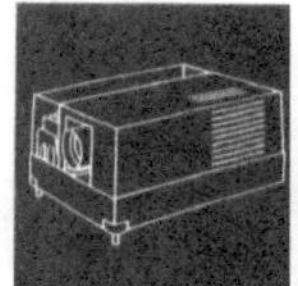

Projectors
PRADOVIT® COLOR automatic 35mm projector
PRADO® UNIVERSAL versatile classroom projector
Episcopes
Epidiascopes
Large lecture hall projectors
Micro projectors
Overhead projectors

ERNST LEITZ GMBH D-6330 WETZLAR
Subsidiary: ERNST LEITZ CANADA LIMITED, MIDLAND, ONTARIO

Generalagency for Denmark:

NOCO   Nordisk Optisk Co. A/S
Industriholmen 17-19, 2650 Hvidovre

LØVENS KEMISKE FABRIK
LEO PHARMACEUTICAL PRODUCTS
BALLERUP – DENMARK

A new result
of Sandoz-research:

Calcitonin-Sandoz synthetic salmon-calcitonin

Ampoules of
1 ml (20 µg / ml)

10.75

Calcified Tissue Research

This international journal publishes research into the structure and function of bone and other mineralised systems in living organisms. It includes reports and reviews of connective tissues and cells, ion metabolism and transport, hormones, nutrition, ultrastructure and molecular biology.

50 reprints of each paper are provided free of charge. Additional copies may be ordered at cost price; this must be done when the page proofs are returned to the publisher.

Manuscripts and inquiries may be directed to one of the following addresses:

The Editorial Secretaries, Calcified Tissue Research,
The Medical School, Leeds University,
Leeds, 2, Yorkshire, Great Britain

The Editorial Secretaries, Calcified Tissue Research,
Case Western Reserve University, School of Medicine, 511 Wearn Bldg.
2065 Adelbert Road
Cleveland, Ohio 44106, USA

It is a fundamental condition that submitted manuscripts have not been and will not simultaneously be submitted or published elsewhere. With the acceptance of a manuscript for publication, the publishers acquire the sole copyright for all languages and countries. Unless special permission has been granted by the publishers, no photographic reproductions, microform or any other reproductions of a similar nature may be made of the journal, of individual contributions contained therein or of extracts therefrom.

The use of registered names, trademarks, etc. in this publication does not imply, even in the absence of a specific statement, that such names are exempt from the relevant protective laws and regulations and therefore free for general use.

Subscription Information

Volumes 20—22 (3 issues each) will appear in 1976. The publisher reserves the right to issue additional volumes during the calender year. (Information about obtaining back volumes and microform editions available upon request.)

All Countries (Except North America). Subscription rate: DM 414.—, plus postage and handling. Orders can either be placed with your bookdealer or sent directly to: Springer-Verlag, Heidelberger Platz 3, D-1000 Berlin 33, Tel. (030) 822001, Telex 01-83319.

North America. Subscription rate: $173.90, including postage and handling. Subscriptions are entered with prepayment only. Orders should be addressed to: Springer-Verlag New York Inc., 175 Fifth Avenue, New York, N.Y. 10010, USA. Telex 0023232235.

Publishing Office. Springer-Verlag, Journal Production Department, Postfach 105280, D-6900 Heidelberg 1, Tel. (06221)487331, Telex 04-61690.

Responsible for advertisements: L. Siegel, Springer-Verlag, Kurfürstendamm 237, D-1000 Berlin 15
Telephone (030)8821031, Telex: 01-85411
© by Springer-Verlag Berlin · Heidelberg 1976
Printed in Germany by Universitätsdruckerei H. Stürtz AG, Würzburg

---